Distillation Diagnostics

Distillation Diagnostics

An Engineer's Guidebook

Henry Z. Kister

Fluor Corporation, USA

WILEY

Library of Congress Cataloging-in-Publication Data:

Names: Kister, Henry Z., author. | John Wiley & Sons, publisher.
Title: Distillation diagnostics : an engineer's guidebook / Henry Z Kister.

Description: Hoboken, New Jersey : Wiley, [2025] | Includes index.
Identifiers: LCCN 2024023262 (print) | LCCN 2024023263 (ebook) | ISBN
    9781119640110 (hardback) | ISBN 9781119640158 (adobe pdf) | ISBN
    9781119640127 (epub)
Subjects: LCSH: Distillation. | Distillation apparatus–Maintenance and
    repair.
Classification: LCC TP156.D5 K557 2025 (print) | LCC TP156.D5 (ebook) |
    DDC 665.5/32–dc23/eng/20240604
LC record available at https://lccn.loc.gov/2024023262
LC ebook record available at https://lccn.loc.gov/2024023263

Cover Design: Wiley

Set in 10/12pt TimesLTSTd by Straive, Chennai, India

SKY10094149_121824

To my son, Abraham, and my daughter, Helen, who have been my love, blessings, and the lighthouses illuminating my path

Contents

Preface

For over a century, distillation, the "king of separations," has been by far the most common separation technique in refineries, petrochemical, chemical, and natural gas plants and will remain the prominent separation technique in the foreseeable future. Every chemical process has a reaction section and a separation section, with distillation usually dominating the latter. Three percent of the world's energy is tied in distillation. The dominance of distillation stems from its low capital compared to other techniques, scaling up well, and not introducing an extra agent that requires removal later.

The past half-century has seen tremendous advances in distillation technology. High-speed computers revolutionized the design, control, and operation of distillation towers. Invention and innovation in tower internals enhanced tower capacity and efficiency beyond previously conceived limits. Despite all these advances, the distillation failure rate has been on the rise and growing.

The great progress has not bypassed the distillation troubleshooting tools. Reliable pressure drop measurements, gamma scans, laser-guided pyrometers, and thermal cameras are tools that not-so-long-ago troubleshooters would only dream of. But again, despite this great progress, the ability of engineers to effectively diagnose and solve plant problems appears to be on the decline.

Why are we losing the problem-solving war? Despite the prominence of distillation in industry, academics prefer to focus on more trendy "cool" fields. The *Chemical Processing* editorial from March 2018 decries the lack of practical research on distillation in chemical engineering departments. In industry, mergers, cost-cuts, and the pandemic have retired most of the experienced troubleshooters and thinly spread the others. The literature offers little to bridge the experience gap. In the era of information explosion, databases, and computerized searches, finding the appropriate information in due time has become like finding a needle in an ever-growing haystack. To locate a useful reference, one needs to click away a huge volume of wayward leads.

The result is that distillation diagnostics is becoming a lost art. Engineers are given little practical guidance on how to correctly apply and get the most out of the diagnostic tools, so the tools are utilized sparsely and only on a very basic level. Problems are left undiagnosed or misdiagnosed. Solutions based on misdiagnoses are doomed to fail. Lacking a proper diagnosis, some engineers turn to so-called experts and end up being sold a bill of goods or sledge hammers to crack nuts. It is not uncommon to see an expensive new tower solving a problem that can be solved by replacing a faulty feed pipe for nickels and dimes.

The trend away from diagnosis could not have come at a worse time. We are being alarmed by adverse climate and environmental changes, are recognizing the importance of saving our planet, and are striving for a cleaner tomorrow. The engineering community is engaged in developing new technologies for reducing the carbon footprint, aspiring to "zero carbon" energy transition, carbon capture, and replacement of fossil fuels by renewables. Unfortunately, many of these technologies require massive new equipment and additional energy usage, neither of which helps achieve the prescribed goals of reducing carbon footprints.

One simple technology is being forgotten: correct diagnostics. In his book *Process Engineering for a Small Planet* (Wiley, 2010), Norman Lieberman demonstrates how poor troubleshooting and wasteful practices guzzle energy, generate carbon dioxide, and waste the earth's precious resources. In Chapter 1 of his book, Lieberman describes a case in which a correct diagnosis would have led to modifying trays and downcomers in a fractionator, and adding mist injection to the overhead compressor, which could have circumvented erecting a giant new fractionator with a new oversized overhead compressor. Just the energy usage due to the compressor oversizing was estimated to waste the amount of crude oil that 400 families use daily. Fabricating the new steelwork and structures consumed additional immense amounts of energy and emitted tons of carbon dioxide, all of which were unnecessary.

In another example (Chapter 2), four uninstalled tray manways induced the escape of a large quantity of diesel in the bottom of a crude fractionator. To vaporize it took 40 MM Btu/h, which is equivalent to the fuel consumption of a quarter of a million 400 horsepower motor cars driving around the clock. Lieberman diagnosed the problem and recommended re-installing the manways, but the implemented solution was to build a new larger heater, wasting not only energy but also concrete, steel, nickel, and chromium needed for the heater and its tubes, the mining and production of which only adds to the carbon footprint. Lieberman presents a book full of similar experiences.

This book builds up on Lieberman's initiative with a multitude of distillation examples. In one (Chapter 8), a packed stripper was erected to turn a wastewater stream into cooling water makeup. The entering wastewater poured above the oversized hats of the liquid distributor. In the little space between the hats, the vapor sped up and carried the liquid over. Two different consultants studied the problem. The first offered an incorrect diagnosis and a failed fix. The second performed an extensive simulation study, incorrectly concluding that a larger-diameter tower was needed. The plant gave up and junked the tower, with the water still going to the sewer. Good troubleshooting and hat or feed pipe modifications costing nickels and dimes would have saved precious water, reduced sewage, and prevented turning a good tower into scrap metal.

My previous book, *Distillation Troubleshooting*, focused on case studies and the lessons learned. This book goes further, providing the tools to correctly diagnose and solve operating problems, avoid the wasteful practices, and make the most of existing equipment. It travels through the multitude of invaluable tools available for diagnosing tower problems, providing application guidelines derived from the school of hard knocks. The book describes what each tool can and cannot do, provides insight into how different ideas and theories can be tested using these tools, shares the experience of how to correctly interpret the results provided by each technique, and guides troubleshooters to a multitude of additional tests that they can perform to get closer to a correct diagnosis and an effective fix. The only other place where its information is available is in the heads of experienced troubleshooters.

The techniques are illustrated with real case studies from my experience of 45 years as well as from the experiences of many of my colleagues and those presented in the literature. Every technique is accompanied with an extensive list of "do and don'ts" based on the author's and industry's experience. This guidance can make the difference between success and failure, and between good and bad results.

The case studies and lessons learned are described to the best of the author's and the contributors' knowledge and in good faith, but may not always correctly reflect the problems and solutions. Many times I thought I had the answer, only to be humbled by a new light or another experience later. The experiences and lessons in this book are not meant to be followed blindly. They are meant to be stories told in good faith and to the best of knowledge and understanding of the author or contributors, which hopefully give troubleshooters ideas to think about. We welcome any comments that either affirm or challenge our perception or understanding.

If an incident that happened in your plant is described, you may notice that some details could have changed. Sometimes, this was done to make it more difficult to tell where the incident occurred. At times, it was done to simplify the story without affecting the key lessons. Sometimes, it happened long ago and memories of some details faded away. Sometimes, and most likely, this incident did not happen in your plant at all. Another plant had a similar incident.

The book emphasizes the "do-it-yourself" approach. It endeavors to place the tools and good application guides in the practitioner's hands, eliminating reliance on the so-called experts. People who attend my seminar classes come out with knowledge on how to apply the tools and express concern about forgetting this knowledge over the years. They express wishes to have a book like this so they can refresh their memories in the future. This is that book. It is my hope that the book will lead the readers to correct diagnostics and effective fixes to tower problems, and to trouble-free operation.

If you can successfully solve only one problem using the techniques in this book, the savings to your company will be enough to pay for thousands of copies of the book. But far beyond the savings, you will be taking a step in the right direction of saving our small planet and ushering in a cleaner tomorrow.

HENRY Z. KISTER

May 2024

List of Acronyms

AIChE	American Institute of Chemical Engineers
AGO	Atmospheric gas oil
API	American Petroleum Institute
ASTM	American Society for Testing and Materials
ATB	Atmospheric tower bottoms
BPA	Bottom pumparound
BPD	Barrels per day
BPH	Barrels per hour
BPSD	Barrels per stream day
BTM	Bottom
CAT	Computer aided tomography
CFD	Computational fluid dynamics
COND	Condensate
COT	Coil outlet temperature
COV	Coefficient of variation
CW	Cooling water
dP	Differential pressure (pressure drop)
ED	Extractive distillation
FCC	Fluid catalytic cracking
FI	Flow indicator
FRI	Fractionation Research Inc (or Institute)
FRN	Full range naphtha
HAZOP	Hazard and operability analysis
HCO	Heavy cycle oil
HCU	Hydrocracker Unit
HETP	Height equivalent to a theoretical plate
HF	Hydrogen fluoride
HN	Heavy naphtha
HP	High pressure
HSE	Health, Safety, and Environment
HVB	Hot vapor bypass
HVGO	Heavy vacuum gas oil
ID	Internal diameter
IPA	Intermediate pumparound
IR	Infrared
LCO	Light cycle oil
LED	Light Emitting Diode
LI	Level indicator

LN	Light naphtha
LNG	Liquefied natural gas
LV	Liquid volume
LVGO	Light vacuum gas oil
MBPD	Thousand BPD
MCB	Main Column Bottoms
MDEA	Methyldiethanol amine
MEA	Monoethanol amine
MP	Medium pressure
MPA	Mid pumparound
MRI	Magnetic resonance imaging
MSCFH	Standard cubic feet per hour $\times$ 1000
OSHA	The Occupational Safety and Health Administration
OVHD	Overhead
P&ID	Process and instrumentation diagram
PA	Pumparound
PFD	Process flow diagram
PI	Pressure indicator
PPR	Polypropylene return
STM	Steam
TAR	Turnaround
TI	Temperature indicator
TPA	Top pumparound
TRC	Temperature recorder/controller
VCFC	Vapor crossflow channeling
VLE	Vapor–liquid equilibrium
VLLE	Vapor–liquid-liquid equilibrium
VTB	Vacuum tower bottoms

Acknowledgments

Many of the experiences and case histories reported in this book have been invaluable contributions from colleagues and friends who kindly and enthusiastically supported this book. Many of the contributors elected to remain anonymous. Kind thanks are due to all contributors, you have done a magnificent service to our beloved science of distillation and to the industry. Special thanks to those whose names do not appear in print. To those behind-the-scenes friends, the author extends special appreciation and gratitude.

Writing this book required breaking away from everyday work and family demands. Special thanks are due to Fluor Corporation, particularly to my supervisors, Curt Graham, Maureen Price, Jeff Scherffius, Jennifer Foelske, and Bill Parente, for their backing, support, and encouragement of this book-writing endeavor, going to great lengths to make it happen. Warmhearted gratitude to my wife Susana for her love, dedication, and devotion to our family, guiding our children to walk in G-d's path and do the right thing, and encouraging them to excellence.

Recognition is due to my mentors who, over the years, encouraged my work; immensely contributed to my achievements; and taught me much about distillation and engineering: to my life-long mentor, Dr. Walter Stupin, who warmly mentored and encouraged my work, throughout my career at C F Braun and later at Fluor, being a ceaseless source of inspiration behind my books and technical achievements, and who taught me a lot of what I know about distillation; Curt Graham and Paul Walker, Fluor, whose warm encouragement, guidance, and support were the perfect motivators for professional excellence and achievement; Professor Ian Doig, University of NSW, who inspired me over the years, showed me the practical side of distillation, and guided me over a crisis early in my career; Professor James Fair, distillation guru, University of Texas, who warmly encouraged my books and distillation work; Reno Zack, who enthusiastically encouraged and inspired my achievements throughout my career at C F Braun; Dick Harris and Trevor Whalley, who taught me about practical distillation and encouraged my work and professional pursuits at ICI Australia. Thanks are due to my colleagues Garry Jacobs and Carlos Trompiz (Fluor), Ron Olsson (Celanese), Daryl Hanson (Valero), Christine Long (Chevron), and troubleshooting guru Norman Lieberman, all of whom taught me a lot by sharing their vast experiences and expertise. The list could go on, and I express special thanks to all who encouraged, inspired, and contributed to my work over the years. Much of my mentors' and colleagues' teachings found its way to the following pages.

Special thanks are due to my mother, Dr. Helen Kister, and my father, Dr. John Kister, may their memories be blessed, who dedicated their lives to me, guided me to do things right, enthusiastically supported my work throughout my life, and have been there for me every moment of the day. Their prayers have remained with me even after they left this world. Thanks also for always encouraging me to excel and teaching me to diagnose and critically evaluate problems.

Special thanks are due to Tracerco, in particular to the dedication of Margaret Bletsch and Lowell Pless, who contributed wonderful expertise, experiences, and images for the radioactive troubleshooting sections of this book and assisted me in fitting them to the audience.

Chapter 1

Troubleshooting Steps

"In our industry we have many more troublemakers than troubleshooters"
—(Professor Zarko Olujic, presentation to AIChE)

Troubleshooting, revamping, and eliminating waste offer huge benefits in reducing capital investment, downtime, carbon dioxide emissions, and energy consumption. Unfortunately, the attention paid to this resource in the energy-transition era has been too little to reflect its tremendous potential.

A multitude of examples in Norman Lieberman's book *Process Engineering for a Small Planet* (287) demonstrate how poor troubleshooting leads to wasteful practices that guzzle energy, increase the carbon footprint, and deplete the earth's precious resources. The chemical process industry (CPI) has been paying dearly for downtime, lost production, substandard product quality, raw material problems, safety and environmental issues, and excessive energy consumption due to ineffective troubleshooting of abnormal situations (16). In 1997, a consortium estimated that the annual loss for the CPI due to ineffective abnormal situation management was US $10 billion (16).

Correct diagnosis is at the heart of problem identification and implementing a correct, cost-effective solution. An incorrect diagnosis breeds ineffective solutions that prolong the agony and escalate the costs. This chapter focuses on the systematic diagnostic steps. The following chapters will describe the multitude of techniques that have been found effective for diagnosing distillation problems.

A well-known sales axiom states that 20% of the customers bring in 80% of the business. A sales strategy tailored for this axiom concentrates the effort on these 20% without neglecting the others. Distillation diagnostics follow an analogous axiom. A person engaged in diagnosing column problems must develop a good understanding of the factors that cause the vast majority of column malfunctions and the techniques available for narrowing in on their root causes. While a good knowledge and understanding of the broader field of distillation is beneficial, the diagnosis often requires only a shallow knowledge of this broader field.

It is well accepted that diagnosing problems is a primary job function of operating engineers, supervisors, and process operators. Far too few realize that distillation diagnostics start at the design phase. Any designer wishing to achieve a trouble-free column design and operation must be as familiar with diagnostic techniques, many of which are applicable during the design (and more so at the revamp) phase.

Expansive surveys of the causes of column malfunctions were described in previous studies (196, 198, 201). Abundant resources are available to distinguish good from poor practices and to

Distillation Diagnostics: An Engineer's Guidebook, First Edition. Henry Z. Kister.
© 2025 American Institute of Chemical Engineers, Inc. Published 2025 by John Wiley & Sons, Inc.

avoid and overcome troublesome design and operations (e.g., 192, 285, 286, 293). What is often missing is the connect. How does one link field observations with the known tower malfunctions in order to develop an effective remedy? This book is all about the link: translating field observations into diagnoses and cures.

Following a brief survey of the primary causes of tower malfunctions, this chapter looks at the basic diagnostic: the systematic strategy for diagnosing distillation problems and the dos and don'ts for formulating and testing theories. Finally, it reviews the techniques for testing these theories and for focusing on the most likely root cause.

1.1   CAUSES OF COLUMN MALFUNCTIONS

Close to 1500 case histories of malfunctioning columns were extracted from the literature and abstracted in Ref. 201. Most of these malfunctions were analyzed in Ref. 198 and classified according to their principal causes. A summary of the common causes of column malfunctions is provided in Table 1.1. If one assumes that these case histories make up a representative sample, then the analysis presented below has statistical significance. Accordingly, Table 1.1 can provide a useful guide to the factors most likely to cause column malfunctions and can direct troubleshooters toward the most likely problem areas.

The general guidelines in Table 1.1 often do not apply to a specific column or even plant. For instance, foaming is not high up in the table; however, in amine absorbers, it is a very common trouble spot. The author therefore warns against blindly applying these guidelines in any specific situation.

The total number of cases in each category is shown in the column headed "Cases." The other three columns show the split of these cases according to industry categories, namely refining, chemicals, and olefins/gas plants.

An analysis of Table 1.1 suggests the following:

- Plugging, tower base, tower internals damage, instrument and control problems, startup and/or shutdown difficulties, points of transition (tower base, packing distributors, intermediate draws, feeds), and assembly mishaps are the major causes of column malfunctions. They make up nearly two-thirds of the reported incidents. Familiarity with these problems, therefore, constitutes the "bread and butter" of distillation and absorption troubleshooters.

- Primary design is a very wide topic, encompassing vapor–liquid equilibrium, stage-to-stage calculations, reflux-stages relationship, unique features of multicomponent distillation, tray and packing capacities and efficiencies, scale-up, column diameter and height determination, type of tray, and size and material of packing. This topic represents most of our distillation know-how and occupies the bulk of most of the distillation texts (e.g., 34, 193, 299, 451, 492). While this topic is paramount for designing and optimizing distillation columns, it plays only a minor role in operations and troubleshooting in distillation. As shown in Table 1.1, only one column malfunction among fourteen is incurred in the primary design. The actual figure is probably higher for a first-of-its-kind separation, but lower for an established separation. Due to the bulkiness of this topic and its low likelihood to cause malfunctions, and due to the excellent coverage that the topic receives in several texts (e.g., 34, 193, 299, 451, 492), it is only lightly touched upon in this book.

- The above statements must not be interpreted to suggest that troubleshooters need not be familiar with the primary design. Quite the contrary. A good troubleshooter must have a solid understanding of primary design because it provides the foundation of distillation

Table 1.1 Most common causes of column malfunctions. (From Kister, H. Z., Transactions of the Institution of Chemical Engineers, 81, Part A, p. 5, January 2003. Reprinted Courtesy of the Institution of Chemical Engineers in the UK.)

No.	Cause	Total cases	Refinery cases	Chemical cases	Olefins/ gas cases
1	Plugging, coking	**121**	68	32	16
2	Tower base and reboiler return	**103**	51	22	11
3	Tower internals damage (excluding explosion, fire, implosion)	**84**	35	33	6
4	Abnormal operation incidents (startup, shutdown, commissioning)	**84**	35	31	12
5	Assembly mishaps	**75**	23	16	11
6	Packing liquid distributors	**74**	18	40	6
7	Intermediate draws (including chimney trays)	**68**	50	10	3
8	Misleading measurements	**64**	31	9	13
9	Reboilers	**62**	28	13	15
10	Chemical explosions	**53**	11	34	9
11	Foaming	**51**	19	11	15
12	Simulations	**47**	13	28	6
13	Leaks	**41**	13	19	7
14	Composition control difficulties	**33**	11	17	5
15	Condensers that did not work	**31**	14	13	2
16	Control assembly	**29**	7	14	7
17	Pressure and condenser controls	**29**	18	3	2
18	Overpressure relief	**24**	10	7	2
19	Feed inlets to tray towers	**18**	11	3	3
20	Fires (excluding explosions)	**18**	11	3	4
21	Intermediate component accumulation	**17**	6	4	7
22	Chemicals release to the atmosphere	**17**	6	10	1
23	Subcooling problems	**16**	8	5	1
24	Low liquid loads in tray towers	**14**	6	2	3
25	Reboiler and preheater controls	**14**	6	–	5
26	Two liquid phases	**13**	3	9	1
27	Heat integration issues	**13**	5	2	6
28	Poor packing efficiency (excluding maldistribution/support/hold-down)	**12**	4	3	2
29	Troublesome tray layouts	**12**	5	2	–
30	Tray weep	**11**	6	1	3
31	Packing supports and hold-downs	**11**	4	2	2

know-how. However, the above statements do suggest that in general, when troubleshooters examine the primary design for the cause of a column malfunction, they have less than one chance out of ten of finding it there.

1.2 COLUMN TROUBLESHOOTING – A CASE HISTORY

In Sections 1.3 and 1.4, the systematic approach recommended for diagnosing distillation problems is presented. The recommended sequence of steps is illustrated with reference to the case history described below.[1]

The following story is not a myth; it really happened. One morning as I sat quietly at my desk in corporate headquarters, the boss dropped by to see me. He had some unpleasant news. One of the company's refinery managers was planning to visit our office to discuss the quality of some of the new plants that had been built in his refinery. As an example of how not to design a unit, he had chosen a new gas plant for which I had done the process design. The refinery manager had but one complaint: "The gas plant would not operate."

I was immediately dispatched to the refinery to determine which aspect of my design was at fault. If nothing else, I should learn what I did wrong so as not to repeat the error.

Upon arriving at the refinery, I met with the operating supervisors. They informed me that, while the process design was fine, the gas plant's operation was unstable because of faulty instrumentation. However, the refinery's lead instrument engineer would soon have the problem resolved.

Later, I met with unit operating personnel. They were more specific. They observed that the pumparound circulating pump (see Figure 1.1a) was defective. Whenever they raised hot oil flow to the debutanizer reboiler, the gas plant would become destabilized. Reboiler heat-duty and reflux rates would become erratic. Most noticeably, the hot-oil circulating pump's discharge pressure would fluctuate wildly. They felt that a new pump requiring less net positive suction head was needed.

Both these contradictory reports left me cold. Anyway, the key to successful troubleshooting is personal observation. So I decided to make a field test.

When I arrived at the gas plant, both the absorber and debutanizer towers were running smoothly but not well. Figure 1.1b shows the configuration of the gas plant. The debutanizer reflux rate was so low it precluded significant fractionation. Also, the debutanizer pressure was about 100 psi below design. Only a small amount of vapor, but no liquid, was being produced from the reflux drum. Since the purpose of the gas plant was to recover propane and butane as a liquid, the refinery manager's statement that the gas plant would not operate was accurate.

As a first step, I introduced myself to the chief operator and explained the purpose of my visit. Having received permission to run my test, I switched all instruments on the gas-plant control panel from automatic over to local manual. In sequence, I then increased the lean oil flow to the absorber, the debutanizer reflux rate, and the hot-oil flow to the debutanizer reboiler.

1 Reproduced from Norman P. Lieberman, *Troubleshooting Process Operations* (4th ed., PennWell Books, Tulsa 2009). This case history is a classic example of how to perform a systematic troubleshooting investigation. The permission of PennWell Books and Norman P. Lieberman for reproducing this material is gratefully acknowledged.

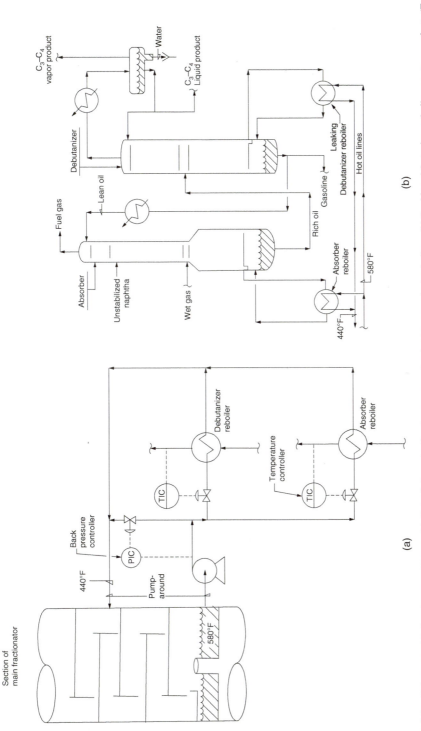

Figure 1.1 Column troubleshooting case history. (a) Hot oil from the fractionator supplies heat to gas plant reboilers. (b) Leaking debutanizer reboiler upsets gas plant. (From Lieberman, N. P., "Troubleshooting Process Operations," Ed. 4, PennWell Books, Tulsa, OK, 2009. Reprinted with permission.)

The gas plant began to behave properly. The hot-oil circulating pump was putting out a steady flow and pressure. Still, the plant was only producing a vapor product from the debutanizer reflux drum. This was because the debutanizer operating pressure was too low to condense the C_3–C_4 product. By slowly closing the reflux drum vapor vent valve, I gradually increased the debutanizer pressure from 100 psig toward its design operating pressure of 200 psig.

Suddenly, at 130 psig, the hot-oil flow to the debutanizer's reboiler began to waiver. At 135 psig the debutanizer pressure and the hot-oil flow plummeted. This made absolutely no sense. How could the debutanizer pressure influence hot-oil flow?

To regain control of the gas plant, I cut reflux to the debutanizer and lean-oil flow to the absorber. I was now back where I started. The thought of impending failure loomed.

I repeated this sequence twice more. On each occasion, all went well until the debutanizer pressure increased. By this time it was 3 a.m. Was it also time to give up and go home?

Just then, I noticed a commotion at the main fractionator control panel. The operators there stated that the fractionator was flooding again—for the third time that night. The naphtha production from the fractionator had just doubled for no apparent reason.

In every troubleshooting assignment there always occurs that special moment, the moment of insight. All of the bits and pieces fall into place, and the truth is revealed in its stark simplicity.

I cut the debutanizer pressure back to 100 psig and immediately the flooding in the main fractionator subsided. The operators then closed the inlet block valve to the hot-oil side of the reboiler and opened up a drain. Naphtha poured out instead of gas oil. This showed that the debutanizer reboiler had a tube leak.

Whenever the debutanizer pressure reached 130 psig, the reboiler pressure exceeded the hot-oil pressure. The relatively low-boiling naphtha then flowed into the hot oil and flashed. This generated a large volume of vapor that then backed hot oil out of the reboiler. The naphtha vapors passed on into the main fractionator and flooded this tower. Thus, the cause of the gas plant instability was neither a process design error, instrument malfunction, nor pumping deficiency. It was a quite ordinary reboiler tube failure.

1.3   STRATEGY FOR TROUBLESHOOTING DISTILLATION PROBLEMS

In almost any troubleshooting assignment, it is desirable to solve a problem as fast as possible with the least expense. Surprisingly, this objective is often only partially achieved, the obstacle being a poor (often nonexistent) strategy for diagnosing the problem.

While devising a troubleshooting strategy, it is useful to think in terms of a "doctor and patient" analogy. The doctor's strategy for diagnosing an illness is well established and easily understood by most people. Applying a similar strategy to diagnosing distillation problems often constitutes the most effective and least expensive course of action.

The sequence of steps listed below is often considered optimum for tackling a troubleshooting problem. It is based on the author's experience as well as the experience of others (112, 116, 161, 166, 286, 293, 318, 411, 498). The step headings refer to the doctor and patient analogy. Actions described in Lieberman's case history (Section 1.2) are referenced to demonstrate the optimum sequence of steps. A good troubleshooting strategy always proceeds stepwise, starting with the simple and obvious.

1. *Save Patient and Prevent Disease from Spreading to Others*

    Assess the safety, health, and/or environmental hazards that the problem can create. If a hazard exists, an emergency action is required prior to troubleshooting. In terms of the medical analogy, measures to save the patient or prevent the patient's illness from spreading to others have priority over investigating the cause of the problem.

2. *Temporary Strategy: Hospitalization, Bed Rest, Special Diet*

    Implement a temporary strategy to live with the problem. Problem identification, troubleshooting, and implementing the solution take time. Meanwhile, negative effects on safety, health, the environment, and plant profitability must be minimized. The strategy should be as conducive for troubleshooting as practicable. The strategy, and the adverse effects that need to be temporarily tolerated (e.g., lost production, off-spec product, instability, higher utility costs), usually set the pace of the troubleshooting investigation.

    In the debutanizer case history, the short-term strategy was to operate the column at a pressure low enough to eliminate instability and to tolerate an off-spec bottom product. In the medical analogy, the short-term strategy is hospitalization, bed rest, special diet, anti-inflammation drugs, or just "taking it easy."

3. *Determine Urgency of Treatment: Is it Life-Threatening?*

    The extent to which the temporary solution (step 2) can be tolerated and, to a lesser degree, the complexity of the problem and resources required and available need to be considered next, which will set the pace of the troubleshooting investigation. "Crisis" urgency is assigned when there are significant adverse safety, health, or environmental concerns, if the column cannot produce an on-spec valuable product, or if there is a major impact on plant profitability. "Medium" urgency is typical when the column falls short of producing the desired capacity or product quality but can still operate and make acceptable products. "Low" urgency is usually assigned to instability or operating nuisance or when the cost effects are not major. In the debutanizer example, the urgency was medium to crisis, as the plant could not make the overhead product and the bottom product was off-spec. In the medical analogy, life threat sets a fast pace of treatment, while minor pain sets a slow pace.

    The urgency of treatment is affected by the proximity of the next scheduled plant outage or turnaround. Urgency is likely to be stepped up when there is an opportunity to attempt a fix in a forthcoming outage or when there is a prospect of having the tower limp along for a lengthy period. In contrast, the urgency is reduced if cleaning, major repairs, or column replacement is planned for a forthcoming outage and is likely to eliminate the problem. In the medical analogy, there may be no need to do some tests when the patient is about to have a major checkup.

    The urgency may change, sometimes quickly, with changing market conditions. For instance, a rise in product demand or price often steps up the urgency.

4. *Doctor Obtains Problem Definition (Detailed Symptoms) from Patient*

    Obtain a clear, factual definition of the symptoms. A poor problem definition is one of the most common diagnostics pitfalls. In the debutanizer case history above, the definitions used by different people to describe the symptoms of a reboiler tube leak problem were:

    • "The gas plant would not operate."

    • "The gas plant's operation is unstable because of faulty instrumentation. However, the problem will soon be resolved by the instrument engineer."

- "The oil circulating pump is defective. Whenever the oil flow to the reboiler is raised, reboiler heat duty and reflux rate would become erratic, and the pump's discharge pressure would fluctuate wildly. A new pump requiring less net positive suction head is needed."

- "The column was running smoothly but not well. Reflux rate was too low, so it precluded significant fractionation. The column pressure was 100 psi below design. Only a small amount of vapor, but no liquid, was being produced from the reflux drum, which should have produced mainly liquid. Other problems noticed by plant personnel are as described above."

The above represents a typical problem definitions spectrum. The last definition, provided by a troubleshooting expert, can clearly be distinguished from the others. The first two definitions were nonspecific and insufficiently detailed. The third described one symptom, but left the other symptoms out. The first three definitions also contained implied diagnoses, none of which turned out to be correct.

Listening to the people involved helps formulate a good definition. It is easy to miss or overlook crucial details. Different people focus on different details, so talking can bring out hidden details. In the debutanizer case, the observation by the plant personnel became part of the problem definition.

The doctor–patient equivalents to the first three definitions are statements such as "I feel I am going to die," "I am feeling a bit off, but I will be OK soon," and "I do have a sharp headache" (without mentioning other pains and having a fever as well). These statements do not provide the doctor with the entire story.

5. *Doctor Becomes Familiar with Patient*

Familiarize yourself with the column (if not familiar yet). What is the main purpose of the column? What are the key performance criteria? What are the operating temperatures, pressures, product purities, and mass balance that the column is attempting to achieve? Have there been any recent modifications? Familiarize yourself with the physical arrangement of the column and its instrumentation. What instruments are available? Do not overdo this at this point; there will be opportunities to learn the finer details as the investigation proceeds. In the medical analogy, upon entry to the doctor's office, a nurse often measures height, weight, temperature, and blood pressure for the doctor to review.

6. *Doctor Examines Patient*

Examine the column behavior yourself. This is imperative if the problem definition is poor. In the debutanizer example, had the troubleshooter based his investigation merely on other people's observations, he would have missed a major part of the problem definition. Communication gaps are often hard to bridge. In a similar manner, a doctor always needs to examine the patient before coming up with a diagnosis.

In some situations, it may be impractical or too costly for the troubleshooter to visit the site (e.g., a column located on another continent). In this case, the troubleshooter must be in close direct video communication with the operating person, who should be entirely familiar with the column, its history, and its operation.

7. *Doctor Looks for Swelling, Rashes, or Sounds*

Walk around the column, looking for outside signs. Check all lines containing valves that cross-connect products and measure surface temperatures on each side of each valve. Valves often leak or are inadvertently left open (Section 3.2.1 has some case studies). Survey the column piping for any unusual features, poor piping arrangement, leaking valves, "sticking" control valves, and valves partially shut. In one case (166), a cooling water valve left quarter-opened from a previous campaign constrained the vacuum and led to an off-spec product. Filter baskets may have not been adequately reinstalled after cleaning. Check that meters are correctly installed and samples are taken from the correct lines. For instance, a thermocouple near the junction of two streams may give an unreliable indication of the temperature of either. Listen to sounds coming from the tower. These may indicate vibrations, eruptions, sloshing, loose nuts, or pump cavitation. Inspect liquid lines entering and leaving. Crackling, vibration, and hammering indicate vapor presence in the lines. Column sways may indicate a high base level.

8. *Doctor Obtains Health History of Patient*

Learn about the column history. The question "What are we doing wrong now that we did right before?" is one of the most powerful diagnostic tools available. Closely examine differences between the column and columns used for identical or at least similar services. Examine any differences between the expected and the actual performance. Each difference can provide a major clue. A tabular listing of the similarities and differences (possibly using a spreadsheet) may provide useful leads. In search of clues, doctors always ask patients about their health histories. In the debutanizer example, the troubleshooter compared the operation to the design performance (he was working with a new column).

Dig into past issues. Had plugging, precipitation, scaling, corrosion, tray, or packing damage previously occurred? Review the maintenance records. A pump, filter, or control valve that required an abnormal amount of servicing may give a useful lead.

Digging into the past may reveal a recurring ("chronic") problem. Search for the correct link between the past and present circumstances. Be cautious: a new problem may give the same symptoms as a past problem, but be caused by an entirely different mechanism.

A history search may also unveil a hidden flaw. In one case (161), a column modification reduced column efficiency. The reduced efficiency was unnoticed, and the poor performance became the norm. The problem was noticed several years later.

9. *Doctor Asks Patient What Hurt First*

Search and identify events that occurred around the time the problem started. Carefully review operating charts, trends, computer records, and operator logs. Establish event timing to differentiate an initial problem from its consequences. Chapter 9 describes several case histories with actual operating charts that demonstrate the value of analyzing the event timing. In terms of the medical analogy, doctors always ask patients what happened first and what they did differently about the time when the trouble started.

Do not exclude events that may appear completely unrelated, as these may be linked in an obscure manner to the problem. In the debutanizer example, it was the observation that the onset of debutanizer instability coincided with flooding in the fractionator that gave the troubleshooter the vital clue. At first glance, the two appeared completely unrelated.

10. *Doctor Learns about Family Health History*

Do not restrict the investigation to the column. Many column problems initiate in upstream equipment. Doctors frequently seek clues by asking patients about people they have been in contact with or their family health history. Question the designers, equipment suppliers, installers, and others that are familiar with the column or worked on the column.

Listen to shift operators and supervisors. Take good notes; details that appear unworthy of remembering may turn out to provide vital clues. Experienced people often spot problems even if they cannot fully define or explain them. Operators and supervisors know the behavior of the column better than anyone, and familiarity with this behavior can lead to the correct diagnosis. Different operators may offer different observations, all of which may be useful. Listening to all of them can provide vital clues.

When something operators or supervisors say sounds odd, finding out why they say so is crucial and is likely to provide a major clue. In the debutanizer example, some of the key observations were supplied by the shift team. Hanson et al. (157, 158) described two cases, and Hasbrouck et al. (166) another case, where observations by operators led to the root cause. Achoundong et al. (2) showed how systematically listing and understanding the reasons for a variety of comments by the operating team became central to a correct diagnosis. In one crude tower, the author calculated a feed capacity of 85,000 BPD, but the operating team swore they could not go past 70,000 BPD. Understanding why the operating team was saying this revealed a previously unknown bottleneck. A similar experience was reported by Litzen and Bravo (298).

Language spoken by operators and supervisors may differ from process engineers' language. For instance, an operator may state that increasing the column reflux makes the trays more efficient. For process engineers, tray efficiency is the number of theoretical stages achieved by a tray, and this number is usually independent of the reflux (193, 245). For operators, the tray efficiency criterion is separation, which does improve with higher reflux. Effective listening requires understanding why the operator is making each statement.

Operating personnel often tell you their conclusions first (166). You need to find out how they reached these conclusions. For example, when they tell you that a pump is defective (as they did in the debutanizer example), asking what led to this conclusion revealed the key observation about instability setting in when the oil flow to the reboiler was raised.

During all discussions, be on the same team with the operators and supervisors. Do not be afraid to ask questions, even those that may appear stupid. The attitude that you want to learn more about the issue and work with the operating team to seek a solution is productive and will win cooperation. If you suspect an operator is incorrect, a polite reply like "I guess that is possible" allows saving their face (your face if the operator turns out to be correct!). The attitude that you have the answers is counterproductive, and you will lose the much-needed cooperation.

11. *Doctor Checks Patient's Responses* (*"Take a Deep Breath"*)

Study the column behavior by making small, inexpensive changes. These are central for refining the definition of the symptoms. Record all observations and collect data; these may contain a major clue, which can become forgotten or hidden as the investigation progresses. In the debutanizer example, the expert watched the column response to raising pressure. This led to the observation that the debutanizer pressure affected the oil flow – an unexpected occurrence that became a major clue.

In the doctor–patient analogy, this is similar to the doctor asking the patient to take a deep breath or momentarily stop breathing during a medical examination.

12. *Doctor Obtains Lab Tests and Bloodwork*

Take a good set of readings on the column and its auxiliaries, including laboratory analyses. This step is equivalent to laboratory tests prescribed by a doctor to the patient.

To avoid misleading information, suspect instrument readings and laboratory analyses, and make as many validation cross-checks as possible. An instrument may lie even when the instrument technician swears it is correct. One expert (112) stated, "I've spent a good deal of company money, and a lot of time, chasing a perceived operating problem because of an improperly calibrated instrument." In one example (498, Case 25.2 in Ref. 201), an erroneous reading of a reflux flow meter, resulting from an incorrect pipe design, led to unnecessary, costly shutdowns.

At the same time, do not disbelieve your instruments. They may be trying to tell you something. You need not trust them, but you should listen to them. In one case (213), an apparently defective flow meter reading led to the identification of incorrectly connected cooling water pipes.

Compile mass, component, and energy balances; these provide a valuable check on the consistency of instrument readings and the possibility of leakage. Carefully review the column drawings for unusual features. Verify that the column drawings used are the latest update and explore the possibility of past modifications evading the updates. Review the column internals for violations of good design practices. If identified, examine the consequences of such violation and its consistency with the symptoms. Perform a hydraulic calculation under test conditions to determine whether any operating limits are approached or exceeded. For a separation problem, carry out a computer simulation of the column; check against test samples, temperature readings, and exchanger heat loads. Chapters 2 and 3 elaborate on the various checks.

13. *Doctor Uses Ultrasound, X-Rays, Other Noninvasive Tests*

If more information is needed, like looking inside the tower, there are a large number of noninvasive techniques, some of them high-tech, that can give close insights. These include gamma scans, neutron backscatter, surface temperature surveys, thermal scans, computer-aided tomography (CAT) scans, tracer injection, quantitative multi-chordal gamma scans, and others. These are described in detail in Chapters 5–7. These are equivalent to ultrasound, X-rays, magnetic resonance imaging (MRI), and CAT scans used in medicine.

14. *Doctor Watches Out for Selective Attention Bias*

"Selective attention bias" refers to strongly concentrating on one diagnosis to the exclusion of other observations one should catch (328). Under pressure and stress, narrowing of focus is natural, but can exclude important and relevant information. A well-known example illustrating this bias is discussed in the book *The Invisible Gorilla and Other Ways Our Intuition Deceives Us* (62). In the medical analogy, experienced X-ray technicians looking for a particular diagnosis on an X-ray missed obvious, serious problems not related to the looked-at diagnosis (62, 328). In the debutanizer example, selective attention bias is likely to have played a role in the incomplete problem definitions (step 4) and was overcome by the expert's definition that combined all the relevant information. The doctor needs to step back and look at the bigger picture to counteract this bias.

15. *Doctor Critically Reconciles Results from Different Sources*

The data obtained from one technique should be consistent with those of others. For instance, in flood testing, check that alternative techniques such as gamma scans and differential pressure measurements give consistent results. Investigate any inconsistencies; these may provide a vital clue. Repeat measurements as necessary. Doctors check that the X-ray results are consistent with their examination results and with the bloodwork.

16. *Doctor Compares Sick Patient with Healthy Individual*

A simulation of the current column operation can be compared to one with good performance. A tabular listing of the similarities and differences (possibly using a spreadsheet) may highlight the issues. Comparing the patient with a healthy individual can provide the doctor with invaluable clues.

The simulation can also provide the internal vapor and liquid loads and the physical properties needed for hydraulic checks. These can determine whether a hydraulic issue is expected or whether it occurs prematurely. The simulation can also provide the feed flashes and properties needed to rate-check distributors and inlet pipes.

17. *Doctor Critically Determines Need to Receive Input from Other Disciplines*

Input from other disciplines may be critical to problem diagnosis (455). For instance, mechanical engineers will be the best to determine whether a tray was dislodged upward, or downward, or vibrated loose and why. Doctors often enlist help from specialists.

Beware of placing undue reliance on expert opinion. While experts have vast knowledge, they are less familiar with the problem and data than you. Closely scrutinize their comments in light of your findings, and do not hesitate to challenge them. As an advisor, my opinion had often been correctly challenged by members of a process team, sending me in a different direction. Doctors often go back and forth with a specialist until the correct diagnosis is established. For maximum effectiveness, make the expert a part of the troubleshooting team, working closely together.

18. *Doctor Calms down Patient to Get the Best Information*

Tests conducted under upset conditions can be relied on only as a preliminary indication (112), and their data should be suspect and treated with caution. Backing off from the upset to calm conditions eliminates interactions, narrows in on the key variables, and minimizes bad leads.

19. *Doctor Gets it Right versus Doctor Gets it Fast*

The huge costs of lost production, off-spec products, or idle equipment often tilt the scale to "How fast can we fix the problem?" However, this need for speed should not be allowed to overpower good testing and adequate analysis (16). A study by Swain (16, 470) showed that the probability of coming up with a correct diagnosis initially rises rapidly with the time spent and then flattens off. The author's father, an old-school doctor, would spend 45 minutes examining a patient before coming up with a diagnosis that was always correct. I have been examined by doctors that took less than two minutes, often with an incorrect diagnosis, and have never returned to them.

1.4 DOS AND DON'TS FOR FORMULATING AND TESTING THEORIES

Following the previous steps, a good problem definition should now be available. In some cases (e.g., the debutanizer), the root cause may be identified. If not, there will be sufficient information to narrow down the possible causes and to formulate a theory. In general, when problems emerge, everyone will have a theory. In the next phase of the investigation, these theories are tested by experimentation or by trial and error. The following guidelines apply to this phase:

1. *Get Your Facts Right*

Sherlock Holmes once stated, "The difficulty is to detach the framework of fact – of absolute undeniable fact – from the embellishments of theorists and reporters." This detachment is central to obtaining a correct diagnosis. Check and recheck the validity of your data until you are positive that they are correct. Never assume anything. See step 12 in Section 1.3 for typical validation checks. Incorrect data support the wrong theories and deny the correct ones. Look for independent ways of confirming or denying the validity of measurements and observations. Any theory must be consistent with adequately validated data. Adequately validated data form a strong basis for formulating theories.

2. *Theory versus Data*

Logic is wonderful as long as it is consistent with the facts and the information is good. Clearly distinguish facts from theories and interpretations. The pitfall to avoid is "Don't let the data get in the way of a good theory." Follow the data. There are no "impossible" data. If data appear "impossible," perform additional validation checks to confirm or deny them. When you have conclusive data, adhere to them.

Make sure that the data are good. Do not trust instruments and drawings without verifying their correctness (356). Look for anything that does not make sense. Doubt everything you are told, no matter how much you trust and respect the person (356). Seek to positively confirm all data and facts.

3. *Theory versus Laws of Physics and Chemistry*

Critically check that the theory does not violate the laws of physics and chemistry (356). Closing a valve cannot increase the flow. An exothermic reaction does not reduce temperatures. Any theories that violate the laws of physics or chemistry need to be disposed of.

4. *Learn from Past Experiences*

Distillation failures are repetitive (see Section 1.5). Therefore, learning from past experiences in similar systems is invaluable for formulating a good theory. Look for something that happened in the past rather than to large molecular-weight protein molecules wreaking havoc in your system. Talk to people that operate similar columns or check experiences in the literature (see Section 1.5).

At the same time, beware of being biased by experience in past cases. Treat a theory based on past experiences as one more theory that needs to be tested.

5. *Beware of Biases*

It is important to prevent biases from steering theory formulation in the wrong direction. Mostia (328) elaborates on the variety of biases. Foremost is tending to see what one wants to see. Other biases include being swayed by presentation quality (speech, visuals, color), initial piece of information, group thinking ("bandwagon effect"), or an authority in the field. To counter biases, it is essential to be aware that they exist (no one is immune) and to watch out for them when formulating theories.

6. *Visualize What is Happening*

When formulating a theory, attempt to map the paths of liquid and vapor travel inside the column. Imagine yourself as a pocket of liquid or vapor traveling through the column internals. Which way will you travel? Remember that this pocket will always look for the easiest path.

7. *Closely Examine Relevant Points of Transition Using Simplified Sketches*

Table 1.1 shows that points of transition (tower base, feed points, draws, distributors) are common bottlenecks. The relevant points of transition can be effectively troubleshot by preparing simplified sketches and addressing the question "Would it work like it should?" This technique is described in detail in Chapter 8. When drawings are not clear or miss important information, do not guess. Check with the supplier, designer, or mechanical engineers. If needed, prepare a cardboard model.

8. *For Chemical Systems, Brush up on the Chemistry and Solubilities of the Components*

Focus on the main components, but do not overlook any foreign material or unexpected byproducts. There have been cases (e.g., 401) in which an apparent poor separation was caused by the presence of an unexpected component. This subject is discussed in detail in Section 3.2.12.

9. *Think of Everyday Analogies*

Another useful technique is to think of everyday analogies. The processes that occur inside the column are no different from those that occur in the kitchen, bathroom, or yard. For instance, blowing air into a straw while sipping a drink will make the drink splash all over; similarly, a reboiler return nozzle submerged in liquid will cause excessive entrainment and premature flooding.

10. *Do Not Overlook the Obvious*

In most cases, the simpler the theory, the more likely it is to be correct.
Very few problems are really random. If something happens more than once, there is likely a root cause (318).
The most likely problem is probably the problem (488).
The key you lost is often in the last place you look (356).

11. *Beware of the "Obvious Flaw" Pitfall*

An obvious flaw is not necessarily the root cause of the problem. One of the most common troubleshooting pitfalls is retarding or discontinuing further investigation once an obvious flaw is identified. Often, this flaw fits in with most theories, and all are sure that the flaw is the root cause of the problem. The author has seen many cases where correcting an obvious flaw neither solved the problem nor improved the performance. Once an obvious flaw is detected, it is best to treat it as another theory and continue troubleshooting.

12. *Avoid Tunnel Vision*

Start with an empty sheet of paper (150). Beware of an obvious cause blinding the team to other causes. It is common to blame new trays or packings for poor performance initiating immediately after a tower revamp; in reality, the root cause is often unrelated to the new trays or packings. In one case (481), premature flooding following a tower repack turned out to be due to a maintenance replacement of a leaking plug valve by a lower-pressure-drop ball valve. This led to a false level indication that in turn induced excessive tower base level and premature flooding. In Case 20.2 in Ref. 201 as well as in the debutanizer example, an exchanger leak rather than the new trays was the root cause. In yet another case (237), a 40-year-old liquid draw issue bottlenecked a tower following a retray. There were other cases where the existing draw rather than the new trays turned out to be the issue (e.g., 150). In all these cases, the retray or repack was initially suspected. Systematic testing identified the real root cause.

13. *Ask "Why?"*

In the debutanizer example, it was the troubleshooter's asking "Why did the tower pressure influence the hot oil flow to the reboiler?" that was invaluable in connecting the dots.

14. *Calculation is Better than Speculation*

Premises on which theories are based can often be easily supported or disproved by calculation. In one case, it was argued that liquid entrainment was an issue. A simple calculation showed that at the upward velocities involved, the rise of any liquid drops would be reversed by gravity within less than 1 in. This totally invalidated the theory.

Calculations are only as good as their basis. Closely review any assumptions and that all the dimensions used for tray internals are correct. Request the supplier to provide any missing dimensions.

15. *Test Theories Effectively*

Testing theories should begin with those easiest to prove or disprove, almost irrespective of how likely or unlikely these theories are. If shutting the tower down is expensive (which is almost always the case) but is required for testing a leading theory, it is worthwhile to first cater to alternative theories that require less drastic actions even if they are longer shots. In the medical analogy, surgery should not be performed before a blood test that may identify a less likely cause.

16. *Use One-Variable Reversible Changes to Test Theories*

Test the response of the column to changes in variables such as vapor flow rate or liquid flow rate. Compare the results with predictions from the various theories. For instance, if a column flood responds to changes in the vapor load but not in the liquid load, any theory that argues that the flood is due to excessive liquid load is invalidated. In one case (385), determining that the tower responded to changes in the vapor load but not to changes in the liquid load invalidated the leading theory and identified the correct root cause. Change one variable at a time. If several variables are allowed to change simultaneously, the result is likely to be inconclusive.

Take time to plan every step of your test and consider all possible outcomes (356). Tests that reveal very little not only are a waste of time and effort, but also undermine people's confidence in the investigation. Discussing the plans with knowledgeable individuals can help avoid this trap.

Refrain from making any permanent changes until all practical tests are done.

17. *Can the System be Simplified?*

Look for possibilities of simplifying the system. For instance, if it is uncertain whether an undesirable component enters the column from outside or is generated inside the column, consider operating at total reflux to check it out.

In one amine absorber (225), where foaming was suspected, a field trial was conducted in which the amine solution was replaced by nonfoaming clean water. The water trial showed an identical capacity limitation, conclusively denying foaming. A small-diameter chemical tower (481) flooded after packing replacement. The flood persisted in a trial where the old packings were reinstalled, ruling out the new packing as the root cause.

Be concise in drawing conclusions from simplified tests. For instance, in the amine absorber trial discussed above, the plant took special care to check the water quality before the trial. Had there been doubts about water cleanliness, foaming could not have been ruled out.

18. *What Works in Other Towers may not Work in Yours*

Critically examine the argument that the same equipment has been performing flawlessly in a similar/identical column for many years (456). No two columns are the "same." Small variations may make large differences in performance. Closely compare the variations; they may lead you to the root cause.

19. *Do not Overlook Human Factors*

People act based on their reasoning, which is likely to differ from yours. People often have their own agendas, especially when they have a lot to lose in an unfavorable outcome. The more thoroughly you question their operating or design philosophy, the closer you will be able to reconstruct the sequence of events. Their replies may also reveal considerations you are not aware of. Be cautious in your questioning. The attitude that you want to learn more about the system and what can be improved will win cooperation. Pointing fingers or implying that someone screwed up is a sure way of getting noncooperation (411).

20. *Avoid Confirmation Bias or Wishful Thinking*

It is human nature to give preference to data or information that favors one's beliefs over conflicting data. Such bias can steer an investigation away from the correct theory. Be open to new ideas and beware of human nature to rely on the initial diagnostic impression.

To maintain an unbiased attitude when working with conflicting data, request a "cold eyes review" by knowledgeable colleagues or use an alternative analysis method (456). When working with experts, keep in mind step 17 in Section 1.3.

21. *Ensure that Management is Supportive*

Ensure that management is well informed of what is being done and is receptive to it (454, 498). Otherwise, important nontechnical considerations may be overlooked. Further, management is far less likely to become frustrated with a slow-moving investigation when it is convinced that the best course of action is being taken.

Often, management is done by technical people with expertise that can contribute ideas. Moreover, such technical people often expect that their ideas are incorporated into the testing.

22. *Involve the Supervisors and Operators in Each "Fix"*

Whenever possible, give supervisors and operators detailed guidelines for the fix attempt and leave them with some freedom to make the system work. The author was involved in several cases where the actions of a motivated operator made a fix work and had seen other cases where a correct fix was unsuccessful because of an unmotivated effort by the operators.

23. *Promote Teamwork and Prevent the "Us Against Them" Division*

With different people having different ideas and theories, it is important to assemble all these ideas into constructive teamwork and to suppress any confrontations. Some people will have a stake in their theory being correct. They will feel that they win when their theory is pushed ahead and feel rejected when their theory is dismissed. Good troubleshooting leadership needs to encourage all ideas, treat all respectfully, recognize that even the ideas that are disproved contribute in the path to solving the problem, and acknowledge their initiators accordingly.

24. *Do not be Afraid to Admit when You are Wrong*

Admitting that you are wrong is inherently difficult to do. Nonetheless, recognize that the investigation is not about who is right and who is wrong, but about finding the correct technical solution. Everyone serves on the same team, and all will win when the correct solution is found. Accepting the truth, or accepting that others' ideas are superior to yours, sends the message of cooperation and the dominance of technical validity. This will promote idea exchange, productivity, and teamwork.

25. *Beware of Poor Communication While Implementing a "Fix"*

Verbal instruction, multidiscipline personnel involvement, and rush generate an atmosphere ripe for miscommunication (161). Ensure any instructions are clear, concise, easy to follow, and sufficiently detailed. When leaving a shift team to implement a fix by themselves, provide them with written instructions. Be reachable and encourage communication should questions or problems arise. Call in at the beginning of the shift to check whether the shift team understood your instructions and are good with them.

26. *Recognize that Modifications are Hazardous*

Unforeseen side effects of even seemingly minor modifications have been the root cause of many accidents. Disallow "back of an envelope" modifications, as their side effects can generate hazards. Properly document any planned modification, and have a team systematically review it with the aid of a "HAZOP" or similar checklist. Prior to completion, inspect to ensure the modification was implemented correctly and as intended.

27. *Properly Document Fixes*

Document any fix attempt, the reasons for it, and the results. This information will be useful for future fixes. In many cases, a sudden change in plant conditions lowers the priority of a troubleshooting endeavor, and it is discontinued. At a later time, the endeavor is resumed. Good documentation of the initial endeavor gives the resumed endeavor a much better starting point. At one time, we designed and built baffles to prevent vortexing near a feed inlet, only to find that similar baffles had already existed, but did not show on the drawings and no one in the plant knew about them.

28. *Be on the Look for and Capitalize on Testing Opportunities*

Tests that require column shutdown or low-rate operation are usually impractical. Nonetheless, unforeseen circumstances such as plant slowdown or crash shutdown may open an opportunity for performing them if one moves fast. Be alert for and capitalize on such opportunities. In one large column where a tray malfunction was suspected, a crash shutdown opened an opportunity for installing 10 well-designed trays as a trial fix. A subsequent gamma scan showed that this fix would solve the problem.

29. *Take Advantage of Shutdowns to Fill in the Gaps*

If a plant shutdown occurs before the tower problem is diagnosed, do not miss the opportunity to enter the column and investigate. Inspecting column internals often reveals unexpected features that contribute to or are the root cause of the problem (Chapter 10). Opportunities for measuring and taking photographs of the internals will not return once the tower is back in service and may turn out crucial. In a recent experience, the question of whether 40 welded-in bubble cap trays should be cut out and replaced (a mammoth task) hinged on the distance between the vapor riser and the cap, and this dimension was unavailable.

30. *Avoid a Chaotic Approach when a Deadline Approaches*

When a deadline approaches (e.g., a plant turnaround), there is pressure on the team to suspend troubleshooting efforts and to proceed with an arbitrary fix, often imposed by the more vocal or emotional members of the team. This may not coincide with the best engineering fix (455). Anticipate this possibility well before the deadline, and work with the team to promote the best engineering fix while there is still plenty of time.

1.5   LEARNING TO TROUBLESHOOT

Troubleshooting is not magic, nor is it performed by magicians. It is a learned art. Unfortunately, not much of it is taught at school, although a few university courses on troubleshooting exist. It is learned in the school of hard knocks. You can avoid most of these hard knocks by learning from other people's experiences. The objective of Refs. 196, 198, 201 was to put these experiences in the hands of every engineer, supervisor, or operator who is interested. Failures are repetitive, and learning from the past can solve today's problems and avoid tomorrow's.

Three elements have been listed as critical to successful troubleshooting (16): knowledge of the process and equipment, experience with operations and solving problems, and using an effective method to solve the problem. Training programs should provide an understanding of the process and equipment, be based on a large number of experiences, include examples and exercises based on these experiences, proceed stepwise from simple to complex, address interactions with other units, and be accompanied by a relevant manual (14, 16).

There are many other resources. Talk to the experienced people in your plant and organization and to fellow workers in professional meetings, and attend their presentations. Get involved in startup, shutdown, and commissioning work. Get involved in incident investigations. Inspect equipment and participate in equipment testing.

Consider supplementing the above with self-training. After three years in technical services, I was transferred to operations on startup duties, needing to become an overnight troubleshooting expert. I picked up a notebook and talked to experienced people, taking notes at each stop and collecting lessons, guidelines, and advice. I combed the literature in search of other people's experiences, often writing to the authors. Their stories and wisdom, together with many other lessons my colleagues and I learned in the school of hard knocks over the years, are among the pages of this book. It is my hope that all these invaluable lessons will be useful to future students of the art of troubleshooting.

1.6 CLASSIFICATION OF COLUMN PROBLEMS

The problems usually experienced in distillation columns can be classified as follows:

Capacity problems. the column cannot achieve the required feed or product flow rates at the design reflux/boilup rates or incurs excessive pressure drops.

Separation problems. the column cannot achieve the required or design separation at the design reflux/boilup rates. In some cases, the column cannot achieve product specs even when reflux and boilup are raised. In other cases, the column works well at maximum rates but unexpectedly loses efficiency upon turndown.

Instability problems. the column cannot operate under stable conditions or can become touchy and sensitive to small changes in operating conditions.

Often, a separation problem may show up as a capacity or instability problem. The reason is that due to poor separation, operators increase the reflux and boilup to maintain the product on-spec. This hydraulically loads up the column, and the problem shows up as a capacity limit or, when operating right near the limit, as an instability. Conversely, a capacity problem may produce premature flood, which shows up as poor separation or instability. Likewise, pressure or temperature deviations may be a reflection of premature flood or poor separation or may cause them.

Pressure or temperature deviations. the column cannot attain the expected or design temperatures and pressures. Many times, these reflect reboiler/condenser limitations or the presence/absence of a reaction, unexpected components, or a second liquid phase.

Startup/shutdown/commissioning problems. the column operates normally at the steady state and turndown, but problems occur during abnormal operation.

Safety and environmental issues. these strongly depend on the materials processed in the tower and may arise due to design or operation flaws, or unforeseen conditions, either during normal or during abnormal operation.

Excessive downtime. this may be due to fouling, foaming, or damage episodes.

Non-optimum issues. these may lead to purity issues, capacity limitations, or excessive energy usage.

Excessive operator attention. this may be due to instability or operation deviating from the design.

The troubleshooter's challenge is to distinguish the cause from the result. The following chapters will cover the primary techniques available to narrow the root cause down.

Chapter 2

Troubleshooting for Flood

"Troubleshooting, revamping, and eliminating waste offer huge benefits in reducing capital investment, downtime, carbon dioxide emissions, and energy consumption. Unfortunately, the attention paid to this resource in the energy-transition era has been too little to reflect its tremendous potential."

—(Henry Kister, author)

2.1 FLOODING: THE MOST COMMON TOWER THROUGHPUT LIMITATION

Flooding is accumulation of liquid in the tower. This accumulation propagates upward from the flood initiation zone, filling the regions above with frothy liquid. The accumulation may build up all the way to the top of the tower, or to a point where an abrupt change in hardware design or flow conditions (e.g., feed point) interrupts further accumulation. The flood may or may not propagate above that point.

The capacity of most distillation and absorption towers is limited by flooding. Flooding is a common cause of deterioration in tray or packing efficiency, separation falling off, and product purity going off-spec. When the accumulation builds up to the top of the tower, massive liquid entrainment may occur from the tower and reach downstream units, causing contamination and sometimes equipment damage. Flooding may generate intermittent steps of liquid buildup and dumping, which destabilize the tower. This instability may propagate to downstream or thermally coupled units. Liquid accumulation in the tower lowers the bottom flow rate, sometimes starving the bottom pump and causing its cavitation. To prevent the adverse effects of flooding, operators cut throughput, and the plant loses capacity.

The diameter of new towers is set to sustain operation a comfortable margin (typically 20%) below the flood point. In existing towers, throughput is limited by the onset of flooding, which may bottleneck the entire plant. The maximum throughput for tower revamp designs is usually restricted by the approach to (typically 10%) the flood point.

Troubleshooting for flood is mandatory for correctly diagnosing the root cause of poor tower performance and/or debottlenecking it. There are many alternative causes of poor separation, including inefficient trays or packing, poor liquid distribution, excessive weeping, or damage to internals. Liquid entrainment from the top of the tower may be caused by poor reflux piping. Instability may result from oscillating control loops, poor controller tuning, or feed rate

Distillation Diagnostics: An Engineer's Guidebook, First Edition. Henry Z. Kister.
© 2025 American Institute of Chemical Engineers, Inc. Published 2025 by John Wiley & Sons, Inc.

fluctuations. Pump cavitation may result from air leak into the pump suction (under vacuum) or from undersized suction piping. If flooding can be ruled out, the troubleshooter can focus on one of these alternative causes.

It is equally important to correctly identify the location and nature of the flood. The author has seen many cases where misdiagnosis of the root cause or the location of flooding led to ineffective solutions or debottlenecks. When misdiagnosed, the flooding persists and even worsens, and engineers lose their credibility, if not their jobs.

Once the flood, its root cause, and location are correctly diagnosed, an adequate solution can be devised. The solution usually involves hardware modifications to tower internals or auxiliary equipment, or process re-optimization (e.g., changing feed preheat, tower pressure).

This chapter aims to provide engineers with the tools needed to distinguish flooding from other issues and to diagnose the nature of the flood, its location, and its likely mechanism. Only nonradioactive techniques for flood diagnosis are covered in this chapter. Radioactive techniques such as gamma scans and neutron backscatter are highly effective for diagnosing flood and are discussed at length in Chapters 5 and 6.

2.2 FLOOD MECHANISMS IN TRAY AND PACKED TOWERS

As stated in Section 2.1, flooding is accumulation of liquid in the tower. This accumulation propagates upward from the flood initiation zone. This applies regardless of the mechanism described below.

In tray towers, there are four different major mechanisms that lead to this liquid accumulation:

1. *Entrainment (Jet) Flood.* Figure 2.1 shows an unflooded and a flooded tray. As the vapor velocity is increased, froth or spray height on trays rises. When the froth or spray approaches the tray above, some of the froth or spray is aspirated into the tray above as entrainment. Upon further increase in vapor velocity, massive entrainment of the froth or spray begins, causing liquid to accumulate on the tray above and flood it.

   Entrainment (*jet*) flood is the most common flood mechanism for towers operating below 100 psia. The active areas of trays are sized to keep tower operation within a comfortable margin from jet flood (typically keeping the design vapor rate 10–20% below the jet flood vapor rate at a constant liquid to vapor ratio). Entrainment flood is always

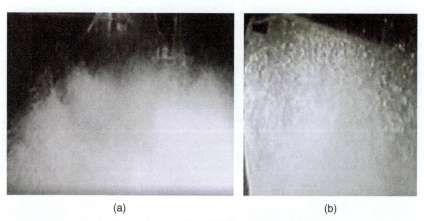

(a)   (b)

Figure 2.1 Unflooded and Flooded Tray (a) Unflooded, froth height about 12 inches; (b) Jet-flooded, froth height exceeding tray spacing. (Copyright © FRI. Reprinted Courtesy of FRI.)

primarily a vapor flood. If liquid has an effect, it is very secondary. The key parameter for jet flood is the C-factor, given by

$$C_{active} = U_{active}\sqrt{\frac{\rho_G}{\rho_L - \rho_G}} \tag{2.1}$$

where C_{active} is the C-factor (ft/s), i.e., the superficial vapor velocity U_{active} (ft/s) based on the active area of the tray, adjusted for vapor and liquid densities (ρ_G and ρ_L). The flood point occurs at a maximum C-factor that is calculated from jet flood correlations that can be found in most standard texts (34, 193, 245, 299, 451, 492).

2. *Downcomer Backup Flood*. Tray pressure drop, liquid height on the tray, and frictional losses in the downcomer apron (clearance head loss) cause aerated liquid to back up (or build up, or stack up) in the downcomer, as illustrated in Figure 2.2. Note that the relevant pressure drop is that across the tray above, as the pressure difference between the vapor space above the lower tray and that above the upper tray pushes the liquid up in the downcomer.

   All of these increase with *higher* liquid rates; tray pressure drop also increases with higher vapor rates. When the backup of aerated liquid fills up the downcomer (i.e., exceeds the [tray spacing + weir height]), it will start accumulating on the tray above, causing downcomer backup flooding.

   In the absence of fouling or foaming, aerated downcomer backup usually does not rise above 16–18 in. Downcomer backup is therefore an uncommon flood mechanism at tray spacings of 24 in. and higher. The lower the spacing, the more likely it is to run into a downcomer backup flood. When the tray spacing is 18–20 in. or less, it is important to closely check for downcomer backup.

   When tray holes or valves plug up, downcomer backup shoots up parabolically. The dry pressure drop (vapor friction pressure drop incurred by the vapor flow through the tray openings in the absence of liquid) is proportional to the hole velocity **squared**. The dry pressure drop across a tray is typically 2–3 in., which incurs the same clear liquid backup in the downcomer. This would translate to about 5 in. of aerated liquid. If half the holes

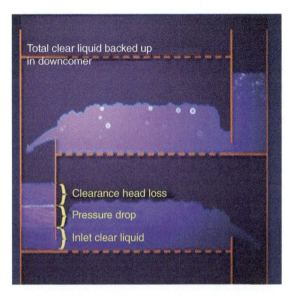

Figure 2.2   Downcomer backup in heads of clear liquid. (Copyright © FRI. Reprinted Courtesy of FRI.)

plug, the dry pressure drop will escalate by a factor of two **squared**, from 2 to 8 in. The additional 6 in. liquid head will raise the downcomer froth by about 10 in., which can easily rise to the tray above and initiate accumulation. A plugging/fouling flood is almost always a downcomer backup flood because of the parabolic rise in vapor friction (dry) pressure drop as holes or valves plug up.

Downcomer backup flood is also common in foaming services. Foam formation increases the aeration (i.e., the vapor content) of the downcomer liquid, which in turn raises the height of the foamy dispersion in the downcomer to the tray above, initiating accumulation there.

Criteria for downcomer backup flood calculation are available in most distillation texts (34, 193, 245, 299, 451, 492).

3. *Downcomer Choke Flood (also called downcomer entrance flood or downcomer velocity flood)*. The downcomer entrance area must be sufficient to transport all of the liquid downflow. The material entering the downcomer is the dispersion on the tray, which is typically only 20–30% liquid by volume, the balance being vapor. In the downcomer, the vapor disengages from the liquid and returns to the vapor space above the tray. As the vapor–liquid mixture enters the downcomer, it is countered by the vapor venting back from the downcomer liquid. If the entrance area is not large enough, the venting vapor will prevent the liquid from descending ("choke the downcomer"). Some of the liquid will then accumulate, initiating a downcomer choke flood on the tray above.

This phenomenon is analogous to liquid pouring out of a bottle or a cup. When a water-filled open bottle is turned upside down, air entering upward to take the place of the outflowing water impedes the water downflow rate at the neck of the bottle, and the bottle will empty slowly. In contrast, if a water-filled glass is turned upside down, the area at the neck will be too large for the air to impede the water downflow, and it will empty instantaneously.

Just like in the analogy, downcomer choke flood is primarily dependent on the cross section area at the entrance to the downcomer. Downcomer choke floods are always primarily liquid floods. If vapor has an effect, it is very secondary. The key parameter is the liquid velocity at the downcomer entrance. Criteria for the maximum downcomer inlet velocities can be found in many distillation texts (34, 193, 299, 451, 492), with the most updated list in Ref. 245.

Downcomer choke flood is common in high-pressure distillation where liquid loads are high and the liquid is foamy. Over 90% of the floods experienced by the author at pressures 165 psia and above were downcomer choke floods. Downcomer choke floods are also common with foaming systems, where the foam has difficulty entering the downcomer and disengaging from the liquid. Finally, downcomer choke floods are commonly experienced with high liquid rates.

4. *System Limit Flood (also called ultimate capacity flood)*. This is an ultimate jet flood, which occurs when the vapor *velocity* acting to lift the large liquid drops above the tray exceeds the terminal velocity of the drops. Differently expressed, system limit occurs when the momentum force pushing the large drops upward exceeds the gravitational force trying to bring them down. System limit flood is independent of tray geometry and tray spacing. The C-factor used for jet flood (Eq. 2.1) is also used to characterize the system limit. References 245 and 463 provide a good correlation for system limit.

In packed towers, there are three flood mechanisms:

1. *Flood in the vapor-rich region*. As vapor velocity is raised, a point is reached where the high vapor velocity interferes with the free drainage of liquid. The packed bed begins to

load up with liquid. Upon further increase in vapor velocity, rapid liquid accumulation takes place, initiating a flood.

This is by *far* the most common packing flood mechanism. Again, it is characterized by the C-factor (Eq. 2.1), but here the C-factor is based on the entire tower cross section area rather than the tray active area. Theoretically, the cross section area occupied by the packing should be subtracted from the tower cross section area, but with modern metal or plastic packing this area is small (typically 1–3% of the tower cross section area) and is neglected. A procedure for calculating the flood velocity is described elsewhere (193, 245).

2. *Flood in the Liquid-rich region.* Liquid holdup in the packing directly increases with liquid *load*, becoming much higher at high liquid loads. In high-pressure distillation, the higher vapor densities slow down the liquid descent, and the lower surface tensions increase frothiness. All these factors make liquid drainage more difficult. As vapor or liquid loads are raised, a point is reached where the drainage of liquid is impeded. The bed starts accumulating liquid. Flood initiates when the accumulation becomes large. The same procedure applied for flood calculation in the vapor-rich region (193, 245) can be used for this region as well.

3. *System Limit Flood (also* called *ultimate capacity flood*). This is the same as in tray towers.

2.3 FLOOD AND FLOOD MECHANISM DETERMINATION: HYDRAULIC ANALYSIS

Calculating the proximity of flood limits is invaluable in diagnosing the root cause and location of a tower flood or capacity problem.

Hydraulic procedures are available in most distillation texts (e.g., Refs. 34, 193, 299, 451, 492), with the most updated in Ref. 245, for calculating the percent of flood, i.e., the proximity of each tray or packed bed in the tower to the points of initiation of each of the various flood types. Hydraulic calculation software is also available from technology suppliers like Fractionation Research Inc. (FRI), from equipment vendors, and in tower simulation software.

First, a tower simulation is prepared per the guidelines in Chapter 3 to provide valid internal vapor and liquid loads and physical properties for each stage in the tower. The simulation should be based on the highest throughput before the tower runs into trouble. Adherence to the procedure in Chapter 3 is important, as internal vapor and liquid loads based on a simulation that incorrectly represents plant data will lead to invalid hydraulic analysis and a misleading diagnosis. The highest internal vapor and liquid loads in each section of tower are then used in the hydraulic calculations together with the relevant physical properties.

A case where the hydraulic analysis was sufficient to diagnose the root cause of a tower flood (24) is illustrated in Table 2.1. This tower was an olefins plant demethanizer, 6 ft diameter at the bottom, swaging to 3 ft at the top. When the feed rates were raised, flooding was observed just above the swage. The tower was simulated based on plant data at the maximum throughput. Capacity limits were calculated using the FRI software on the basis of the simulation and tray/downcomer geometry.

The upper six rows of Table 2.1 list the geometrical parameters. The next three rows list the key calculated hydraulic parameters. The next three rows present the calculated results as "Jet flood, %," "Froth in the downcomer, %," and "Downcomer choke velocity, %." These describe the proximity of the operating parameters to main capacity limits. A value approaching or exceeding 100% for any of these parameters indicates likely flooding by this mechanism.

Table 2.1 Demethanizer hydraulic analysis that diagnosed tower problem. (From Bellner, S.P.,W. Ege, and H. Z. Kister, Oil & Gas Journal, November 22, 2004. Reprinted Courtesy Oil & Gas Journal.)

	Top section		Bottom section		
	Trays 1–6	Trays 7–14	Trays 15–25	Trays 26–28	Trays 29–73
Tower diameter, ft	3	3	6	6	6
Tray spacing, inches	18	18	18	21	23
Hole area, % of active area	6.5	6.5	3.2	5.6	7
Downcomer inlet area, % of tower area	14	14	30	30	30
Downcomer clearance, inches	1.3	1.3	1.5	2.8	2.8
Outlet weir height, inches	1.5	1.5	1.5	2	2
Vapor C-factor, based on net area, ft/s	0.14	0.13	0.08	0.11	0.14
Liquid load, GPM/inch of outlet weir	2.1	2.8	6.3	11.0	13.5
Downcomer inlet velocity, ft/s	0.13	0.18	0.11	0.20	0.24
% Jet flood	47	46	62	65	70
% Froth in downcomer	58	53	106	94	87
% Downcomer choke velocity	51	53	38	65	78
Total downcomer backup, inches of liquid	6.0	6.0	11.8	12.3	12.3
Tray inlet clear liquid height, inches of liquid	2.4	2.5	3.1	4.4	4.4
Head loss under downcomer, inches of liquid	0.2	0.3	1.2	1.1	1.7
Tray pressure drop, inches of liquid	3.4	3.2	7.5	6.8	6.2
Tray dry pressure drop, inches of liquid	2.3	1.9	6.4	4.7	3.7

Table 2.1 shows all the trays to operate a comfortable margin away from jet and downcomer choke floods. In the top section, froths in the downcomers were also a large margin away from flood. In contrast, froth heights in the downcomers of trays 15 to 25, immediately below the swage, exceeded 100%, suggesting a likely flood due to excessive downcomer froth heights. On trays 26–73 below, the froth heights in the downcomers approached flood but did not get there yet. This analysis concluded that the observed flood most likely originated just below the swage (trays 15–25), and definitely not above it as was previously thought, even though the symptoms appeared above the swage.

To diagnose the root cause, the bottom five rows of Table 2.1 list the calculated hydraulic factors that contribute to the downcomer backup. The downcomer backup is the sum of the inlet clear liquid height, the tray pressure drop, and the head loss under the downcomer (34, 193, 245, 451, 492), as illustrated in Figure 2.2. Table 2.1 shows that below the swage, most of the downcomer backup is due to the pressure drop, particularly the dry pressure drop. The dry pressure drop is the friction head incurred by the vapor flow through the tray openings when no liquid is present. To give high dry tray pressure drop takes a small open area on the tray. So the hydraulic analysis identified low open area as the root cause. The hole areas below the swage were indeed low, 3.2% of the tray active area, compared to typical values of about 8–12%. Low hole areas may or may not lead to flood, and this is why it is important to compile a hydraulic analysis like in Table 2.1. The Table showed that the low hole areas led to flood on trays 15–25. Based on this finding, the sections below the swage were retrofitted with trays containing larger hole areas. This eliminated the flood and debottlenecked the tower.

2.4 OPERATING WINDOW (OR STABILITY) DIAGRAMS

These graphics, also referred to as "performance diagrams," plot the operating limits for a given tray design and operation. Within these limits, a region ("window") of satisfactory operation is defined. They were introduced by Bolles (38), and offer visualization of the proximity of the tray operation to the surrounding limits. Operating window diagrams were advocated by many authors (e.g., Refs. 28, 99, 305, 464–466) and extensively discussed in Refs. 99 and 464–466.

Figure 2.3 is an example of an operating window diagram for a tower operating at a slight positive pressure, containing trays at 24 in. tray spacing. The x-axis is the liquid weir load, which is the tray liquid flow rate (gpm) divided by the outlet weir length (inches). The y-axis is the C-factor from Eq. 2.1, representing the vapor load. To maintain constant product purity, towers usually operate at a constant reflux ratio, i.e., a constant ratio of liquid to vapor. For instance, if the vapor rate is raised, the liquid rate will be raised proportionally to keep a constant liquid-to-vapor ratio. The operating line is the long-dashed sloped line in Figure 2.3. The hard limits are plotted in Figure 2.3 as continuous lines, while the softer limits as short-dashed lines. The shaded region is the region of stable operation for the system and tray design, and is often referred to as "stable tray operation window." The limits plotted are as follows:

1. SL, the system limit.
2. JF, the jet flood.
   Figure 2.3 shows that both the SL and JF limits primarily depend on the vapor load and are only slightly affected by the liquid loads. This is typical of these limits. Figure 2.3 also shows that the jet flood limit at 24 in. tray spacing is not too far below the system limit.
3. DB, the downcomer backup limit. As stated earlier, in the absence of fouling or foaming, aerated downcomer backup usually does not rise above 16–18 in. and is therefore an uncommon flood mechanism at tray spacing 24 in. or higher. At lower spacing, or in the

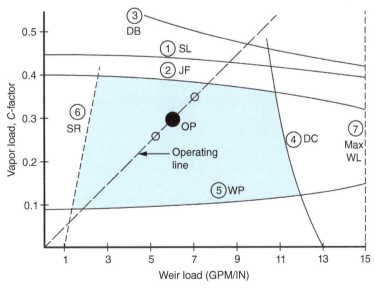

Figure 2.3 Example of a Tray Operating Window Diagram, showing the hard limits as continuous lines, and the softer limits as short-dashed lines. The shaded region is the region of stable operation. (From Kister, H. Z., "Practical Distillation Technology," Course Manual. Copyright © 2013; reprinted with permission.)

presence of fouling or foaming, the DB curve will descend and fall below the SL, and in many cases, also below the JF limits.

4. DC, the downcomer choke limit. This limit strongly depends on the liquid load, is only slightly affected by the vapor load, and strongly depends on the downcomer entrance area. A 50% increase in downcomer inlet area would move this limit way to the right.

5. WP, excess weeping. This is a turndown limit, which primarily depends on the vapor rate. WP is usually considered a hard limit, but when some efficiency loss at turndown can be tolerated, operation below this limit is possible.

6. SR, spray regime. This regime is discussed in Section 2.11. Many towers are designed to operate inside the spray regime. Getting into the spray regime is only a limit when the trays are not adequately designed to handle the spray regime.

7. MAX WL, the maximum weir load. This is another soft limit. Operation beyond this limit remains efficient. The only problem with it is that at the high weir loads, considerable capacity is lost, as depicted by the downward-bending jet flood curve in this region (Figure 2.3). Reaching this limit provides incentive to increase the number of tray passes.

Additional limits may apply. A minimum liquid rate may apply for specialty trays containing truncated downcomers that do not have positive liquid seals (28). Below this limit vapor may enter downcomers. A dumping limit, in which all the liquid weeps through the tray and nothing gets to the downcomer, is also sometimes included; this is associated with severe loss of fractionation and often with instability. A pressure drop limit may apply in some cases (99). Even reboiler and condenser limitations can be included (29).

Bernard (28) demonstrates the use of this technique to depict effects of changing the operating line, feedstock, design variables, and fouling. Summers (464, 466) and Engel (99) demonstrate application of this technique to optimize tray design.

While recognizing their merit of providing a graphic illustration of the relevant limits and the stable operating region, the author's experience has been that for the nonexpert user, these diagrams may be misleading. The limits are often specific to one set of operating conditions. The diagrams often make very poor distinction between "hard" limits and those that can be violated at a minor or no penalty. Some limits depend on inaccurate correlations. Some limits like "80% of jet flood" may be important for design, but not for operation or troubleshooting, where "100% (or 95%)" of jet flood is more meaningful. Limits like channeling are seldom shown. Foaming and fouling are often not represented, poorly represented, or based on guesswork. The author recommends caution with the application and interpretation of operating window diagrams for any specific situation.

Startup Stability Diagrams   One situation where the author and others (26, 28) found the stability diagrams particularly useful is for addressing startup conditions. Here, the above-listed limitations are usually not an issue, as the limits are few and high accuracy is not required.

In many startups, vapor is introduced to the tower before the liquid. This vapor will flow both through the trays and the downcomers. If the vapor velocity through the downcomers exceeds the system limit (based on the downcomer area), it will not allow liquid descent into the downcomer and cause flooding in the tower. This is a downcomer unsealing flood, as described in Section 2.11. An attempt to introduce liquid before the vapor is unlikely to help, because in the absence of vapor the liquid will dump through the tray openings and bypass the downcomers without sealing them. The startup stability diagram, showing both the critical vapor velocity (downcomer unsealing flood limit) at which the vapor will prevent liquid descent in the downcomer and the liquid dumping limit, is invaluable for defining the stable region on the vapor–liquid map at which neither limit will be restrictive and the tower can be started up. Once started up, the tray liquid will seal the downcomers, and the critical vapor velocity will no longer be a limit. Successful applications of this technique have been reported (26, 28, 190).

2.5 FLOOD POINT DETERMINATION: FIELD TESTING

To determine the flood point, the tower is first set in non-flooded conditions at high rates. Then the tower internal vapor or liquid flow rates or both are raised. Usually both are raised, because raising one without the other will adversely affect the tower material balance, generating a poor-purity product before flooding conditions are reached. The vapor and liquid rates during flood testing are typically raised by one of the following procedures:

1. *Raising reflux and reboil rates while keeping feed rate constant.* This is the most popular procedure. Only two variables (rather than three in the procedures below) need to be changed. Product compositions do not change until actual flooding starts. This procedure is independent of the capacities available in upstream or downstream units, making it simpler and easier to implement. Almost always, the data provided by this procedure can be easily and reliably extrapolated to predict the maximum column feed rate.

2. *Raising feed rate while simultaneously increasing reflux and reboil rates in proportion, or in a way that keeps constant product purities.* This technique provides the most direct measurement of the maximum feed rate that can be handled by the tower but can only be applied when upstream and downstream units have sufficient capacity to handle the additional feed. It may be adversely affected if the feed rate increase causes an upset in an upstream unit.

3. *Changing tower pressure.* Usually, when tower pressure is below 50 psig, this procedure lowers the pressure; when tower pressure exceeds 150 psig, it raises the pressure (219). Product purities are kept constant by manipulating reflux and reboil. This procedure is often limited by reboiler or condenser capacities or by the relief valve setting. Evaluation of the data derived from this procedure needs to take into account the effect of varying the pressure on feed vaporization.

4. *Varying preheater or precooler duty while adjusting reflux and reboil.* This procedure is only applicable when the feed is preheated or precooled. This procedure is the least popular, and is often restricted by the exchanger capacity. In some multicomponent distillations, the results may be misleading because it may induce accumulation of an intermediate impurity in one section of the column (192). When the preheater or precooler is a feed-bottom or feed-overhead interchanger, it may recycle disturbances and cause instability well before the flood condition is reached.

With all the above procedures, reflux and reboil flow rates are varied. The procedure of varying these rates influences the results and must take the tower control system into account.

In most tower control schemes, the composition (or temperature) controller manipulates either the reflux or reboil, directly or indirectly. The stream that is not controlled is regarded as "free," i.e., on flow control. During flood testing, this "free" stream flow rate is usually raised, with the temperature control automatically adjusting the controlled stream to maintain product purity. Figure 2.4 shows a tower with the temperature control on the boilup and the "free" stream the reflux rate. Flood testing in this tower raises the reflux rate, which cools the control tray, so the temperature controller will raise boilup. The net result is that both boilup and reflux are raised, and a new steady state is established.

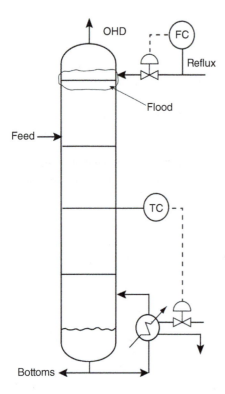

Figure 2.4 Distillation tower with reflux on flow control and temperature control manipulating boilup. Flood initiates near the top in this example. (From Kister, H. Z., "Practical Distillation Technology," Course Manual. Copyright © 2013; reprinted with permission.)

This procedure may overshoot the flood point. The higher liquid rate may take some time to reach the tray where flooding initiates, to fill the packing voids, or to accumulate on trays and downcomers to a detectable extent. So following the change, the tower may appear stable for some time, even though the flood reflux and boilup rates have been exceeded. This issue is most common in towers containing a large number of trays or in towers where the volume is large and the internal liquid flow rate is small. In the meantime, the test may progress, and the vapor and liquid rates may be raised further. The flood point thus determined will be higher than the actual one.

"Undershooting" the flood point is just as common. Its occurrence depends on the tower dynamics. For instance, increasing boilup can raise the froth height and thus the inventory of liquid on the trays (or in the packings). This extra liquid holdup may take up enough of the void space in the packing or the disengagement space between the trays to initiate premature flooding. The flood point thus determined will be lower than the actual one.

To avoid overshooting and undershooting, vapor (or liquid) rates should be raised in small steps, followed by long stabilization periods between steps. This is critical in towers with a large number of trays. A preliminary flooding test, with relatively large and fast steps, will indicate when to slow down in the final test. Typically, vapor (or liquid) rates are raised by 5–10% increments at 15- to 30-minute intervals during the preliminary test (4, 192). Increments as small as 1–2% are preferable, even at the preliminary test. It was found (187) that frequent small increases in vapor (or liquid) rates are less likely to prematurely upset the tower and generally require shorter stabilization periods (4, 187).

Although the preliminary test may suffer from overshooting or undershooting, it is likely to determine the flood point within ±10% and often within ±5% (4, 192). The results of this test provide a good starting point for the slow test. The preliminary test was found effective both for improving accuracy and reducing time requirements for flood point determination (187).

Accurate material and energy balances are required for flood point determination, should close within 3% and 5%, respectively (4, 187), and be checked before and during the test. The author encountered one case where an energy balance error of 25% made it impossible to determine whether the tower was flooding prematurely or at the intended hydraulic loads. Generally, there is no need for accurate component balances, as these have a relatively small impact on the tower hydraulic loads.

Ambient conditions should be recorded during the flood test (4). Flood tests should be avoided under extreme weather conditions, as instability may set in and heat losses (especially in smaller, poorly insulated towers) may affect the results.

Many of the guidelines described in Section 3.1.3, particularly those pertaining to material and energy balances, also extend to flood testing. However, flood tests are far less sensitive to analytical errors than efficiency tests and therefore require a much lower level of effort.

2.6   FLOOD POINT DETERMINATION IN THE FIELD: THE SYMPTOMS

Recapping the key concepts, flooding is accumulation of liquid in the tower. Almost always, this accumulation propagates from the lowest flooded region upward. Accumulating liquid backs up into the tray (or packed section) above, until the whole column fills with liquid or until an abrupt change in flow conditions (e.g., feed point) or tray/packing design is reached. Flooding may or may not propagate above that point. One or more of the following symptoms indicates flooding:

1. Excessive column differential pressure
2. Sharp rise in column differential pressure
3. Onset of fluctuations in pressure drop, bottoms level, and tray temperature
4. Reduction in bottoms flow rate
5. Rapid rise of entrainment from the top of the tower
6. Deterioration of separation (as can be detected by temperature profile or product analysis)
7. Flooded appearance on gamma scans (Chapters 5 and 6).
8. Onset of instability

Pressure drop measurements across various tower sections are the primary tool for flood detection.

Excessive Column Pressure Drop   The liquid accumulation that characterizes flooding incurs high pressure drop. In general, a normal tray pressure drop is 4–5 in. of liquid. For most organic and hydrocarbon systems whose specific gravity is around 0.7–0.8, this gives 0.1–0.14 psi per tray. If the measured pressure drop per tray is 0.15 psi per tray with an organic or hydrocarbon system, possible flooding should be suspected. Once the measured pressure drop exceeds 0.2 psi per tray and the measurement is correct, flooding is highly likely.

For aqueous systems, as well as other higher-density systems like chlorosilanes with a specific gravity of about 1.0, the normal pressure drop can be somewhat higher (about 0.18 psi/tray), possible flooding occurs when the measured pressure drop is above 0.2 psi/tray, becoming highly likely when the measured pressure drop exceeds 0.25 psi per tray, and the measurement is correct.

With packed towers, the flood pressure drop is given by the Kister and Gill Equation (193, 220, 245, 462)

$$\Delta P_{\text{flood}} = 0.115 \, F_p^{0.7} \tag{2.2}$$

where

ΔP_{flood} = flood pressure drop, inches water per ft of packing
F_p = packing factor, ft^{-1}

The values of packing factors to be used in Eq. 2.2 should be taken from the 9th Edition of Perry's Handbook (245). Packing factors originating from other sources may not be compatible with Eq. 2.2 and can lead to inaccurate or even incorrect predictions. Measured pressure drops significantly higher than the value predicted from Eq. 2.2 indicate flooding.

While a high pressure drop (per the above criteria) almost always signifies flooding, many towers flood, yet the pressure drop remains low. The pressure drop rise is due to liquid accumulation. When the liquid accumulation is small, the pressure drop rise is small. Typical scenarios include flooding near the top of a tower (only a short packed length or a few trays accumulate liquid), flooding with channeling (the channeled vapor bypasses the accumulated liquid), flooding in vacuum-packed towers (accumulation is channeled and the vapor bypasses the accumulation zone), and flooding at low liquid rates (slow rate of liquid accumulation).

Techniques for pressure drop measurements are discussed elsewhere (4, 54, 192, 410).

Watch out for the dynamic velocity head of a vapor feed or reboiler return adding to the pressure drop reading. In Case 4 in Section 8.16, the dynamic velocity head of the vapor–liquid reboiler return mixture converted into static pressure at the tap, giving a reading of 165 psig, compared to 160 psig at a neighboring tap. Likewise, the pressure drop across the wash bed in a refinery vacuum tower (335) declined from 3.1 mmHg (indicating flood) to 0.7 mmHg (indicating normal operation) when the pressure measurement tap was switched from the feed (flash) zone to a point above the chimney tray just below the bed. The high pressure at the feed zone tap was due to impingement of the high-velocity vapor/liquid feed.

Watch out for reranged differential pressure transmitters in towers where trays were replaced by low pressure drop internals like packings (166). The accuracy of a reading is a fixed percentage of the range of the sensing element. Reranging the output signal to go full scale at a lower value increases the fractional error. Lower-range transmitters should be used instead.

Closely audit the location of the pressure drop points, especially when trays are replaced by packings. In one revamped tower (442), a pressure drop of 1.1 psi was measured across the top packed bed, more than double the design. The lower pressure tap was in the overhead vapor line. The real pressure drop across the bed was only 0.4 psi, while the remaining 0.7 psi was the entrance loss into the overhead vapor pipe.

In high-pressure (>50–100 psig) packed towers, depending on the measurement method, the vapor static head between the lower and upper taps may need to be subtracted from the pressure drop reading. This pressure drop may exceed the packing dynamic pressure drop (4, 54, 512). In dP cells, the gas filling the measurement leg may have a different density than the vapor in the tower, and the difference would affect the pressure drop measurement. A detailed procedure allowing for both of these effects is very well illustrated by Xu et al. (512) and by Cai and Resetarits (54).

Sharp Rate of Rise of Pressure Drop   A sharp rate of rise of pressure drop with vapor load is a more sensitive flooding indicator than the magnitude of pressure drop. Pressure drop rises with increasing vapor loads. Upon flooding, the pressure drop rise is largely enhanced by the liquid accumulation. In many cases, once started, the pressure drop will keep rising even when vapor loads are no further raised.

The flood point can be inferred from a plot of pressure drop against vapor flow rate. It is the point where the slope of the curve steepens (Figures 2.5 and 2.6). In tray columns, the

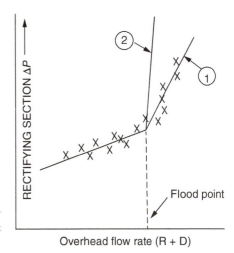

Figure 2.5   Plots of pressure drop versus vapor flow rate for tray column. (From Kister, H. Z., "Distillation Operation," Copyright © McGraw-Hill, 1990. Used with permission of McGraw-Hill.)

slope change can be relatively mild (curve 1), or relatively steep (curve 2), as illustrated in Figure 2.5. A mild slope is more common for jet (entrainment) flooding or a small number of flooded trays, while downcomer flooding or flooding that propagates throughout several trays tends to give a steeper slope. It is not unusual to find a vertical slope once the flood point is reached (119, 185).

In packed towers, inferring the flood point from a pressure drop versus load curve (Figure 2.6) is generally more difficult, because the slope begins changing at the loading point (point *B*), and

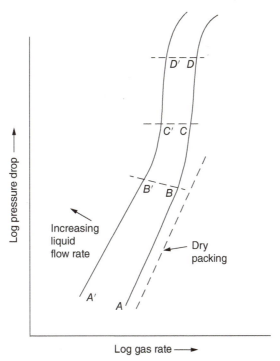

Figure 2.6   Typical pressure drop versus vapor flow rate for packed tower. (From Kister, H. Z., "Distillation Operation," Copyright © McGraw-Hill, 1990. Used with permission of McGraw-Hill.)

the change may be continuous (curve *BCD*), rather than sharp. Further, many packed towers experience a rapid drop in efficiency well before reaching the hydraulic flood point. Here, the throughput limit is the loss of separation, and the hydraulic flood point may be of lesser practical significance. There are some cases of flood, especially in vacuum distillation, where no point of inflection is observed (232).

For the best definition of the rate of pressure drop rise, it has been recommended (187, 192, 319) to install differential pressure recorders across each section of the tower prior to troubleshooting and flood testing. This technique (Figure 2.7) can clearly identify the region of flooding initiation. In one case (204), applying this technique to identify the flood initiation zone prevented misdiagnosis and unnecessary shutdown. In another case (363, 383), it conclusively confirmed that the flood initiated in the bottom section of a column.

The open circles in Figure 2.8 show the multitude of points where pressure transmitters can be installed. Pressure drops between judiciously selected pairs of points can concisely identify the flood location and nature. In one case (21), at upset conditions, the top packed bed pressure drop fluctuated between 0.24 and 1.75 psi, while the pressure drop between the bottom tray and the top of the tower cycled at the same frequency between 3 and 4.7 psi, roughly the same amount. This proved that the pressure drop increase was in the top packed bed, not on the trays below. There was a pressure measurement point (not marked on Figure 2.8) in the top P/A (pumparound) liquid draw line from the collector tray, which gave a pressure drop of over 2 ft of liquid across the collector tray (the risers were only 10 in. tall), proof that the collector tray was liquid-full and backing liquid into the packing. The collector tray would fill up only if the downcomers of the upper trays below were plugged and backing up. This concise identification of the flood location and nature led to a solution that eliminated the flood without tower shutdown.

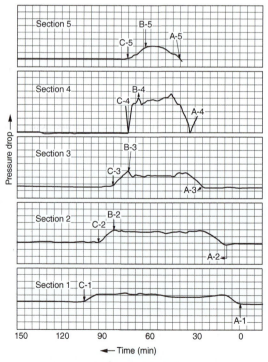

Figure 2.7   Pressure drop profile obtained with high-speed multichannel strip-chart recorder. (From McLaren, D. B., and J. C. Upchurch, Chemical Engineering, p. 139, June 1, 1970. Reprinted with permission.)

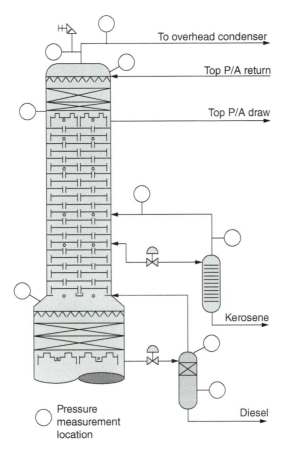

Figure 2.8   Pressure drop measurement locations in one complex tower (the upper part of an atmospheric crude tower). dP measurements can be conducted between any two and helped identify flood location and nature of flood. (From Barletta, T., J. Nigg, S. Ruoss, J. Mayfield, and W. Landry, Hydrocarbon Processing, p. 71, July 2001. Reprinted with permission.)

A similar technique using a multitude of pressure measurement points enabled Öztürk et al. (354) to diagnose a flood zone between the kero draw and the top in another atmospheric crude tower. Similarly, Barletta and Kurzym (20) used two high-accuracy electronic gauges to identify the top pumparound and the heavy naphtha pumparound to be the two flooded zones out of six zones in an atmospheric crude tower. They took pressure readings simultaneously to ensure column pressure variations would not influence the measured pressure drops.

If it can be afforded, a multichannel pressure drop recorder has been highly recommended (166, 319) and successfully applied (119, 319) for flood testing and diagnosis. This instrument can trace the sequence of events, zero in on the location of flood initiation, the conditions when it occurs, how the flooding propagates, and which remedial action is working. This instrument is invaluable for identifying flooding at the top tray (e.g., due to plugging), when the pressure drop rise may be too small to be noticed on the section or tower differential pressure. Figure 2.7 (319) shows an application of this technique, depicting step-by-step initiation and propagation of flood and, later, of the tower getting out of flood, clearly identifying the location and timing of each step.

Onset of Fluctuations in Pressure Drop, Bottoms Level, and Tray Temperature   This symptom is less common than the previous two, and may be caused by factors other than flooding.

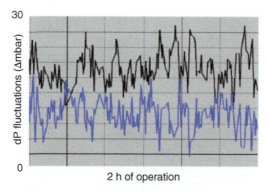

Figure 2.9   Pressure drop fluctuations due to foam-flooding of a chemical tower. Lower curve, pressure drop of packed bed above the feed, upper curve, pressure drop of trays below the feed. (From Yarbrough, B., T. Cooper, and C. Davidson, "Extractive Column Troubleshooting after Internals Revamp," Paper presented in the Distillation Symposium, AIChE Spring Meeting, Virtual, August 18, 2020. Reprinted with permission.)

It signifies liquid accumulation above a certain point, and when enough liquid builds up, the liquid rapidly dumps. In the Figure 2.8 case cited above (21), at stable operation the top packed bed pressure drop was 0.2 psi, but upon an upset fluctuated between 0.24 and 1.75 psi. The dumping step is usually faster than the buildup step and is accompanied by cooling down below the flooded zone.

Figure 2.9 shows pressure drop fluctuations due to foam flooding of a chemical tower (516). The tower contained a packed bed above the feed, with trays below the feed. Both the trays and packing pressure drops fluctuated, the fluctuation amplitudes varied 6–12 mbars (0.1–0.2 psi). The average tray pressure drop aligned with the design, while the packing pressure drop was higher than expected. Gamma scans showed intermittent loading and unloading of the packed bed, no tray flooding, and signs of foaming throughout. A thorough investigation determined definite foaming in the tower (516). Senger and Wozney (422) reported similar pressure drop behavior "pulsating flood" in some of their tests with another chemical foaming system in a 12 in. ID tower that flooded from the random-packing support up.

Pressure drop fluctuations are not unique to foam-flooding. Hanson and Lee (148) report cycles of pressure drop dumping and building resulting from partial trays plugging in a non-foaming service. These were accompanied by temperature excursions and bottom level loss. The author experienced similar cycles in a similar service.

The pressure drop fluctuations are often accompanied by bottom level fluctuations that tend to be out-of-phase with the pressure drop fluctuations. A pressure drop peak represents liquid accumulation and will correspond to a bottom level valley. Similarly, a pressure drop dive represents a liquid dump and will coincide with a bottom level rise. In packed towers, the pressure drop peaks tend to occur simultaneously with the bottom level valleys, and vice versa. Trays have time constants of the order of 0.1 minutes per tray, so the pressure drop and bottom level peaks tend to be out of phase.

Figure 2.10 shows operating charts from a large chemical column (298). Here, the bottom sump level and tray temperature 10 trays above it ("tray N") showed very repeatable oscillations at a fixed amplitude and a period of several minutes. The amplitude was small at incipient flood, steeply rising upon the onset of full flood. The plant added an incipient flooding indicator that detected the small-amplitude fluctuations and warned the operators. Adding this indicator reduced the average number of flood episodes per year from 12 to 2 and raised tower capacity by 5%.

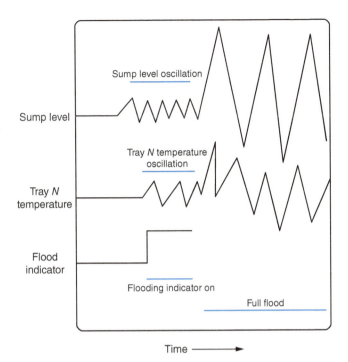

Figure 2.10   Bottom level and Tray 10 temperature fluctuations upon incipient flooding, intensifying upon the onset of full flood. (From Litzen, D. B., and J. L. Bravo, Chemical Engineering Progress, p. 25, March 1999. Reprinted Courtesy of Chemical Engineering Progress (CEP). Copyright © American Institute of Chemical Engineers (AIChE).)

Reduction in Bottoms Flow Rate   Reduction in bottom flow is a common symptom of flood, and a primary criterion to determine the flood point (192, 293, 437). Upon flooding, liquid accumulates in the tower, so less liquid reaches the bottoms, and the bottom level falls. Most frequently, the bottom level is controlled by manipulating the bottom flow rate, so the level may stay constant while the bottom flow rate is reduced.

While a reduction in bottom flow indicates flooding, many towers flood with no significant decline in bottom level or flow rate. For instance, if the feed is mostly liquid and flooding occurs above the feed, the bottom section may continue to operate normally with no significant decline of bottom level or flow rate. Also, if the flood location is well above the bottom, the time delay from the onset of flooding to the reduction in bottom level or flow may make accurate measurement of the flooding conditions difficult.

Generally, reduction of bottoms flow rate is a good indicator of flooding in towers that flood near the bottom and in towers that are relatively short (437).

Rapid Rise in Entrainment   A rapid rise in entrainment is another common flood symptom (4, 192, 284). The liquid accumulating in the tower builds up to the top and is entrained in the tower overhead vapor stream.

When the tower overhead vapor stream goes to a knockout drum (e.g., absorbers), or to the bottom of another tower, this entrainment can be recognized as buildup or rapid rise of a liquid level in the drum or bottom of the downstream tower. In most distillation towers, the overhead vapor stream goes to a condenser, and from there to a reflux drum. The reflux drum level control usually manipulates either the distillate rate (Figure 2.11a) or the reflux rate (Figure 2.11b).

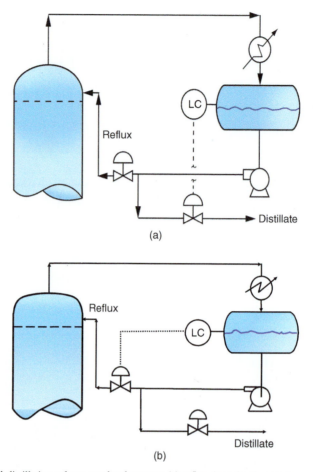

Figure 2.11 Typical distillation column overhead systems (a) reflux drum level control connected to the distillate valve (b) reflux drum level control connected to the reflux valve. (From Kister, H. Z., "Practical Distillation Technology," Course Manual. Copyright © 2013. Reprinted with permission.)

When the drum level manipulates the distillate rate (Figure 2.11a), the entrainment rise is often indicated as a significant increase in distillate rate for no apparent reason. When the drum level manipulates the reflux rate (Figure 2.11b), the entrainment is often indicated as an increase in reflux for no corresponding increase in boilup rate. The flooding prevents most of the increased reflux from descending down the tower, so it entrains back into the tower overhead, further increasing the drum level and reflux rate. The reflux flow meter reads high and the reflux valve often opens widely due to the recirculation of entrainment around the tower overhead loop. Alternatively, the entrainment is often indicated as a rise in reflux flow rate that does not raise the heat input required to maintain the same tower temperatures. In one case (280), the primary flood symptom was a normal reflux rate even though the boilup rate was extremely low.

In some cases, as in many refinery fractionators, tower overhead temperature (instead of drum level) manipulates the reflux rate (Figure 2.12). Upon flooding, entrainment from the top raises the heavies content of the tower overhead, and therefore the overhead temperature. The temperature controller raises reflux. The flooding prevents the increased reflux from descending down the tower, so it entrains back into the overhead, further increasing the overhead temperature. Here too, the reflux flow meter reads high, and the reflux valve often opens widely due to the recirculation of

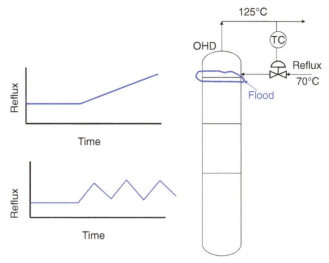

Figure 2.12   Typical refinery fractionator distillation column overhead with top temperature controlling reflux flow rate. Operating charts show the different responses that have been experienced upon top section flooding. (From Kister, H. Z., "Practical Distillation Technology," Course Workbook. Copyright © Henry Z. Kister. Reprinted with permission.)

entrainment around the tower overhead loop (upper operating chart in Figure 2.12). Alternatively and less frequently, with highly subcooled refluxes, the cold reflux entrainment reduces the tower overhead temperature, and the controller cuts back reflux, which in turn heats up the overhead, causing the controller to increase reflux. In this case, the entrainment rise shows up as temperature and reflux flow rate fluctuations (lower operating chart in Figure 2.12).

The entrainment from the tower overhead may overwhelm the condenser gravity drain lines, which are generally sized for the normal process flow rate (reflux plus distillate) and not for the massive entrainment incurred by flooding. Unable to drain, the entrained liquid builds up in the condenser, partially flooding it, thus reducing the area available for condensation. This in turn causes the column pressure to rise. Pressure differential builds up between the column and reflux drum, and eventually becomes large enough to quickly push and empty the liquid out of the condenser. The area available for condensation quickly rises, and with it the condensation rate, causing the tower pressure to dive. Following the fast liquid-emptying step, drainage from the condenser returns to a gravity mode, liquid accumulates, and the process repeats. The end result is severe pressure fluctuations in the tower, which can be as high as 7 psi in a few minutes. So a scenario of steady tower pressure breaking into severe fluctuations is often a symptom of flooding near the top with the entrainment interacting with the condenser drainage.

The rapid rise in entrainment indicator is particularly useful when flooding occurs near the top. However, this indicator will fail to indicate a stripping section flood that does not propagate to the rectifying section.

Loss of Separation   As flooding approaches, the amount of liquid entrained by the vapor sharply rises. At high pressures and/or high liquid rates, the amount of vapor entrained in the downflowing liquid also rises. As either type of entrainment rapidly escalates in the vicinity of the flood point, efficiency and separation decline (Figure 2.13). The decline in efficiency tends to occur closer to the flood point with downcomer floods than with entrainment floods. In Figure 2.13, the limiting mechanism for the isobutane–normal butane system is believed (513) to have been downcomer flood, while that for the cyclohexane–heptane system was

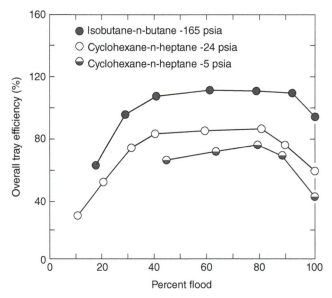

Figure 2.13   Overall tray efficiency, four foot ID tower, at total reflux, illustrating drop of efficiency as the flood point is approached and also at low loads. (Reprinted with permission from Yanagi, T., and M. Sakata, Industrial and Engineering Chemistry Process Design and Development, 21, p. 712. Copyright (1982) American Chemical Society.)

entrainment flood. In packed towers, the decline in efficiency tends to take place closer to the flood point with smaller packings than with larger packings. With large packings (e.g., 2 in. or larger random packings, and of surface area per unit volume below $60\,ft^2/ft^3$ for structured packings), and/or under vacuum, the decline in efficiency may begin at rates well below the hydraulic flood point.

Since loss of separation starts before the tower is hydraulically flooded, using it gives a lower flood point than other indicators. This is hardly a disadvantage, because the point at which separation begins to rapidly deteriorate is of greater practical importance than the hydraulic flood point. The point at which the separation begins to deteriorate is often referred to as "the maximum useful capacity," or "maximum operational capacity," and the vapor loads at which it occurs are typically about 0–20% below the hydraulic flood point. In most atmospheric and superatmospheric tray towers, this point occurs at flow rates of 5% or less below the hydraulic flood point.

The loss of separation is usually inferred from laboratory analyses of column products. A plot of tray or packing efficiency (Figure 2.13) against flow rate at a constant reflux ratio can identify the point where the separation begins to deteriorate as flood is approached.

The tower temperature profile is an alternative indicator of the loss of separation. Temperature often rises above the flooded region because the accumulating liquid rising from below is richer in heavies, and also because the flooded region no longer performs an efficient separation. Temperature may also rise below the flooded region because the smaller amount of liquid downflowing from the flooded region leads to heating up below, and also because the higher pressure drop raises the boiling points below.

A good knowledge of the normal and flooded temperature profiles under similar feed conditions is invaluable for the successful application of temperature profiles for flood indication. Figure 2.14 shows temperature profiles under normal and flooded conditions (231). The points are pyrometer measurements of wall temperatures taken from the access ladder of an uninsulated tower. All the temperatures are for the even-numbered trays, as the ladder was on the left side of

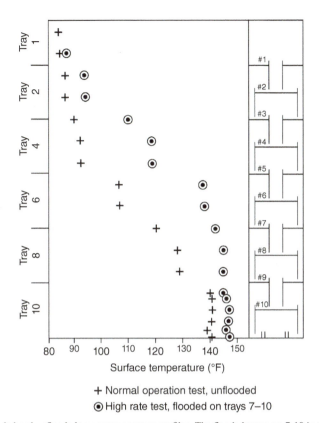

+ Normal operation test, unflooded
⊙ High rate test, flooded on trays 7–10

Figure 2.14   Flooded and unflooded tower temperature profiles. The flooded trays are 7-10 in the high-rate test. Temperatures measured by surface pyrometers. (From Kister, H. Z., K. F. Larson, J. M. Burke, R. J. Callejas, and F. Dunbar, "Troubleshooting a Water Quench Tower," Proceedings of the 7th Annual Ethylene Producers Conference, Houston, Texas, March 1995. Reprinted with permission. Copyright © American Institute of Chemical Engineers (AIChE).)

the tower in Figure 2.14, and the side downcomers obscured the odd-numbered trays. The crosses represent the normal temperature profile, showing a discrete reduction in temperature every two trays as one ascends the tower. The circles represent the temperature profile when the bottom three to four trays were flooded. The temperature spread across the bottom four trays completely disappeared, indicating loss of separation. The circles also show hotter temperatures above, due to movement of heavier components up induced by the poor separation on the bottom three to four trays. A similar application of wall-measured temperature profiles to successfully identify flood was reported by Lieberman (284).

Section 2.9.1 presents an example of a packed tower under deep vacuum in which neither the pressure drop nor gamma scans provided conclusive evidence for flooding. It was the loss of temperature spreads in the packed bed that provided the conclusive flooding diagnosis.

A shortcut for the Figure 2.14 procedure is to look at the differential between two suitable temperatures in the relevant tower section. For instance, in the tower in Figure 2.14, the normal temperature difference between tray 10 and tray 6 was $(140 - 108) = 32\,°F$. Upon flooding, this temperature difference fell to $(147 - 137) = 10\,°F$. A differential temperature indicator between trays 10 and 6 will indicate the onset of flood. It is important to select the correct trays in this procedure. For instance, the temperature differential between trays 6 and 4 in Figure 2.14 will be a poor indicator of flood. A flood-detection method using temperature differentials changing with time was proposed by Dzyacky et al. (91, 92). Generally, temperature differentials change faster than differential pressure and can give an earlier detection of the onset of flood.

Caution must be exercised with temperature profile interpretation, because unchanging temperatures may also indicate a "pinch" condition (i.e., poor separation due to insufficient reflux or reboil). In order to tell the difference, reflux and reboil can be raised. If this improves separation, pinching is indicated. If this worsens separation, or does not affect it, flooding is indicated. Other flood symptoms can help the interpretation. In one tower (24), flood was noted by a rise in pressure drop, but the separation stayed good. Reducing reflux got the tower out of flood, but the separation deteriorated. The tower had more trays than needed and operated near a pinch, so its separation was unaffected by the loss of stages upon flooding but deteriorated when reflux was cut.

Caution is also required when interpreting packed tower temperature profiles, as it may be difficult to distinguish poor separation due to vapor or liquid maldistribution from one due to flooding. Again, tests of raising reflux and reboil can shed light. If separation improves, it will argue against flooding.

In order to reliably establish the column temperature profile, a large number of temperature measurement points is needed. If unavailable, a vertical temperature survey can be conducted. Temperature surveys are discussed in detail in Chapter 7.

Temperature profiles provide an effective, low-cost method of detecting flood, but its success depends on the existence of a large enough number of measurement points and on the tower having a large enough temperature gradient under normal operating conditions. If the normal temperature difference from tray to tray is small, as in close separations, the flooded temperature profile does not vary significantly from the normal profile, and temperature profiles will be poor indicators of flooding.

Gamma Scanning   Gamma scanning is one of the best techniques for detecting flood. It is powerful for diagnosing flood, identifying flooded regions, and gaining insight into the nature of the flood. A detailed discussion is in Chapters 5 and 6.

Bleeders   Use of vapor bleeders was found effective for flood testing (187) in atmospheric and pressure towers. Each bleeder is located in the vapor space above a tray, or more frequently in the overhead line upstream of the condenser. Upon bleeder opening, only vapor coming out indicates no flooding; liquid spraying out indicates flooding. When the tower is above its atmospheric boiling point, bleeding flashes and chills the liquid. The presence of liquid in the bleeder can therefore be detected as a sharp temperature drop of gas issuing from the bleeder (187), or "weeping" or icing up downstream of the bleeder.

The bleeder technique is not very popular and is often hazardous or environmentally unacceptable. Other disadvantages are a shortage or lack of valved bleed points inside the tower, the need to know which trays are most likely to flood, and a lack of indication of an approaching flood condition. Of its few applications, most are bleeders in tower overhead lines that discharge into closed systems.

Sight Glasses (Viewing Ports)   These provide superb visual indication of flooding (187, 192). Sight glasses are expensive, increase the leakage potential, and upon glass breakage may lead to a chemical release. Supplying a light source that permits adequate viewing can also be an issue. For these reasons, this technique is uncommon in commercial towers, mainly used for towers that process nonhazardous materials at near ambient pressure.

Noise Increase   This technique was demonstrated in a pilot packed column at the Separation Research facility at the University of Austin, Texas, but not commercially. A sharp increase in the noise signal from the packing indicates the onset of flooding (365), remaining high when the packing is flooded. This is analogous to the fizzling noise of a carbonated drink.

Pattern Recognition   In many cases, as flood is approached, certain variables were observed to oscillate in a brief, repeatable, systematic pattern (91, 92). For example, liquid inventories on trays rise momentarily, resulting in a temporary decrease of the bottom level simultaneously with a temporary increase in the tower dP, or the onset of erratic movement of either or both. Since prior to flood these variables remained within the normal operating limits, it may be difficult to recognize the pattern from observing the operating charts. In contrast, it was reported (91, 92) that this pattern can be readily recognized empirically using first and second derivatives with time, which change a lot more than the variables. Operators often have expertise in identifying such patterns, and should be consulted. This field-tested pattern recognition technique can provide early detection of flood and permit tower operation closer to the flood limit (91, 92).

2.7   FLOOD MECHANISM DETERMINATION: VAPOR AND LIQUID SENSITIVITY TESTS

In a troubleshooting investigation, there are usually many theories. The troubleshooter's challenge is to reject invalid theories and to narrow down the remaining theories to a manageable number that can be catered for. With flooding, one of the best methods of invalidating incorrect theories is by field tests that determine whether the flood is sensitive to vapor, liquid, or both. For instance, when the test proves the flood to be sensitive to vapor but not to liquid, any theory that argues a liquid-sensitive flood is invalidated. In the author's experience, good vapor and liquid sensitivity tests discard about half the theories. In one case (Section 9.2.2), vapor and liquid sensitivity tests invalidated 7 out of 12 theories, reducing the number of theories from 12 to 5, a much more manageable number, and later led to the root cause. In another case (385), vapor and liquid sensitivity tests ruled out a theory that was considered a certainty, thus preventing an incorrect diagnosis and ineffective solution.

The flood mechanism discussion in Section 2.2 teaches that entrainment (jet) floods, packing vapor-rich floods, and system limit floods are induced by excessive vapor loads and are therefore highly vapor-sensitive. If there is any sensitivity of these floods to liquid loads, it is small. To this list, Sections 2.11 and 2.12 will add downcomer unsealing flood and flood induced by channeling. In contrast, downcomer choke floods and downcomer side draw bottlenecks (Section 2.15) are caused by excessive liquid loads, and therefore are highly liquid-sensitive. If there is any sensitivity to vapor loads, it is small.

Packing liquid-rich floods, and floods due to packing distributor overflows, are caused by excessive liquid loads and are therefore highly liquid-sensitive. However, they can show significant sensitivity to vapor loads as well. Downcomer backup flood can be induced by either excessive vapor load or excessive liquid load and can be sensitive to either, depending on the dominant term causing the backup. The example in Table 2.1 and Section 2.3 had the dry vapor pressure drop as the dominant term, so the downcomer backup flood was a vapor flood.

Section 2.5 describes common flood test procedures. The most popular of these starts with the tower in normal operation. In the tower in Figure 2.4, reflux is gradually increased until symptoms of flood are observed. This test cannot tell whether the flood is vapor-sensitive, liquid-sensitive, or both. The increase in reflux rate raises the liquid load, but at the same time, it also cools the tower. The temperature controller responds to the cooling by increasing boilup, thus raising the tower vapor load. It is impossible to tell whether the observed flood is due to the initial increase in reflux (therefore, liquid-sensitive), the subsequent increase in boilup (therefore, vapor-sensitive), or due to both.

To determine the flood sensitivity to vapor and liquid, the temperature control needs to be disconnected, so the boilup is kept constant (on flow control or in manual). The reflux is raised. If the tower floods, the flood is liquid-sensitive. An issue with this test is that since reflux is raised without a matching increase in boilup, lights are induced into the bottom product, making it off-spec.

Similarly, for a vapor-sensitivity test, the reflux is kept constant (in Figure 2.4, it is on flow control, and therefore already constant) and the boilup is raised. In this test, the temperature controller can be kept in auto and the temperature set point is raised. Flooding in this test is vapor-sensitive. This test induces heavies into the top product and makes it off-spec.

The good news is that these tests yield quick results; if correctly performed, each test normally will provide the answer within two to three hours. Once both the vapor and liquid sensitivity tests are performed, all the theories are checked against their findings, and those that do not predict the observed sensitivities are invalidated and crossed off the troubleshooting list.

There may be other tests that can be useful for testing theories. The key is to test them by changing one variable at a time in the correct direction. In one case (237), a theory postulated that an observed tower instability was caused by losing the downcomer seal. To test it, the tower was set at stable operation, and the liquid flow rate was reduced without changing the vapor flow rate. Neither flood nor instability resulted, invalidating this theory.

Overall, the trick for sensitivity tests is to monitor the response of the tower to changing one variable at a time in the correct direction.

2.7.1   Extension of Vapor/Liquid Sensitivity Tests to Complex Fractionators

This extension is mainly applicable to refinery complex fractionators, although it may apply to other complex towers with side draws and heat addition/removal.

The trick is the same as in simple sensitivity tests: in one test raise the liquid rate without changing the vapor rate, and in the other raise the vapor rate without changing the liquid rate.

Figure 2.15 shows a typical refinery fractionator with naphtha top product, kerosene and diesel side draws, and residue bottom product. There are two pumparounds (PAs), mid-PA (MPA) and bottom PA (BPA), which are direct-contact heat-removal sections. Tray liquid is drawn from the tower, cooled, and returned to the tower a few trays up. Direct-contact condensation of part of the tower vapor by the cooled streams generates most of the reflux to the section below and some of the product.

In the example used here, there appears to be a flood bottleneck between the kerosene and diesel draw. It is desired to determine whether the flood is liquid- or vapor-sensitive.

For the liquid sensitivity test (Figure 2.15a), the liquid rate below the kerosene draw needs to be raised without affecting the vapor flow rate in this section. Either cutting the kerosene draw rate or increasing the MPA duty will increase the liquid flow rate into the section below. The higher liquid rate should be prevented from continuing below the diesel draw, as changes there may unpredictably affect the vapor flow up the tower. For this reason, the additional liquid rate needs to be drawn as diesel product. Just like in simple sensitivity tests, the liquid sensitivity test here will induce lights (kerosene components) into the diesel product.

To keep the vapor rate in the kerosene–diesel separation section constant, the heater and the BPA heat duties need to be kept constant. Keeping the BPA duty constant can be challenging, because letting some kerosene into the diesel product will cool the BPA draw temperature, which will tend to reduce the BPA heat duty. If the BPA duty is allowed to drop, the vapor rate in the section above will increase, which will defeat the test. It is therefore imperative to closely monitor the BPA duty and counter any tendency to fall off (e.g., by closing exchanger bypasses).

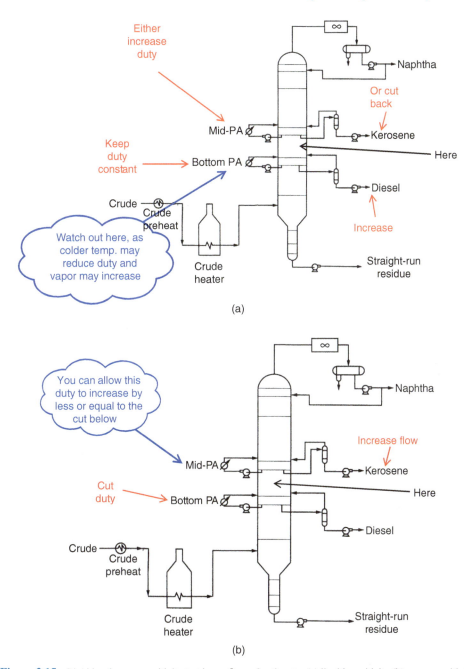

Figure 2.15   Liquid and vapor sensitivity test in a refinery fractionator (a) liquid sensitivity (b) vapor sensitivity.

For the vapor sensitivity test (Figure 2.15b), the vapor rate above the diesel draw needs to be raised without affecting the liquid flow rate in this section. Either cutting the BPA duty or raising the heater duty will increase the vapor flow rate into the section above the BPA. Generally, the BPA duty cut is preferred because raising the heater duty may cause adverse changes below the diesel draw. The additional vapor needs to be condensed either by increasing the MPA duty (preferred, as it will minimize impact on the section above) or by raising the overhead condenser duty.

To keep the liquid rate in the kerosene–diesel separation section constant, the diesel product flow rate needs to be kept constant. As more vapor will be condensed at the MPA or overhead condenser, this additional condensate needs to be removed and not allowed to reflux below the kerosene draw. This means increasing the kerosene draw rate and, in some cases, the naphtha draw rate. Just like in simple sensitivity tests, the vapor sensitivity test here will induce heavies (diesel components) into the kerosene product.

Section 9.2.5 presents a case study in which vapor and liquid sensitivity tests were extensively used and turned invaluable in determining the location and nature of a fractionator flood. Other such cases are described in Refs. 18, 52, 129, and 180.

2.8   GAINING INSIGHT INTO THE CAUSE OF FLOOD FROM dP VERSUS VAPOR RATE PLOTS

Pressure drop in a column is usually sensitive to only three factors: vapor load, liquid accumulation (flooding), and abnormalities like plugging or damage. It follows that a plot of dP against the vapor load, similar to those in Figures 2.5 and 2.6, can decouple the effect of vapor load on the dP from that of liquid accumulation. If this plot is generated at different times during the run, it can also decouple the effect of plugging and damage on the dP.

It is important to note that while the dP depends on the liquid load, it is not sensitive to the liquid load. The dP depends on the liquid load via the liquid head on the tray. The liquid head equals the weir height plus the head over the weir (34, 193, 245, 451, 492). The weir height is independent of the liquid load, while the head over the weir is only dependent on the liquid load to the 2/3 according to the Francis weir formula (34, 193, 245, 451, 492). Overall, the dependence of the dP on the liquid load is generally very weak. In contrast, the friction (dry) pressure drop, which typically makes up about a third to two-thirds of the total pressure drop, is proportional to the square of the vapor load, making the total pressure drop sensitive to the vapor load.

Similarly, the dP in packed towers in the vapor-rich regime is insensitive to the liquid load (34, 193, 245). In this regime, the liquid occupies only a small portion of the bed volume, so its effect on the pressure drop is small. In the liquid-rich regime, the pressure drop becomes sensitive to the liquid load. Criteria for the regime are presented in Section 2.16.

In each plot, the measured dP should be plotted against the characteristic vapor flow rate as derived from plant data. For the top section of a simple column, the vapor flow rate equals the sum of the reflux and distillate meters (as in Figure 2.5) plus any vent gas. For subcooled refluxes, the vapor condensed by the subcooling should be added. In the bottom of the tower, the reboiler duty or steam meter gives a measure of the characteristic vapor load.

Figure 2.16 shows two applications of this technique. Figure 2.16a, derived from actual plant data, positively affirmed flooding at a steam flow rate of about 60 klb/h. Here, the flood correlated with the steam flow rate. Figure 2.16b is for one C_3 Splitter tower (242). In this tower the pressure drop was low, about 0.09 psi per tray, which argued against flood per the guidelines in Section 2.6, but gamma scans showed many flooded trays. To reconcile the two conflicting observations, the measured tower pressure drop was plotted against the tower internal vapor flow rate (Figure 2.16b) calculated from the plant meters. The upper curve is the dP for the entire tower, and the lower curve is the dP for the rectifying section. Both curves show a point of inflection at vapor flow rate just above 2000 MPH. This affirmed that at this operating vapor load, the trays were right at incipient flooding. While some flooding could have started earlier (as seen by the gamma scans), the significant accumulation of liquid (the dP rise) started above this load. Mystery explained.

Figure 2.17 shows how dP versus vapor rate plots can track the condition of the tower internals. A reference dP versus vapor load plot is prepared from plant data in the first month or so

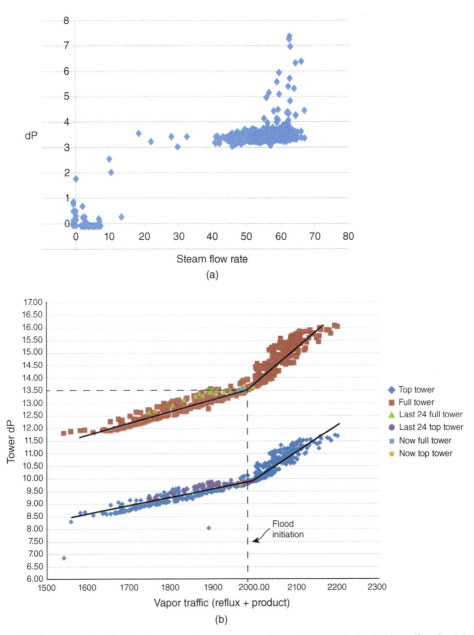

Figure 2.16   Application of plots of pressure drop versus vapor flow rate to identify floods (a) to affirm flood (b) to reconcile two conflicting observations. (Part (b): From Kister, H. Z., B. Clancy-Jundt, and R. Miller, PTQ, Q4, p. 97, 2014. Reprinted with permission.)

after the tower returns to service following a turnaround in which the trays were cleaned. During this period the internals are usually clean and non-corroded. If flooding in the tower is due to plugging, the reference curve may show no point of inflection and no flooding. If after a year or two in operation a point of inflection appears, or occurs at lower vapor loads (Figure 2.17a), plugging is indicated. If only a few trays are plugged, or if the plugging is in the downcomers and

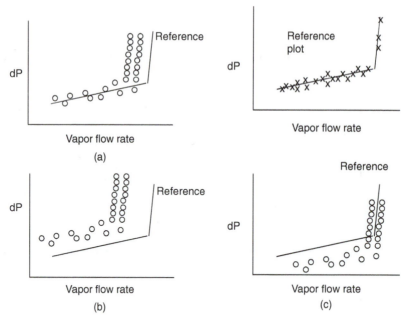

Figure 2.17   Use of dP logs to diagnose tower problems (a) plugged downcomer, or a few plugged trays (b) many plugged trays (c) missing or corroded trays. (From Kister, H. Z., "Practical Distillation Technology," course workbook. Copyright © Henry Z. Kister. Reprinted with permission.)

not the trays, the pressure drops below the point of inflection (at the lower vapor loads) will stay the same as in the reference curve. A vapor load at the point of inflection that gradually declines (moves to the left in Figure 2.17a) over a period of time indicates plugging worsening with time.

The vapor load at which the point of inflection occurs does not indicate how many trays are plugged, but how badly the trays or packings are plugged. It is enough to have one badly plugged tray to lower the point of inflection severely. If the pressure drops below the point of inflection also rise (e.g., Figure 2.17b), or the slope at the lower vapor loads increases, it is likely that many trays are plugged.

Figure 2.17c shows a scenario in which damage occurred, either due to dislodged trays or due to corroded trays or packings.

When plotting dP versus vapor rates over lengthy periods, watch out for dP recalibrations. Audit the dP measurement over the period, looking for zero dPs, which would indicate recalibrations. Then compare the dP versus vapor rate curves before and after. Most of the time a recalibration will only give a small bump, but there are other times when it needs to be taken into account.

In one tower, flooding was observed after six months of operation. According to the vendor hydraulic calculations, the flooding should not have occurred. The question was whether the trays were plugged or the vendor calculations were optimistic. Figure 2.18a plots the dP against the vapor load in the top of the tower and does not show a clear point of inflection. Plotting the pressure drop against the reboiler duty (Figure 2.18b) gave a much better correlation, suggesting the limit was below the feed. The point of inflection occurred at a reboiler duty of 16 units, which was well above the duties used in the first three months of reference period. Therefore, this plot could not tell whether the limit was plugging or a tray limit. Plotting additional data over the next year showed that the limit did not change, suggesting that plugging was not the issue.

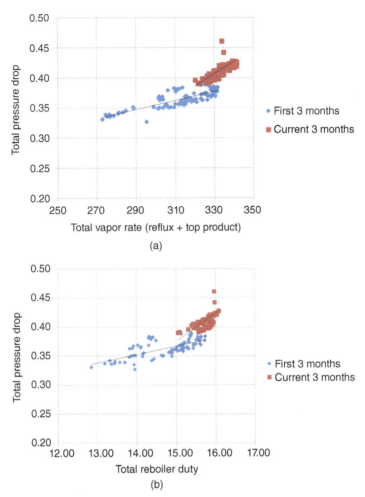

Figure 2.18   Application of plots of pressure drop versus vapor flow rate to identify flood nature and location (a) *x*-axis is the overhead vapor rate (b) *x*-axis is the reboiler duty.

Figure 2.19 (226) shows application of this technique to track changes throughout the run of the top fractionation section of a refinery crude fractionator equipped with moving valve trays. In the winter through summer of 2016, no flooding was observed. In the spring of 2016, the section experienced severe corrosion. In early autumn 2016, the dP shot up, indicating flooding, probably due to corrosion products and popped-out valve floats plugging trays. By the winter of 2016–2017, the dP came down. This was because the corrosion products and the remnants of the popped valve floats were washed away or disappeared due to corrosion. The dP versus vapor load plot enabled tracking this sequence of events, and by doing so, explaining a mystery as discussed in detail elsewhere (226).

Figure 2.20 (265) plots the dP of a condensate stripper column against the feed rate. The feed rate is usually not a good representation of the vapor load in the tower and is therefore not the ideal variable to plot on the *x*-axis. In this case, however, the tower vapor rate varied with the condensate feed rate, making the feed rate a reasonable representation of the vapor traffic. The plot shows high dP at multiple vapor loads and erratic dP rises before the turnaround (TAR). This indicates premature and erratic flood, arguing against a natural flood limit. In this case, the

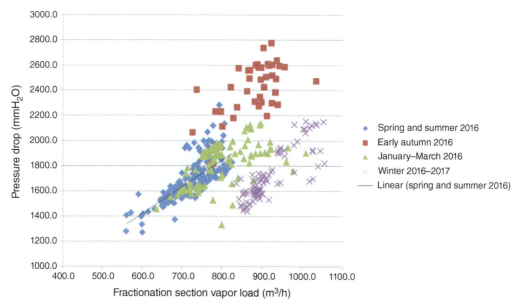

Figure 2.19   Application of pressure drop versus vapor flow rate plot to track changes in tower performance during the run. (From Kister, H. Z., and M. Olsson, Chemical Engineering, p. 29, January 2019. Reprinted with permission.)

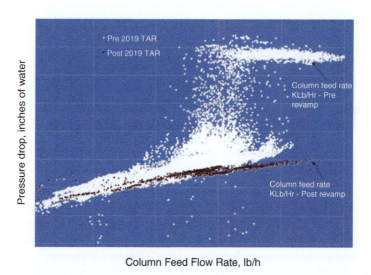

Column Feed Flow Rate, lb/h

Figure 2.20   Application of pressure drop versus vapor flow rate plot to a tower experiencing erratic premature flooding due to foaming and a point of transition issue (Contributed courtesy of Timothy Zygula P.E., BASF Petrochemicals).

flood was due to foaming combined with a point of transition issue (Section 8.7, Figure 8.7). Foam-flooding (Section 2.10) often gives similar plots. Note the post-TAR curve, after the bottleneck was eliminated showing no flood anywhere. A plot similar to the pre-TAR curve was produced by Kumar et al. (266) in a C2 Splitter tower due to intermittent hydrates (similar to freeze-ups) that caused premature flooding.

In Figure 2.21, a pressure drop versus vapor flow plot was applied to diagnose the root cause of separation fall-off in a packed tower. Over the preceding six months, tower feed flow rate

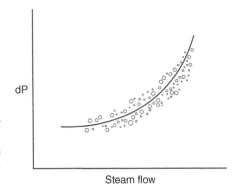

Figure 2.21 Application of plots of pressure drop versus vapor flow rate to diagnose poor performance in a packed vacuum tower. (From Kister, H. Z., "Practical Distillation Technology," course workbook. Copyright © Henry Z. Kister. Reprinted with permission.)

dropped by 10% for the same reflux, steam rate, and product purity. There were several theories on the table, including plugged packings, onset of packed bed foaming, and deterioration of liquid distribution.

The large circles in Figure 2.21 are measurements before the performance deterioration, and the small circles after. Figure 2.21 shows the before and after values to fall on the same curve, invalidating the plugged packings or onset of foaming theories. Had any of these occurred, the recent points would have lied at higher dP's than the older points. This left distributor plugging the most likely theory. Distributor plugging (with no liquid overflow) will not lead to higher pressure drop but would maldistribute the liquid, causing separation issues, which were countered by increasing the reflux-to-feed and stripping-to-feed ratios to keep the same product purity.

A somewhat similar observation led Lenfeld and Buttridge (281) to state "Drop in column efficiency before there is any noticeable increase in column pressure drop is a sign that the root cause is the liquid distributors rather than the packings."

2.9   DIAGNOSING FLOODS THAT GIVE SMALL dP OR NO dP RISE

The rise in pressure drop discussed in Sections 2.6 and 2.8 is caused by liquid accumulation in the tower. When this accumulation is small, or when it does not interfere much with the vapor rise, the pressure drop rise can be small, even unnoticeable.

Flooding Near the Top of the Tower   A small pressure drop rise, often of the order of 0.3–0.5 psi, is typical when the flooding occurs near the top of a tower. Since only one or very few trays flood, the liquid accumulation is not high, which incurs only a small pressure drop rise. When the tower dP is 3–4 psi, a small dP rise of 0.3–0.5 psi remains indistinguishable from the general noise in the dP measurement. Figure 2.22 shows actual operating charts, with Figure 2.22a being the most common. The chart in Figure 2.22b is from a tray tower in Figure 2.4 with vapor flood at the top. An increase in vapor rate generates flood (the vertical rise in dP). Once liquid accumulates at the top, the liquid downflow drops, and the control temperature rises. The controller cuts back on boilup, reducing vapor upflow, and the upper trays gradually unflood (the sloped downward trend in dP). The unflooding re-instates the normal liquid downflow, the control temperature declines, increasing boilup, and the flood returns (the vertical rise in dP). The typical cycle amplitude is about 0.3–0.5 psi with a period of three to five minutes.

The ability of the dP versus vapor rate plot (as described in Section 2.8) to decouple the dP changes due to flooding from regular dP changes makes this plot invaluable for diagnosing flooding near the top of the tower. One 20-trays high-pressure (200 psig) hydrocarbon absorber in

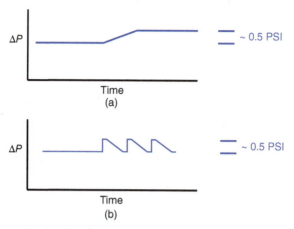

Figure 2.22 Typical operating charts for flood near the top of a tower (a) most common, a small rise in pressure drop (b) column swings in and out of flood. (From Kister, H. Z., "Practical Distillation Technology," course workbook. Copyright © Henry Z. Kister. Reprinted with permission.)

foaming service experienced premature flooding at about 67% of the design lean oil rate. Flooding was identified by liquid carryover from the tower top to the knockout drum downstream. The drum normally operated with no liquid in it. The plant always identified flooding by the appearance of a liquid level in the drum. When the operators were asked whether they saw a high dP upon flooding, their answer was "Never. The tower operates with 2 psi pressure drop both under normal and flooded conditions."

Figure 2.23 plots data collected over a two-month operating period. The hollow circles are periods of four hours or longer in which there was no liquid level in the drum, i.e., definite no-floods. The filled circles are periods of four hours or longer in which there was continuous level in the drum, i.e., definite floods. The points fall on two distinct lines, with the difference

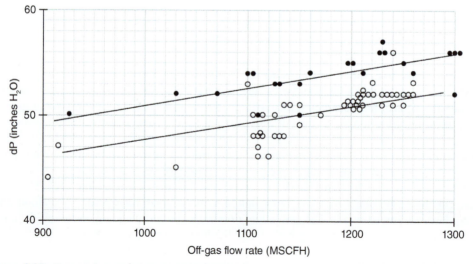

Figure 2.23 Pressure drop versus gas rate data collected over two months for hydrocarbon absorber. Hollow circles, no liquid in downstream knockout drum, filled circles, >4 hours of liquid level in drum. (From Kister, H. Z., "Practical Distillation Technology," course workbook. Copyright © Henry Z. Kister. Reprinted with permission.)

between them being 3 in. of water (0.11 psi) pressure drop. So the dP versus vapor load curve was sensitive enough to identify such a small pressure drop change.

An interesting feature of Figure 2.23 is absence of a point of inflection like in Figures 2.5 and 2.16–2.19. The absence of a point of inflection indicates a flood that is insensitive to vapor load, i.e., a liquid flood. Indeed, in this case it was a downcomer choke flood.

Channeling Allows Liquid Accumulation Without Interfering with Vapor Rise
Another situation where a tower floods with little or no pressure drop rise is when channeling occurs, and the channeling permits vapor to rise, bypassing the liquid accumulation. This sometimes occurs in packed vacuum towers. One such case is described in detail in Section 2.9.1. In another vacuum tower (65), a CAT scan (Section 6.4) showed that upon flooding a bed of structured packings, liquid accumulated along the periphery, allowing vapor to channel through the center.

Grobenrichter and Stichlmair (142) studied pressure drop rise due to crystallization of salt by passing dry air through saturated aqueous solutions in a 6-in. column. With wire gauze packings, there was always a large pressure drop increase due to liquid accumulation. With corrugated sheet packings, the pressure drop rise was high at higher liquid rates, but much smaller at low liquid loads (<12 gpm/ft^2 in their tests), suggesting vapor channeling around zones of local flooding.

A similar scenario may also occur in tray towers. In the tower described in Section 2.8 (Figure 2.19), after the popped-out valve floats disappeared, the pressure drop came down (the winter 2016–2017 curve). However, the flood experienced earlier (the early autumn 2016 curve) did not go away, as evidenced by gamma scans and off-spec products. Popping out the valve floats induced vapor channeling, which generated a rise path for the vapor without incurring a high pressure drop. A similar experience of flooding without pressure drop rise was reported (226) in a chemical column whose manways were left uninstalled at a turnaround, allowing vapor to channel through the open spaces.

2.9.1   Flood Diagnosis in a Chemical Vacuum Tower with no dP Rise

The bottom section of a chemical deep-vacuum tower had two beds containing high-efficiency structured packings, with a vapor side draw between. Above the feed, there was another packed bed. The problem experienced was that the side draw product had excessive concentration of the heavy key component. The tower was operated at about 60% of flood.

To gain insight into the bed hydraulics, the reboiler duty and reflux rates were raised, and the pressure drop versus vapor load (represented as the C-factor, Eq. 2.1) plot for the test was constructed (Figure 2.24a). The C-factor was highest in the bottom section, became lower above the side draw, and stayed uniform above that.

The hollow circles in Figure 2.24a are pressure drops measured across the bottom bed. The filled circles are the pressure drops measured across the two upper beds. The rate of pressure drop rise with C-factor was the same for both, which is not surprising since all beds contained the same packings. This, however, suggests lack of large-scale plugging in the beds, because such plugging would have affected one bed more than the others, and the difference would show on the plot. This is one theory put to rest.

The plot shows neither a point of inflection nor a sharp rise in pressure drop, giving no indication of flood. The diagram also shows a comparison to calculated values. Pressure drops were somewhat higher than calculated, but tracked the calculated values quite well.

The Kister and Gill equation (2.2) predicted flood at a pressure drop of 1 in. of water per foot of packing for this packing. In Figure 2.24a, this pressure drop was measured in the bottom section at a C-factor of 0.37 ft/s. The pressure drop calculation gave a slightly higher C-factor of

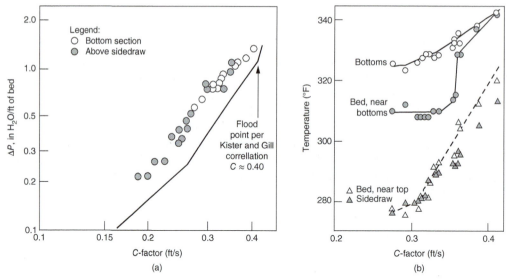

Figure 2.24 Data analysis to diagnose flood in a packed deep vacuum tower (a) application of a plot of pressure drop versus vapor flow rate (b) loss of temperature spreads near the bottom of the bottom bed at C-factor of 0.36 ft/s, indicating the onset of flood. (From Kister, H. Z., R. Rhoad, and K. A. Hoyt, Chemical Engineering Progress, p. 36, September 1996. Reprinted Courtesy of Chemical Engineering Progress (CEP). Copyright © American Institute of Chemical Engineers (AIChE).)

0.4 ft/s. However, Figure 2.24a showed that the 1 in./ft pressure drop was exceeded still without any sharp rise in pressure drop.

Gamma scans were shot and are described in detail in Section 5.5.15. At C-factors just above 0.36 ft/s and at 0.3 ft/s, the scans showed good liquid distribution throughout the bottom bed except for the lowest 2 ft or so. In these bottom 2 ft, the distribution was good in the 0.3 ft/s scan, but maldistribution occurred in the 0.36 ft/s scan, with two of the four chords shot showing possible accumulation. This suggested channeling that permits vapor rise to bypass the liquid accumulation.

Figure 2.24b is a plot of the temperature profile of the bottom bed against the C-factor. There was a good temperature spread at the lowest C-factors. Raising vapor and liquid loads reduced the temperature spreads, indicating lower packing efficiency. This behavior is experienced with wire gauze packings, like the packings in this bed (193). At a C-factor of about 0.36 ft/s, the "near bottoms" temperature abruptly shot up, becoming the same as the bottom temperature, indicating loss of separation between the two points. The packing efficiency sharply dropped in this region. This coincided with the point where the accumulation in the bottom 2 ft was seen on the gamma scans and when the flood pressure drop from Eq. (2.2) was measured. The efficiency drop was due to the channeling or entrainment induced by the liquid accumulation.

Overall, the test showed that the flood occurred where expected, and at the low operating C-factor of 0.3 ft/s or less, hydraulic issues were unlikely to have caused the high impurity issue. Further testing showed that the separation issue was due to a very non-forgiving concentration profile (232).

2.9.2 Flood Diagnosis in Packed Pumparound (PA) Beds

Flooding in PA beds can be difficult to diagnose. Being typically used as direct-contact condensers, these have high vapor loads at the bottom of the bed, tapering as one travels up the bed.

So the bed may flood at the bottom, but the flood may not propagate past a certain height. Often, flooding may lead to the formation of some stable channeled pattern in the PA bed.

Reference 21, a case study of flooding near the top of a refinery atmospheric crude fractionator, gives an excellent description of symptoms typical of PA bed flooding. Operation of the top PA bed (shown in Figure 2.8) was stable, with a pressure drop of 0.16 psi across 9 ft of structured packings (the measurement in Ref. 21, 0.22 psi, includes 0.06 psi in the overhead nozzle) compared to 0.17 that the author calculates from Eq. (2.2). This suggests operation near flood, as at the operating rates the pressure drop should have been significantly less (21). When a PA bed mildly floods, liquid stacks up until there is enough head to push it down in a channeled mode. An equilibrium is established, with liquid stacking at one part of the bed and vapor channeling through another, giving a mild pressure drop rise. This channeled mode hurts heat transfer. This can be inferred from a reduced temperature difference between the PA liquid draw at the bottom of the bed and the vapor exiting from the bed. In the case study in Ref. 21, the difference was 10 °F for two years, where it should have been 25–35 °F had the bed not flooded.

In this tower, the flooding initiated at the collector under the bed (Figure 2.8) due to plugging in its downcomers. When the rates through the tower were raised, the "upset conditions" described in Section 2.6 were initiated, in which the bed pressure drop swung between 0.24 and 1.75 psi, indicating the channeled mode transitioned into a "buildup and dump" mode. This mode indicated flooding coming from underneath. The dump takes place through the vapor openings of a collector or distributor, or through the open area of trays. The author is not aware of any PA beds experiencing the "buildup and dump" mode when the flood initiated in the bed itself, i.e., at the bottom layers of packing.

2.10   FOAM FLOODING SYMPTOMS AND TESTING

A foam (Figure 2.25) forms when the films surrounding vapor bubbles are stabilized to resist gravity drainage. The stabilization could be provided by surface forces, surfactants, or by chemicals helping to form a gelatinous surface layer. The bubbles then persist without coalescence with one another, or without rupture into the vapor space. The lifespan of foams varies. A quickly breaking foam may take five seconds to break, and a moderately stable foam can last two to three minutes (122, 192). The science of foams is extensively discussed in the literature (e.g., Refs. 32, 39, 170, 192, 263, 299, 366, 402, 403, 496).

Foams are generated during vapor–liquid contact, especially when the contact is turbulent. High-turbulence regions and impact of liquid jets on liquid surfaces promote foams. In tray towers, foams increase aeration in downcomers, leading to excessive froth backup in the downcomers. Foams also make vapor disengagement from the vapor–liquid mixture entering the

Figure 2.25   Foam in Structured Packing. (Courtesy of Prof. G. Wozny, TU Berlin.)

downcomer more difficult, requiring lower velocities in the downcomers to disengage the vapor and vent it out. This is analogous to the lengthy draining of a washing machine (after stopping it) during the wash cycle compared to the quick draining during the rinse cycle. Compared to non-foaming systems, foams cause premature downcomer backup and/or downcomer choke flood. Usually foams do not bottleneck trays, they bottleneck downcomers.

In packed towers, foams cause premature floods as well. This is most severe in the liquid-rich regime. Random packings allow lateral movement of the foam bubbles, and therefore can handle foams well, even in the liquid-rich regime, but will still flood prematurely under foaming conditions. With structured packings, the corrugated sheets restrict the sideways movement of the bubbles, and also support and often stabilize the foam. The inter-layer channels are therefore prone to choking, some earlier than others, and premature flooding results, with greater severity in the liquid-rich regime.

Symptoms   A foam flood gives similar symptoms to other floods, namely high pressure drop, entrainment from the top of the tower, reduction in bottom flow rate, and poor separation. However, in addition, it may give one or more of the following symptoms:

1. An erratic rise in pressure drop, sometimes with no apparent reason (Figure 2.26a).
2. Fluctuating tower pressure drop (Figure 2.9).
3. Nonreproducible flood data. The points in Figure 2.26b depict different vapor and liquid loads at which flooding occurred at different times. All occurred prematurely.
4. Flood sensitive to temperature (Figure 2.26c; also, Ref. 469). At the hotter temperature, the foaming did not occur.
5. An abnormal temperature profile. For example, in amine absorbers foaming will cause the reaction to occur further up in the tower (15, 325). The rich solution will become cooler, while the lean gas will warm up. A significant drop in temperature difference between the inlet and outlet gases, as well as between the rich and lean amine, will suggest foaming.
6. Flooding goes away when antifoam is added.
7. Gamma scans and downcomer neutron scans effectively identify foams (Section 5.5.5).
8. Downcomer inlet velocities exceeding the criteria in Section 2.16.

Where to Look for Foaming   A survey of published case histories (198, 201) found foaming to be a service-specific phenomenon. Ethanolamine absorbers that absorb acid gases such as H_2S and/or CO_2 from predominantly hydrocarbon gases, and their regenerators, account for about one-third of the published cases. Another 10% were also in acid gas absorption/regeneration service, but using alternative solvents such as caustic, hot potassium carbonate (hot pot), and sulfinol. Another 10% were in absorbers of gasoline and LPG from hydrocarbon gases using hydrocarbon solvents.

Case histories of foaming were also reported (198, 201) in each of the following services:

Chemical: aldehydes, soapy water/polyalcohol oligomer, solutions close to their plait points (solubility limits), extractive distillations, solvent residue batch still, ammonia stripper, DMF absorber (mono-olefins separation from diolefins), cold water H_2S contactor (heavy water GS process). Recent work added to this list include distillation of organic acids and alcohols (369), heterogeneous azeotropic systems (369), and butanol–water (422).

Refinery: crude preflash, crude stripping, visbreaker fractionator, coker fractionator, solvent deasphalting, hydrocracker depropanizer.

Gas: glycol contactor.

Olefins: high-pressure condensate stripper.

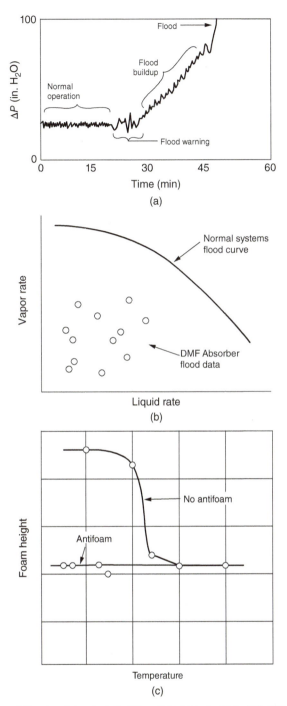

Figure 2.26 Symptoms of foaming (a) a typical differential pressure chart, indicating foam-induced flooding; (b) non-reproducible flood data, indicating foaming; (c) effect of temperature on foam height. (From Bolles, W. L., Chemical Engineering Progress, 63(9), p. 48, 1967. Reprinted Courtesy of Chemical Engineering Progress (CEP). Copyright © American Institute of Chemical Engineers (AIChE).)

There is a rich literature on experiences with foaming problems, with most of the pre-2006 reported cases described on pages 543–557 of *Distillation Troubleshooting* (201). The discussion below briefly describes some of the highlights. The reader troubleshooting a potential foaming problem is strongly encouraged to explore these experiences and see whether one or more can be beneficial to the problem at hand.

In about a third of the reported cases (198, 201), suspended solids, corrosion products, or polymers catalyzed foaming. Although not specifically reported, solids could have catalyzed foaming in many of the other cases. The survey by Le Grange et al. (272–274) of 108 foaming incidents in ethanolamine absorber–regenerator systems found that half of them had foam promoting contaminants entering the plant with the feed. Good filtration can often alleviate, even mitigate the foaming, as described by many of the cases abstracted in Ref. 201 and detailed in Refs. 98, 100, and 316. However, solvents may react with the material used in filters, and the reaction products may cause foaming (15, 81, 341, 357).

In ethanolamine service, or other alkaline absorber–regenerator systems, chemical analysis or coloration of the lean solvent showing a high solids or degradation product content indicates a high potential for foaming. Solids and degradation products catalyze foams. A qualitative guide correlating appearance with foaming tendency in ethanolamine solutions was presented by Lieberman (286). Engel (98, 100) provides an in-depth discussion of the various root causes of foaming in ethanolamine services.

In many cases, the foaming was caused or catalyzed by an additive such as a corrosion inhibitor, in some cases in ppm concentrations. In one reported case (452), as well as many others experienced by the author, severe foaming occurred in amine or caustic absorbers due to a corrosion inhibitor contained in the steam condensate used for solution makeup. Another case was reported (516) of a heavy contaminant coming into the feed of an extractive distillation tower and causing foaming. The author is also familiar with a case where an anti-foulant induced foaming.

Ross and Nishioka (403) found that the potential for foaming is maximized near the plait point (near the solubility limit), i.e., where a solution is still homogenous but is near the point of breaking into two liquid phases. Once the plait point is reached, and two phases are formed, one phase may serve as a foam inhibitor and destroy the foam. Alternatively, they may form a foaming emulsion. Hydrocarbon condensation into aqueous solutions (6, 81, 181, 273, 274, 331, 341, 452, 473), or small amount of water in a hydrocarbon system (Case 16.7 in Ref. 201, also Ref. 265, both Olefins condensate strippers), is known to promote foams. One refinery preflash tower (287) foam flooded because water was refluxed into the tower due to malfunction of the reflux drum boot. The flooding stopped when the boot again removed the water. A similar foam flood was experienced in a butyraldehyde column (469) due to refluxing small quantities of water; the problem was solved by adding a water-removal boot to the reflux drum.

High pH stabilizes foaming emulsions. The author experienced foaming in one depentanizer and one de-isobutanizer, normally non-foaming services, when small quantities of caustic entered the towers.

A foaming problem may occur in one installation but not in another even though both handle the same materials. Often a foaming problem remains "hidden" because a tower or downcomers are oversized, or may not occur in one installation because a surface-active impurity is absent. In one case (39), a new absorption unit experienced foaming while a similar older unit did not. This was because the older unit had oversized downcomers that adequately handled the foam. In another case (1), foaming was observed only after a capacity expansion. Tests by Senger and Wozney (422) showed that foaming with the same chemical system may proceed differently depending on the liquid flow rates and the packings

Avoiding local high-turbulence regions may help prevent foaming. In one case (250), a tower foamed when the reflux entering via a downward elbow dropped 12 in. into a liquid pool behind

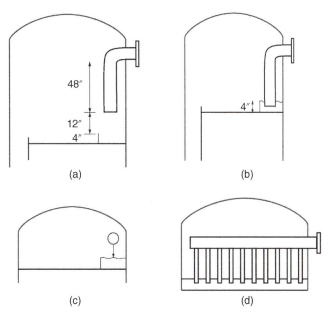

Figure 2.27   Reflux entry into towers with foaming systems (a) dropping the reflux from a pipe 12 inches into a liquid pool produced tower foaming (b) extending the pipe into the pool eliminated the foaming (c) another foaming system, reflux introduced via a pipe distributor with holes pointing downwards into the pool, foaming produced (d) foaming eliminated by adding dip tubes that brought the reflux below the sump liquid level. (From Kister, H. Z., G. Jacobs, and A. A. Kister, Chemical Engineering Research Design, 191, p. 313, 2023. Reprinted Courtesy of IChemE.)

an inlet weir (Figure 2.27a). A gamma scan showed foam on a few trays, but severe foaming only on the top tray. Extending the pipe into the pool (Figure 2.27b) eliminated the foaming. In another case (250), liquid to the regenerator of a hot pot absorber entered via a pipe distributor with holes pointing downwards into a sump behind an inlet weir (Figure 2.27c). Again, gamma scans showed foam on the top tray. To cure, the pipe was fitted with dip tubes that were submerged in the sump liquid (Figure 2.27d). There was no more foaming after this.

In one amine absorber (223), a retray with low pressure drop trays led to channeling of the inlet gas jet to the region opposite the gas inlet nozzle, generating a high-gas velocity region, and the increased turbulence produced premature foam flood. Figures 2.28 and 2.29 show how increased gas velocities raise foam height and foam decay time in test equipment.

Testing   Foaming tests were reported in many literature sources (1, 32, 39, 79, 122, 192, 201, 225, 422, 438, 490, 516), and these have been applied with various degrees of success. Experience has shown (39, 122, 222) that when not performed under actual plant operating conditions, foam tests can be inconclusive. Generally, methods 3–7 below reproduce actual plant conditions, and at least one of them is often simple to set up. Common test methods include:

1. The "bottle shake" test. A clean bottle is charged with a sample of the solution. It is left open for some time to allow all the entrapped gas to escape. The bottle is capped and shaken vigorously. It is then set on a table. The foam height and the time it takes for the foam to settle to the liquid level are recorded. Settling times exceeding five seconds indicate foaming; the longer, the more severe. Successful application of this test for amine absorbers was reported (15, 81). This method is easy, simple, and reliable when it indicates foaming. However, it often fails to detect foaming.

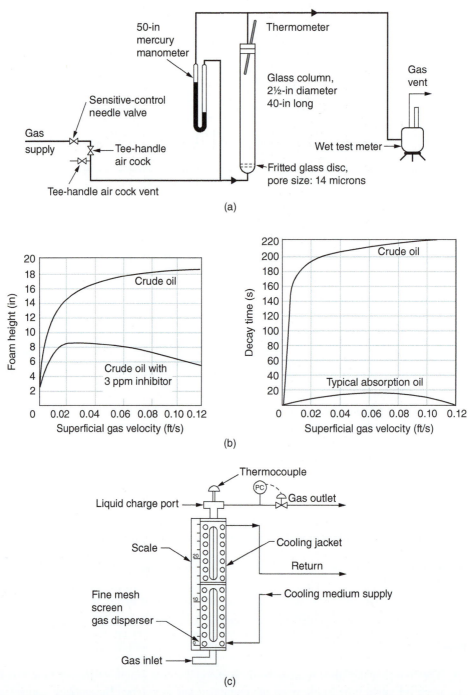

Figure 2.28   Testing for foam (a) an apparatus for dispersing gas through a solution sample; (b) a typical foam height and decay time from the test apparatus; (c) an armored level glass technique. (Parts (a) and (b): From Glausser, W. E., *Chemical Engineering Progress*, 60(10), p. 67, 1964. Part (c): From Bolles, W. L., *Chemical Engineering Progress*, 63(9), p. 48, 1967. All Parts Reprinted Courtesy of Chemical Engineering Progress (CEP). Copyright © American Institute of Chemical Engineers (AIChE).)

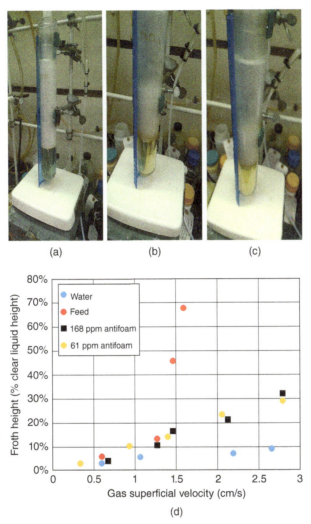

Figure 2.29 Foaming tests, antifoams screened and tested and a suitable one was found. (a) column feed 1.8 l/min, no antifoam, tall foam height; (b) column feed 2.1 l/min, 168 ppm antifoam medium foam height; (c) column feed 1.2 l/min, 61 ppm antifoam, low foam height; (d) an excellent result presentation as a plot of foam height versus gas velocity in apparatus, including results for a non-foaming substance (water) as reference. (Contributed Courtesy of Dr. Carine Saha Kuete Achoundong, Dr. Claudio Ribeiro, and Dr. Lihua Liu, BASF.)

2. Dispersing gas (e.g., air, nitrogen) into a solution sample in a small vessel. Foam height and foam decay time give a measure of the foaming tendency. This is again a simple method and reliable when it indicates foaming. One watchout is that foaming cases were reported, with this method failing to indicate foaming (39, 192, 201, 222, 225). It was recommended to conduct this test at the process temperature, but sometimes the temperature makes little difference (452). Refs. 452 and 504 provide detailed procedures that have been successfully applied for ethanolamine absorber–regenerator systems.

3. An apparatus that disperses gas (e.g., air, nitrogen) through a solution sample (e.g., Figures 2.28a and 2.29). To achieve good dispersion, the gas is often dispersed via a fritted filter. Details are described elsewhere (1, 122, 452). In principle, this test may

suffer from the same problem as in 2 above, but the experience reported with it has been excellent (1, 122, 273, 438, 452, 516). The good experience and its simplicity have made this test the most popular.

Figures 2.28b and 2.29 show how this method should be applied. For best results, the foam heights and decay times should be plotted against the superficial gas velocity. The test should first be performed on a non-foaming solution like demineralized water (blue dots in Figure 2.29d). The test is then repeated with the test solution (orange dots in Figure 2.29d). Different antifoam agents, at different concentrations, can then be tested (Figure 2.28b; yellow and dark brown dots in Figure 2.29d).

4. Oldershaw or pilot-scale towers have been successfully used for diagnosing foaming (490, Case 2.23 in Ref. 201). With glass Oldershaws or pilot towers with observation windows, this method can also provide invaluable information on the nature of the foam and where in the tower it is likely to foam. This technique is most suitable for atmospheric and vacuum towers but may have difficulty dealing with solids and high-pressure applications.

5. An armored level glass (Figure 2.28c) can be set either in the field (222) or in the laboratory (39), and nitrogen is dispersed into the sample at actual plant temperature and pressure. This procedure has special merit when column temperatures and pressure cannot be reproduced by other tests and has been successful where methods 1 and 2 failed (39, 222).

6. Antifoam tests. If a tower responds to antifoam injection, it means there is foaming. However, if the tower does not respond to antifoam injection, it does not necessarily deny foaming. Some foams respond to one antifoam agent but not to others. The antifoam concentration also matters, too much antifoam may produce more severe foaming than none (177, 201). In one case (81), an antifoam was able to break down the foam, but the foam would return less than an hour later due to degradation by heating and cooling cycles. Excessive use of some antifoams can produce fouling and accumulation (6, 473).

When the tower responds to antifoam, these tests can help find the optimum agent and concentration, as described in one case (81), see also Figure 2.29. To rule out foaming, try a variety of agents and concentrations. It would be a good idea to experiment with them in the lab before injecting into the tower to determine what the best agents and concentrations are to try. If several agents and several concentrations have been tried and the tower has not responded, one can conclude that foaming is unlikely, as described in one case (225) and many other experiences by the author.

7. Field tests. Testing a system where foaming is suspected with a non-foaming solution can conclusively confirm or deny foaming. In one amine absorber (225), where foaming was on the suspect list, a field trial was conducted in which the amine solution was replaced by clean water. While an amine system often foams, a clean water system does not. The water trial showed an identical capacity limitation as experienced with the amine solution, conclusively denying foaming.

In the abovementioned case, it was necessary to interrupt operation and perform the water test. In other cases, the trial requires no process interruption. The author is familiar with cases in which a steam condensate makeup was used for caustic or amine solutions, and foaming was experienced. Foaming disappeared within a few days after a trial in which demineralized water replaced the condensate makeup.

In addition, there are tests that can help in specific circumstances:

1. Foam tests on different samples. In one ethanolamine absorber (273), foaming caused severe carry-under of gas into the rich amine, resulting in a venting problem downstream. The foaming source diagnosis was established by performing foam tests on

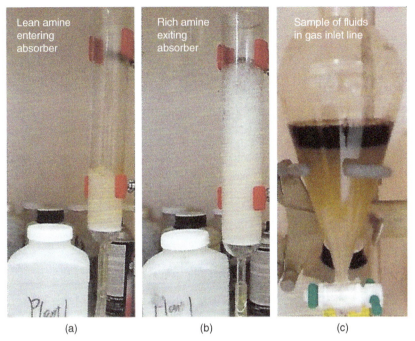

Figure 2.30   Inlet contamination foaming in an amine absorber (a) lean amine entering absorber, little foaming; (b) rich amine leaving absorber, severe foaming; (c) sample of liquids in the inlet line. (From Le Grange, P., M. Sheilan, and B. Spooner, "Why Amine Systems Fail in Sour Service", Sulphur 368, p. 42, January–February 2017. Reprinted with permission.)

both the lean amine and rich amine (Figure 2.30a,b). The foaming was induced by contaminant-containing liquids in the inlet gas line (Figure 2.30c).

2. Stepping up the bottoms flow rate can provide inconclusive supporting evidence for diagnosing foaming. A large step-up in bottoms flow rate that does not produce a significant change in bottoms level supports a foaming diagnosis. When the foam height exceeds the upper tap of the level transmitter, the transmitter measures the foam density, not the liquid level. Detecting liquid when opening the vent valve on the upper tap of the level glass or transmitter also supports foaming. A step-up in bottoms flow accompanied by a drop in bottoms level does not deny foaming; it just means that the level has not risen above the upper tap.

3. A discrepancy between staggered level gauges in the bottom sump can indicate foaming (504). Observation of the level glass can also show a hydrocarbon layer above an aqueous solution (6), which may be linked to foaming.

Cures   Three cures have been successful for foaming problems: eliminating the foam-causing chemical (e.g., by eliminating the additive or by filtering out the solids that catalyzed the foams), injecting antifoam, and debottlenecking downcomers (either by using larger downcomers, or by reducing downcomer backup, or by replacing trays by random packings). Injecting antifoam has not always been effective, and some experimentation with the inhibitor type and concentration are often required.

These cures are discussed at length in Ref. 192. Cures specific to ethanolamine absorber–regenerator systems are discussed by Le Grange et al. (273, 274).

2.11   DOWNCOMER UNSEALING FLOODS AT LOW LIQUID LOADS

The challenges of handling low liquid loads in tray towers are keeping the little liquid on the tray without drying up, and keeping the downcomers liquid-sealed (198). When a downcomer loses its liquid seal, vapor rises up through it and interferes with liquid descent. This interference has led to downcomer unsealing floods, capacity bottlenecks, poor separation, instability, and total inability to start towers up.

On conventional trays, at liquid flow rates exceeding about 2–3 gpm/in. of outlet weir length, the outlet weir generates a pool of frothy liquid (Figure 2.31a). As long as the downcomer clearance is lower than the liquid head in the pool, this liquid pool will seal the downcomer. In this regime, liquid is the continuous phase and vapor is dispersed in the liquid. In contrast, at low liquid flow rates (below about 2–3 gpm/in. of outlet weir), and especially at high vapor velocities, there is a phase inversion. The vapor atomizes the tray liquid, forming a vapor-continuous dispersion known as the spray regime (190, 193) or drop regime. In this regime, there is no frothy pool, and the liquid is dispersed as atomized drops in the inter-tray vapor space (Figure 2.31b). The continuous vapor phase reaches the downcomer outlet. There is no static seal preventing vapor from rising up the downcomer. In the spray regime, the downcomers are prone to seal loss.

Many high-capacity trays contain truncated downcomers, which terminate a few inches above the tray floor. These downcomers discharge their liquid through slots or holes at the bottom of the downcomers. Blowby (vapor breakthrough into downcomers, or downcomer seal breakage) is a major potential issue in truncated downcomers because they are not sealed by the tray liquid. Vapor breakthrough into truncated downcomers reverses tray action, inducing some liquid to weep through the trays and some vapor to channel up the downcomers. Severe blowby can lead to a sharp drop in tray efficiency. If the velocity of vapor rise in the truncated downcomer is high, it too can lead to downcomer seal flood similar to that described for the spray regime.

Figure 2.32 illustrates how downcomer unsealing can initiate tower flood in the spray regime. Consider a tray with very little liquid on it. The vapor, looking for the easiest path, ascends through both the trays and downcomers (Figure 2.32a). Upon the introduction of liquid, some of it reaches the downcomer (Figure 2.32b). If the vapor velocity through the downcomer is too high, it will not allow the liquid to descend into the downcomer (Figure 2.32c). Unable to descend, the liquid will accumulate on the tray above. This will lead to flooding of the tray, with liquid carry over from the top of the tower (Figure 2.32d). A very similar mechanism occurs in high-capacity trays with truncated downcomers.

An alternative scenario frequently occurs. The accumulating liquid head may lead to significant weeping. This weeping may then provide enough liquid to seal the downcomer on the tray

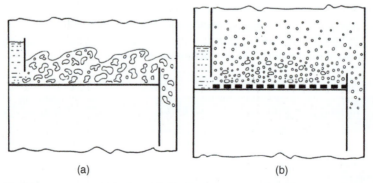

(a)   (b)

Figure 2.31   Flow Patterns on trays (a) froth regime (liquid phase is continuous) (b) spray regime (gas phase is continuous). (From Henry Z. Kister, Chemical Engineering, September 8, 1980. Reprinted with permission.)

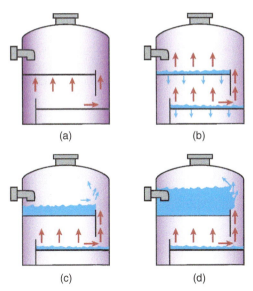

(a)   (b)

(c)   (d)

Figure 2.32   Downcomer Unsealing Flood in the spray regime (a) vapor distributes between tray and downcomer; (b) liquid introduced, reaches the downcomer, attempts to descend; (c) when vapor velocity up the downcomer is too high, the vapor does not permit liquid descent; (d) unable to descend, liquid accumulates on the upper tray, generating flood. (From Kister, H. Z., "Practical Distillation Technology," course workbook. Copyright © Henry Z. Kister. Reprinted with permission.)

below. Once this downcomer is sealed, there is nothing to stop the liquid from descending, and it will pour right down. This will propagate through the trays and dump into the tower bottom. Once the liquid pulse goes through, the downcomer becomes unsealed again, and the process repeats. The end result will be a sequence of liquid buildups and dumps, generating oscillations and instability. Operation personnel usually overcome this instability by increasing the liquid rate, so that the downcomer seal is not lost.

The phenomenon of vapor in the downcomer impeding the liquid descent is similar to that of the system limit discussed earlier (Section 2.2). It is therefore a vapor flood. Lowering the vapor rate is a common cure for downcomer unsealing flood. At the reduced vapor rate, the seal will still be broken, and some vapor will still rise up the downcomer, but the lower vapor velocity will permit the liquid to descend, so no flood will occur. The outcome of changing the liquid rate depends on the vapor rate. Below the flood point, adding liquid helps establish the downcomer seal, thereby preventing the unsealing altogether. Above the flood point, adding liquid will aggravate the flood, as this liquid will be unable to descend and will accumulate.

The symptoms of this type of flood resemble those of other types of flood: high pressure drop, entrainment from the top of the tower, poor separation, and instability. What makes it unique is that it occurs only in the spray regime or with truncated downcomers, well below the flood point obtained by hydraulic calculations (vendors and technology providers seldom include downcomer unsealing in their hydraulic calculations). One useful indicator is calculating the liquid head loss at the downcomer exit. Head loss values below 0.4 in. of liquid, and especially below 0.2 in. of liquid, suggest that downcomer unsealing is likely to occur (but not necessarily initiate flood, as with large downcomer areas the vapor rise velocity in the downcomer may be too low to initiate flood). Applying the downcomer unsealing model by Kister (190), with its revised equations published in Refs. 192 and 224, or the model by Bernard (26, 28), can shed more light on this issue.

An excellent method of diagnosing downcomer seal loss is by neutron backscatter, as described in Section 6.3, and/or downcomer gamma scans (Section 5.3).

In multipass trays, it is possible that due to partial plugging, geometry, or maldistribution, the liquid seal is lost on one downcomer but not on the others. Gamma scans or neutron backscatter are the best guides. In the wash trays at the top of one amine regenerator, the gamma scans showed no liquid on one panel and some liquid on the other. Section 6.3.2 and Figure 6.20 (227) describe a similar phenomenon.

Downcomer unsealing is unlikely to occur if the downcomer is sealed by an inlet weir or a seal pan, unless the recess area behind it leaks or an instability causes the level in the sump to fall below the downcomer clearance.

2.12   CHANNELING-INDUCED PREMATURE FLOODS AT HIGH LIQUID LOADS

Large open areas on trays can lead to vapor channeling on the trays. Such channeling leads to the formation of a high vapor velocity zone, accompanied by excessive entrainment and premature flooding, and vapor-deficient zones with excessive weeping. On sieve and valve trays, the most common type of channeling, known as vapor cross-flow channeling (VCFC), was shown to occur when the following four factors come together (194, 228, 245, 406):

1. A high fractional hole area (hole or open slot area/active area) on the tray (>11% with sieve trays, >12% with fixed valve trays, >13–14% with sharp-orifice moving valve trays, any venturi (smooth orifice) moving valve trays);
2. A ratio of liquid flow path length to tray spacing above 2;
3. A weir load exceeding 5–6 gpm/in. of outlet weir;
4. Pressure below 120 psia.

Hartman (162) showed that at very high ratios of flow path length to tray spacing (>3–4), VCFC can occur even at slightly lower fractional hole areas than in 1 above. Hennigan (169) presented a case of sieve trays performing at about half the expected efficiency, in which all the above limits were violated except for the hole area that as installed was 9%. However, plant data showed the tower dP to have declined steadily during the run, and it is likely that corrosion (which was known to occur in this service) increased the hole area. Resetarits and Papademos (397) show that push valves on the trays may help alleviate the channeling to some degree. Experience by Ruffert et al. (406) teaches that channeling may persist if the forward push is not well-adjusted.

VCFC (Figure 2.33) is encountered when, due to a hydraulic gradient, vapor preferentially channels through the tray outlet and middle, generating a high-vapor-velocity region with high entrainment and premature flood. At the same time, the tray inlet region remains vapor-deficient, which promotes excessive weeping and poor tray efficiency.

Davies (80) provided a mechanistic explanation for VCFC. The channeling induced by the tray hydraulic gradient is countered by the dry pressure drop (the friction pressure drop through the holes or valves) that tends to keep the vapor distribution uniform. The criterion proposed by Davies (80) is that to avoid VCFC, the hydraulic gradient needs to be kept below 40% of the dry pressure drop.

A rule of thumb for sieve and valve trays (203) states that the hydraulic gradient equals 0.2 in. per foot of flow path length for weir loads (gpm per inch of outlet weir length) of 6 and higher, and is zero for weir loads of 2 gpm/in. and less. For weir loads between 2 gpm/in and 6 gpm/in the rule recommends linear interpolation in weir loads.

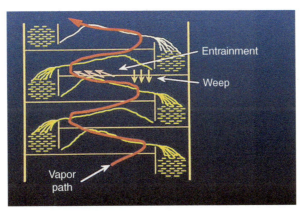

Figure 2.33   Vapor cross flow channeling (VCFC) on trays. (From Kister, H. Z., K. F. Larson and P. E. Madsen, *Chemical Engineering Progress*, p. 86, November 1992. Reprinted Courtesy of Chemical Engineering Progress (CEP). Copyright © American Institute of Chemical Engineers (AIChE).)

VCFC is not the only form of channeling experienced on distillation trays. Other forms of channeling may be due to vapor maldistribution or multipass tray maldistribution (223, 231, 242, 244, 252). One thing they have in common is that they have only been experienced at large tray open areas, high ratios of flow path lengths to tray spacing, and high weir loads.

Vapor channeling floods are vapor floods. The symptoms of this type of flood resemble those of other types of floods: high pressure drop, entrainment from the top of the tower, poor separation, and instability. The pressure drop may not increase too much, as the channeling gives the vapor a path bypassing the accumulating liquid.

What makes the channeling flood unique is that it occurs prematurely, well below the flood point obtained from hydraulic calculations (vendors and technology providers seldom include VCFC in their hydraulic calculations). It occurs only when the abovementioned four criteria come together; if any one of these factors falls significantly outside the range, channeling flood is unlikely. Another symptom is that once out of flood, turning the vapor and liquid loading down will cause a loss of efficiency due to the intensification of weeping. Finally, multi-chordal gamma scanning with quantitative analysis (Section 6.1) is a superb method for diagnosing channeling.

In all channeling, the best prevention is to boost the dry pressure drop.

2.13   FLOODS BY HIGH BASE LEVEL OR ENTRAINMENT FROM THE TOWER BASE

Layton Kitterman, one of the all-time great experts in distillation troubleshooting, estimated that 50% of the problems in the tower originate in the tower base region (253). Over 100 reported case histories out of 900 surveyed (198) testify that indeed, more problems initiate at the tower base than in any other tower region, although the actual percentage may be lower than 50. The survey found little improvement in tower base incidents over the years (198), so it will continue to be a major trouble spot. Numerous case studies are described in Ref. 201.

In half of the case studies reported (198), liquid levels rose above the reboiler return or gas inlet. When this happens, the reboiler return vapor entrains the liquid above the nozzle into the bottom tray or packing, causing it to flood, as depicted in Figure 2.34a.

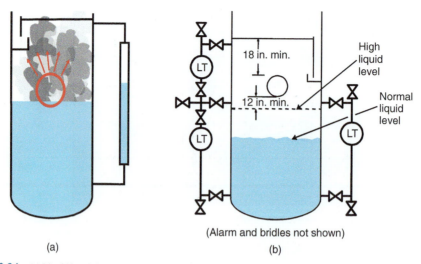

(a)   (b)

Figure 2.34   (a) Liquid level above reboiler return, but level instrument fails to see it due to incorrect upper tap location; (b) Good liquid measurement practice. (Part (a): From Kister, H. Z., "Practical Distillation Technology", course workbook. Copyright © Henry Z. Kister. Reprinted with permission. Part (b): From Kister, H. Z., Chemical Engineering Progress, p. 16, March 2006. Reprinted Courtesy of Chemical Engineering Progress (CEP). Copyright © American Institute of Chemical Engineers (AIChE).).

Faulty level measurement or control was the primary cause of these high levels. A multitude of case studies have reported that faulty base level measurement led to premature flooding (e.g., in Refs. 192, 201, 285, 286, 410, 450, 509). Restriction in the outlet line (including loss of bottom pump, obstruction by debris, and undersized outlets) is another cause. A third major cause is excessive pressure drop in a kettle reboiler circuit, which backed liquid in the tower base beyond the reboiler return inlet.

In the majority of cases, high base levels caused tower flooding, instability, and poor separation. Less frequently, vapor slugging through the liquid also uplifted or damaged trays or packing. In even fewer cases, the tower was overfilled with liquid to the top, pouring into the overhead line and reflux drum, at times causing safety and environmental hazards and accidents. In some cases the entire tower shook and appeared like it could collapse.

To prevent excessive tower base level, the author endorses the measures recommended by Ellingsen (95): reliable level monitoring, often with redundant instrumentation, and good sump design. One added measure (198, 201) is avoiding excessive pressure drop in a kettle reboiler circuit. These measures are discussed at length in Ref. 192 and are briefly summarized below:

Reliable Level Monitoring   Some of the lessons learned are as follows:

- Do not rely on a single level indicator. Always have a cross-check by another, ideally using a different technology (if reliable). In key fractionators, some plants add a third level indicator with a voting mechanism (2/3).

- Ensure that the upper tap of the level transmitter is beneath the reboiler return or vapor inlet. Figure 2.34a illustrates the reason. Once the reboiler vapor blows into the liquid, the liquid becomes a frothy mixture, typically with half the specific gravity of the pure liquid. Therefore, a column of froth that is x ft tall will only show up as $x/2$ ft of liquid in the level transmitter. The tower can be liquid-full, with the transmitter only reading 70% level.

- Ensure that the reboiler return or gas inlet does not impinge on the level taps or the liquid level.
- Closely check that the correct liquid density is used in the calibration. An out-of-calibration level transmitter has been one of the most common causes to false level indications.
- Under startup and abnormal operation conditions, liquid compositions and temperatures, and therefore liquid densities, are likely to differ from those of normal operation. Guide the operators on how to interpret the level transmitter readings (calibrated for normal operation) to the startup and abnormal operation.
- Pay attention to control during startup. Most towers manipulate the bottoms flow rate to control the base level. However, at startup the bottom flow rate is yet to be established. Work with the operating team on how to control the level at startup without encountering high liquid level issues. This may require rigging lines from the tower bottom to an off-spec tank, or from a product tank to the tower base, to supply liquid for the reboiler startup.
- Ensure that the dP transmitter in the tower is operational. A flood initiated by a high base level will show up as an increase in dP.

Figure 2.34b shows a good level measurement practice. It features two independent level transmitters mounted on two separate bridles. The upper tap of each transmitter is beneath the reboiler return nozzle. There is a third level transmitter coming from one of the upper level taps and going to another tap just below the bottom tray or packed bed. This transmitter should always read zero. A reading on this transmitter (which can be set to trigger an alarm) indicates a rise of the liquid level above the reboiler return inlet and likely flood initiation.

Good Sump Design Some of the lessons learned are as follows:

- Never introduce the reboiler return or vapor feed below the liquid level (see Ref. 192 for a possible, very infrequent exception).
- Always have the bottom of the reboiler return nozzle at least 12 in. above the high liquid level (HLL), as shown in Figure 2.34b. In one case (141), the bottom of the nozzle was only 4 in. above the HLL, causing liquid entrainment from the bottom sump into the first tray, which not only reduced tower capacity but also perpetuated foaming and severe fouling on the tray.
- Avoid impingement.

Impingement by the reboiler return or incoming gas on the liquid level generates entrainment that can travel up the tower and flood trays or packings. Reboiler return or inlet gas impingement on instruments gives false readings, which may cause high liquid levels to go unnoticed. Premature floods can also occur due to impingement of the reboiler return vapor on the bottom tray, the seal pan overflow, and the inlet from a second reboiler. Some case studies (201) also reported severe local corrosion due to gas flinging liquid at the tower shell in alkaline absorbers fed with CO_2-rich gas, mostly in ammonia plants.

Figure 2.35 shows poor arrangements that generated impingement of reboiler return or inlet gas on the liquid level, leading to premature floods or entrainment that traveled right up the tower. Sections 8.14–8.16 contain more cases.

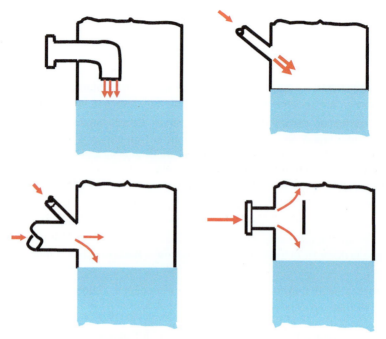

Figure 2.35   Practices to be avoided. Each of these can lead to premature flooding starting from the tower base. (From Kister, H. Z., Chemical Engineering, May 19, 1980. Reprinted with permission.)

Diagnosis   The following can help diagnose flood due to a high liquid level or an impingement issue:

1. Check the level measurement by an alternative technique. This may involve another transmitter. Gamma scanning or neutron backscatter is particularly powerful and informative for this purpose. Section 5.5.18 shows the application of gamma scans to detect flooding due to excessive pressure drop in a kettle reboiler circuit.

2. A useful test is increasing the bottoms flow rate to see if the level comes down (like it should). If the level does not come down, liquid is likely stacked above the upper level transmitter tap, in which case the transmitter measures the froth density rather than the liquid level. Keep raising the bottom flow rate until the level starts declining or another constraint is reached. Has this alleviated the flood? Observe the tower dP when performing this test. If the flood is caused by a high liquid level, the tower dP should decline as the bottom flow rate is increased. Incorporating gamma scans or neutron backscatter with this test can make its results even more conclusive.

3. A useful check is to open the vent valve at the upper level tap (if this can be done safely). Normally, it should vent vapor. Liquid coming out is an indication of a liquid level above the transmitter.

4. Draw a to-scale sketch of the bottom sump. Look for impingement of the reboiler return vapor on the liquid level, instrument taps, and the seal pan overflow. See Sections 8.14–8.16 for many cases in which such a sketch was key to a correct diagnosis.

5. If the reboiler is a kettle reboiler, compile a pressure balance on the kettle circuit. Reference 217 provides a case study of column flood due to excessive pressure drop in a kettle reboiler circuit. The reference contains a detailed description of the pressure balance that was central to the diagnosis.

2.14   TROUBLESHOOTING INTERMEDIATE COMPONENT ACCUMULATION

Column feed often contains components whose boiling points are intermediate between the light and heavy key components. In some cases, the column top temperature is too cold and the column bottom temperature is too hot to allow those components to leave the column at a rate sufficiently high to match their feed rate into the column. Having nowhere to go, these components accumulate in the column, causing flooding, cycling, and slugging. If the intermediate component is water or acidic, it may also cause accelerated corrosion; in refrigerated columns, it may cause hydrates. A high temperature difference between column top and bottom, a large concentration of the component in the feed, a large number of stages, and a high tendency to form azeotropes or two liquid phases are conducive to intermediate component accumulation.

A high activity coefficient at the tower bottom can make an intermediate component, even a heavy component, particularly volatile. For instance, when a tower bottoms is mainly water, organic components such as n-butanol are repelled from the liquid into the vapor and become highly volatile. The same organic can become highly nonvolatile if the top of the tower is rich in methanol or ethanol. For a methanol–water separation column, n-butanol in the feed will tend to concentrate near the middle of the tower. This effect has been discussed at length by Saletan (408) and Kister (199).

Figure 3.1 is an example of a column that separates a 50/50 mixture of methanol and heptanol. The reboiler receives a feed of butanol at a rate of about 0.1% of the feed rate. The tiny amount of butanol builds up to 80–90% in the liquid at the middle of the tower, dominating the tower composition (and temperature) profile. The butanol accumulates in the column over a period of time, during which the column will be in an unsteady state. Such an unsteady state period is common to accumulation incidents. Figure 2.36 (the "no decant" curves) is another example. Here, a minor high-boiling impurity in the feed, 2 ppm, formed a heterogenous azeotrope with water and built up to >40% in the tower, generating a two-liquid phase zone from stage 9 to 16.

The component builds up toward the equilibrium concentration that reinstates the component balance in the tower. However, when the volatility of the intermediate component near the bottom is high and near the top is low, and/or the number of stages is high, and/or the concentration of the intermediate component in the feed is high, the equilibrium concentration may not be reached. Instead, unsteady state cycling may set in, as described below.

Intermediate component buildup most frequently takes place over the entire tower, but there are times when it is confined to the section below the feed or the section above the feed. Excessively subcooled feed (or fouling of a feed preheater) can lead to accumulation of an intermediate component between the feed and the bottom. Similarly, excessive preheat (oversurfaced or clean preheater) may lead to an accumulation between the feed and the tower top.

Hiccups (Also Referred to as Burping, Puking, Cyclic Slugging)   A typical symptom of nonsteady state accumulation is cyclic slugging, which tends to be self-correcting. The intermediate component builds up in the column, typically over hours or days. Eventually, the column floods, or a slug rich in the offending component exits either from the top or from the bottom. The column end from which the slug leaves often varies unpredictably or according to the operator's action. Once a slug departs, normal column operation resumes fairly quickly, often without operator intervention. The cycle will then repeat itself. Several experiences have been reported in the literature, many of them described in Refs. 94, 172, 199, 201, 389, and many others cited by the mentioned references.

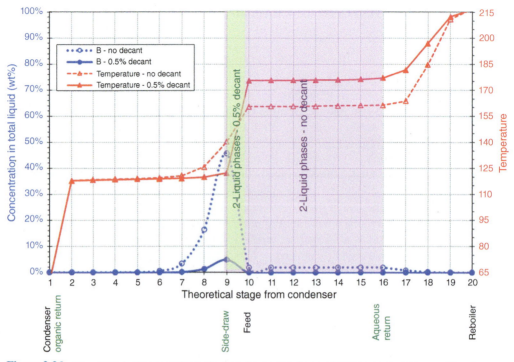

Figure 2.36 Minor high-boiling impurity in the feed, 2 ppm, that forms a heterogeneous azeotrope with water, can build up to >40% in the tower, generating a two-liquid phase zone from stage 9 to 16. Removing a side draw 0.5% of the product take off and decanting to remove the component cuts the two-liquid phase zone to two stages and eliminates the bulge. (From Schon, S. G., "A New technique to Prevent Minor Component Trapping in Distillation", Paper presented at the Distillation Symposium, AIChE Spring Meeting, San Antonio, Texas, April 2022. Reprinted Courtesy of Steven G. Schon, Arkema Inc.)

Intermediate component accumulation may interfere with the control system. For instance, a component trapped in the upper part of the column may warm up the control tray. The controller will increase reflux, which pushes the component down. As the component continues to accumulate, the control tray will warm up again, and reflux will increase again; eventually, the column will flood. Cases describing a similar sequence have been reported (Case 2.13 in Ref. 201, also 94, 389).

One of the most fertile breeding grounds for intermediate component accumulation is a condition in which at least some section in the tower operates close to the point of separation of a second liquid phase (201). Either the accumulation itself, a condition in which phase separation equipment in the feed route (a decanting drum, a coalescer) malfunctions, or simply a change in feed composition can induce separation of a second liquid phase inside the tower. The introduction of a second liquid phase changes volatilities and azeotroping, which can aggravate hiccups.

Hiccups can cause major upsets in downstream equipment. A slug of lighter material escaping from the bottom of one tower can quickly vaporize in the hotter downstream tower, causing a pressure surge there and possibly lifting its relief valve.

Diagnosis Periodic cycling that tends to be self-correcting, with a long time period, is the key symptom for hiccups. The time period is seldom less than one hour, which distinguishes hiccups from other, shorter cycles such as those associated with flooding, hydraulic, or control issues, whose period is typically a few minutes. Typical cycle period for hiccups ranges from about once every two hours to once every week. The cycles are often regular, but if the tower feed or product flows and compositions vary, the cycles may be irregular.

Drawing internal samples from a tower over the cycle (or over a period of time) is invaluable for diagnosing accumulation. Such samples will show high and increasing concentrations of the intermediate component over the cycle. Even a single snapshot analysis can detect accumulation of a component. Unfortunately, safety, environmental, and equipment constraints often preclude drawing internal samples.

Closely tracking and monitoring temperature changes is also invaluable. Temperature changes reflect composition changes. Accumulation of an intermediate component tends to warm the top of the tower and cool the bottom of the tower. The response of the control system to temperature changes and any rise in pressure needs to be considered when interpreting temperature trends. One symptom is common: at the initiation stage, temperature deviations from normal are small, often negligible. As the accumulation intensifies, systematic temperature deviations ("excursions") from normal become apparent. Close to the hiccup point, temperature deviations become relatively large.

Figure 2.36 illustrates the variations in the temperature profile. The "0.5% decant" curve shows a temperature profile close to normal, practically unaffected by the accumulation. As the component builds up, the curves move closer to the "no decant" curves, first at stage 11 and then progressively to stage 17. In this case, the temperature declines by a sharp step due to the formation of the second liquid phase on these stages. Had there been no second liquid phase, the accumulation would have lowered the temperatures in the accumulation zone below the feed in a more gradual (rather than stepped) manner.

Steady state bulges can be readily simulated and recognized on a plot of component concentration against stage numbers, as in Figures 3.1 and 2.36. The simulation needs to be able to correctly simulate a second liquid phase, which may form during component accumulation. With hiccups, steady state simulations are likely to run into convergence problems, reflecting the reality that steady state is not reached. To get around this, "sneak into the solution" by studying a related system that can converge, e.g., input less intermediate component in the feed, or draw a side product. Once a converged solution is reached, slowly proceed toward the actual system (by raising the intermediate component in the feed, or cutting back on the side draw). Track the changes by plotting a concentration versus stages diagram for each step. In the case of a second liquid phase, many stages in the simulation may alternate between a single liquid phase and two liquid phases, making convergence particularly problematic. Either increasing or decreasing the concentration of the second liquid phase can help reach a solution.

Gamma scans and dP measurements can detect the intermittent flooding. Gamma scans typically show normal operation at the beginning of the cycle and mid-column flooding when a hiccup occurs. If enough pressure taps are available on the tower, dP transmitters can be just as informative. Finally, sight glasses are extremely useful when safety requirements permit.

Cures There are three cures that can effectively mitigate the hiccups and two "band-aid" solutions that can reduce the frequency and severity of the hiccups. These cures are discussed in detail with several examples elsewhere (199, 201, 419):

- Reducing the column temperature difference. This allows the accumulating component to exit with one of the products but will increase the impurity in this product. It is low-cost, easy to implement, and therefore often used.

- Removing the accumulated component from the tower using a side draw in the accumulation zone. This is often accompanied by using decanting or another separation technique to separate the intermediate component from others. Figure 2.36 shows the effectiveness of this technique for mitigating the accumulation. The bulge went down from 40% to 5%, and the number of stages with two liquid phases went down from 7 to 2.
- Removing the accumulated component from the feed.
- Modifying tower internals—this is a band-aid solution.
- Improving controls and living with the problem—another band-aid solution.

2.15   TROUBLESHOOTING LIQUID SIDE DRAW BOTTLENECKS

This includes side draw lines to products or reboilers. Bottlenecks at a liquid side draw can cause one or more of the following:

Initiate flooding that propagates up the tower. This occurs when the bottleneck backs liquid up in the tower. It is most likely when the product is taken from a sump under the downcomer with a restricted path for the excess liquid to flow down the tower.

Cause tower instability, initiating at the draw point. This occurs when the bottleneck backs liquid up in the tower, but once the liquid reaches a certain height or accumulation, it can dump down the tower or even down the draw line.

Cause a loss of product yield and, at the same time, lead to a product lower down in the tower going off-spec. This is most likely to occur when the excess liquid has an easy path down the tower.

Downcomer Draw Boxes   Column flooding is likely to initiate in the draw arrangement in Figure 2.37a whenever the liquid flow rate to the draw exceeds the draw hydraulic limit. The excess liquid has nowhere to go, so it will stack up in the downcomer and flood the next tray. From there, the flood will propagate upward. Column flooding is unlikely to initiate in the draw arrangement in Figure 2.37b because the generous clearance under the downcomer will allow excess liquid to drain onto the tray when the draw encounters a hydraulic limit.

Depending on the dimensions, column flooding may or may not initiate in the draw arrangement in Figure 2.37c when the draw reaches its hydraulic limit. At the limit, liquid will stack up in the seal pan and in the downcomer up to the seal pan overflow elevation. If the overflow weir is tall enough to back up liquid in the downcomer all the way to the tray above, flooding is likely. It is important to note that the seal pan liquid is likely to be degassed, while the downcomer liquid is likely to be frothy and therefore have a significantly lower specific gravity, typically about half, so the seal pan can back downcomer liquid to about twice the submergence of the downcomer. On the other hand, if there is ample vertical height to allow overflow without excessive downcomer backup, flooding is unlikely.

A potential problem with the arrangement shown in Figure 2.37b is vapor entry into the downcomer during excessive draw rates. Should the draw rate exceed the rate of incoming liquid, vapor will be sucked into the downcomer and the draw via the downcomer clearance. The entering vapor may choke the downcomer, the draw, or both. Intermittent flooding or cycling may initiate, producing instability. With arrangements shown in Figure 2.37a and c the downcomers are always liquid-sealed, so vapor entry is unlikely.

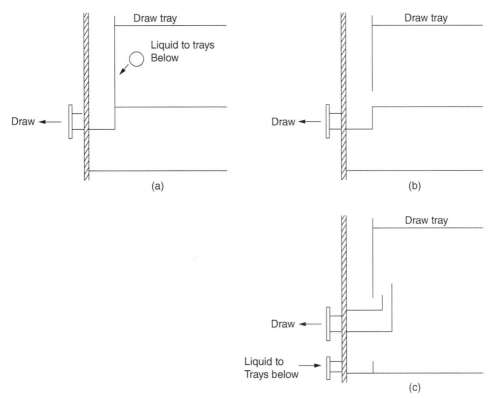

Figure 2.37   Side draw-off arrangements (a) likely to initiate flooding at excess draw rates; (b) generous clearance under the downcomer allows liquid drainage to the tray, but may be prone to vapor entry into the downcomer; (c) downcomer sealed, preventing vapor entry.

Should a side draw bottleneck be suspected, the following checks and field tests can help diagnose:

1. *Check for self-venting flow in the draw line.* Liquid in the column is aerated, i.e., contains vapor bubbles, unless it has been degassed. Degassing a liquid takes a minimum residence time of 30–60 seconds based on the sump volume and the total flow rate through it, from the point of liquid collection to the top of the draw nozzle. Downcomers are usually designed for a residence time of 3–7 seconds, which may increase to about 10 seconds after adding the volume of the sump underneath. So the fluid leaving in the draws from any of the arrangements in Figure 2.37 is usually not liquid; it is aerated liquid.

   Aerated liquid does not follow the liquid laws; it follows the aerated liquid laws, which state that all the pipes downstream of the draw, all the way to the lowest point, including any horizontal sections, and including the draw nozzle, need to be sized for self-venting flow. A self-venting flow is where liquid descends and any entrapped vapor bubbles rise back up to be vented back through the line and draw nozzle. Just like they do in a downcomer. If the line is not large enough for self-venting flow, the bubbles will accumulate in the line and choke it, initiating flooding or instability.

   Figure 4.9 is a graphical version of an excellent correlation for self-venting flow by Simpson (439) and later by Sewell (425). Above the diagonal, the flow is self-venting.

Below the diagonal the downpipes are likely to choke, backing up liquid in the tower. The correlation straight line extrapolates to larger and smaller pipe diameters.

Cases 10.1 and 10.2 in Distillation Troubleshooting (201) describe case studies of premature tower floods due to product lines undersized for self-venting flow. Many more were described in the literature (8, 18, 198, 280, 489). Some of these resulted in poor product quality or instability rather than flood.

There are a number of situations where one can often get away with draw lines undersized for self-venting flows:

- When the line originates in a sump with more than 30–60 seconds residence time. This is a common situation when the line originates in a tower bottom sump (typical residence time three to five minutes) or in a chimney tray that holds liquid level above the nozzle (typical residence time of about one minute).

- When the line feeds a low-head centrifugal pump. A typical example is pumparound pumps. It appears that these pumps suck in the bubbles. This may not extend to high-head pumps in which vapor bubbles may lead to cavitation.

- When a subcooled stream enters the sump and quenches the bubbles.

2. Observe what happens when increasing the draw rate (e.g., by opening the control valve). If opening the control valve does not increase the flow rate, or if the draw pump cavitates, it means that no more liquid is available from the draw sump to the draw. For best results, install a static pressure gauge at the bottom of the line, just before the draw pump (if pumped), or just before the control valve if by gravity. The difference between the pressure gauge reading and the column pressure at the draw point (which can be estimated from the column pressure transmitters) should equal the static head in the line. From this reading one can infer how liquid-full the line is at different moments during the test. Likewise, gamma scans (Chapter 5) or neutron backscatter measurements (Section 6.3) can estimate the fluid density in the line. When practical, gamma scan or neutron scan the line.

Chimney Tray draws are not immune to side-draw bottlenecks. When they hold liquid, chimney trays usually, but not always, provide sufficient degassing time to degas the liquid (see criterion in item 1 above). In Figure 2.38, the pumpdown draw will benefit from the degassing time on the chimney tray, and if sufficient (>30–60 seconds), there will be no need for self-venting flow. Not

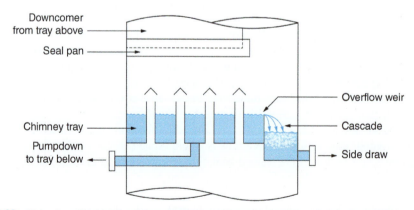

Figure 2.38   Examples of liquid draws from a chimney tray. The residence time will degas the pumpdown liquid, but not the side draw liquid. (From Kister, H. Z., G. Jacobs, and A. A. Kister, Chemical Engineering Research Design, 191, p. 313, 2023. Reprinted Courtesy of IChemE.)

so with the side draw in Figure 2.38; the cascade over the chimney tray weir re-entrains vapor into the side draw takeoff sump (250). If the side draw line is undersized for self-venting flow, liquid will back up onto the chimney tray, possibly resulting in instability, and/or overflow and/or a premature tower flood.

In another case (150), liquid to circulating thermosiphon side reboiler was drawn from a chimney tray, with the reboiler effluent returning to the same chimney tray. Overflow pipes transported the reflux to the tray below. Circulation through a thermosiphon reboiler generated a very large liquid flow through the chimney tray, depriving it from the residence time needed for degassing. The reboiler draw lines were undersized for self-venting flow of the huge recirculating liquid flow and choked. The excess liquid overwhelmed the overflow pipes, so it accumulated in the tower and flooded it, with gamma scans showing 15 ft of liquid piling up above the chimney tray. This scan observation was the key to the diagnosis. The problem was resolved primarily by returning the reboiler effluent to the tray below the reboiler (in essence, converting the circulating thermosiphon reboiler to a once-through thermosiphon reboiler), which eliminated the excessive circulation through the chimney tray and the reboiler draw line.

Chimney tray overflow or premature flood due to downpipes undersized for self-venting flow are discussed with reference to packed tower case studies under "collectors" in Section 4.6. Tray case studies are found in Section 9.2.5 and in Refs. 118 and 388.

2.16   TWELVE USEFUL RULES OF THUMB

Rules of thumb should always be taken for what they are: rough criteria that reflect the experience of practitioners. They are not meant to replace detailed calculations such as those presented in most distillation texts (34, 193, 245, 299, 451, 492). Their merit is in providing troubleshooters with a preliminary orientation of which path is more likely to lead to problem identification.

Most of these rules of thumb apply to systems where there is a lot of experience. A rule may not extend to other systems where the experience is limited. Many systems are not covered by these rules. In addition, every rule of thumb has exceptions. Some rules of thumb useful for distillation troubleshooting are:

1. Flood C-factor, ft/s (C-factor from Eq. 2.1), pressures <100 psia, liquid loads <20 gpm/ft^2)

   0.45 – System limit for hydrocarbon and organic systems

   0.35–0.42 – Jet flood for trays at 24–36″ spacing and vapor flood for large structured packings (<250 m^2/m^3 surface area per unit volume) and large random packings (<100 m^2/m^3 surface area per unit volume).

   0.25–0.35 – Jet flood for trays at 18–24″ spacing and for medium structured packings (250–350 m^s/m^3 surface area per unit volume) and medium random packings (100–150 m^2/m^3 surface area per unit volume).

   Flood C-factor, ft/s, is likely to be lower at pressures >100 psia and liquid loads >20 gpm/ft^2.

   Detailed calculation methods are available in most distillation texts (34, 193, 245, 299, 451, 492), with Ref. 245 the most updated.

2. Maximum downcomer entrance velocities, ft^3 per second clear liquid/downcomer cross section area at top, ft^2 (pressures <100 psia)

   0.45 – Organic and hydrocarbon non-foaming systems

   0.2–0.25 – Foaming circulating systems that use antifoam

   0.15 – Once-through foaming systems and others that minimize or avoid antifoam

   Detailed table is in Refs. 193 and 245 (245 is updated).

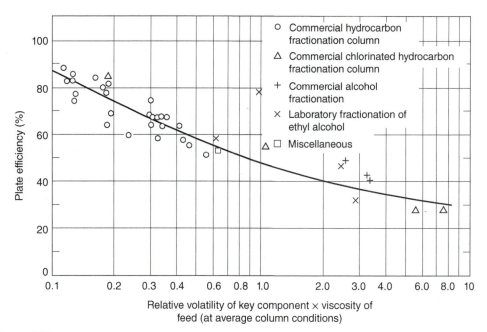

Figure 2.39   The O'Connel plot for tray efficiency. (From O'Connell, H. E., *Transactions of the American Institute of Chemical Engineers*, 42, p. 741, 1946. Copyright © American Institute of Chemical Engineers (AIChE). Reprinted with permission.)

3. Tray efficiency. The O'Connel plot is provided in Figure 2.39. Predictions are usually within plus or minus 25%.

4. Packing HETP in meters, as a function of a_p, the packing surface area per unit volume, m^2/m^3, for organic and hydrocarbon systems, surface tension <25 dynes/cm:
   HETP = $93/a_p$ Metal random packings, $1''$ or larger
   HETP = $100/a_p + 0.1$ Y-type or third-generation structured packings
   HETP = $145/a_p + 0.1$ X-type structured packings, $a_p < 300 \, m^2/m^3$
   Multiply the above values by two for water due to its higher surface tension (70 dynes/cm). For intermediate surface tensions, interpolate.
   These rules of thumb are generally plus or minus 25% and assume good industrial-grade liquid and vapor distribution (not perfect like in test columns).

5. Tray flow regime, expressed in weir loads (gpm liquid per inch of outlet weir length):
   Spray regime occurs at weir loads <2–3 gpm/in. Downcomer unsealing in conventional trays is unlikely to occur at higher weir loads.
   A downcomer is prone to losing its liquid seal in the spray regime when the downcomer exit liquid head loss at the downcomer exit is <0.4 in. of liquid, and especially below 0.2 in. of liquid. The downcomer exit head loss can be estimated from

$$h_{DC \, exit} \text{ (inch of liquid)} = 0.604 \, u_{exit}^2 \qquad (2.3)$$

   u_{exit} is in ft/s. The area used in the calculation is based on the most restrictive area for liquid flow at the downcomer exit.

6. Channeling is likely to occur on trays when the four factors listed in Section 2.12 come together.

7. Packing flow regimes are set according to the flow parameter F_{LG}

$$F_{LG} = \frac{L}{G}\left(\frac{\rho_G}{\rho_L}\right)^{0.5} \tag{2.4}$$

where L and G are the liquid and vapor mass flow rates, respectively, consistent units, and ρ_G, ρ_L are gas and liquid densities, consistent units, respectively. F_{LG} is dimensionless. The vapor-rich region occurs at $F_{LG} < 0.1$. The liquid-rich region occurs at $F_{LG} > 0.3$.

8. On trays with organic and hydrocarbon liquids, normal pressure drop is about 0.1 psi per tray. If measuring 0.15 psi per tray, flooding should be suspected and checked out. If measuring 0.2 psi per tray and the measurement is good, flooding is highly likely. The numbers may be somewhat higher for aqueous and high-liquid density systems, as described in Section 2.6.

9. In packed towers, the flood pressure drop can be calculated from Eq. (2.2). Significantly higher values are likely to indicate packing floods, assuming the measurements are good.

10. The head–flow relationship for packing orifice distributors is given by the Torricelli Orifice Equation:

$$Q = 5.67K_D n_D d_h^2 h^{0.5} \tag{2.5}$$

where Q is the liquid flow rate, gpm; K_D is the orifice discharge coefficient, with a recommended value of 0.707 (68, 161, 192, 245); n_D is the number of holes; d_h is the hole diameter, inches; and h is the net liquid head, inches. See Section 4.6.2 for pressure drop and buoyancy corrections.

11. The Summers et al. (468) tray stability factor η (dimensionless) provides a measure of tray stability

$$\eta = (\Delta PD/H_{CL})^{0.5} \tag{2.6}$$

where ΔPD is the dry pressure drop and H_{CL} is the clear liquid height, both in inches of liquid. H_{CL} can be calculated from the difference between the total and dry pressure drops or from the Colwell equation (73, 193, 245). When the stability factor exceeds 0.6, weeping is less than 30%, and the tray is stable. Weeping and pass maldistribution rapidly escalate at lower stability factors, with 50% weeping or more by the time the stability factor declines to 0.4.

    Summers has recently published (467) an updated criterion. The author has extensive good experience with the 2010 version, but only limited experience with the updated criterion.

12. Avoid excessive inlet velocities. Good practices are:
    Liquid feeds to towers (192) $V < 5$ ft/s
    where V is the liquid velocity at the inlet, ft/s.
    Flashing and vapor feeds to tray towers (37): inlet $\rho_M V_M^2 < 2500$,
    Where ρ_M is the mixed density (or vapor density when all vapor), lb/ft³, and V_M is the mixed (or vapor when all vapor) velocity, ft/s
    Flashing and vapor feeds, packed towers (192, 326): inlet $\rho_M^{0.5} V_M < 52.4\,(\Delta P)^{0.5}$
    where V_M and ρ_M are as defined above and ΔP is the packed bed pressure drop, inches of water per foot of packing.

Chapter 3

Efficiency Testing and Separation Troubleshooting

"A theory without a test is B.S."

—(Norman P. Lieberman and E. T. Lieberman, Ref. 295)

The first part of this chapter deals with efficiency testing. The second part deals with various sources of poor separation, other than those covered in Chapter 2 (flooding, foaming, channeling, loss of downcomer seal), those in in Chapter 4 for packed towers (maldistribution of liquid or vapor), and those associated with control instability that are covered in distillation control texts (50, 192, 431, 453).

3.1 EFFICIENCY TESTING FOR TROUBLESHOOTING

3.1.1 Purpose and Strategy of Efficiency Testing for Troubleshooting

The incentive for efficiency-testing a poorly performing column is obvious. The simulation with the expected efficiency shows how the tower should perform, while that based on the measured plant data provides the real efficiency. In the absence of flooding, a real efficiency below the expected one signifies issues such as maldistribution or channeling in trays and packings, poor hydraulic design, fouling, corrosion, or damage. In many cases (e.g., 195, 197, 234, 286, 351), a mismatch between simulation and plant data, or a low efficiency obtained from good plant data, was the key for correct diagnosis and an effective solution to an operating problem.

A well-documented performance/efficiency test report is essential even for a well-performing unit, for the following reasons:

1. This report provides an excellent starting point for almost every troubleshooting assignment, present or future, and highlights unit limitations and operating deficiencies.

2. When the tower malfunctions at a future date, availability of good operation test data dramatically reduces the troubleshooting effort.

3. A tower may appear to perform well, even while running at non-optimum conditions; for instance, energy usage may be excessive. Because of overdesign, these factors may be hidden. Yarborough et al. (515) presented several case histories (not all distillation-related) where plant testing directly led to major improvements in profitability.

Distillation Diagnostics: An Engineer's Guidebook, First Edition. Henry Z. Kister.
© 2025 American Institute of Chemical Engineers, Inc. Published 2025 by John Wiley & Sons, Inc.

4. Excess reflux and reboil due to poor efficiency may go unnoticed while generating excessive hydraulic loads and a premature capacity bottleneck. When unappreciated, an incorrect, or ineffective, or unnecessarily expensive debottleneck action may result.

5. Simulations used to determine the best running conditions and assess the effectiveness of the proposed modifications may be misleading unless tested against reliable data.

6. A discrepancy between the simulation and the tower performance (e.g., differences between the simulated and measured temperature profiles) may identify a hidden problem (351).

7. Basing a revamp on a reliable simulation, with efficiency tuned to correctly match the plant data, can be an enormous cost-saver. In two cases (277, 289), an expensive new tower was (or was to be) constructed based on a simulation that incorrectly reflected tower performance. In both cases, a correct simulation achieved (or would have achieved) the revamp by an inexpensive, minor modification. The author is familiar with many other similar cases.

8. A reliable efficiency determination provides a good basis for simulation and future design or optimization of similar towers and valuable input for the company database.

9. It provides a check that the contractor and/or internals vendor meets their guarantees.

Good performance/efficiency testing is rigorous and effort- and time-consuming. The cost-effectiveness of the rigorous procedure is often questioned, and a shortcut version is advocated. A suitable shortcut procedure can evolve from the rigorous procedure outlined here and in Refs. 4 and 192 by skipping over guidelines that are considered less critical.

The best procedure to adopt depends on the objective of the test. A shortcut test is best suited for detecting gross abnormalities and is often performed as a part of a troubleshooting effort. When investigating a gross malfunction, rigorous testing may slow down the identification and rectification of the fault. When a column appears to perform well, there is no urgency, so the focus should be on obtaining reliable data for future reference, favoring a rigorous test.

A shortcut test is unsuitable and often misleading for detecting subtle abnormalities, for obtaining column efficiency, for checking the design, for optimization, and as a basis for debottlenecking modifications. The author is familiar with many cases where shortcut tests applied for these purposes needed repeats, yielded conflicting data, provided inconclusive results, and led to ineffective modifications. In most of these cases, the test objectives were not met even though the total time, effort, and expense spent were severalfold those that would have been spent had a single rigorous test been performed. The author therefore strongly warns against applying a shortcut procedure for these purposes. This recommendation is shared by others (515).

Hanson, Golden, and Martin (126, 153) strongly warn against using shortcut tests, as well as using the plant management information system (MIS) or plant information (PI), as the sole basis for performing a revamp. They question the validity of some of these data, the ability of the statistical methods to detect faulty instruments, and the absence of key data not gathered by the plant systems. Sloley and Fraser (446) add concerns about MIS data gathering being often designed to analyze data trends, most suited for plant operation, rather than the absolute values needed for evaluation and revamps. Rigorous testing can fully take care of these concerns. "The surest way to have a successful, low investment revamp is to carefully (rigorously and correctly) define your current operating basis" (126). The author endorses their recommendation.

Shortcut tests range from those that do little more than taking a set of readings and samples from the column to those that incorporate checks of material, component and energy balances, and key instrumentation. Even for shortcut tests, the author recommends incorporating the above checks and spending time on preparations to ensure that key indicators provide adequate data. The key items can be extracted from the list of preparations and checks recommended here and in Refs. 4, 192.

3.1.2 Planning and Execution of Efficiency Testing for Troubleshooting

General It is best to carry out a test encompassing the whole unit. Individual testing of columns, one at a time, increases the total effort and time consumed and reduces the reliability of measurements. Testing the entire unit provides several material balance cross-checks and enables better identification of erroneous meters and lab analyses. For instance, if the column feed analysis is off, the column component balance may not be sufficient to reveal which analysis is erroneous; but if data from a component balance on upstream and downstream equipment are also available, the incorrect analysis can readily be identified.

A shortcut to the above recommendation may be acceptable when the plant inlet rate and compositions have not significantly changed since the last plant test and the problem areas are well known. In such cases, it may suffice to test only the specific column area (192, 515).

Another shortcut is often acceptable when the column is near the end of a processing train that yields reasonably pure products. Product analysis and metering tend to be far more reliable than intermediate-stream measurements, and there is usually more to be gained from cross-checks with downstream measurements than with upstream measurements. In such cases, it may suffice to restrict the testing to the column and downstream equipment.

Safety, health, and environment Test procedures must conform to all statutory and company safety, health, and environmental regulations. The test plan should be reviewed with persons familiar with safety, health, and environmental requirements and amended as necessary to fully conform to these.

Duration "Snapshots" are never a suitable basis for efficiency tests and validating simulations. Plants seldom run in the steady state. There are always peaks and valleys in key variables. Over a short duration, test data are likely to be biased toward such peaks or valleys. Over a longer time interval, these deviations average themselves out. The key to success in efficiency tests is collecting data over a *long enough period of steady and stable operation*.

This raises the question of what is *a long enough period*. In the author's experience, for smaller columns (<40 trays, <10 ft diameters, <3 packed beds), one shift (eight hours) is long enough. In larger fractionators (refinery main fractionators, superfractionators such as C_2 or C_3 splitters, ethylbenzene–styrene towers), a day or two minimum is needed.

In either case, the key is *steady operation*. This again raises the question of what is *steady operation*. If a reflux or feed flow rate bounces between 80 and 120 units, it is not steady. On the other hand, if this flow varies between 99 and 101 units, it is very steady. The criterion should be deviations and standard deviations from the average. Compositions, too. If product impurity varies between 0.5% and 5%, it is not steady. In contrast, if it varies between 0.8% and 1.2%, it is usually considered steady. At least three successive samples should show essentially constant composition (4). Finally, temperatures. The temperatures in the sections that have the highest temperature gradients are the most sensitive and best indicators of steady operation (4).

The author has had success with two alternative options based on the above discussion:

Option 1. The traditional test option (4, 192). Carry out a performance test over a two- to three-day period. This option is ideal when testing several towers at the same time, as advocated earlier. This option is also strongly preferred for towers that experience fluctuations in key variables. During shorter periods, variations in plant conditions may introduce serious errors. Over a period of two days, fluctuations are averaged out. Further, column control problems may make it difficult to obtain a sufficiently long period of steady operation if the test is short. Over a 2-day period, the column should be running continuously under steady conditions at least for one shift or so.

If product or charge tank levels are to be gauged to obtain accurate flow rates, the test period should be long enough to measure tank displacements within 1–2% (4).

One drawback of this option is that if the plant experiences an upset during this test, it is likely that the test will need to be terminated and later repeated.

Option 2. This option is most suitable for towers that usually do not experience fluctuations in key variables and is often favored by operation personnel. It requires computerized data collection for all key process variables. A commitment needs to be made to operate the tower steadily over three to four days for smaller columns or over a week for larger columns. Minute-by-minute data are recorded for the tower and other relevant units. For each meter, averages, mean deviations, and standard deviations are compiled for portions of the test periods. From these, a period of minimum deviations is selected as the test period. The meters are averaged over this period. This period needs to be at least 8 hours for smaller towers and 24 hours for larger towers and can be as long as one wishes, providing the deviations remain small. Longer periods include more lab analyses. One can also select two steady periods and compare them.

With this option, it is important to collect the minute-by-minute data. If the time interval is much longer, it may hide important swings and counteracting conditions.

This option can handle a major upset during the test. Data collected over the upset will show large deviations from the average and will be discarded. However, this option requires enhanced effort from the lab and those taking local meter readings, as frequent lab analyses and local readings are needed throughout the total longer test duration.

Table 3.1 is a summary of the recorded test data for a simple C_3 stripper tower, showing minimum and maximum values and standard deviations. Using similar summary tabulations was also recommended by other studies (4, 418, 465).

A rule of thumb by Summers (465) for the length of time it takes to establish the steady state following an operational change is to multiply the liquid holdup of the tower (including the sump and reflux drum holdup) in weight units by the reflux ratio and divide this by the mass flow rate of the feed.

Consideration should be given to the possibility of unsteady state accumulation ("bulging") of intermediate components (139, 199, Section 2.14). Figure 3.1 is an example of a column that separates a 50/50 mixture of methanol and heptanol. The reboiler receives a feed of butanol at a rate of about 0.1% of the feed rate. The tiny amount of butanol builds up to 80–90% in the liquid at the middle of the tower, dominating the tower composition (and temperature) profile. It accumulates in the column over a period of time, during which the column will be in an unsteady state. Data taken under the accumulation conditions are likely to be of little value for determining tray efficiency. Unsteady-state accumulation of an intermediate component must be avoided during the test, either by changing top or bottom temperatures to allow it to escape or by drawing it via a side draw or by eliminating it from the feed during the test (199).

Timing   The best time to carry out a performance test is when the plant is stable. In most plant situations, weekends are ideal, as changes due to fluctuations in upstream units are minimized.

Table 3.1 Data collection summary for a C_3 stripper tower.

Start time 20-Mar-18 15:00

End time 20-Mar-18 13:00

	Description	Tag No.	Units	Min	Value	Max	Std Dev
Feeds							
	C_3 from FCC + PPR return	ecfc03	ton/hr	22.96	23.21	24.64	0.26
Products							
	Stripper bottom flow rate (liquid)	efc04	ton/hr	21.91	23.05	24.75	0.51
	Stripper bottom temperature	eti02	C	60.99	61.28	61.57	0.11
	Stripper bottom C_2H_6 concentration	eac01a	ppm	80.04	97.35	126.23	14.52
	Stripper bottom C_3H_6 concentration	eai02b	%	79.08	79.88	80.62	0.38
	Stripper bottom CH_4 concentration	eai01c	ppm	1.75	3.56	11.73	2.50
	Stripper purge flow rate	efc07	ton/hr	0.17	0.49	0.81	0.08
	Stripper condenser outlet temperature	eti08	C	45.10	46.19	47.00	0.53
General							
	Stripper reflux flow rate	efc05	ton/hr	23.19	24.52	25.32	0.45
	Stripper LP steam to reboiler flow rate	efc06	ton/hr	3.40	3.65	3.86	0.04
	Reboiler return to stripper temperature	eti03	C	60.75	61.16	61.57	0.13
	Stripper overhead pressure	epc01	kg/cm^2	24.4	24.48	24.60	0.03
	Stripper pressure drop	epi02	kg/cm^2	0.36	0.36	0.36	0.00
	Stripper overhead temperature	eti01	C	50.89	51.73	52.63	0.39
	LP steam temperature	eti09	C	194.25	198.22	201.7	1.81
	LP steam pressure	epc11	kg/cm^2	1.69	1.72	1.74	0.01
Local	Temperature after FC-07	local T1	C		10.0		
	Pressure before FC-06	local P1	barg		2.0		
	Pressure after FC-06	local P2	barg		0.0		
	Pressure after vent condenser	local P3	barg		23.9		

3.1.3 Preparations for Efficiency Testing

This is the most important phase of the test. An erroneous meter, a leaking block valve, or an incorrect lab analysis during the main test can dramatically reduce the reliability of the results, defeating the purpose of the test. This is the time to eliminate all potential problems and complications. The following considerations are important and are based on the author's experience as well as many literature sources (4, 47, 121, 139, 153, 286, 319, 418, 426, 447, 465, 515).

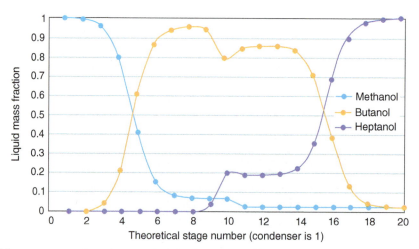

Figure 3.1 Relatively small amount of intermediate component can build up to large concentrations within a distillation column, interfering with the separation and generating an unsteady-state accumulation that makes tray efficiency very difficult to determine. (From Graham, G., P. Pednekar, and D. Bunning, Chemical Engineering, p. 30, February 2018. Reprinted with permission.)

1. *Prepare a needs list for work required.*

    This should include a detailed list of instruments to be checked and required laboratory analyses. Consider the following:

    • Thoroughly check for leaks. Closely study the P&ID and look at any lines connected to the tower, or cross-connecting products. Check that all the block valves on lines connecting different products with each other are shut. Use a temperature gun to make sure they are at ambient temperatures and not passing fluids. Look for any unmetered connections to the tower or its heat exchangers and make sure they are fully isolated and not leaking. See Section 3.2.1 for how unexpected leakages affect product purities.

    • Thoroughly check that all streams entering the tower are accounted for. Audit any purges entering or leaving. Devise ways of accounting for them or, if practical, have them blocked in for the test period. In one case (41), a small product stream was cooled, used as seal oil to the vacuum pump, and then returned to the column. The cooling of the stream had a significant impact on the column heat balance. In another case, a seemingly benign unmetered vapor equalizing line from a column overhead to the feed drum accounted for a 10–12% error in the tower mass balance (136).

    • Ensure that no important meters are left out and that the meters included are sufficient to entirely define unit performance. Add temporary instrumentation to fill any gaps. If in doubt, include the meter. It is much easier to ignore data that were collected but not needed than to generate needed data that were not collected (418). Include control valve openings as well as pump suction and discharge pressures that can be used to verify important flow rates. Seemingly unimportant information may turn out to be the deciding factor between contradicting measurements.

    • Closely review the literature on column instrumentation basics and pitfalls. Excellent coverage by Sands (410) and other resources (4, 192) should be closely studied, and any uncertainties that may adversely affect the validity and accuracy of key instrumentation used in the test should be addressed. Proper flow meter corrections

Figure 3.2   Ultrasonic flow meter installed too close to pipe fittings. (From Sloley, A. W., "Reuse and Relocation: A Case history Part III: Troubleshooting Approach and Technique", in Kister Distillation Symposium Proceedings, AIChE Spring Meeting, Virtual, April 18–22, 2021. Reprinted with permission.)

should be applied. Ref. 361 has superb guidelines for and listing of flow corrections for a wide variety of flow meters.

- Prepare a list of all handheld monitoring devices (e.g., timers, pyrometers, ultrasonic flow meters, electronic pressure gauges) required. Check their availability and have them ordered as necessary. Review any special requirements for using them. For instance, an ultrasonic flow meter should not be placed too close to an elbow as flow patterns created by the elbow may reduce reading accuracy. Unlike the installation in Figure 3.2, it was recommended (444) that it should be installed >15 pipe diameters from a single elbow, >40 pipe diameters downstream of a tee, >5 pipe diameters upstream of an elbow, and >10 diameters upstream of a tee. Confirm that the transducer is properly fixed and gelled to the pipe (435). Seek ways of verifying the measurement. A thorough testing program was shown to lead to high confidence in the ultrasonic meter readings (159). For split lines or manifolds, consider whether it is practical and desirable to measure the flows in each arm using an ultrasonic flow meter, as was demonstrated in one case (471).

- Check whether any scaffolding is required to obtain the desired readings and have them ordered.

- Ensure all meters are operational and read on-scale. Look for oddities in field instrument installation. A flow meter will read inaccurately when an orifice plate is incorrectly sized or damaged or when a tap is plugged. Ensure that all orifice plates are of the proper size; if practical, inspect them to ensure good condition. Failure to check orifice plates has ruined the results of many test runs (4, 192, 465). Audit meters on steam condensate lines; they may become erratic (e.g., 330) when flashing occurs in the condensate lines. All orifice taps should be blown. Inspect pressure indicator installations and check their taps for blockage. A pressure gauge located near an elbow or a tee will read the impact pressure rather than the fluid pressure. In Cases 25.1 and 25.3 in Ref. 201, poor calibration or installation of pressure transmitters misled troubleshooting. A vapor feed or reboiler return blowing into a pressure tap will add the velocity head to the pressure measurement; in one case (Case 4, Section 8.16), this made a difference of 5 psi. Inspect thermocouples and thermowells; they may read incorrectly if too short (as a thermowell described in Ref. 442 did) or in the presence of deposits. Check that all recorded temperatures for your test are actual, not pressure-compensated, temperatures. Inspect impulse lines, heat-tracing equipment, and vibrations near transmitters. Correct any oddities prior to the test.

- Ensure all flow and differential pressure indicators are zeroed and calibrated. All dial thermometers and outside pressure gauges that can be safely removed with no leakage should be removed, and their accuracy checked. The unsatisfactory ones should be either replaced or removed before the test, and an alternative way of taking the measurement should be considered. Many unit thermowells are installed without a temperature-sensing element. These wells can be probed with a handheld device, or a temporary thermocouple can be installed in them to provide valuable data.

- Whenever possible, verify flow rates by tank gauging. A reliable determination requires liquid density as well as correct height readings at the beginning, end, and additional time intervals. Identify additional sampling and temperature measurements needed to obtain liquid density, and check that you have accurate dimensions of the tanks.

- Identify the location of sample points to be used in the test. Add cooling coils where required. Samples should be taken where a stream is either all liquid or all vapor and from a location where the stream is well mixed. Liquid samples normally give better accuracy except when flashing of lights is an issue. For sampling vapor lines, consider the possibility of atmospheric condensation, and for sampling liquid lines, consider that they may contain vapor bubbles.

- Check whether samples can be safely drawn from the column; these can become very useful. For samples drawn from the column, phase separation may be required, because some vapor is likely to be present even in the bottom of the downcomer, while some liquid is likely to be present in the vapor space between the trays. Avoid drawing samples at the feed tray (465); such samples are likely to be dominated by the feed and are seldom representative of the tower composition. In packed towers, drawing samples from a collector right above the feed zone is usually good and informative.

- Ensure all sampling lines are operable, free of blockage and comply with the relevant safety, health, and environmental requirements. Valuable sets of guidelines for sampling are available elsewhere (4, 319).

- Prepare a list of all samples required for the test and their frequency. Together with the lab supervisors, determine the sample containers suitable for each analysis (e.g., bombs, glass bottles, plastic bottles) and the best way of sealing the containers (plugs, corks, screwed tops). Samples to be transported to outside labs may require special containers. Review the sampling procedure with the lab supervisor. The procedure should minimize flashing of lights from liquid samples. If vapor is sampled under pressure, the purge valve should be at the outlet to avoid Joule–Thompson condensation in the sample bottle. Determine which samples need special handling (e.g., refrigeration, stabilization, settling out of a second liquid phase). Attempt to use as few types of sample containers as possible to minimize errors. It has been recommended (447) that each sample taken should be large enough to run the analysis at least three times.

- Check that the on-line analyzers are adequately calibrated and read the same as the lab analyzers. Review the records of these analyzers over the past three months. Straight lines and nonsensical numbers indicate a bad analyzer. Investigate the reason and ensure the problem was fixed and is unlikely to reoccur during the test. Check that the trends indicated by the analyzer make sense and that it properly responds to process changes.

- Verify that the basis for the compositions reported by the lab analysis is valid and applicable throughout the expected composition range. Check that none of the reported compositions relies on extrapolation of data. This is addressed in detail by Kunz (270) for ethanol–water and methanol–water systems. For instance, inferring composition from liquid density measurements for ethanol–water mixtures above 60°C is based on extrapolation rather than data. Also, maxima occur in the refractive index–composition plots for these systems, so a measured refractive index near the maxima can indicate two different compositions.

- Check that the range of component concentrations of the important components in all samples falls within the ranges of the analyzers available in the lab. A measurement of "< 0.1%" for an important component can make the entire test futile. Should any important components fall outside the measurable range, consider sending the samples to an outside lab or altering the test conditions to increase this concentration.

- Heavy component groups are often reported as "C_x plus," where x is some number between 5 and 12. Evaluate whether it is important to have an analysis of the split of these components. In one aromatic tower, it was found that assumptions regarding the split of the heavy non-key C_9+ components varied simulation results significantly, and analysis was necessary (278).

- Refinery distillation product analyses are usually carried out either by chromatographic simulated distillation (ASTM D2887, D7169, and D3710) or by laboratory distillation (ASTM D86 and D1160). Occasionally, the laborious true boiling point (TBP) laboratory distillation (ASTM 2892) is used. Consider the pros and cons of each method for the test, and choose the preferred method for each tower. Ref. 154 provides some useful guidelines. Most commercial simulators use a TBP curve to define a series of individual pseudo-components. This is done by breaking down the TBP curve into small ranges so that the volumetric average temperature for each range corresponds to a defined TBP cut point. Simulations also calculate pseudo-component gravity from an API gravity curve, which can be obtained from lab analysis. If unavailable, the simulators use a less accurate approximation. Therefore, it is important to include the API gravity analyses (154).

- In refinery gas plant towers as well as some petrochemical towers (debutanizers, depentanizers, naphtha splitters, aromatic towers), whenever practical, obtain full component analyses of the feeds and products (235, 278). As described in step 17 of Section 3.1.6, these can provide better characterization than laboratory distillation or simulated distillation tests.

- Check with lab supervisors, safety supervisors, production supervisors, health and environmental officers what measures and equipment (e.g., protective clothing, access, training) are needed for sampling and for other tests duties. Order any unavailable items.

- Contact the research department for input. Decide whether the designer should be contacted.

2. *Set a tentative date.*

   Check with the various departments how long it will take to carry out the preparations for the test. Add a "Murphy's law" factor and set the earliest date at which the test can be carried out. A good "Murphy's law" factor is half the longest preparation period, or 1 week, whichever is longer. For instance, if the instrument department advises that it will take four weeks to calibrate all instruments, allow another two weeks for delays. Better Murphy's law factors can be determined by experience in each situation.

Once the preparations are completed, set the time most convenient for the plant operation personnel. Stable operation during the test is essential, and the operation personnel can anticipate periods of potential instability best. An efficiency test usually means increasing the lab workload, and the lab supervisor will need to schedule additional personnel. Obtain a commitment from the lab supervisor to complete the analysis by a reasonable date. The longer a sample sits on the shelf, the lower the reliability of its analysis because of possible leakage or chemical reaction and the higher its chance of being lost.

Inform all parties involved about the proposed dates and duration of the test. Check for scheduled maintenance work or unusual feed conditions that can interfere with the test. The author has seen cases when minor maintenance such as filter cleaning interrupted a test. Do not forget to inform upstream and downstream units and check for potential bottlenecks.

Troubleshoot for any linked operations that may adversely affect the test. In one case (444), the reboiler heating liquid also heated rail cars. High heating loads on the rail cars would rob the reboiler of some of its heating duty.

3. *Become familiar with the data acquisition system.*

Check the reliability of the data acquisition system and to what extent you can use it without interfering with normal operation. Check the time intervals at which key readings are updated and ensure they fit the test needs. Compare acquired data to control board readings. Investigate any discrepancies. When more than one system is used to acquire data (e.g., a distributed control system and a separate on-line process analytical system), the system clocks should be synchronized and any offsets noted so that data acquired from different systems can be matched to each other.

It is frequently possible to have a control computer or data acquisition system programmed to prepare a special report of data, to store key data points more frequently, to trend key variables, to add to its records normally unmonitored data, and to calculate deviations and standard deviations for selected time periods. These options should be considered and utilized as needed.

4. *Prepare data collection sheets for the test.*

For readings of computer data logs, it is a good idea to offload 1-minute data from the computer and to later transform them into an Excel spreadsheet, which calculates the maximum, minimum, average, and standard deviation for each measurement over the test period or selected increments of the test period. For other readings, it is best to use the modified versions of the log sheets used by the shift personnel. This is particularly important if the shift team is requested to take readings over the night shift. A new data sheet may confuse them and reduce their cooperation.

Readings such as levels, control valve positions, and times at which readings are taken may not appear important but later turn out to be valuable in analyzing results. For instance, a reboiler control valve setting can provide useful information for determining whether the steam side is flooded with condensate. These should be included in the data collection sheets.

Include ambient measurements (temperature, barometric pressure, precipitation). They may be valuable for test result interpretation. Case 25.1 in Ref. 201 describes how the correct barometric pressure was central to vacuum tower evaluation.

Aim to include as many sets of data as possible on each data sheet to reduce the volume of paper. An attractive flow sheet enhances cooperation from those taking readings.

5. *Carry out a preliminary test run.*

This validates the items on the data sheets and identifies the measurements that were left out. It is too late to discover those during the test. Eliminate any unnecessary readings, but be careful in doing this. For instance, a pressure gauge reading that appears unnecessary may turn out to be the deciding factor between two conflicting readings. Do temperatures, pressures, and flows move in sensible directions? If not, investigate and resolve.

6. *Based on the preliminary test run, perform a mass balance on the unit and each column.*

Ensure that these close within ±5% (4). Leaking valves and exchanger tubes affect the mass balance, and these leaks need to be identified prior to the final test. Meters have read incorrectly even when the instrument technicians were sure they were right. Ensure that density corrections are adequately applied for all flow meters; correction equations are described elsewhere (4). Reference 361 has superb guidelines for and listing of flow corrections for a wide variety of flow meters. Verify the accuracy of flow meters; use the orifice diameter and pressure drop across it to calculate the flow using the orifice equation as described by Summers (465). Ensure the valves on the lines connecting the orifice taps to the flow meters are correctly set. A material and component balance closure within ±3–5% and an energy balance closure within ±5% are advocated for the final test (4, 187, 192).

Some experts (465) accept closure within ±10%, while preferring ±5%. Others (153, 447) target a stricter criterion of ±2% for the mass balance, but recognize that this may not be achievable. When the closure of any of the balances exceeds ±5%, the author strongly recommends performing a sensitivity analysis to define the impact of the nonclosure on the efficiency and only accepting the data when the nonclosure makes little difference to the result or when there is an independent verification of the key meter readings. Sensitivity analysis is discussed in steps 10–14 in Section 3.1.6.

In one case, the energy balance on a giant tower closed within ±8%. The sensitivity analysis found this error to be too high to provide a suitable basis for the intended revamp. The plant process engineer solved the problem by diverting the reboiler condensate into a large rectangular trash bin ("Dempsey dumpster") and timing its filling to give the condensate flow rate. After a couple of trials, the engineer got repeatable results, which accurately matched the steam orifice meter. This test successfully established the reboiler duty as the correct basis for the simulation and revamp.

Pay attention to flow measurements of streams near their bubble point or dew point. Flashing of liquid or condensation of vapor can lead to major errors. Any "noisy" or heavily dampened flow meter signal may indicate unexpected flashing or condensation.

Whenever practical, cross-check the mass balance against direct weight or volume measurements (e.g., from product tank levels). Make sure you can determine the density of liquid in the tank; you would most likely need to have temperature and composition measurements of the tank liquid. Table 3.2 is a checklist for troubleshooting the nonclosure of material, component, and energy balances.

Follow the guidelines in the first two points under step 1 above. Check relief valve bypasses and drains for leakage. Use a pyrometer to find whether these are warm. Leakage will affect the material balance. If it cannot be prevented, check whether it can be estimated from flare or vent header analysis for certain components. Adjust the nitrogen flow to the flare or vent header accordingly.

Table 3.2 Checklist for troubleshooting nonclosure of material, component, and energy balances.

1. Are any valves leaking (check vents, drains, bypasses, crossovers, relief valves)?

2. Are there any signs of exchanger leaks (cooling water or steam chemicals in the process, process components in the cooling water or steam condensate)?

3. Are any exchanger or control valve bypasses open or leaking? Would any of these leaks negatively impact the mass or energy balance?

4. Are all lines connected to the tower accounted for (look for purges, pressure-equalizing lines, liquid to/from pump seals, relief valve bypasses)?

5. Are the meters included sufficient to entirely define unit performance? Are any handheld meters needed?

6. Are the orifice plates of the correct size? Are their flow factors correct?

7. Are the orifice plates in good condition?

8. Are any orifice taps plugged?

9. Are the valves on the lines connecting the orifice taps to the flow meters correctly set?

10. Are there any flow meters measuring two-phase flow?

11. Can any condensation occur in measured vapor streams?

12. Can any vaporization occur in measured liquid streams?

13. Is there sufficient pipe run free of obstruction upstream and downstream of all flow meters?

14. Are there any spiral wound (seamed) pipes used? Are they discontinued in the vicinity of all meters?

15. Are the appropriate density corrections applied to adjust the flow meter reading?

16. Are all meters zeroed and calibrated?

17. For mass measurements from tank level gauging, are tank liquid temperature and composition metered to determine liquid density?

18. Do flow measurements match volume or mass measurements?

19. Are metered flow rates consistent with those derived from control valve settings and pump curves?

20. Is an ultrasonic flow meter needed? Is there a suitable and accessible measurement point?

21. Is the reboiler return or vapor feed blowing into any pressure tap?

22. Are all the needed samples included?

23. Can all samples be withdrawn in compliance with HSE requirements?

24. Are all the sample lines operable? Any blockages?[a]

25. Are the sampling procedures adequate?[a]

26. Can there be any liquid in vapor samples or vapor in liquid samples?[a]

27. Are the correct sample containers used? Are they adequately purged?[a]

28. Do the sample containers leak?[a]

29. Is the preferred analysis method(s) used?[a]

30. Are the lab analyses correct?[a]

31. Do any important components fall outside the lab analyzer measurable range?[a]

32. Are the on-line analyzers calibrated and read the same as the lab?[a]

33. Did any on-line analyzer produce any nonsensical values in the last three months? Why?[a]

34. Are the analyzers functioning properly? Are the sample handling loops adequately purged?[a]

35. Does a log–log plot of distillate/bottom components' concentration versus relative volatility produce a straight line or a smooth curve?[a]

36. Does a reaction or accumulation occur in the column?[a]

37. Can any samples be taken from the column?[a]

(Continued)

Table 3.2   (Continued)

38. For a partial condenser, does a flash calculation match the measurement?[a]

39. Are the subcooled condensate and superheated steam temperatures measured?[b]

40. Is an exchanger duty calculated using a small temperature difference ($< 10°F$) between inlet and outlet streams?[b]

41. Can the cooling water flow rate be measured with an ultrasonic meter?[b]

42. Are the air temperatures into and out of the bays of an air condenser and fan pressure drop measured?[b]

43. Are temperature gauges accurate?[b]

44. Do thermocouples adequately fit in the thermowells? Are there deposits in the thermowells?[b]

45. Is feed enthalpy based on the temperature measurement of a two-phase flow?[b]

46. Does the steam trap leak?[b]

47. Is there a possibility of water in the saturated steam inlet to the reboiler?[b]

48. Do heat losses from the column need to be allowed for?[b]

49. Are the ambient conditions recorded?

50. Are all the HSE requirements (e.g., protective clothing, access, training) needed for sampling and for other test duties taken care of?

[a] Check for separation quality and nonclosure of component balances only.
[b] Check for nonclosure of energy balances only.
  All unmarked items affect mass, component, and energy balances.

A flow meter on a liquid stream leaving the column or drum can sometimes be checked by heavily throttling the control valve on the stream for a measured short time period and timing the level rise in the column or drum. The mass accumulated in the tower or drum can be calculated from the level rise and can be used to check the change on the flow meter.

In case of a gross mass balance nonclosure, compare the measured flow rates to those calculated from control valve opening, pump curves, and/or ultrasonic flow meters. If practical, perform this comparison over a range of flow rates rather than a single flow rate. These checks are not accurate, but can often identify the meter that is grossly out of line.

7. *Based on the preliminary test run, perform component balances on the column.*

Component balances for the main components, as well as other components for which the separation in the column is important, should close within ±3–5% as prescribed in step 6 above.

For minor components, the lower the concentration of the component, the higher the percentage error in the component balance is likely to be (4), which will easily exceed the 3–5% closure. At low concentrations, these components are difficult to analyze, and they may azeotrope with others. Usually, these small components do not affect efficiency determination, and one can get away with even large nonclosure of their component balances. The author encountered situations where even a factor of 2 or 3 in the component balance of a minor component made little difference to the column and its efficiency determination.

In cases where the split of the minor components is important, the AIChE Column Testing Guide (4) suggests estimating their concentration from downstream units, where they are concentrated and can be analyzed more reliably.

Other cases where the component balance closure may be relaxed are pseudo-components of petroleum fractions, or dilute components, where a 10% closure is considered realistic (4).

A useful check for component balances (4, 192) is preparing a log–log plot of the ratio of distillate to bottoms concentrations for each component against its relative volatility in relation to the heavy key component. Relative volatility is calculated under average column conditions. Except for columns that are primarily stripping or rectifying, this plot should give a straight line or a smooth curve approximating a smooth line. This plot can help resolve discrepancies in component balances.

8. *Carry out an energy balance.*

An incorrect reflux meter or a leaking steam trap will not show up in the material balance. Ensure closure is within ±5% (4, 166, 192), but see step 6 above for a possible exception.

Ensure that the temperatures of subcooled condensate leaving the reboiler and of superheated vapor (or steam) entering the reboiler are monitored. It has been stated (278) that this is critical for reliably establishing column internal flows. Often, there is no temperature indicator on either. Either add indicators or measure using a pyrometer.

As a general rule, plant steam is inherently wet (399, 400). The common belief that letting down steam from a higher to a lower pressure through valves can create saturated (or even superheated) steam is often fallacious. Risko (399) shows that letting down 5% wet steam from 200 psig to 50 psig will only reduce the wetness to 2–3%. Audit the steam line to the meter for sources of condensate such as poor condensate trapping (check all steam traps), damaged insulation (check with a temperature gun), and other heat sinks such as uninsulated flanges. Be on the watch when saturated steam or vapor is used to heat a reboiler or preheater. Any water present in the steam will generate a significant error in the heat duty. Check whether the steam temperature can be raised for the test to prevent condensation or whether the condensate can be trapped ahead of the flow meter.

For reboilers heated by heat transfer fluids, check the system for contamination, leaks, and degradation products. These may affect liquid physical properties (115), introducing errors in heat duty calculations. Seek guidance from the fluid supplier as needed.

Pay special attention to exchanger duties calculated from small temperature differences (e.g., from the cooling water flow and the difference between inlet and outlet temperatures, where the temperatures are less than 10°F apart). Often, the flow can be throttled to increase the temperature difference; if this is impractical, high-accuracy temperature indicators and/or pyrometer checks may be required. Often, cooling water flow measurement is unavailable, but can be achieved using an ultrasonic flow meter.

For air condensers, measure the air temperatures into and out of each bay, as well as the pressure difference across the fan. These measurements may prove crucial for evaluating and checking the condenser duty and tower heat balance.

For flashing feeds, determining feed enthalpy from feed temperature is often inaccurate, particularly if their boiling range is narrow. Further, the pressure at the feed temperature measurement point is often unknown due to the pressure drop between that point and the column feed zone. A better estimate can often be obtained from the temperature at the last point where the feed exists as a single phase (prior to flashing). This procedure affords an additional validation check when the feed prior to flashing is at its bubble point (e.g., when it comes from the bottom of a previous tower). Using its temperature and composition (per component balance in step 7 above), the bubble point pressure can be calculated and compared with the measured bottom pressure in the previous column.

If the feed is preheated, calculate the preheater duty from the heating side (steam side duty or, if liquid heating, the temperature difference between the inlet and outlet temperatures multiplied by the specific heat and multiplied by the flow rate). Similar considerations apply to feed precoolers.

If the temperature of a vapor–liquid mixture needs to be measured, an effort should be made to measure pressure as close to that point as possible. Alternatively, the temperature measurement should be as close to the column as possible where the pressure is known.

Use a thermal camera to identify any regions of major heat losses. In some high-temperature vacuum columns, columns with small (<2 ft) diameters where the surface-to-volume ratio is high, or uninsulated columns, or where insulation is damaged, radiation and convection heat losses may be significant. In these cases, heat losses need to be estimated or measured and accounted for. In most cases, these losses are negligible (4).

Check temperature increases along lines that are heat-traced. This can be achieved by cutting holes in the insulation at the beginning and end of the line and measuring the temperature difference using a pyrometer. There have been instances (240) where heat tracing caused flashing in a reflux line.

Check for steam leaks through reboiler and preheater steam traps. These can be detected by comparing the temperatures entering and leaving the trap with a pyrometer.

Heat-integrated towers should be evaluated simultaneously for proper checking of the energy balances (465).

9. *Prepare the simulation program and spreadsheet for carrying out the mass, component, and energy balances.*

Once they are ready, data can be quickly processed and inconsistencies easily detected before or during the test, before it is too late.

10. *Check whether the tower operates near a pinch or an azeotrope.* See steps 10–14 in Section 3.1.6. The efficiency of towers operating near a pinch, or right near an azeotrope, cannot be reliably determined. Efficiencies determined under pinched conditions can be out by a factor of 2 of 3. If close to a pinch, consider increasing the reflux ratio. If too close to an azeotrope, consider reducing the reflux ratio to move away from the azeotrope. For example, if an ethanol–water column produces a 96.1% product, right near the azeotrope, consider reducing the reflux ratio to make a 95% ethanol product. Apply the simulation to guide your move.

If operating near a pinch, a "pinch test" is worth considering. Immediately after completing the main test, the column can be retested at another two or three reflux ratios. The reflux ratios should be higher if no azeotrope and lower if one product is near the azeotrope. If the column is near a capacity limit, before raising the reflux for this test, it may be necessary to reduce the feed rate.

11. *Monitor and enhance the reliability of laboratory analyses.*

The AIChE Guide (4) recommends that at least three successive samples of each stream should show essentially constant composition. Below are some actions that help make this happen:

A useful trick to monitor lab reliability is to take several samples of the same stream. If possible, take them yourself. Give one sample to the lab tester and request an analysis. Plug the other sample containers, ensuring no leakage. Give the same lab tester the second container several hours later, requesting another analysis (do not forget to remove the plugs first – otherwise the tester may become suspicious). Give the other containers to the lab testers on afternoon and night shifts, leaving one for the tester the next day. Compare the analyses. You may surprise yourself with the difference in results! If necessary, discuss with the lab supervisor, but avoid getting personal. Don't forget, your objective is reliable data, and bad personal feelings will not help.

Another useful check is to obtain a number of test containers early, fill them with samples, then close and plug them to prevent leakage, and record their weight. Reweigh them a number of days later; a weight loss will indicate leakage despite all the precautions. Alert the laboratory supervisor as needed.

Check the sampling technique used by the lab tester for each sample. Volatile samples need to be taken in bombs, and these must be adequately purged.

Compare the lab analyses against design data, previous performance test results, or by component balances. Look for anything that does not make sense.

12. *Check product analyses using bubble point or dew point calculations.*

The tower bottoms stream is at its bubble point, so a bubble point calculation using the bottom composition and the tower bottom temperature should yield the measured tower bottom pressure. Similarly, the tower overhead product should be saturated liquid or vapor at the condensing temperature. Physical property measurement (e.g., liquid viscosity) may also be available and should be checked against the calculated values.

13. *Consider wall temperature measurements.*

Holes can be cut in the insulation in strategic locations around the tower, and wall temperatures measured by a pyrometer or a thermal camera. Wall temperatures add invaluable data for validating simulations, testing theories, detecting packed tower maldistribution, spotting accumulation ("bulges") of intermediate components, and identifying the presence or absence of a second liquid phase. In tray towers, wall temperatures can provide a detailed tray-by-tray measurement of the temperature profile. For rigorous efficiency tests, wall temperature measurements can become the deciding factor (234) between two conflicting models both matching most of the test data. Chapter 7 has many examples in which these wall temperatures were invaluable.

Some engineers use contact thermocouples when measuring wall temperature. The author's experience with contact thermocouples has not been good. They often take a long time to reach equilibrium and are sensitive to surface irregularities.

When inserting handheld thermocouples into thermowells, make sure to allow sufficient time to reach equilibrium at the final temperature. This may be lengthy, especially when measuring hot temperatures.

14. *When there is a partial condenser, perform a flash calculation on the reflux drum.*

If the calculation and measurements do not match, investigate the reason. The condenser may provide more or less than a single stage. In Case 1.13 in Ref. 201, the capacity of an entire refinery was limited by the gas rate exiting what should have been a single-stage total condenser, but became partial due to Rayleigh condensation providing more than one stage. Should the flash calculation mismatch the measurements, include additional composition, pressure and temperature measurements in the region to fully characterize the split. The author recalls a very painful exercise one time when he skipped this check.

15. *Check feed and draw points where there are multiple valved locations.*

The author had been in many situations where the operating team thought the feed entered one point but temperature measurement confirmed it entered another, or the closed valve was leaking and it entered both. A pyrometer survey is invaluable. Then change to the desired feed point.

16. *Check control valve and exchanger bypasses for leakage.*

Such leakage will affect the mass and energy balances. Here too a pyrometer survey is valuable.

17. *In deep vacuum towers, attempt to estimate the air leakage rate.*

    Air leakage reduces the partial pressure of the components, therefore affecting the mixture's boiling temperature. Attempt to also estimate the point of air entry, as the boiling temperature will only be affected by the leakage from that point up.

18. *Explain the importance of the efficiency test to the shift and lab personnel by informal discussions.*

    Their cooperation is crucial for obtaining reliable results. Pay special attention to shift supervisors, chief operators, panel operators, and lab testers that will be working with you; they are the key to success. Explain to them the importance of the test, seek their advice, and incorporate their input into your plans.

19. *Tag and label instruments and sample points with weatherproof tags.*

    Some of the participants are only partially familiar with the test zone. Assign areas of readings for each participant. Request each participant to perform a "dry run" and to point out any instruments they cannot locate.

20. *Prepare a "flag sheet" of the unit.*

    This is a simplified process flow diagram showing items of equipment, temperature, pressure, and key component compositions on it (Figure 3.3). Use it to take key readings during the test and to troubleshoot for gross deviations. On the sheet, show expected values and next to them leave empty boxes to enter measured values. Perform a "dry run" using this sheet before the test.

21. *Unmetered reboiler steam flow can often be measured by running the condensate to a drum of known volume and timing it.*

    Check whether this technique can be applied. Remember that steam condensate flashes, giving rise to incorrect readings. It may be necessary to cool the condensate or run it into a drum partially filled with cold water. Check for special safety, health, and environmental precautions associated with such measurements.

22. *Compare exchanger duties with exchanger design.*

    An exchanger duty grossly exceeding its heat transfer capability based on reasonable coefficients should raise a flag and is indicative of an issue that needs resolving (277).

23. *Ensure sufficient operating margin is available in filters, coalescers, and storage tank levels.*

    If they reach a limit, they may interrupt the test.

24. *If the material and component balances are to be carried out using a computer.*

    Preparing the program and simulation at this stage may prove very beneficial.

25. *If involving outsiders such as the designers, consultants, or service providers.*

    Make sure that they are planning to attend, are aware of the safety training requirements, and are willing and able to comply.

3.1.4 Last-Minute Preparations

Last-minute preparations are critical for a successful efficiency test. Problems that arise at the last minute most likely persist during the test and may lead to meaningless test results. The guidelines below are based on the author's experience, supplemented by Refs. 4, 47, 121, 192, 286, 319.

1. Verify that all test supplies ordered have been received and are adequately located. Check that the required types and numbers of sample containers and plugs needed are ready at the lab. Tag them with weatherproof tags. These tags should be different from those

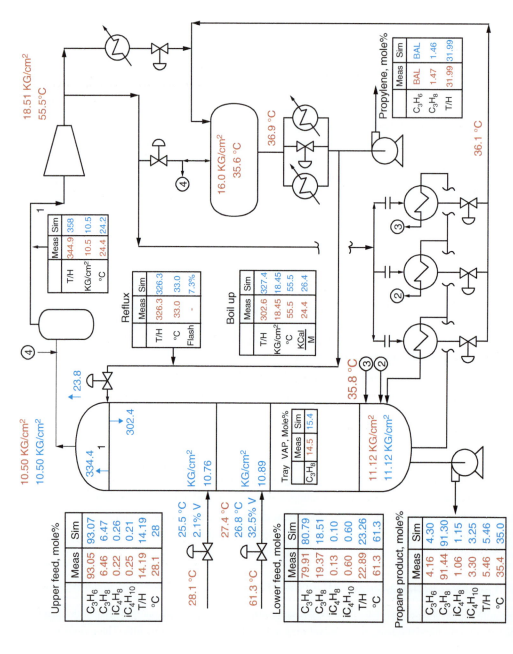

Figure 3.3 Flag sheet comparing measured test data (red) with simulated values (blue) for a C₃ splitter tower. This flag sheet was prepared following the test, after a test simulation became available. A similar flag sheet should be used to keep track of the test, but in the absence of a test data simulation, the blue values show expected or design values.

used for normal analysis. Verify that all the required personnel protective equipment and safety clothing has been received by the participants.

2. Verify that all sample points are unplugged and operable.

3. Check that the shift supervisors and laboratory personnel are aware of the test and any special safety, health, and environmental requirements. They will tend to be more cooperative if they appreciate the importance of the test.

4. Ensure the availability of a sufficient number of blank data sheets.

5. Difficulties encountered during data collection may lower the quality of the data and the cooperation of the participants. Make use of tricks that make data recording easier. These include (121):

   • Supply waterproof dark ballpoint pens for recording data. Inked readings run, while penciled readings are hard to record when the paper is wet.

   • Provide each participant with a clipboard for holding the data sheets and taking readings and plastic/nylon covers to protect them from rain.

   • Supply light, sturdy ropes for draping the clipboard or handheld devices over the shoulders while climbing ladders.

   • Provide pocket notebooks so there is no need to carry clipboards while climbing up tall towers.

6. Do a final check to confirm that the required instrument, mechanical, and lab personnel are available. Ensure all participants are aware of their duties and of any special safety requirements.

7. Do a final check on relief valve bypasses and drains and steam traps for leakage (see Section 3.1.3, steps 6 and 8 and the first two points in step 1). Leakage may have developed since the preparation phase.

8. Perform a last-minute check using the flag sheet. Look for any inconsistencies.

9. Check that tank levels have sufficient margins and that pressure drops across filters and coalescers are low, so that they do not interrupt the test. Deoil refrigerated heat exchangers as necessary. Inject methanol into columns susceptible to hydrate formation. Backflush strainers.

10. Check that there is a sufficient margin between the plant operating rates and the maximum limit. A popping relief valve is the last thing you want during a test.

11. Make a last-minute check with upstream units, preferably with the shift supervisors as well as the plant superintendents, and ensure that they are aware of the test. Also check that other units will not constrain the availability of heating or cooling media.

12. Postpone the test if an extreme weather condition is approached. Operation personnel are likely to focus on keeping the plant on-line, instability is likely, outside participants will rush to get out of the bad weather, and meters may be adversely affected. In one unit test run (153), cold ambient temperatures (-15 to $-30°F$) prevented much of the desired data from being collected, leading to poor closure of mass and energy balances and significant errors.

3.1.5 The Test Day(s)

When preparations are adequately carried out, the test should proceed smoothly. The main watchouts in this period are as follows:

1. Ensure that all the safety procedures are adhered to and that protective clothing is worn as required. Unsafe practices are not only hazardous, but can also generate friction with operating supervisors and may lead to the termination of the test.

2. At the beginning of the day, verify that the column has been running stably and steadily for several hours.

3. Closely watch the flag sheet (Figure 3.3). Look for any major inconsistencies.

4. Stay in close touch with the plant superintendent, shift supervisor, panel operators, and upstream unit control rooms. Quickly question any decisions that may affect the test. A nightmare scenario is someone forgetting that the test is in progress and making a move that destabilizes the tower and adversely affects the data.

5. Closely watch the sampling technique. Attend personally when key samples are taken. A common issue is the sampler allowing insufficient purge time through the sampling line and sample bomb, especially on stormy days and when sample lines are long. Obtain key samples in duplicate, and keep one in case the other is lost. Check that the sample labeling is correct.

6. Ensure that the lab tester can find all the requested sample points. Tests often call for nonstandard samples from spots with which the lab tester is unfamiliar. If the tester cannot find the spot, the sample may come from a wrong spot or not be taken at all.

7. It is best to have test readings taken around the clock. If duties are delegated to the shift team overnight, check that they are fully aware of what you want. A phone call to the plant does wonders to avoid problems and to affirm the importance of the test to night-shift personnel.

Tidy-up The test is incomplete until all special equipment (e.g., instruments, tags, labels, ladders) is properly removed. Any temporary equipment remaining may turn into a safety hazard or a nuisance. Temporary instruments may become lost or damaged. Labels may confuse the operators. Poor tidy-up is one of the most frequent sources of complaints against test organizers.

It is best to first obtain instructions from operation supervisors as to which items need removing and how soon and which items (e.g., some instruments, labeling) constitute a beneficial addition to the unit. Retaining items (e.g., instruments) should be approved both by the operation team and by the department supplying them. The rest of the items should be promptly removed.

3.1.6 Processing the Results

The first step is compiling material, energy, and component balances for the unit based on the test data. These should close per the guidelines in steps 6–8 in Section 3.1.3. Verify laboratory analyses using dew point and bubble point calculations. Some flows and compositions may need readjustment to satisfy the balance equations. Any inconsistencies must be resolved before proceeding with result processing.

Prepare a master flag sheet with the test data. This master flag sheet will serve as the test data input to the simulation.

Column efficiency determination   To determine column efficiency, follow the steps below:

1. *Verify that the correct VLE data are used in the simulation program.*

   Thermodynamic properties are critical for the validity and success of a simulation model (4, 139, 171, 176, 313, 414, 418, 446, 465, 522). Addressing column design, one expert (315) went as far as stating "The choice of the thermodynamic option for any process simulation is invariably the most critical of all the choices you will make. This is true for any process simulator. A mistake at this stage will be fatal to the accuracy of your results." Another expert (314) stated (and demonstrated) that "correlations from commercial simulators need to be evaluated and validated."

   Errors in relative volatility (α) are the most underrated factor affecting tray efficiency. *Direct* effects of errors in volatility are independent of the reflux ratio, while *indirect* effects are additional effects that depend on the reflux ratio.

   Figure 3.4 shows the *direct* effect of the errors (82, 193, 202, 404). At very low relative volatilities ($\alpha < 1.2$), small errors in volatilities have a huge impact on tray efficiency. For instance, at $\alpha = 1.1$, a 3% error gives a tray efficiency 40–50% higher than its true value. Since the accuracy of relative volatility is seldom better than $\pm 2\%$ to 3%, tray efficiencies of low-volatility systems become meaningless unless accompanied by VLE data. Likewise, comparing efficiencies derived for a low-volatility system by different sources is misleading unless using identical VLE. This applies to both trayed and packed towers.

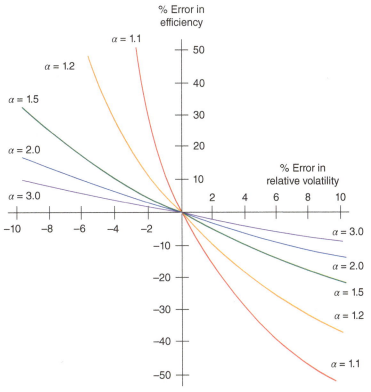

Figure 3.4   Direct effect of errors in relative volatility on errors in tray efficiency. (From Kister, H. Z., "Distillation Design," Copyright © McGraw-Hill, 1992. Used with permission of McGraw-Hill.)

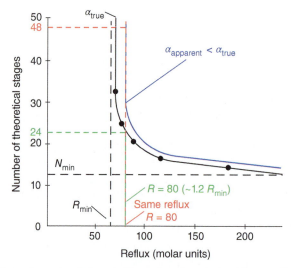

Figure 3.5   Indirect effects of errors in relative volatility on the value calculated for tray efficiency from test data. The apparent calculated efficiency is double its true value due to a relatively small error in relative volatility. (From Kister, H. Z., Chemical Engineering Progress, p. 39, June 2008. Reprinted Courtesy of Chemical Engineering Progress (CEP). Copyright © American Institute of Chemical Engineers (AIChE).)

Figure 3.4 also shows that direct errors in relative volatility are a problem only at low relative volatilities. For $\alpha > 1.5$–2.0, VLE errors have a small direct impact on tray efficiency.

For measurements at the finite reflux ratios used in commercial towers, the *indirect* effects of VLE errors (below) compound the direct effect of Figure 3.4. The indirect effects are enhanced errors in tray efficiency due to the enhancement of the VLE errors by the reflux ratio. Consider a case where $\alpha_{apparent} < \alpha_{true}$ and test data at a finite reflux are analyzed to calculate tray efficiency (Figure 3.5). Consider the minimum reflux R_{min} and actual reflux R (molar units). Due to the volatility difference, $R_{min,apparent} > R_{min,true}$. Since the test was conducted at a fixed reflux flow rate, $(R/R_{min})_{apparent} < (R/R_{min})_{true}$. A calculation based on the apparent R/R_{min} will give more theoretical stages than a calculation based on the true R/R_{min}. This means a higher apparent efficiency than the true value. In Figure 3.5, this effect doubles the number of calculated stages from 24 to 48.

The indirect effects add to those of Figure 3.4, widening the gap between the true and apparent efficiencies. The indirect effects exponentially escalate as minimum reflux is approached. Small inaccuracies in VLE or reflux and mass balance measurement alter R/R_{min}. Near minimum reflux, even small R/R_{min} errors induce huge errors in the number of stages and therefore in tray efficiency. Efficiency data obtained near minimum reflux are therefore meaningless and potentially misleading.

Verification of VLE data should include reviews by thermodynamic specialists, as well as comparisons with in-house or published VLE data. Blind use of commercial simulator VLE has led to many invalid simulations. The checks should focus not only on the quality of the data and data regression, but also on how well the data extrapolate to the tower temperature, pressure, and composition ranges. With high-purity products, the data available often do not extend to very low concentrations, causing significant – even major – errors (76, 171, 311, Case 1.6 in Ref. 201), or to different temperature conditions (76, 313, 315, 494).

With nonideal systems, it is common to have a low-volatility zone ("pinch") near an azeotrope or near a pure component (139, 311), with enhanced sensitivity to VLE errors. Again, when $\alpha_{apparent} < \alpha_{true}$, it can lead to a gross overestimate of the number of stages from test data. Many nonideal systems experience the formation of a second liquid phase, and inaccurate liquid–liquid equilibrium (LLE) values can largely reduce simulation reliability (139). Lack of information about the process chemicals and their interactions, unexpected column chemistry, and possible product degradation lead to uncertainty in the reliability of the VLE, and experimental verification may be necessary (139).

The author could not emphasize the importance of correct VLE more, especially when dealing with low relative volatilities. In one case (289), a new C_3 splitter tower (these are large and expensive towers) was constructed based on a simulation that implied that the existing tower was too small. A slightly different relative volatility would have yielded the conclusion that the existing tower was adequately sized, and its through-put was constrained by the undersized 3-in. steam condensate drain line backing up condensate into the reboiler and restricting boilup.

2. *Verify that the correct thermal properties are used in the simulation program.*

This includes the physical properties (such as densities and latent heats) that convert tower external mass flow measurements into the internal molar flows used in the simulation. When the measured apparent molar reflux ratio is less than the true reflux ratio, a calculation based on the measured reflux ratio will give a higher apparent efficiency than the true value. As with the indirect VLE effects, the effect of errors in mass balance and reflux measurements on tray efficiency exponentially escalates as minimum reflux is approached. Again, efficiency data obtained near minimum reflux are therefore meaningless and potentially misleading.

Based on test data, it was concluded that the HETP of one packed tower was twice what it should have been. It was intended to split the bed and change distributors. A later thorough review of the thermal properties found the latent heats in the simulation program to be low, making the true molar reflux and boilup rates significantly lower than those simulated. Revising these data brought up the number of stages and reversed the conclusion. The packings performed well; the separation problem was due to insufficient reflux and boilup.

In many absorption, stripping, and reactive systems, rate-based simulation methods have given better results than the stage model. In addition to the VLE data, it is important to correctly model the interfacial area. Interfacial area correlations often yield widely different results or fail badly outside the correlation range (314). Reaction kinetics may also come into play. With these systems, it is very important to have lab or pilot plant data to tune the model (314). In common systems like absorption of acid gases with ethanol amines, commercial simulations tuned with plant data are available (501, 502), but they need to be checked and often require some adjustments for specific situations (314). Verification of the model VLE data should include reviews by thermodynamic specialists and close comparisons with data.

3. *Verify the mass, component, and energy balances and analysis validity.*

This is a repeat of steps 6–8, 11 and 12 in Section 3.1.3. Also pay attention to the possibility of intermediate components' bulging, as described in Section 3.1.2. If not adequately allowed for, the simulation may not match the data (176).

4. *The number of components in the simulation should be enough to accurately characterize tower performance, but not unnecessarily excessive to generate convergence and run-time problems.*

For instance, if there are 10 C_7 components and the C_7 is not a key, they can easily be lumped into the 3–4 main components. Be sure to include intermediate components that may bulge in the middle of the column. In petroleum fractions, flat regions on the temperature–percent distilled curve require fewer pseudo-components, while steep regions as well as regions near cut points require more for good characterization.

5. *Make sure the feed enters the column in the simulation the same way that it does in the actual column.*

Feed is usually specified to enter at a desired stage, with the simulation choosing the mixing pattern at the feed stage. When there are no side draws in the vicinity of the feed and no chemical reactions and the feed is a single stream, this is usually satisfactory. In contrast, in the situations listed below, the feed entry needs to be modeled as a series of flashes that mimics the physical way in which the feed enters:

- *When a fast vapor- or liquid-phase reaction occurs at the feed zone.*
  The reaction block in the simulation needs to work with the correct concentrations, and these depend on how the feed liquid and column liquid mix at the feed zone.

- *When two or more feed streams enter the same stage.*
  Figure 3.6 is inspired by Hanson and Hartman's (147) crude tower case study where a 300°F vapor feed from a flash drum and a 695°F flashing feed from a heater both entered the feed (flash) zone. A special distributor on the flashing heater effluent feed separated the liquid from the vapor and directed the liquid to the stripping section below. Figure 3.6a shows a common, but incorrect, simulation, with both feeds mixed at the feed zone. In reality, the cold vapor does not contact the flash liquid due to the feed distributor. Figure 3.6b shows the correct simulation model that matched plant data (147), with the flash drum vapor not mixing with the heater effluent liquid. In Figure 3.6b, the liquid from the wash bed descended via pipes from a collector to the stripping section and therefore bypassed the flash zone; had it rained from the bed, that liquid would have needed to go into the mixer rather than the stripping section.

- *There is a major side draw within 1–2 stages of the feed.*
  A typical example of this is a refinery vacuum tower, where the major side product, the heavy vacuum gas oil (HVGO), is drawn one or two stages above the feed. In this application, modeling by a series of flashes has been highly recommended (19, 131, 135).

6. *Make an initial guess of the column efficiency.*

Unless there is a reason not to, it is common to assume uniform efficiency throughout a tray column (465). For columns containing sections packed with different packing types or sizes, it is common to apply rule 4 in Section 2.16 for the efficiency of each section. This initial guess needs not be accurate. From knowledge of the actual number of trays and the assumed tray efficiency, or packed height and assumed HETP, estimate the number of theoretical stages.

The author and others (48, 169, 465) strongly recommend analyzing test results using theoretical stages and avoiding the efficiency options in the simulations. The efficiency options in the simulation are Murphree or vaporization efficiencies, and these are complex concepts that tend to be variable in the tower. Recently, Braswell et al. (48) showed that these efficiency options are often incorrectly applied in major simulators, violating thermodynamic principles, and leading to incorrect temperature

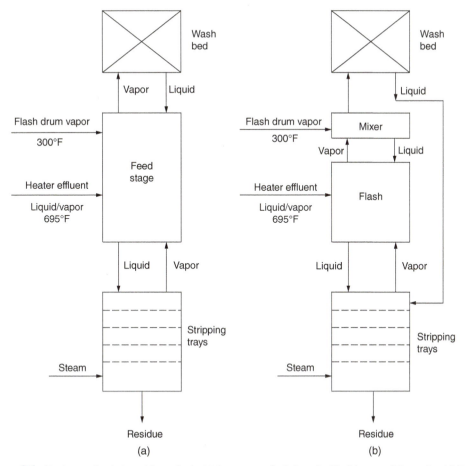

Figure 3.6 Feed entry simulation with two feeds. (a) Incorrect, as flash drum feed in this tower did not mix with liquid in heater effluent; (b) correct, no mixing of flash drum vapor with liquid effluent from heater.

and composition profiles. There is also experience of these options producing erroneous results at low efficiencies (465). In some situations, especially when dealing with absorptions or stripping with chemical reactions, or with chemical reactions inside the tower, the stage model is not suitable and a rate-based model should be preferred (474, 501, 502).

7. *Using the test material balance and the estimated theoretical number of stages, run a computer simulation for the test conditions.*

Useful hints for good architecture and organization of simulation blocks, inputs, and outputs are discussed elsewhere (414).

In simple columns (two products, no side draws), it is common to specify the test-measured top and bottom product purities in the simulation, making the reflux and boilup dependent variables. The number of stages is then increased or decreased until the measured reflux and boilup in the simulation match those measured in the test.

Alternatively, and often for easier convergence, the reflux and one product purity; or boilup and one product purity; or a product rate and purity; or a product rate and a temperature are specified, and the number of stages is altered to match the other product purity.

When developing a simulation, seek simplicity. Simplicity reduces calculation time, makes it easier to make changes later, but most importantly, makes it easier to debug and spot errors that can adversely affect the results.

Compare all simulated dependent variables (reflux rate, exchanger duties, and product purities) with measured values. If any mismatch, investigate the cause.

8. *Compare the temperature profile measured in the test with the simulated temperature profile.*

If significant discrepancies exist, alter the number of stages in the relevant section to improve the match. Similarly, if internal column samples were obtained, check them against stage compositions predicted by the simulation and adjust the number of stages accordingly.

A useful technique by Krishnamoorthy and Yang (264) plots temperature and concentration profiles showing their split among sections of the tower. Figure 3.7 shows such a plot for an ethylbenzene–styrene splitter with six packed beds. Measured temperatures (and if available, also compositions) can be superimposed on the plot to determine to what extent the number of theoretical stages in each bed is achieved. They expressed the results for each bed as a percentage of the observed number of stages to the expected number of stages at good liquid and vapor distribution. A value of 100% indicates that the expected efficiency is achieved, and less than 70–80% indicates problems. They demonstrate success with this method despite the challenge of a small temperature change (<0.2°C per stage) for this close-boiling application. The results may not be accurate, but they successfully identified the trouble areas.

9. If the number of stages in one column section was significantly altered in steps 7 or 8 above, the total number of stages in the column may also need adjustment. Repeat steps 6

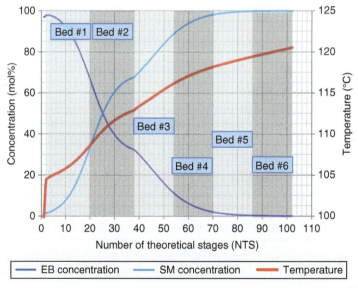

Figure 3.7   Temperature and concentration profile in an ethylbenzene–styrene splitter operating at 400 mbars, showing the expected change in each packed bed assuming all operate as expected. When compared with actual measurements, such a diagram can help identify low-efficiency sections. (From Krishnamoorthy, S., and L. M. Yang, "Troubleshooting EB/SM Splitters: How Can a Maldistribution Analysis Help?," in Kister Distillation Symposium, Topical Conference Proceedings, p. 43, AIChE Spring Meeting, San Antonio, Texas, March 26–30, 2017. Reprinted with permission.)

through 8 until the simulation matches both the component balance and the temperature or internal composition profile.

10. *Apply the simulation to examine the sensitivity of product purity to changes in the number of stages.*

This is perhaps the most critical step in processing test data; overlooking it has been a primary source of grossly misinterpreted test data. It is not uncommon to find that column efficiency was overestimated by a factor of 2, and even more, in columns where product purity is insensitive to the number of stages. Scale-up of such misinterpreted data has proven disastrous on many occasions.

Figure 3.8 shows an application of the purity versus stages sensitivity analysis to a depentanizer column. It was critical to determine the number of stages (i.e., the efficiency) in the stripping section. Figure 3.8 shows that the only number of stages that will match the measured bottom composition is 5 or 6. Any other number of stages (4 or 7) will take the composition measurement error to over 50%, which was highly unlikely.

The dashed curve in Figure 3.8 examines the sensitivity of the number of stripping stages to the reflux by arbitrarily raising the reflux ratio by 50%; 50% was chosen because this test had no energy balance (the reboil meter was nonfunctional) and the only validation checks on the reflux measurement were the pump and control valve curves, which are typically of ±30% accuracy. Since Figure 3.8 showed a negligible effect of reflux on the number of stages, the lack of energy balance could be tolerated. This confirmed the conclusion that there are 5–6 stages in the stripping section.

Figure 3.8 also looks at changing the sensitivity of the efficiency in the stripping section to the number of rectifying stages. The number of rectifying stages was varied widely in the simulation with no effect on the bottom composition. In this tower, there was no way of determining the rectifying section efficiency, as the product compositions were totally insensitive to the number of rectifying stages.

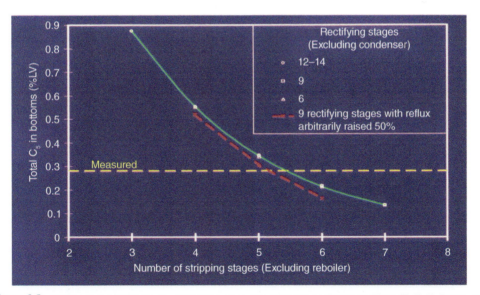

Figure 3.8   Efficiency sensitivity checks, stripping section of a depentanizer. (Based on Kister, H. Z., E. Brown, and K. Sorensen, Hydrocarbon Processing, p. 124, October 1998.)

The reason for the strange behavior in this tower was that the tower feed point was too low, so the number of stages in the bottom was near the minimum stages, where the composition depends on the number of stages but is independent of the reflux ratio.

More commonly, a loss of stages in one section (e.g., bottoms) can be compensated by better efficiency in the other section (e.g., top). In a case reported by Geipel (120), poor stripping section efficiency due to fouling was partially compensated by better efficiency above the feed; in another case reported by Lee (276), it was compensated by raising the feed point. In both cases, this was established by a sensitivity study.

11. When a column operates near minimum reflux, contains an "excess" of trays, or operates under other pinched conditions, product purity is often insensitive to the number of stages. The author is familiar with one tower operating close to minimum reflux where product purity remained practically unchanged when the number of stages was halved. When the column was simulated with less than three stages, product purity was sensitive to the number of stages. In the same column, changing the number of stages between 5 and 15 yielded product purity changes far smaller than those that could be detected by the lab analysis. A similar experience was reported in the top section of the column case study featured in step 10 above; per Figure 3.8, the number of rectifying stages had a negligible effect on product purity. Pinching (either due to a mislocated feed, proximity to minimum reflux, or a tangent pinch) is commonly implicated by the above insensitivity. A McCabe–Thiele diagram and a key ratio plot can help identify the cause; the application of these techniques for this purpose is described elsewhere (193, 195, 276).

12. Similar to step 10 above, check the sensitivity of the number of stages to errors in the reflux rate. Prepare a stages versus reflux plot like Figure 3.5 or 3.9 around the measured reflux rate. Near minimum reflux or a pinched condition, minor changes (equivalent to typical flow meter errors) can have a greater effect on product purity than doubling (or halving) the number of stages in the column. Figure 3.5 shows that when the number of

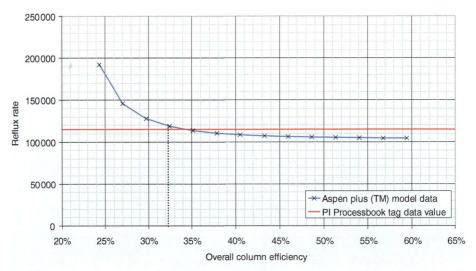

Figure 3.9 As minimum reflux is approached, the column efficiency cannot be obtained with confidence. In this tower, efficiencies higher than 40% are impossible to quantify. (From Hennigan, S., "Full Scale Plant Column Efficiency – Measurement and Management," in Distillation 2011: The Dr. Jim Fair Heritage Distillation Symposium, Topical Conference Proceedings, p. 151, AIChE Spring National Meeting, Chicago, IL, March 13–17, 2011. Reprinted Courtesy of S. Hennigan, INEOS Acetyls.)

stages in that tower exceeded about 30, the number of stages was extremely sensitive to the reflux ratio, and any small errors in reflux measurement, energy balances, or latent heats could double or half the number of stages. In contrast, near the minimum number of stages (say about 15 in Figure 3.5), it is possible to get reliable efficiency determination even when there is a poor closure of the energy balances.

The difficulty of determining the tray efficiency as minimum reflux is approached is further illustrated in Figure 3.9. The minimum reflux can be inferred from the asymptote to which the model (connected dots) curve approached, which is about 103,000 units. The measured reflux is about 114,000 units. Assuming a 3% maximum error in the energy balance, the tray efficiency is between 32% and 37%, with 35% most likely. However, had the measured reflux been 106,000 units, the tray efficiency would have been anywhere between 40% and infinity. In this tower, efficiencies above 40% are impossible to quantify.

Going the other direction, had the measured reflux been about 130,000 units, the efficiency would have been between 28.5% and 30.5% (assuming a 3% maximum error in the energy balance) with 29.5% being most likely, so the efficiency error would have been ±3%, very minor. A plot like Figure 3.5 or 3.9 also gives an evaluation of how errors in the energy balance affect the efficiency. In Figures 3.5 and 3.9, the reflux ratio or reboiler duty can be plotted instead of the reflux rate.

13. Similar to the checks in steps 10 and 12 above, one more useful plot is the product purity against reflux ratio. Figure 3.10 plots product composition against reflux ratios for two columns. Plant data are the red points, and simulation results are the continuous blue curves. Since the efficiency test is held at a constant product composition, it will produce only a single red point. The blue simulation curve will shed light on the sensitivity. For example, the simulation curve for the deethanizer in Figure 3.10a shows a variation in the reflux ratio from 0.42 to 0.48 changing the C_3 in the overhead from 0.2% to 0.1%. This is a good sensitivity to reflux. The debutanizer simulation curve in Figure 3.10b shows a much higher sensitivity, with the C_5 ppm in the top doubling from 100 to 200 ppm with a reflux ratio change from 0.9 to 0.85. In this case, small errors in reflux and reboil can lead to large errors in efficiency determination.

Purity versus reflux ratio plots are most informative when plant operating data (not just the efficiency test data, which, as stated, only produce a single point) are added as shown in Figure 3.10. Neglecting the outlying points, Figure 3.10a verifies an excellent simulation that predicts the plant data very well. In contrast, Figure 3.10b shows a much poorer correlation, and since most data points fall to the right of the simulation curve, that an optimistic tray efficiency is used in the simulation. It also shows a much larger scatter of data, suggesting that other variables may play a role in the performance of the tower.

14. Check the sensitivity of the simulation to a reduction in the number of stages in each section. Simulating a gas plant demethanizer with a side reboiler, Shah and Stucky (426) found that examining the sensitivity of the side and bottom reboiler duties and product rates to the number of stages in each section was one of the most critical steps in finalizing the tray efficiencies. In multicomponent distillation, examine the effect of reducing the number of stages on the key component ratios (ratios of light key to heavy key concentration) in the top and bottom products. The author experienced one case where using an efficiency ranging from 40% to 60% in the stripping section matched test data quite well; in that case, the key component ratios helped choose the most likely efficiency value.

When the design, hardware, internal loads, and operating conditions of the top and bottom sections are not widely different and there are no issues with packed tower liquid

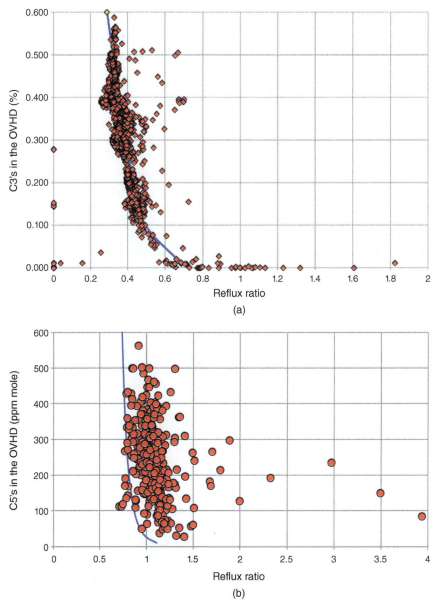

Figure 3.10 Invaluable checks of the validity and sensitivity of efficiency determination by plotting product composition against reflux ratios over a wide range of product compositions. Plant data are red points, and simulation is the continuous blue curve. (a) For a deethanizer column, showing a small data scatter, an efficiency determination that closely matches the data, and good sensitivity of composition to reflux ratio. (b) For a debutanizer column, showing a much larger scatter. The data fall to the right of the simulation curve, showing that the efficiency used is optimistic. The product compositions are quite insensitive to the reflux ratios. (From Bernard, A., "Practical Approach to Distillation Tower Revamp," in Kister Distillation Symposium, AIChE Spring Meeting, New Orleans, Louisiana, March 31–April 4, 2019. Reprinted Courtesy of A. Brernard.)

or vapor distribution, column efficiency should be reasonably uniform throughout the column. If the simulation indicates wide variation from top to bottom, be suspicious. It may suggest an error in the simulation or an actual performance problem.

15. When a tower has an alternate feed point, consider repeating the test while feeding the alternate feed point. Check that the simulation correctly predicts both. In one case (436), a contractor's insistence on switching to the alternate feed point provided an invaluable test for the simulation that correctly showed that the switch would worsen the separation.

16. For towers operating in deep vacuum, check the simulation sensitivity to air leakage. In one case (436), since the actual air leakage rate could not be measured, it became an adjustable parameter in the simulation used to match the simulation to test data.

17. In refinery gas plant towers (debutanizers, depentanizers, naphtha splitters), whenever practical, obtain full component analyses of the feeds and products, and use them in high preference to the ASTM D86 curves (or even the simulated distillation curves) for validating the simulation. One of the lessons from the depentanizer case in Ref. 235 is the importance of adhering to this guideline. The number of stages determined from ASTM D86 analyses can become sensitive to the flashing of lights from the samples and cannot be validated by a component balance. In the case study, going by the ASTM D86 analyses incorrectly gave double the number of stages in the tower simulation.

18. In refinery fractionators (other than those in gas plants), overlay the measured and simulated TBP distillation or ASTM distillation curves, and ensure a good match. Poor match may be caused by mass balance deviations, entrainment, leakage, plugging, maldistribution, or damage not adequately modeled in the simulation.

    Figure 3.11 compares TBP distillation measurements with the simulation for a refinery vacuum tower. The process flowsheet is identical to that in Figure 7.2a, with heavy vacuum gas oil (HVGO) being the upper (lower boiling range) side product,

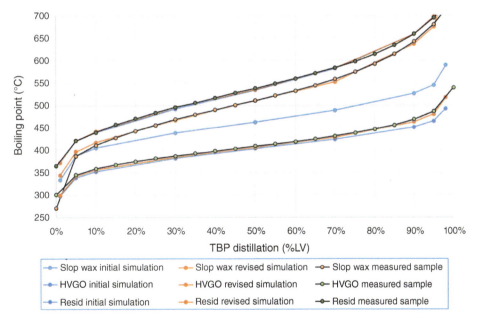

Figure 3.11 Comparing simulated TBP to test data for side products HVGO, slop wax, and bottom resid. (From Kister, H. Z., "Practical Distillation Technology," Course Manual. Copyright © Henry Z. Kister. Reprinted with permission.)

the slop wax the intermediate (middle boiling) side cut, and the resid (or residue) the bottom, heavier product.

The initial simulation (blue pen) gave an excellent match to the measurements for the resid. For the HVGO, the match was good below 70% LV, but there were some deviations above 70% (measurement 10–25°C above simulation). The large discrepancy was with the slop wax, where the measured TBP temperatures exceeded the simulated, especially at the higher end. The difference gradually rose from zero at 10% LV to about 110°C at 90% LV. This suggested that the heavier material (resid) is likely to be entrained into the slop wax. The simulation was revised by incorporating additional resid entrainment into the slop wax until the measured and simulated TBP curves for the slop wax matched. Likewise, to match the heavy ends of the HVGO, the reflux was slightly reduced, lowering the net draw rate of the slop wax (i.e., not including the entrainment). This pushed some heavier material into the HVGO. The revised simulated TBPs (orange pen in Figure 3.11) are very close to the measurements.

19. Allow for stages contributed by reboiler, condenser, inter-reboiler, and intercondenser. Common rules of thumb are as follows:

   • A single stage is allowed for a once-through reboiler or a kettle reboiler or when the bottoms draw compartment is separated by a preferential baffle from the reboiler compartment (192). Half or zero theoretical stages are allowed for an unbaffled recirculating reboiler arrangement.

   • A single stage is allowed for a partial condenser, and none for a total condenser. Note, however, that most computer simulations count a total condenser as a stage in which there is no separation.

   • Determine whether an inter-reboiler or an intercondenser approximates a theoretical stage. If so, allow for it.

   Subtract the total number of stages contributed by these devices from the number of stages calculated by the simulation. The difference is the number of stages in the column.

20. Compare the test run efficiency with the design efficiency or the efficiency of other towers in the same service. As per step 1 above, a comparison with the design can be misleading when relative volatility is low (<1.5, and especially <1.2), unless the designer used identical VLE data to those used in the test run simulation. Differences in VLE values will be reflected as differences in column efficiency when relative volatility is low. In one case experienced by the author, a 2% difference between the relative volatility used by a designer and an operator in a low-volatility ($\alpha = 1.1$) system was sufficient to account for a difference of 50% in efficiency. In low-volatility systems, it is best to simulate the design conditions using identical VLE data to those used for test run simulation.

21. A good test simulation has predictive capabilities. If practical, check these. In one ethylene oxide absorber–stripper system (358), a simulation tuned with plant data was further validated by demonstrating that it correctly predicted the plant-measured effect of reducing lean absorbent flow and lean absorbent temperature. In Case 4.8 in Ref. 201, a simulation was further validated by correctly predicting revamp product purities following packing changes.

22. In multicomponent distillation, each binary pair of components has different separation efficiencies (33, 193, 345, 474). For example, if both water (light non-key) and a C_6 hydrocarbon (light key) are separated from a C_7 hydrocarbon (heavy key), the efficiency of the high-volatility pair water–C_7 will be significantly lower than the efficiency of the lower-volatility C_6–C_7 pair. For tray towers, the efficiency of each pair can be estimated using the O'Connell plot (344, Figure 2.39). Most ideal-stage simulations do not permit

different efficiencies for different binary pairs. Usually, only the separation of the key pair is important, so the simulation only needs to match it, and the optimistic prediction it gives for the high-volatility non-key component separation is not an issue.

Occasionally, the non-key separation becomes important, like when investigating the presence of 100 ppm of water in the bottom of a C_6–C_7 splitter. The test simulation was prepared to match the test data for C_6–C_7 separation, giving an optimistic separation for the water. To investigate the water separation, it was rerun with a lower efficiency. The rerun was only good for predicting the water–C_7 separation, but was incorrect for the main C_6–C_7 separation.

With some nonideal systems, especially when operating near the azeotrope point or when a second liquid phase is involved, the need to use equal efficiencies for the various binary pairs may be prohibitive, and rate-based (nonequilibrium) simulation models need to be applied (474). Rate-based models are usually preferred also in reactive distillation and gas absorption/stripping applications (474, 501, 502).

23. Document your analysis in a detailed report, outlining the data used, steps made, and the basis for the various choices (193, 286, 465). This is often overlooked yet crucial for retaining your analysis and hard labor for future use. The author has seen far too many cases in which extensive effort put into testing and analyzing data became worthless a few years later just because no one knew where some data or findings came from. In contrast, the author has seen cases when test data and simulations obtained years earlier were readily applicable and turned out invaluable for a revamp or evaluation many years later.

3.1.7 Determining Hydraulic Loads

To perform hydraulic calculations on the tower, tower internals, and distributors, the hydraulic loads are inferred from the simulation. This determination is not always straightforward. The following watchouts apply.

Table 3.3 provides a simplified internal flows output, directly derived from a real tower simulation, for rating trays for a tower revamp. The tower was simulated as 16 stages, numbered from top (stage 1) to bottom (stage 16). There was a major flashing feed entering between stages 14 and 15. In this case, the task was to rate the internals **below** the feed (the stripping trays) and check whether they could handle the maximum simulated revamp loads shown.

At first glance, the maximum stripping section vapor flow rate is 19,600 kg/h and the maximum liquid rate is 259,600 kg/h. These were the numbers initially sent to the tray supplier. Based

Table 3.3 Internal flow rates from tower simulation. (From Kister, H. Z., "Practical Distillation Technology," Course Manual. Copyright © 2013. Reprinted with permission.)

Liquid profile (from tray)					
Stage	12	13	**14**	**15**	16
Mass flow (kg/h)	155,400	33,300	12,800	259,600	244,500
Volume flow (m³/h)	242	52	20	375	347
Density (kg/m³)	641	635	644	693	705
Vapor profile (to tray)					
Mass flow (kg/h)	225,000	215,000	35,300	19,600	4,500
Volume flow (m³/h)	29,500	29,000	8,300	6,600	4,158
Density (kg/m³)	7.633	7.415	4.258	2.963	1.082

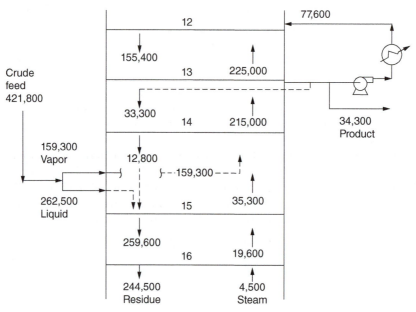

Figure 3.12   A sketch leads to the correct internal flow rates from the tower simulation. (From Kister, H. Z., "Practical Distillation Technology," Course Manual. Copyright © 2013. Reprinted with permission.)

on these, the supplier saw no need to change the stripping trays. The client was not happy, as they believed they observed a limitation at current rates and the expected revamp rates were higher.

It is imperative to appreciate that the listing of the vapor and liquid internal loads in the simulation output is often user-unfriendly, especially near the entry point of flashing, subcooled, or superheated feeds or refluxes. Directly using the numbers in the output can lead to incorrect conclusions, as it did in this case. A simple yet effective technique can easily identify the correct liquid and vapor loads: put together a sketch like Figure 3.12 and list the flows on it.

In this tower, each of the bottom two stages (15 and 16) modeled three trays (33% efficiency). The internal vapor flow rate for the upper stage, 19,600 kg/h (Table 3.3), is the vapor flow _to_ the stage, i.e., into the lowest of the three trays represented by stage 15. The flow out of stage 15, which is a lot closer to the vapor flow rate through the upper of the three trays, is a much higher 35,300 kg/h. Table 3.3 lists this flow as the vapor entering stage 14. This is nonsense; the sketch shows the flow into stage 14 to be 35,300 kg/h _plus_ the feed flash vapor (159,300 kg/h), a total of 194,600 kg/h.

Table 3.3 gives the liquid rate _from_ stage 15. This is the flow leaving the lowest of the three trays. Figure 3.12 shows that the flow into stage 15 is significantly higher and equals the flash liquid (262,500 kg/h) _plus_ the liquid descending from stage 14 (12,800 kg/h), a total of 275,300 kg/h. This higher number is not in the computer simulation! It is obtainable from the simulation. You need to flash the feed to column entry pressure and then add the flash liquid to the liquid flow from stage 14 (this number is in the simulation), and the result gives the liquid rate to stage 15. So stage 15 needs to be rated for 35,300 kg/h vapor and 275,300 kg/h liquid.

All these numbers need to be checked with mass balances to ensure the correct numbers are plotted. In this case, the material balance for stage 15 gives:

Into stage 15 275,300 (liquid) + 19,600 (vapor) = 294,900 kg/h

Out of stage 15 259,600 (liquid) + 35,300 (vapor) = 294,900 kg/h

The author has experience with another major simulation software in which the vapor flash (equivalent to 159,300 kg/h in Figure 3.12) was combined with the vapor from tray 15 (35,300 kg/h) and the result was presented as vapor from stage 15. This would have required stage 15 to be rated for 194,600 kg/h, which is nonsense. In that case, the feed flash was a lot smaller, about 10% of the vapor load from stage 15, so it was missed. The problem was picked by the tray supplier who could not design trays to satisfy the excessive required vapor load. A sketch like Figure 3.12 showed that the additional 10% vapor from the feed flash should not be included in the stage 15 vapor load, and therefore, the supplier's tray will work.

Caution is also needed with highly superheated feed. The superheat vaporizes some of the tower liquid upon entry. The vapor loading peaks when all the superheat is consumed to generate dew point vapor. The simulation often misses this peak ("the bubble effect"). A classic example is a typical refinery FCC main fractionator, where a highly superheated feed at about 1000°F enters, leaving the stage above it at about 700°F. The maximum vapor rate can be inside the stage and, if not accounted for, may lead to premature flood as reported in one case (286). The author had success by rerunning the simulation with two stages above the feed and using the intermediate vapor rate between them to rate the bottom section. This rerun is only used for the hydraulic loadings, as one stage is realistic for separation.

Similar considerations apply to liquid side draws or pumparound draws. The simulation output may not include the product draw rate, the pumparound draw rate, or both, in the draw tray liquid loadings. Identification of such issues is the same as outlined above: draw a sketch like Figure 3.12 and check the numbers using a material balance as described above.

For packing distributor and parting box ratings, entry piping rating, and pipe distributors bringing feed or reflux, a flash calculation of the feed to tower pressure at the feed entrance point is a necessity. The distributor and piping need to be rated based on the flash vapor and flash liquid rates. The computer simulation often does not report the results of this flash, but would easily include the results upon request by the user. Skipping this flash calculation is a very common gross error in packing distributor ratings. The author has seen many cases where the feed was assumed to be liquid when the flash vapor volume at column entry pressure was more than 80% of the feed. Needless to say, the distributors designed for liquid had no chance of working with a large vapor fraction.

Numbers coming out of the simulation also need to be closely reviewed with the help of sketches like Figure 3.12 with highly subcooled feeds. In one case (113), high ethylene losses occurred after replacing trays by packings in an olefin demethanizer. Inadequate allowance for the highly subcooled (70°F) feed overloaded distributor capacities. In another case (450), inadequate accounting for subcooling in tower revamp design led to premature flooding. In one more case (41), the distributor maximum liquid load exceeded the reflux pump capacity by far. The supplier design was based on the liquid rate *from* the top stage when it should have been based on the reflux feed *to* the top stage.

In some applications, vapor and liquid loads change throughout a trayed section or a packed bed. A common optimizing design strategy in packings is to use the higher-capacity, less efficient packing where the loads are high, and the lower-capacity, more efficient packing where the loading is low. In this case, the vapor and liquid loadings at the demarcation point need to be evaluated, with proper allowance for the efficiency change, making sure the lower-capacity packing can handle the intermediate loads.

3.1.8 Hennigan's Rules

Below are rules of thumb proposed by Hennigan (169, reprinted courtesy of S. Hennigan, INEOS Acetyls) for critically reviewing tray efficiency determinations. While not all may apply to a specific situation, they are all worth considering. Rules of thumb should always be taken for what they are: rough criteria that reflect the experience of practitioners. They are not meant to replace detailed calculations. Their merit is in providing troubleshooters with practical insight that can be useful in their endeavors.

Rule #1 – For a new system: expect and design for 40–70% typical efficiency for a system around the middle of the O'Connell curve (Figure 2.39) (i.e., systems of moderate relative volatility around 1.5–3 and liquid viscosity between 0.1 and 2 cP).

Rule #2 – Don't hope to get much more efficiency by modifying a hydraulically sound tray design across its operating envelope.

Rule #3 – Check efficiency in depth, particularly of older kit – they may reveal a performance deficit that has been lived with for years.

Rule #4 – If tower performance "feels" wrong, this is probably a strong indication of an efficiency issue (which may have arisen from emergent hydraulic problems or bad design).

Rule #5 – Measuring efficiency consistently before and after changes and shutdowns enables sensible conversations between vendor and operator.

Rule #6 – Follow through the actual financial consequences of both low and high efficiency – the answer may not be what you think.

Rule #7 – Look for similar efficiencies in families of similar separations. This is the most powerful way to detect deficits and paves the way for future changes.

3.2 DIAGNOSING POOR SEPARATION

The second part of this chapter deals with various sources of poor separation, other than those covered in Chapter 2 (flooding, foaming, channeling, loss of downcomer seal), those described in detail for packed towers (maldistribution of liquid or vapor, Chapter 4), and those associated with control instability that are discussed in distillation control texts (50, 192, 431, 453).

3.2.1 Troubleshoot for Process Leaks

The first step when evaluating tower separation, and when preparing for efficiency testing, is to eliminate any process leaks (Section 3.1.3, step 1). Any such leaks are likely to misguide troubleshooting and efficiency determination. More importantly, process leaks can be health, safety and environmental hazards. These hazards are dealt with in safety texts and are beyond the scope of this book. Here, we focus on the effects of process leaks on separation and efficiency testing.

Consider also heat exchanger leaks (see Section 7.3.5) and other extraneous materials that may affect tray or packing efficiency. For instance, seal oil leaking into the feed of a heavy water distillation tower covered the tower packings with a thin hydrophobic film, which reduced wettability and halved efficiency (70). In another case experienced by the author, gas oil leaking through the seals of a diesel product pump got the diesel off-spec.

Closely watch out when one of the heat transfer streams is also the product stream. For instance, a reboiler or condenser tube leak may affect the concentration of water in one of the products, as reported in one methanol–water column (333). In another case (284), gas dehydration was poor due to feed–effluent exchanger that leaked cold, wet glycol into the heated lean dry glycol entering the top of the contactor. In one other case (138), a kerosene product from a refinery crude tower showed color because a mid-pumparound heat exchanger leaked crude in. Leakage of crude into its preheating stream (tower overhead, pumparound liquid, distillate product) is a common occurrence that contaminates distillates or plugs trays in the tower. In another case (138), a water leak overloaded the condenser, causing a 40% loss of overhead product. A similar case is presented as Case 5 below. Another case is described in detail in Section 1.2. Useful techniques for exchanger leak detection were described by Pugh et al. (386).

Case 1. Off-spec ethylene product (216) The final step in ethylene plant distillation, the C_2 splitter (Figure 3.13), separates the ethylene product from the ethane recycle. This is a close separation under pressure (typically 200–280 psig), which produces a high-purity ethylene product (typically a few hundred ppm ethane impurity) and requires a large number of trays, with about 100 trays common.

At a turnaround, one C_2 splitter was retrofitted with high-capacity trays to raise throughput. Upon restart, the ethylene purity spec could not be met. Tray efficiency was less than half the design. Eventually, the plant got the ethylene on-spec, but at the price of losing tons of ethylene in the bottoms and operating at very low throughput. For an ethylene plant, this is a performance disaster, costing tens of thousands of dollars, maybe more, per day. The plant recruited experts, troubleshooters, and gamma scans. There were meetings, discussions, brainstorming, theories, and ideas. The gamma scans surprisingly showed no apparent problems, but they were not believed.

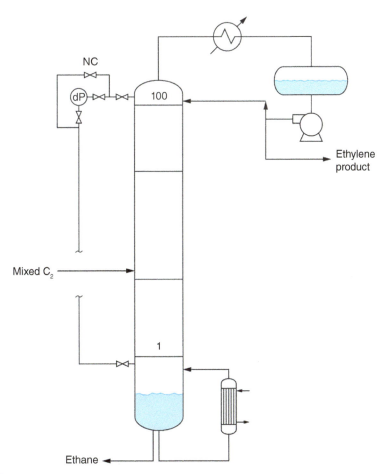

Figure 3.13 C_2 splitter tower, showing the dP bypass. (From Kister, H. Z., "Practical Distillation Technology," Course Manual. Copyright © 2013. Reprinted with permission.)

Two weeks later, an operator climbed up the tower. The C_2 splitter had a differential pressure transmitter (Figure 3.13), mounted right at the top of the tower. This is a good location for minimizing issues with liquid accumulation in the impulse lines (192), but is also an inconvenient spot to look at and maintain. The operator noticed that around the transmitter was a short bypass line with a ¾-in. valve. To his surprise, the valve was open, probably to isolate the transmitter and protect it during some commissioning operation. It should have been closed when the transmitter was reconnected, but this did not happen. The open valve passed a large quantity of ethane from the tower bottom to the top product. A typical differential pressure on this tower was about 8–10 psi, which will push a lot of ethane into the ethylene via this ¾-in. line.

When the operator closed the bypass valve, the ethylene purity came good, the trays operated efficiently, and the consultants were sent home.

The author had another similar experience with an analyzer cross-connection on another C_2 splitter left open. This tower also had problems making on-spec ethylene upon restart.

Case 2. Poor butene separation (216)   The author was involved in consulting a client who built a new butene purification facility. The product tower, which was the largest in the plant, separated two butene isomers. Again, a difficult separation. This was a licensed process. Two licensors bid: one offered a tower with 160 trays, and the other with 100 trays. Our evaluation was that 100 trays would be tight, but would at least get close to achieving the desired separation. The large cost difference gave the client sufficient incentive to go with the shorter tower, and its licensor was chosen.

Upon startup, the worst fears materialized. The tower was unable to make on-spec products without increasing the reflux to the point of significantly losing capacity. "We told you so, you get what you pay for" was heard from the skeptics that favored the taller tower. The author was called in to see whether there was anything that could be salvaged.

Learning from Case 1, troubleshooting began with looking for any possible connections between the products. There was no bypass around the differential pressure transmitter. However, to reduce the number of storage tanks, the design provided the flexibility of sending either product to each of the three storage tanks (Figure 3.13). So there were cross-connections. The author checked each valve on these connections. At the time, the top product was going into tank #1, and the bottoms product to tank #3. All valves marked as "NF" or "SHUT" were shut. As the author was checking them, he put his hand on the lines on each side of the valves (the product temperatures were low enough so there was no hazard in doing so; if the lines are hotter, putting a hand on the line is a burns hazard, and a pyrometer should be used instead). One of the cross-connections felt slightly warm. The author called the operator. "Let me give you another half a turn on the valve and see if it helps," said the operator, but he could only close it another quarter of a turn (see Figure 3.14).

The next day, the operators told the author that overnight they comfortably reached the design throughput with the products on-spec. The reflux was still slightly higher than the design, but not enough to bottleneck the tower capacity.

Case 3. Loss of yield   For years, engineers in a refinery wondered why the diesel yield from the atmospheric crude tower fell short of the simulated value, while the gas oil product yield was higher. Eventually, a process engineer decided to troubleshoot for cross-connections between these products. She identified some pumps that could be used to pump either product, giving flexibility to perform maintenance on the pumps. So there were cross-connections. She checked the valves on these cross-connections; all were shut. She then took readings using a pyrometer. One of the cross-connection lines was hot. Closing the valve another half a turn did not help here.

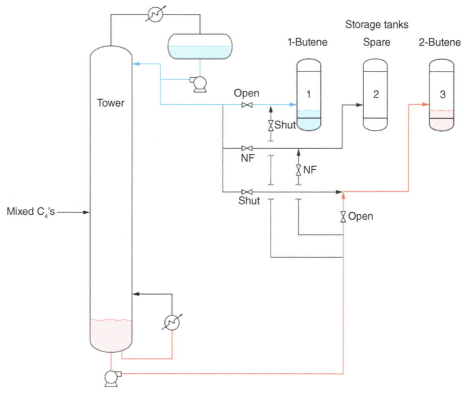

Figure 3.14 C$_4$ splitter tower, showing the connections to the storage tanks. (From Kister, H. Z., "Practical Distillation Technology," Course Manual. Copyright © 2013. Reprinted with permission.)

However, that line could be isolated and the valve replaced. Following the valve replacement, the diesel yield predicted by the simulation was achieved.

Case 4 Sampling from a different point (Case 8.1 in Ref. 201, contributed by Mark Pilling). Kerosene product was off-spec with a high end point. The cause was identified by taking a sample immediately off the column rather than from the product rundown line. This sample was not taken initially as it was hot and awkward to obtain. The sample directly off the column had the kerosene on-spec. Pursuing this lead showed a leaking valve between the diesel and kerosene products upstream of the normal product sample point. This leaking valve was repaired, and the column achieved the design rates and product specifications.

Case 5 Sampling from a different point detects exchanger leak (432) An amine absorber unit (Figure 3.15) was revamped with high-capacity trays. Following startup, the absorber failed to provide the same removal of H$_2$S that was achieved prior to the turnaround. The initial investigation, focused on the new high-capacity trays, did not uncover any issues.

A lean amine sample from a point immediately downstream of the regenerator showed good quality. Later, an operator took a sample of the same lean amine stream but from downstream of the cross-exchanger (Figure 3.15) and noticed a slightly different color. He had it analyzed. The sample showed contamination with rich amine. The leak raised the H$_2$S concentration in the lean amine, leading to excessive H$_2$S in the treated gas. An examination of the exchanger in the next outage showed pin-sized holes in many of the tubes.

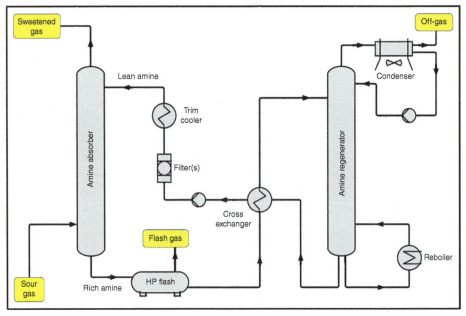

Figure 3.15 Typical amine absorber–regenerator scheme. (From Shiveler, G., and H. Wandke, "Steps for Troubleshooting Amine Sweetening Plants," in Kister Distillation Symposium, p. 96, AIChE Spring Meeting, Austin, Texas, April 26–30, 2015. Reprinted with permission.)

Takeaway from all these stories is that the very first step in any troubleshooting of poor or substandard separations, or in planning for a tower efficiency testing, is to critically look for leaks that can impair product purity across all cross-connections and heat exchangers.

3.2.2 Troubleshoot for Tray Weeping

Weeping refers to liquid descending through tray perforations. Some liquid may also leak at the joints where tray sections are bolted or clamped to supports. Liquid descending through the perforations or at the supports short-circuits the primary contact zone, reducing tray efficiency.

At the tray floor, the static liquid head tends to push the liquid down the perforations. The vapor (dry) friction pressure drop counteracts the downward force, acting to keep the liquid on the tray. Weeping occurs when the static liquid head exceeds the dry pressure drop. For most trays, some weeping takes place under all conditions. Sloshing and oscillations create localities in which the static head exceeds the dry pressure drop, inducing intermittent local weeping.

Some weeping, as much as 20% (106), can be tolerated without an appreciable loss in efficiency (74, 106). Tray weeping may be nonuniform (17, 300). Liquid weeping near the tray inlet drops to the outlet of the tray below, bypassing the contact zones of both trays, making it much more detrimental to tray efficiency than uniform weeping. Liquid weeping near the tray outlet drops to the inlet of the tray below, bypassing very little of the contact zones, with less efficiency loss than uniform or inlet weeping. The maximum amount of weeping that can be tolerated therefore varies with the system and with the location of weeping on the tray, with typical values between 10% and 30%. Excessive tray weeping significantly reduces tray efficiency.

Maldistribution of liquid and/or vapor to multipass trays can also induce nonuniform weeping. In addition, in two-pass and multipass moving valve trays, multiplicity of steady states (348) may create another type of nonuniform weeping, in which some panels weep and others do not.

Amazingly, in some polymerizing services where the polymer forms on dry surfaces, a mild amount of weeping can be beneficial. In such services, a polymerization inhibitor is usually injected into the top reflux, protecting the wet upper side of the trays. If the underside of the trays is unwetted, the inhibitor does not get there, and the polymer will accumulate underneath. In this case, it is beneficial to allow some weeping, which will keep the underside wet.

Hydraulic calculations are the best methods to evaluate weeping. For lower-pressure systems (<150 psig), the author and others (74) have had an excellent experience with the method of Lockett and Banik as modified by Colwell and O'Bara (74). The author has also had good experience with it in high-pressure non-distillation systems like amine absorbers (223). For high-pressure distillation systems, the author and others (74) prefer the Hsieh and McNulty correlation (179). Both of these correlations and their ranges are presented and discussed in detail in Perry's Handbook (245) and *Distillation Design* (193).

The above correlations are based on uniform weeping and do not account for intensification due to nonuniform weeping. In addition, the above correlations make no allowance for the intense inlet weeping generated under channeling conditions as described in Section 2.12, which is detrimental to tray turndown. In one air separation column with sieve trays at 160 mm tray spacing and excessive open area, the products went off-spec at rates below 95% of the design (514).

In some services, corroded tray support rings are replaced by an internal support structure in the tower. The author had an experience with extensive weeping with this technique, which is unaccounted for by the above correlations.

When performing the calculations, audit the uniformity of vapor load in the section under consideration. In some applications, such as steam stripping of hydrocarbons or organics, there is a large variation in vapor loads. This is illustrated in Table 3.3, where the internal vapor flow rates in the stripping section of a refinery crude tower varied from 4500 to 35,300 kg/h, or from 4158 to 8300 m³/h. On a C-factor basis (Eq. (2.1)), this is a variation by a factor of 4–5. If the trays in that section are all of the same design, weeping will be intense in the lower trays. A successful strategy to minimize weeping is to progressively reduce the tray open area from top to bottom in this stripping section (156, 165).

Another useful concept in weeping evaluation is the Summers et al. stability factor (468) defined by Eq. (2.6). When the stability factor exceeds 0.6, weeping is less than 30% and the tray is stable. Weeping and pass maldistribution rapidly escalate at lower stability factors, with 50% weeping or more by the time the stability factor declines to 0.4. Recently, Summers (467) analyzed Fractionation Research Inc. (FRI) data to extend the range of physical and mechanical parameters in the stability factor equation. His revised elaborate equations are presented elsewhere (467). Equation (2.6) is still a good first approximation, but the revised stability factor can have values lower than 0.6 for high-pressure distillation systems and higher than 0.6 at low operating pressures. The author has extensive good experience with the earlier version, but only limited experience with the updated correlation.

To test for weeping, efficiency testing can be repeated at turndown, and the efficiency difference evaluated. Some towers are constantly operated at low rates with poor separation. If excess weeping is confirmed by hydraulic evaluation, these should be tested by increasing reflux and reboil. Separation should improve, as it did in Case 4.4 in Ref. 201.

Leakage, distinct from weeping, refers to the passage of liquid from openings in the tray that are not intended for vapor flow during normal operation. These include (448) leakage around tray support rings, along joints in the tray deck, and through weep holes. Leakage is usually insignificant when a tower operates at high rates, but it may become significant at turndown. It is also important in downsized columns that have been retrayed for internal vapor loads much lower than the original. Seal welding or gasketing and close inspection (Sections 10.3.6 and 10.3.7) are key. Reference 445 contains useful guidelines.

To distinguish tray weeping from other types of internal leakage, the vapor and liquid sensitivity tests described in Section 2.7 may help. The difference is that here one is not looking for flood but for tray efficiency change, so the tests need to be run long enough to afford a reliable efficiency determination. This makes the period of off-spec product longer than in the flood sensitivity tests described in Section 2.7. Regular tray weeping is sensitive to vapor rate but generally insensitive to liquid rate. So for regular tray weeping, the vapor sensitivity test would show a significant efficiency loss, while the liquid sensitivity test would not.

In refinery fractionators, there are often other indicators, such as internal and side product flows, that can be used to monitor the vapor–liquid sensitivity test, circumventing the need for efficiency determination. A good example is Test 4 in Section 9.2.5.

3.2.3   Diagnosing Side Draw Liquid Starvation

A mass balance around the side draw in Figure 3.16a gives

$$D = L - \text{WEEP} - \text{SUMP LEAK} \tag{3.1}$$

where D, L, WEEP, and SUMP LEAK are all mass flows in consistent units.

The side draw will be starved either when the liquid flow rate L from tray 11 above is too low and/or when WEEP from the draw tray (tray 12) and/or SUMP LEAK is excessive. A low liquid flow rate from above can be due to a condenser limitation. In refinery fractionators, it can result from heat transfer limitations in an upper pumparound circuit (e.g., due to exchanger fouling) and/or from excessive condensation in a lower pumparound circuit, as described in some case studies (215, 312).

Starving the draw forces reduction in D. The mass balance will compensate by increasing the flow rate of a lower or bottom product stream. This induces lighter material that belongs in the side draw into the bottom or into a lower product, getting them off-spec.

We have experienced high WEEP due to channeling (228, see Section 2.12), and it has also been reported by Martin et al. (312). Starving the draw is a common issue when the draw tray is damaged (e.g., 310) and/or when the draw sump is leaking (e.g., 61, Cases 10.3 and 10.4 in Ref. 201).

The design in Figure 3.16b allows some liquid to flow back from tray 13 to the sump, so the draw will be starved at a larger weep rate. Using leak-resistant trays, seal-welding the draw sump, and replacing the draw trays by chimney trays are measures that effectively circumvent draw starvation (156, 192).

To test for draw starvation, increase the boilup, the reflux (or internal liquid flow) rate, or both. In the absence of severe damage, both should help increase the draw rate. Raising the internal liquid rate test (at a constant internal vapor rate) can help distinguish a draw starvation from an undersized draw line bottleneck. With a starved draw, the product rate should go up. If the line size is limiting, increasing the liquid flow will not raise the product rate.

With a starved draw, gamma scans will show the draw lines are not liquid-full.

Side Draw Connected to Two Tower Nozzles (Figure 3.16c)   Some towers mix side draw streams from two different elevations to optimize product purity and flexibility. A possible issue described by Lee (276) is that the higher hydraulic head at the upper nozzle can push backflow through the lower nozzle. Symptoms include (276) poor product composition of the upper draw, same temperature at both draws, lower static pressure at the lower draw than at the upper draw, and gamma scans indicating less liquid traffic between the draws than below the lower draw.

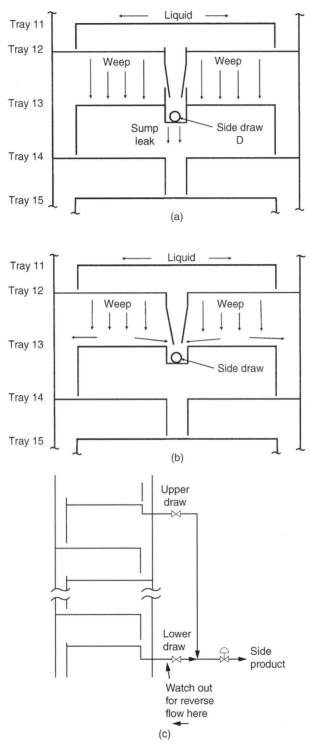

Figure 3.16 Starving a liquid draw. (a) Draw starved when weeping exceeds the intended liquid flow to tray 13. (b) Alleviated with this draw design due to backflow from tray 13. (c) Backflow at lower draw when two nozzles feed the same draw.

3.2.4   Diagnosing Once-Through Reboiler Liquid Starvation

A once-through reboiler uses a total draw via either a downcomer trapout or a chimney (collector) tray per Figure 3.17a or b, respectively, to capture all of the liquid leaving the bottom tray and feed it directly to the reboiler. The reboiler return is directed to the tower bottom sump, from which liquid is drawn as the bottoms product. No bottoms liquid is recycled back to the reboiler, hence the name once-through.

Once-through reboilers can achieve one full theoretical stage of separation (if the trapout draw does not leak) and can also maximize the reboiler temperature difference. Both are advantages over circulating thermosiphon reboilers, in which the recirculation of hot liquid from the bottoms sump into the reboiler reduces the temperature difference and mass transfer stages. Once-through reboiler application is usually limited to systems in which the boilup is less than about 40% of the bottoms product rate, due to the normal limitation of 30% maximum vaporization (by weight) in thermosiphon reboilers (109, 182).

An inherent issue with the trapout arrangement (Figure 3.17a) is weeping (or leakage) from the bottom tray or draw sump. Such leakage has been identified (198) as one of the two most troublesome malfunctions in reboilers. Numerous case studies were described in the literature and surveyed in Ref. 201, together with some of the author's experiences. In one case of severe draw pan leakage (61), the bottoms specifications were hard to achieve, lights were lost to the bottom product, and steam consumption was excessive, costing US $1.5 million/year. The leakage problem is most severe during turndown and startup, where the vapor rate is low. There have been towers that could not be started up due to this leakage. When all the liquid bypasses the reboiler, the reboiler will have nothing to boil, and the lack of vapor will perpetuate the leakage.

In most cases, leakage from the trapout arrangement will give a symptom of reboil shortage with lights in the tower bottoms. It is best diagnosed by measuring the bottoms temperature, the reboiler return temperature, and the temperature of the liquid draw from the bottom tray. A flash calculation based on the temperature measurements can estimate the fraction of liquid leaking. If there is no leak, the bottoms temperature will be the same as the reboiler return temperature. The closer the bottom temperature is to the liquid draw temperature, the larger the leak. If one of these temperatures is not monitored, use the wall temperature measurement as described in Chapter 7. The temperatures in Figure 3.17a show a typical case in which the tower bottoms liquid consisted of about 30% of liquid weeping (about 10% of the weeping liquid flashed).

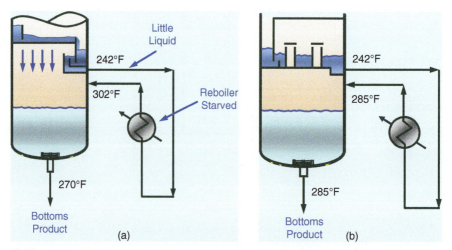

Figure 3.17   Once-through thermosiphon arrangements. (a) Trapout. (b) Chimney tray. (From Kister, H. Z., "Practical Distillation Technology," course workbook. Copyright © Henry Z. Kister. Reprinted with permission.)

As the trapout provides little degassing time, rundown lines to once-through thermosiphons must be sized for self-venting flow (109, 192). When they are undersized, the trapout pan may overflow and again the overflow will bypass the reboiler.

The chimney tray arrangement (Figure 3.17b) can overcome most, even all, of the above problems but is not totally immune. Leakage can occur through the joints of the tray or draw pan (seal welding or good gasketing can prevent this) or through a damaged panel. When the rundown line is undersized for self-venting flow and the liquid residence time on the tray is less than 0.5–1 minutes, overflow may occur. When the vapor velocity through the chimneys is high, it may prevent the descent of the overflowing liquid and entrain it to the next tray, which in turn may initiate flooding that can propagate right up the tower. Figure 3.17b shows the equivalent temperatures after the bottom tray in Figure 3.17a was replaced by a seal-welded chimney tray. The leakage stopped and the bottom temperature increased, indicating less lights in the bottoms. The reboiler return temperature decreased, showing a larger liquid fraction in the reboiler effluent.

3.2.5 Troubleshoot for Liquid in Vapor Side Draws

Liquid can enter a vapor side draw either by entrainment from the tray below or by weeping from the tray or collector above.

To diagnose the source of liquid, it is useful to test at high and low rates. More liquid in the side draw at higher rates indicates entrainment. More liquid in the side draw at lower rates indicates weeping.

When sampling the vapor draw from pressure columns for liquid, beware of condensation due to the Joule–Thomson effect as the vapor expands from the column pressure to atmospheric. It is best to fully open the valve from the draw line to the sample container and to throttle the sample container outlet valve.

Draw a to-scale sketch of the side draw (Figure 3.18). In Figure 3.18, the bottom of the draw nozzles is 18 in. above the tray floor. Research movies (106; see also Figure 2.1a from this movie) show that when a tray operates at about 50% of jet flood, the spray height reaches 12–15 in. So at higher rates, some of the spray is likely to be entrained into the draws. Case studies of liquid entrainment into vapor side draws were presented (201, 367). In packed towers, vapor draws should be placed beneath a total liquid collector. They must not be placed inside or directly beneath a packed bed as these will draw liquid with the vapor.

Check whether the vapor draw is via a bare nozzle or a perforated header. Vapor approaches a bare nozzle from the sides, from above, and from below, and the flow from below may induce entrainment. In packed towers, this flow pattern may lead to vapor maldistribution. Perforated pipe distributors, usually with upward-pointing holes, are common for vapor draws, especially in large towers, but they need to be correctly designed and have the perforations shielded from liquid weeping from above. Invaluable design and checkout guidelines are available in Ref. 441.

Weeping is another source of liquid in vapor side draws. In one case (450), a revamp of a solvent stripper in which bubble-cap trays were replaced by valve trays led to solvent losses in the vapor side draw, which was believed to be caused by the increased weeping from valve trays compared with bubble caps. In Case 10.7 in Ref. 201, weeping from the weep holes of a collector mounted above the vapor draw led to liquid in the draw.

3.2.6 Troubleshoot for Missing or Damaged Trays or Packing

Damage incidents may occur pre-startup (poor installation), during startup, or during the run. These incidents may be sudden (e.g., due to a pressure surge) or gradual (e.g., due to corrosion).

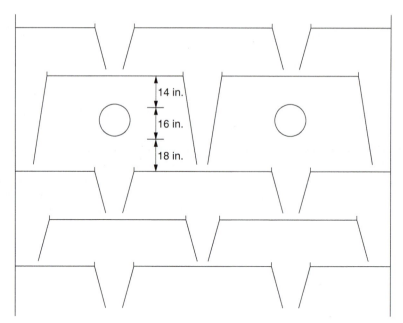

Figure 3.18 A vapor side draw (two nozzles in four-pass trays) that experienced liquid entrainment from the tray below. The bottom of the draw nozzles was close to the spray height on the tray. (From Kister, H. Z., "Distillation Troubleshooting," John Wiley and Sons, NJ, 2006. Reprinted with permission.)

The main indicators of missing trays or packings are historical tracking, pressure drop measurements, and gamma scans. Below are some troubleshooting guidelines:

1. Plot pressure drop measurements against vapor rate, as per Section 2.8, Figures 2.5 and 2.6. Compile a reference plot from the data gathered in the first post-turnaround month, and superimpose later data on the plot. Figure 2.17c indicates damage sometime during the run. A plot like this for discrete time intervals after restart will indicate how sudden or gradual the damage occurred. If sudden, it usually correlates with some upset. This plot also allows close tracking of the events. In some cases, damage may cause pressure drop to rise because the debris block some of the lower trays, downcomers, or packing, initiating flood. The discussion in Section 2.8 around Figure 2.19 and the case described in Ref. 226 are excellent examples of a sequence of steps in which pressure drop first rose due to the blockage of trays by debris from corrosion products and later decreased as the debris were washed away.

2. If the damage incident occurred during startup, there would not be a good reference line. It may be possible to compile a reference line from pressure drop data from the previous run, providing the internals were unaltered, the pressure drop was unaffected by corrosion or fouling, and the instrument was not recalibrated during the outage. The author's experience with compiling a reference line from previous run data has generally not been good. Caution is required.

3. A pressure survey, as described in Section 2.6 and illustrated in Figures 2.7 and 2.8, is invaluable. Unlike the plotting in steps 1 and 2 above, it does not provide a time record, but enables narrowing in on the region where the damage occurred. Regions of very low pressure drop indicate the likelihood of missing or badly damaged trays or packings.

Lieberman (287) describes a classic atmospheric crude tower case in which four uninstalled manways resulted in little separation between diesel and residue, inducing diesel into the residue. To boil off the diesel, the refinery added a new 200-MM Btu/h vacuum heater. Lieberman measured a 0.1-psi pressure drop across the four trays where 0.6 psi was expected and advocated repairing the four trays rather than adding an expensive and wasteful heater, but his recommendation was not implemented. In another tower (312), the field-measured 0.1-psi pressure drop across 11 trays diagnosed their damage.

4. The operation team knows the tower best and will be able to put their fingers on when and how its behavior has changed. They will also be familiar with the tower history and events (some seemingly unrelated) that happened around the time of the damage and could have contributed to it. Incidents from the far past may provide insights which the long-serving operators may recall. All these are invaluable inputs to a damage investigation. Work closely with the operation team.

5. Damaged or missing trays or packings perform little separation, often indicated as off-spec products. A small temperature difference over a section of the tower, where normally there is a significant difference, can indicate damage. The loss of separation discussed in Section 2.6, and Figures 2.14, and 2.24, is adaptable to tray or packing damage troubleshooting.

6. Gamma scans are invaluable for detecting damage. This is discussed at length with many case studies in Chapters 5 and 6.

3.2.7 Poor Material Balance Control Produces High Impurities

Distillation towers are material-balance-controlled (50, 192, 431, 453), i.e., the material flow in and out of the tower is controlled. For this principle to work, a product stream cannot be on an *independent* flow control. Note the word "independent." The flow rate of a stream on flow control, with a cascade (or reset) by another variable such as temperature or level, is *not* independent; it varies as the temperature or level manipulates it. This will not violate the material balance control. In contrast, a stream on a flow control with no resets has a fixed flow rate and does not allow changing the material balance.

The top product in Figure 3.19 is on an independent flow control. Consider 100 units feed, 50/50 lights/heavies, split into 50 lights top product and 50 heavies bottom product, initial steady state. Later, the feed rate is raised to 120 units, the same composition, i.e., 60 units of lights and 60 units of heavies. The flow controller in Figure 3.19 is set at 50, so it would only allow 50 units to leave in the top product. The other 70 units will attempt to leave from the bottom. But since there are only 60 units of heavies in the feed, the other 10 will be lights, and the bottom product will be off-spec for lights, even with high-efficiency tower internals.

The presence of other controllers may alter the response, but will not solve the problem. If a temperature controller manipulates the steam flow rate and a reflux drum level controller manipulates the reflux flow rate, the temperature drop (due to the lights that are unable to leave at the top) will increase the boilup. The additional vapor will be condensed and returned to the tower by the reflux drum level controller. This will cool the tower, and the temperature controller will again raise the boilup, which will again raise the drum level and so on. In reality, the operator will intervene and increase the top draw. But until this happens, one of the product purities will suffer. A case study by Musumeci et al. (330) provides an excellent illustration of how correctly implementing the material balance control tremendously improved product purity. Other case studies have also been reported (192, 319).

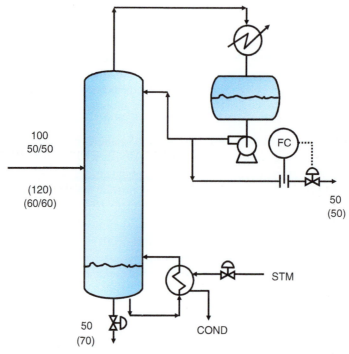

Figure 3.19   A control system violating the material balance control principle. (From Kister, H. Z., "Practical Distillation Technology," course workbook. Copyright © Henry Z. Kister. Reprinted with permission.)

Our malfunctions survey (198) identified violations of the material balance principle to be one of the three leading issues in control system assembly. It should be noted, however, that such violation usually affects the purity of only one of the products, not both. In the above example, the top product will remain on-spec. When the purity of one of the products is not important (e.g., when it is recycled to a reactor without any adverse effects), an independent flow control on a product stream may be acceptable. This is practiced, for example, when a downstream unit is limited in capacity and it is desired to keep a constant feed into it.

Violations of the material balance control principle can be prevented by using the material balance control methods described in Ref. 192 and in distillation control texts (50, 431, 453).

3.2.8   Limitations on Reflux or Reboil Generation

Distillation texts (e.g., 34, 193, 451, 492) teach us that to get a required separation, there is a need to provide sufficient reflux and sufficient boilup. This requirement may seem obvious, but in some multicomponent distillations, this may be difficult to achieve.

A typical scenario is when there is a low-ppm spec (typically <50 ppm) for a volatile component in the bottoms, and at the same time, the feed is either much heavier than the design or deficient in a component of medium volatility. In either case, the bottoms temperature will rise, the reboiler temperature difference will be reduced, and the reboiler may be unable to produce sufficient boilup to strip the volatile component out of the bottoms. Cases 2.2 and 2.3 in Ref. 201 give examples of poor stripping of H_2S from the bottoms of a revamped stripper because the feed was much heavier than the design. In Case 2.3, raising the bottom temperature and increasing the reflux ratio were enough to reinstate good H_2S removal. In Case 2.2, raising the temperature was constrained by a reboiler metallurgical limitation, and it was necessary to supplement the reboil by stripping steam injection.

Excess non-condensables in the feeds, or deficiency of condensable components like light keys, may constrain condensation in the overhead condenser, which in turn may restrict the reflux rate. The insufficient reflux may induce excessive heavies in the overhead product and/or may preclude the generation of enough boilup to remove small concentrations of volatile components from the bottoms product.

Bu-Naiyan and Kamali (51) describe a case in which the bottom product from a hydrocracker debutanizer contained 500 ppmw H_2S (design was <5 ppmw) when operating in the diesel mode with the feed much heavier than the design. A simulation study identified poor fractionation due to insufficient reflux as the root cause. The H_2S concentration was reduced to 10 ppmw by spiking the feed with 2000 BPD of butane (light key, about 5% of the feed) to establish adequate reflux/lift. After the injection, the reflux rate could be increased to more than 6000 BPD, which gave the better purity. A further reduction to <5 ppmw was achieved after another problem (Case 2, Section 8.17) was solved. The author is familiar with other cases in which injection or recycle of light key into the feed solved similar problems.

Excessive preheat or feed cooling can induce similar issues. One debutanizer (286) had 72% C_4's in the feed. A few degrees extra preheat boosted feed vaporization, a drop in stripping rates, and an increase in the C_4's content of the bottoms. Cases of a lack of stripping in a stripper due to excessive subcooling at the feed (149, 286, 449) were also reported.

Incorrect instrument readings can mislead and induce operation at insufficient reflux and/or boilup rates. Reference 396, Tale 6 in Ref. 216, and Case 2.6 in Ref. 201 are examples. Energy balances and differential pressure checks can help diagnose. In other cases (449, and in Case 2.1 in Ref. 201), reflux flow was cut out altogether, resulting in poor separation.

Good simulation based on field data is invaluable for detecting all the above issues. In some cases, the simulation shows almost no vapor. In one case (149), temperature difference and pressure drop measurements across a crude tower stripping section showed insufficient stripping steam; doubling the steam rate was a major improvement. In another case (396), pressure drop measurements helped diagnose insufficient reflux and boilup.

3.2.9 Control Instability Increases Impurities

There is a wide variety of control problems other than those discussed in Section 3.2.7 that can also increase product impurities. In most cases, these problems lead to instability and fluctuations in reflux and boilup, which in turn affects product purities. It will take a book to address these problems in a way that will do them justice. So the author is leaving their discussion to Refs. 208, 192, 201, and 206 and distillation control texts (50, 431, 453).

3.2.10 Impurities and Contaminants Affecting Azeotroping and Product Purities

In chemical systems, azeotroping of compounds often determines which direction certain components will travel in the column and in which product they will end up.

The presence of contaminants, caustic, or salts can change the equilibrium, sending components in an undesirable direction. The author was troubleshooting the presence of water in the bottom of a debutanizer. The hydrocarbon feed contained a minute amount of water, which would normally end up in the top, with none in the bottom. The test that diagnosed the problem was letting a sample of the bottoms settle and then inserting Litmus paper into the water phase.

The dark blue color indicated caustic. Subsequent analysis showed it to be over 50% weight caustic. The tower feed passed through a dry caustic bed to remove acidic components, and caustic fines carried over into the tower.

Similarly, the author has seen many cases where water was supposed to form a minimum boiling azeotrope with an organic low-boiler and leave from the top. Due to salt or caustic entry, a significant concentration of water was often found in the bottoms product.

3.2.11 Absorption of Sparingly Soluble Gases Affecting Product Purity and Downstream Venting

Upon contact, process liquids absorb small quantities of volatile components or sparingly soluble gases. For instance, when nitrogen is used to keep up the reflux drum pressure, a small quantity of nitrogen will be absorbed by the reflux and liquid product. This can make the product off-spec or generate venting problems downstream.

The contact pattern is key. Liquid splashed into the drum vapor space generates a large contact area between the liquid and the nitrogen, and the liquid will absorb a considerable quantity of nitrogen. In contrast, if the liquid line enters the drum well below the liquid level, the liquid will not contact the nitrogen and the absorption will be minimal (if any).

Figure 3.20 describes a case of nitrogen used to maintain pressure in a newly added debutanizer feed drum (210). Following restart, nitrogen needed to be vented from the downstream debutanizer, deisobutanizer, and depropanizer. The plant noticed that the nitrogen control valve was opening almost fully, suggesting that a lot of nitrogen was coming in. This nitrogen had to get out, and the only open route was with the drum outlet liquid. The high absorption rate was due to the feed entry near the liquid surface. This exposed an extensive contact area between the feed liquid and the nitrogen in the drum vapor space. The author is familiar with several similar cases.

Likewise, a freshwater reflux to a tower will absorb a significant quantity of volatile components, while reflux water generated in the overhead system will be saturated with the light components and absorb much less (471).

3.2.12 Reactions and Contaminants Can Affect Product Purity

In chemical systems, reactions can produce undesirable components that end in one of the products. Contaminants can azeotrope with some components and change the direction of their

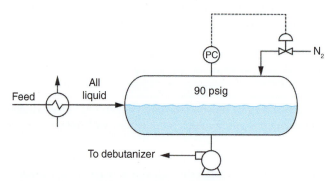

Figure 3.20 Nitrogen absorbed by the liquid in a newly added debutanizer feed drum. (From Kister, H. Z., Chemical Engineering, p. 47, April 2022. Reprinted with permission.)

concentration in the tower. Reactions can also generate components that cause plugging, fouling, corrosion, or foaming. Contaminants soluble in the feed may become insoluble and precipitate out once components to which they have affinity are distilled out of the solution in the tower.

Below are some guidelines that can be useful for troubleshooting reactions that directly (not via plugging or foaming) generate product purity issues:

1. The high temperatures in the bottom of the tower make it a fertile place for reactions. Decomposition and condensation reactions tend to produce light components that end up in the top product. This can be tested by lowering the bottom temperature by about 30–40°F for some time while monitoring the impurity in the top product. Case studies 15.1 and 2.7 in Ref. 201 describe experiences in which temperature-sensitive reactions in the bottom of towers generated light components like water or HCl that contaminated the top product. In one of the cases, the reaction occurred at hot spots in the reboiler. The reaction was stopped by lowering the bottom temperature and eliminating the reboiler hot spots.

   Another test is to operate the tower on total reflux for a short period and monitor the impurity. A growing concentration of the impurity in the top product confirms the reaction. In one tower (401), total reflux testing identified a suspected decomposition of highly chlorinated ethanes to be the root cause of poor separation in the bottom packed bed. In case study 15.1 in Ref. 201, operation at total reflux verified that the product contaminant was formed by a reaction inside the tower.

   When the tower bottom contains thermally unstable compounds that decompose exothermically, excessive temperatures can cause a "runaway" reaction and sometimes lead to an explosion. Some experiences with such explosions have been reported in distillation of nitro compounds, peroxides, hydrocarbon oxides, and acetylenic compounds (10, 86, 87, 255, 309).

   In services where exothermic polymerization occurs, hot spots may form, with the polymer around them hindering cooling by the bulk liquid. This may intensify the above reactions.

   In batch distillation, as the light components are depleted from the residue, the bottom heats up, often initiating decomposition and condensation reactions that generate off-spec products and/or corrosive components. A common cure is to first recover the product from the heavies. Often, this is a quick step that utilizes the high volatility gap.

2. Many reactions are sensitive to acidity. Changing the pH can promote desirable or undesirable reactions. Tight pH control is important. Maintaining sufficient margin from the sensitive zone also helps but is not always practical. One scrubber that used water for absorbing Cl_2 and HCl from off-gas (59) required a pH of at least 3 in the bottom for good absorption. Since pH follows a logarithmic scale, increasing the pH to 3.5 as a margin for upsets required tripling the water rate, which was impractical. The higher pH could only be managed by adding alkali.

3. The presence of oxidants or reducing agents promotes undesirable reactions in the tower or in storage. When contacting alcohols, oxidants often form aldehydes which promote foaming and product contamination or corrosive acids. When contacting unsaturated compounds, oxidants can generate gums, tars, polymers, and plugging. In vacuum towers, air leakage can initiate oxidation reactions in the tower with similar consequences. In one case (116, 117), excessive air leakage into a vacuum tower caused polymerization, which in turn led to plugging and tray damage. In another case, it caused corrosion of structured packings.

4. Corrosion or polymerization inhibitors may be the oxidizers or reducers causing the undesirable reaction. If an adverse reaction is suspected, the inhibitor can be replaced by a more inert inhibitor as a plant test.

5. Catalyst from an upstream reactor may find its way into the column, converting it into a "secondary reactor." Carryover of solids (e.g., metal fines) from upstream may have an identical effect. Both can produce reactions and contaminants that end up as product impurities. In one case (116, 117), carried-over catalyst, together with excessive air leakage into a vacuum column, polymerized the bottom product and caused plugging and damage to trays. In another incident (255), iron carried over from an upstream column is believed to have catalyzed a polymerization reaction in an ethylene oxide column, in turn causing decomposition and explosion.

   If an undesirable reaction is suspected, it is worth checking column products and deposits for the presence of reactor catalyst and metal fines.

6. The materials of construction of the column or internals may influence reactions in the tower. In one case experienced by the author, the replacement of ceramic random packing by metallic packing induced the formation of a compound that contaminated the bottom product. In water peroxide distillation, certain materials of construction catalyze decomposition reactions (87). In hot potassium carbonate (hot pot) absorption–regeneration service, tests (421) showed that ceramic random packings from different suppliers had widely different reactivities toward the solution. Laboratory tests under simulated column conditions can often identify an interaction between the tower chemicals and the materials of construction in the tower.

7. Contaminants leading to product impurities may be generated in storage or during turndown operation due to long residence times. Materials like water, rust, or sludge tend to accumulate in storage and react with the feedstock or in the column, as reported in one case (112). Reactive chemicals left over in the column system from a previous campaign or from commissioning operations may also produce undesirable reactions or contaminants. Reaction products can azeotrope with components in the tower and contaminate products. In Case 15.2 in Ref. 201, reactions at the column feed entry zone generated azeotropes that concentrated in the rectifying section above.

   In one amine absorber (11), aldehyde entering the amine system caused the H_2S in the lean gas to be well above the specification. The aldehyde bonded with the H_2S, which inhibited H_2S stripping in the regenerator. An extensive survey of failures in amine absorber–regenerator systems identified solvent contamination by degradation products or glycols to be a major cause of off-spec lean gas (274). Chemical analysis is the key to diagnosing these contaminants, as demonstrated in Refs. 6 and 432.

   If the contaminant is chemically unstable or explosively reacts with the column chemicals, a detonation may result; some examples were reported (86, 309).

   Chemical analysis and bench-scale tests are key to diagnosis. If practical, it may pay to derive the tower or plant feedstock from an alternative source for a trial period.

8. Long residence times, like those associated with large recycles, can induce a buildup of reaction products or reactive components in the system. Recirculating systems like absorber–regenerators, extractive distillation, or solvent recovery systems are particularly prone to such buildups. In one solvent recovery system (88), the solvent slowly hydrolyzed, with reaction products building up to the point of interrupting separation in the solvent recovery column. The cure was periodically distilling the reaction products out of the solvent. The author is familiar with a similar experience

where acidic products of solvent hydrolysis extensively corroded a solvent recovery distillation unit.

In glycol dehydration systems, slow oxidation of the glycol into organic acids over an extended period of time caused severe corrosion (78). The reaction products should be either removed or rendered harmless, e.g., by neutralization. In one olefin plant caustic absorber (359), the buildup of reactive components when minimum flow bypasses of a compressor were open was theorized to generate additional free radical chains that intensified fouling.

Long residence times in the heated zone at turndown can also promote undesirable reactions like decomposition. Often, the column pressure can be reduced to bring the temperatures down since the condenser and tower are unloaded at turndown.

9. Reprocessing of off-spec streams or streams originating from other units in the tower can bring in contaminants. The off-spec streams typically contain contaminants such as air, water, and rust. Those from other processes can bring in a wider range of contaminants such as catalyst, oxygenates, acids, alkali, foamers, and halides. The author is familiar with a case in which a halogenated contaminant from a downstream plant hydrolyzed to form acid in an olefin plant water-wash column.

10. Excessive concentrations may lead to undesirable reactions at significant rates. An intermediate component may build up in the tower and react by itself or with other components in the tower to form undesirable, corrosive, fouling, or even explosive compounds (199). The buildup can occur in any section of the tower depending on the volatility of the component and its interaction with the main components in the tower. Most buildups occur toward the middle of the tower. When most of the product is liquid, component buildup in the overhead loop is also common. Drawing internal samples from the tower (if valves are available), performing a detailed wall temperature survey (Section 7.1), or running simulations can help detect this buildup.

Increasing reflux to the tower may increase the concentration of a valuable top product component and promote its undergoing secondary reactions accompanied by a loss of product, as has been reported in one case (471).

Several explosions caused by inadvertent concentrations of peroxides in a column or reboiler have been reported (87, 309); similar incidents may occur with other unstable compounds (e.g., 86, 184, 255). Hydrocarbon impurities ingressed in the air intake to air separation plants tend to accumulate as solids in the liquid oxygen at the reboiler of the air separation column. When built up sufficiently, they react explosively with the oxygen (86, 308). The hazard is greatest at shutdowns because the more volatile liquid oxygen vaporizes preferentially, concentrating the impurities.

11. An undesirable reaction may occur only at turndown due to excessive reagent or long residence times. In one case (192), an additive was injected into the feed to eliminate small quantities of aldehydes. At turndown, it also attacked ketones. The problem was overcome by running the column at high rates and shutting it down when product stocks built up. A possible alternative solution would have been to reduce the additive injection at low rates.

12. An unexpected compound may show up as one of the normal components in the routine laboratory analysis of a feed or product stream. Similarly, a component may be present in an unexpected molecular form. In either case, the volatility and solubility characteristics of the unexpected component may cause it to distill toward the wrong end of the column. The routine lab analysis will indicate poor column separation. In one case (401), hydrogen fluoride (HF) was present mainly as carbonyl fluoride (COF_2) in

the feed to an aqueous wash column. While HF can readily be absorbed, COF_2 cannot, giving an apparent symptom of poor absorption of HF. Routine lab analyses should be periodically checked against comprehensive chemical assays. Attention should be paid to concisely defining the components in the streams entering and leaving the column.

13. The presence of even small quantities of water often generates a reaction or affects the course of a reaction and, more importantly, azeotroping of the components.

14. In absorber–regenerator systems, the key to good absorption is to properly strip the lean solvent in the regenerator. Insufficient heat, inefficient stripping by the trays or packings, and contamination of the amine with a strong base have been the main reasons for the poor stripping (272–274).

15. Some reactions are influenced by column pressure. In one case (86), a compound that was nonreactive when heated at atmospheric pressure detonated when heated at column pressure.

16. Material precipitated out of solution or emulsion may initiate an undesirable reaction. In one solvent recovery column (86), explosive-grade nitrocellulose precipitated out of solution and detonated.

17. Unexpected reactions often occur in first-of-its-kind processes or those using a newly developed catalyst. Large-scale commercial processes are often derived from batch or semicontinuous pilot plants, where residence times, impurities, and recycles are different. When an unexpected reaction is suspected, it is often useful to compare key variables in the pilot tests with those in the commercial unit.

Chapter 4

Diagnosing Packed Tower Maldistribution

"The liquid distributor is often considered to be the most important part of a packed bed design. A good distributor cannot make a bad packing perform well, but a bad distributor can certainly make a good packing perform poorly"

—(Dhabalia and Pilling, Ref. 83)

4.1 DIAGNOSING PACKING MALDISTRIBUTION: AN OVERVIEW

Figure 4.1 shows two of the most common basic types of liquid distributors.

An orifice pan distributor (Figure 4.1a) is a pan containing evenly spaced floor perforations, which may be equipped with short tubes, through which the liquid is distributed to the packed bed below, and with round or rectangular riser tubes through which vapor from the bed below passes without contacting the liquid. Liquid is usually fed to the distributor by an elbowed-down pipe above its center or by a pipe with downward-pointing perforations above its center.

In a trough distributor (Figure 4.1b), liquid is fed to a rectangular box called a parting box. The parting box contains holes that apportion the liquid to the troughs below, such that the larger troughs receive more liquid and the smaller ones receive less. The troughs contain evenly spaced openings, either at the bottom (visible in the three troughs in the bottom left of the photograph), or in the side of the troughs near the bottom with downward-deflecting baffles or downflow tubes, through which liquid descends onto the packed bed below. Liquid is fed to the parting box by a central sparger pipe along its length with downward-pointing perforations which may be equipped with downflow tubes.

Many additional features are often incorporated to optimize the performance of either distributor type. Additional common distributor types are notched distributors (Section 4.14) and spray distributors (Section 4.15).

Troubleshooting Our malfunctions survey (198, 201) found liquid distributors to be the second most troublesome internal in distillation towers. The number of malfunctions was higher in chemical than in refinery towers, probably due to the wider spread application of packings in chemical towers. The survey identified liquid distributor issues as the prime malfunction in chemical towers.

Plugging and overflow were the two major liquid distributor issues. Good filtration and fouling-resistant distributor designs were effective cures. While overflows were often caused by

Distillation Diagnostics: An Engineer's Guidebook, First Edition. Henry Z. Kister.

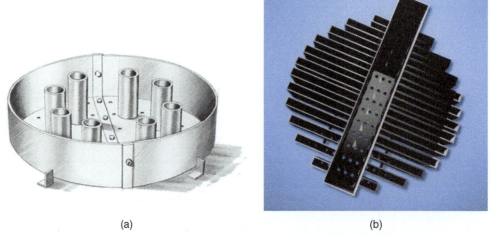

(a)   (b)

Figure 4.1 Two common types of packing liquid distributors (a) an orifice pan distributor, (b) a trough distributor with a parting box metering liquid to the troughs. (Part (a): Courtesy of Koch-Glitsch, LP; Part (b): Copyright © FRI. Reprinted Courtesy of FRI.)

plugging, many more resulted from distributor overloading caused by excessive liquid flows, insufficient orifice area, and hydraulic problems with the feed entry.

Poor irrigation quality, fabrication/installation issues, and poor feed entry configurations were other common issues (198, 201, 239). Poor irrigation quality, which the literature has focused on, accounted for only 20% of the reported liquid distributor malfunctions. The more troublesome items, such as plugging, overflows, and feed entry, received disproportionally less attention.

Feed entry issues, such as excessive velocities, splashing, and poor orientation of the feed pipes, frequently caused poor packing efficiencies. Flashing feed entry into a liquid distributor has been especially troublesome and a fertile source of malfunctions.

Inadequate hole pattern, distributor damage, infrequent redistribution, out-of-levelness, insufficient mixing, and interference with hold-downs constitute the remaining common issues (198, 201, 239). Surprisingly, distributor out-of-levelness was low on the list. Insufficient mixing was mostly troublesome in larger diameter (>10 ft ID) towers and in certain applications like extractive distillation.

Olsson's (346) classic article has been the leading reference for distributor troubleshooting. His key advice is "ASSUME NOTHING." Olsson advocates critically examining the fouling potential and possibility of vaporization in streams entering the distributor, testing distributors by running water through them at the design rates, either in the shop or in situ, and finally, good process inspection. Following Olsson's guidelines would have prevented more than 80–90% of the reported liquid distributor malfunctions in Refs. (198) and (201).

Vapor is easier to distribute than liquid, but since common vapor-distributing devices are far less sophisticated, it is not trouble-free. The prime source of vapor maldistribution has been undersized reboiler return or gas inlet nozzles, which shoot high-velocity jets into the tower (198, 201). These jets persist through low-pressure drop devices like packings. Interference with packing support beams has also been troublesome. Installing vapor distributors, improving vapor distributor designs, and even using inlet baffles have alleviated many of these problems.

This section describes key techniques for diagnosing the nature and cause of maldistribution. Gamma scans (Chapters 5 and 6) and thermal wall measurement by pyrometers or thermal cameras (Chapter 7) provide additional excellent diagnostics. Drawing to-scale sketches is powerful for diagnosing point of transition issues (Chapter 8) at feed and vapor entry and redistribution. For maximum effectiveness, the troubleshooter is advised to combine these techniques.

4.2   EXPECTED PACKING HETPS

A rule of thumb for packing HETP (in inches), as a function of a_p, the packing surface area per unit volume (m^2/m^3), gives:

For organic and hydrocarbon systems, surface tension <25 dynes/cm:

HETP = $93/a_p$ Metal random packings, $1''$ or larger
HETP = $100/a_p + 0.1$ Y-type or third-generation structured packings
HETP = $145/a_p + 0.1$ X-type structured packings, $a_p < 300\,m^2/m^3$

Multiply the above values by two for water due to its higher surface tension (70 dynes/cm). For intermediate surface tensions, interpolate.

These rules of thumb are generally plus or minus 25% and assume good industrial-grade liquid and vapor distribution (not perfect distribution like in test columns).

Measured HETP values that are significantly higher indicate possible or likely maldistribution.

Recognizing maldistribution   The key tool for recognizing maldistribution is the above rules of thumb. Derive the HETP from a simulation that correctly reflects the plant data (follow all the recommendations in Chapter 3), and compare it to the above rules. An HETP >20–30% than the rules of thumb suggests possible maldistribution. An HETP twice that calculated from the rules of thumb signifies a highly likely maldistribution. Closely review the VLE and thermal property data or correlation used in the simulation. Errors in the VLE and thermal properties produce major errors in the HETP determination (Section 3.1.6, items 1 and 2). The author experienced cases where maldistribution was incorrectly identified or not identified due to erroneous VLE or thermal properties in the simulation.

Symptoms of flooding in the bed (Chapter 2) also suggest possible maldistribution. Sometimes flooding causes maldistribution; other times, maldistribution leads to premature flooding. Gamma scans (Chapters 5 and 6) and wall temperature measurements (Sections 7.1.3, 7.2.2, 7.2.3) are especially valuable for detecting flooding and plugging.

Temperature inversions (upper temperatures being hotter than lower temperatures) may also suggest maldistribution. One case study was described (480) and the author experienced others. However, temperature inversions can also suggest other issues such as the presence of a second liquid phase (Section 7.1.5) or a chemical reaction.

4.3   SMALL-SCALE VERSUS LARGE-SCALE MALDISTRIBUTION: DO THEY EQUALLY RAISE HETP?

This topic is discussed at length with extensive literature citation in Kister's book (193) and Perry's Handbook (245). Below is a summary focusing on diagnostics rather than design:

1. Three factors define the effect of maldistribution on efficiency:
   a) *Pinching*. Zonal variations in L/V ratio cause zonal composition pinches. In the rectifying section, the liquid-deficient zones have low L/V ratios, and their composition

approaches a pinch. Upon reaching a pinch, composition change stalls. The converse occurs in stripping zones; here local pinches occur in high L/V zones, with little composition change above.

b) *Lateral mixing.* Liquid and vapor are deflected laterally by packing particles or corrugated layers, promoting mixing of vapor and liquid and counteracting the pinching.

c) *Liquid nonuniformity.* Liquid exits commercial distributors with some nonuniformity. Liquid flows through the packing tend to concentrate at the wall.

2. At small ratios of tower-to-packing diameter ($D_T/D_p < 10$), the lateral mixing offsets the pinch, and a greater degree of maldistribution can be tolerated without a serious efficiency loss. As the D_T/D_p ratio increases, the lateral mixing becomes progressively less effective to offset the pinch, so the loss of efficiency to maldistribution escalates. In large-diameter columns and small-diameter packings, when $D_T/D_p > 100$, lateral mixing becomes ineffective to counter the pinch.

3. Wall flow effects become large only when $D_T/D_p < 10$.

4. Maldistribution tends to be a greater problem at low liquid flow rates than at high liquid flow rates (521). The tendency to pinch and spread unevenly is generally higher at the lower liquid flow rates.

5. Packing HETP has reasonable tolerance for a small-scale maldistribution, i.e., local maldistribution over a small area, and for a random variation. HETP is intolerant of large-scale maldistribution, i.e., large regions deficient of liquid, overirrigated, or zonal variations (267–269, 521). Tests in a commercial-scale column (268, 269) show no significant efficiency loss due to random, small-scale unevenness in the liquid load. The local pinching of small-scale maldistribution is evened out by lateral mixing, and therefore causes few ill effects. In contrast, the lateral mixing is powerless to rectify large-scale maldistribution, and efficiency is lost.

4.4 BY HOW MUCH DOES MALDISTRIBUTION REDUCE PACKING EFFICIENCY?

Figure 4.2 shows the Moore and Rukovena (326) empirical correlation for efficiency loss due to liquid maldistribution in packed towers containing Pall® rings or IMTP® packing. This correlation is simple to use and was shown to work well for several case studies (Figure 4.2), making it valuable, at least as a preliminary guide.

The quality of liquid irrigation is expressed by the Moore and Rukovena distribution quality rating index (Section 4.10). Typical indexes are:

- 10–40% for poor distributors,
- 40–70% for most standard commercial distributors,
- 75–90% for intermediate-quality distributors,
- 90% for high-performance distributors.

Locckett and Billingham (36, 301) modeled maldistribution as two columns in parallel (Figure 4.3a), one receiving more liquid $(1+f)L$ and the other less $(1-f)L$, with the vapor assumed to be equally split between the columns. f is the maldistribution fraction, and L is the liquid flow rate in molar units. At the different L/V ratios, the overall separation is inferior to that at uniform L/V. As the maldistribution fraction f increases, the effective number of stages

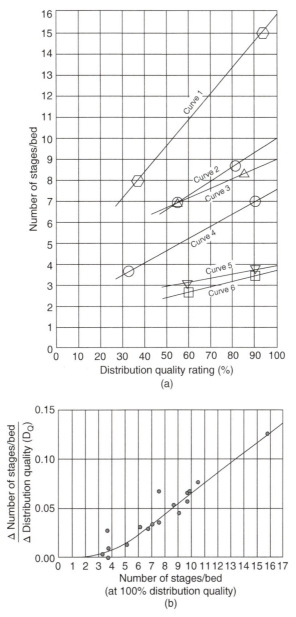

Figure 4.2 Effect of irrigation quality on packing efficiency. (a) Case histories demonstrating efficiency enhancement with higher distribution quality rating. (b) Correlation of the effect of irrigation quality on packing efficiency. (From Moore, F. and F. Rukovena, Paper presented at the 36th Canadian Chemical Engineering Conference, October 5–8, 1986. Reprinted Courtesy of F. Rukovena, Consultant.)

in the combined two-column system declines (Figure 4.3b). The decline is minimal when the maldistribution fraction f is small or in short beds (e.g., 10 theoretical stages), as noted earlier by Strigle's extensive experience (462).

Figure 4.3b shows that beyond a limiting maldistribution fraction f_{max}, the required separation cannot be achieved even if the bed is tall enough to provide many stages. Physically, at f_{max}

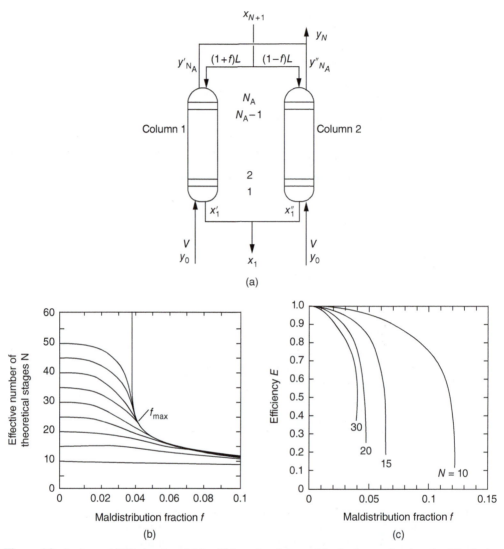

Figure 4.3 Lockett and Billingham's model for efficiency loss due to maldistribution as a function of maldistribution fraction f and the number of stages per bed for a given case study (a) the parallel-column model; (b) $f_{\max}$ and reduction in number of stages. (c) Showing larger efficiency loss at higher number of stages per bed and higher f. The steep drops indicate pinching associated with $f_{\max}$. (From Billingham, J. F., and M. J. Lockett, Transactions of the Institution of Chemical Engineers, 80, Part A, p. 373, May 2002. Reprinted Courtesy of IChemE.)

the liquid-deficient column becomes pinched. Figure 4.3c shows that the steep drop in packing efficiency due to this pinch occurs earlier when the bed contains many stages and that the efficiency loss can be tremendous. For $f_{\max}$ in a binary system, Billingham and Lockett derived the following equation:

$$f_{\max} = \frac{y_{N+1} - y_N}{y_N - y_o} + \frac{x_1 - x_o}{x_{N+1} - x_o} - \left(\frac{y_{N+1} - y_N}{y_N - y_o}\right)\left(\frac{x_1 - x_o}{x_{N+1} - x_o}\right) \tag{4.1}$$

In Eq. (4.1), x and y are mole fractions of the more volatile component in the liquid and vapor leaving a stage, respectively, and the subscripts 0, 1, N, and $N+1$ signify stage numbers, with

0 being the lowest stage and $N + 1$ the top stage. This equation permits the calculation of f_{max} directly without the need for a parallel column model. The terms in Eq. (4.1) can be readily calculated from the output of a steady-state commercial simulation. This method can be extended to multicomponent systems either by lumping components together to form a binary mixture of pseudolight and pseudoheavy components or by normalizing the mole fractions of the two key components. Once f_{max} is calculated, Billingham and Lockett propose the following guidelines:

- f_{max} <0.05, extremely sensitive to maldistribution. The required separation will probably not be achieved.
- 0.05 <f_{max} <0.10, sensitive to maldistribution, but separation can probably be achieved.
- 0.10 <f_{max} <0.20, not particularly sensitive to maldistribution.
- f_{max} >0.20, insensitive to maldistribution.

Figure 4.3c shows that shortening the bed can increase f_{max}. Relative volatility and L/V ratio also affect f_{max}. The bed length and L/V ratio can often be adjusted to render the bed less sensitive to maldistribution.

Analysis of field data using similar principles was discussed by Krishnamoorthy and Yang (264).

Effect of bed length   Both Figures 4.2 and 4.3 show that as a packed bed grows taller, it becomes more sensitive to maldistribution. This is supported by Fractionation Research Institute (FRI) data presented by Shariat and Kunesh (428). In several case studies (178, 201, 285, 301), splitting a bed into shorter beds with redistribution in between largely improved separation. Rules of thumb that recommend limiting packed bed height range from the more conservative that recommend limiting bed heights to <20–25 feet (68, 192, 285, 459), and <10 stages per bed (173, 192) to the less conservative that increase the limits to <30 feet (462) and <15 (53, 462) or <15–20 (262, 458, 459) stages per bed.

Effect of irrigation point density   A minimum of four irrigation points per square foot has been recommended, with 6–10 per square foot being ideal (40, 192, 338, 362, 450, 462). For high purity separations and clean services, a higher number of 10–14 was recommended (40). Commercial-scale tests with both random and structured packings of specific surface areas of $250 \, m^2/m^3$ or less showed no improvement in packing efficiency by increasing the irrigation point density above 4/sq ft (103, 349). So going to higher point densities provides little efficiency benefit while leading to smaller holes, which makes the distributor more prone to plugging. With high surface area structured packings, a larger irrigation point density was shown to help packing efficiency (349). It has been recommended to use irrigation point densities per square ft >12 for $350 \, m^2/m^3$ packings; >16 for $500 \, m^2/m^3$ packing; and >20 for higher areas (262).

4.5   DIAGNOSING PACKING AND DISTRIBUTOR PLUGGING

The following steps have been recommended:

1. Consider all possible sources of solids and identify those likely to be active in your tower (346, 347). Common sources include solids in the feed, catalyst fines, rust in the tower or piping, polymerization, precipitation, crystallization, reaction, tars, antifoam breakdown, gasket disintegration, dirt, debris, poor turnaround cleaning, and commissioning operations. Some solids are living (slime, bacteria). Fouling mechanisms may influence each

other; for example, catalyst fines can induce polymerization. Check if you can make a call whether the solids are formed internally in the tower or externally.

2. Check the potential for solids in the feed. Do not just ask Operations. If possible, sample the feed and closely examine it. Murkiness or coloration may be due to suspended solids. Allow the sample to settle for a few days and observe if it becomes clear. Alternatively, pass the sample through fine filters and see if it clears.

Figure 4.4 shows samples from the feed stream to a chemical column in which trays were replaced by packings (201, 346, 347). Prior to the retrofit, Operations reported no solids in the feed. Samples were drawn and showed coloration (similar to samples 1, 2, and 3 in Figure 4.4) but no solid deposits. After startup, the feed distributor plugged up, causing massive maldistribution, poor packing efficiency, and a capacity bottleneck.

Feed samples 1, 2, and 3 show colorations, but there is no telling whether these indicate suspended solids. The clear sample on the left, drawn after a fine filter was installed on the feed stream, clearly shows that the colorations were due to suspended solids. Deposits were also observed when the color samples were allowed to settle over a week or two. The filter eliminated the plugging problem.

When Operations say they have no problem with solids in the tower, their concise statement should be interpreted as "we have not seen solids in the feed, nor have we experienced fouling issues with our existing (at the time, trayed) internals." Packing distributors and many packing types are far more prone to solids than trays, so a solids issue hidden in a tray tower may become apparent following a packing retrofit.

3. Good filtration should use two full-sized filters in parallel, operating one on one off, without a bypass. The filter openings should be at least 10–20 times smaller than the smallest hole in the distributor. The author is familiar with cases of distributor plugging when the filter openings were only four times smaller than the distributor openings, as in Case 6.4A in Ref. (201). Distributors usually have dead spots where solids can agglomerate, later spalling off and plugging the distributor holes.

Figure 4.4   Samples of a tower feed stream (1, 2, and 3), showing coloration. It is impossible to tell whether the coloration is due to suspended solids or dissolved material. The clear sample on the left, taken after a feed filter was installed, shows that in this case the coloration was due to suspended solids. (From Olsson, F. R., in "Distributors that Worked and those that Did Not," Paper presented at the AIChE Spring National Meeting, Houston, Texas, March 14–18, 1999. Reprinted Courtesy F. R. Olsson.)

Have Mechanical check that the filter screens and baskets are adequately mounted. Liquids can easily find a path around an improperly mounted basket. Check the filter pressure drop and log it over a time period. Loss of pressure drop across a filter may signify a damaged or removed basket.

Check the frequency of filter cleanings. Expect a high frequency during startup and initial run, as the feed moves along residual solids lodged in the line. Closely watch the action of the Operations team. When cleaning is too frequent, especially during startup, baskets are often completely removed, resulting in solids ending up in the distributors or packings (e.g., 216, Case 1). Line up additional personnel to help during these busy periods.

Refer Section 10.3.11 for additional recommended filter checks.

4. It is essential to thoroughly flush all lines and auxiliary equipment from high to low points before connecting to the column to avoid introduction of construction debris, dirt, corrosion products, and other plugging materials. Verify that this has been done. Cases 6.4C, D, and E in Ref. (201), as well as Ref. (282), describe incidents in which construction debris and rust caused distributor plugging.

5. Avoid water entry into carbon steel towers processing water-insoluble components. Water circulating in such towers can become acidic and generate corrosion products that plug distributors.

6. Distributors with small holes often have the holes at the wall of the distributor troughs or in downflow tubes, typically about 1″ or 2″ above the floor. A downflow tube (Figure 4.5) or a baffle directs the liquid vertically downward into the packings. Elevating the holes resists plugging by sediment settling on the floor. In fouling services, distributor run length with wall perforations is typically double that of floor perforations. Perforations of 0.5″ and larger are fouling-resistant, and are usually located in the floor.

With reactive components at high temperatures, elevating holes needs a careful review. Below the perforations, the liquid residence time is high, and when "cooked" by hot vapor, it can undergo undesirable reactions, polymerize, or turn into tars or sludge.

7. Do not underestimate the fouling potential of a service. "A cup of scale can stop up a distributor" (25). Plugging has taken place in seemingly clean systems. In one pan distributor with tubes containing two rows of holes (Figure 4.5), the holes were plugged, the distributor overflowed as evidenced by water marks, and column performance was poor.

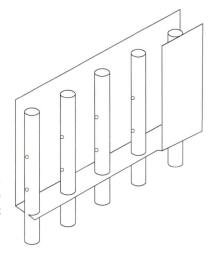

Figure 4.5 Wall perforations in a trough distributor. The distributor liquid issues via two perforations, the lower one a short distance (about 1″ or 2″) above the floor, the upper one higher up. The exiting liquid is directed downwards into the packings by the downflow tubes. (Reprinted Courtesy of Sulzer Chemtech.)

The amount of solids that plugged the holes was minute, would not fill a cup of coffee, and was mainly rust particles. In another case (413), less than 1 lb of solids plugged 80% of the perforations.

8. Plugging-resistant features can be incorporated, often at the price of inferior distribution under clean conditions or at higher costs. Increasing hole diameters greatly improves plugging resistance, usually at the expense of fewer distribution points and lower liquid heads. Experience showed (218, 320) that a lower-quality distributor that works beats a high-irrigation quality distributor that plugs. "Tea strainers" hole-opening guards have successfully prevented plugging in at least some cases. Two-stage distributors, some with rectangular notches instead of holes, are also effective for improving plugging resistance while maintaining irrigation quality (69, 168, 188), but at a higher cost. Notched-trough distributors (Section 4.14) have been successful in fouling services (387, 523), but often provide a lower irrigation quality (192, 245, 267).

9. "Newer is not always better" (282). New state-of-the art distributors can be far more prone to plugging than their low-quality predecessors. In one case, a lower-quality ladder pipe distributor replacement by a "newer and better" trough type introduced severe plugging that was later eliminated by returning to the proven ladder pipe distributor.

10. Corrosion products, scale and salts formed at the top head of a tower can spall off into the distributor troughs below as in Case 6.4B in Ref. (201). Well-designed trough covers or screens can eliminate this.

11. In many services, polymerization inhibitor is injected into the reflux. Outer distribution points are usually about 1–1.5″ away from the tower wall. The open lattice structure of modern random packings does not spread the liquid well, so it may take a foot or two for the liquid to reach the wall. This region remains dry, receives no inhibitor, and may polymerize. First- and second-generation packings are much better liquid spreaders, shortening the dry length at the wall. In one retrofit, changing second- by fourth-generation random packings caused polymer plugging not previously experienced. See Figure 4.16 and Section 4.10 on identifying underirrigated zones near the wall.

12. Beware the possibility of commissioning operations introducing solids that plug distributors. Examples include flushing feed or reflux lines into the tower, washing the tower with solid-containing water, opening the tower to the atmosphere after a water wash without adequate drying (rust). Consider temperature differences between operation and commissioning; in one case (3), residual solids were not removed during a cold flush, but were dislodged later when the system was heated.

13. Consider the possibility of component freezing and plugging distributors when cold reflux or feed enter the column, either during normal or during commissioning operations. O'Brien and Porter (343) provided valuable guidelines.

14. Consider injecting a solvent, either continuously or batchwise, to dissolve the solids. Typical examples include the on-line caustic wash of an aldehyde scrubber in Case 12.3 of Ref. (201), or the pyrolysis gasoline washes to dissolve polymer in olefins plant caustic towers, as described in Ref. (259).

15. Consider the possibility of changing operation conditions to dissolve solids. Examples are raising boilup and cutting reflux in a methanol–water tower, thereby inducing water to the top section to dissolve solids that built up there (285), or using water washes to dissolve salts from the top of refinery towers, described by many case studies in Ref. (201).

16. In vacuum towers processing chemicals that easily polymerize or tar, perform extensive leakage tests before startup to minimize air leakage.

4.6 TROUBLESHOOTING FOR DISTRIBUTOR, COLLECTOR, AND PARTING BOX OVERFLOWS

Overflows are the most common distribution issues, often referred to as "Death to packed beds."

4.6.1 Overflows and Their Impact

An overflow occurs when the liquid in the distributor rises above the top of the distributor troughs or risers and pours out. The overflowing liquid will be maldistributed to the bed below. This is if one is lucky.

If one is not lucky, the vapor flow between the troughs or in the risers will be too high to permit liquid descent. Typically, the vapor rise area of a pan distributor is of the order of 25% of the tower cross-section area, and that of a trough distributor is about 50% of the tower cross-section area. The vapor velocity, and therefore the C-factor (Eq. 2.1), in the vapor risers of a pan distributor, or between the troughs of a trough distributor, will therefore be four or two times, respectively, higher than that in the bed below. In the absence of overflow, the higher vapor velocities will be harmless, and operation will be normal. If an overflow occurs and the system limit C-factor is exceeded in the risers or between the troughs, the overflowing liquid will be blown upward by the vapor and accumulate above the risers or troughs, initiating flood.

The accumulating liquid may be blown into the bed above, propagating the flood up the tower, or when enough head builds up, the liquid may dump through some of the vapor passages, initiating alternating steps of buildup and dump, which destabilize the tower. Section 8.11 describes case studies of system limit floods initiated by a similar mechanism. Many of these were due to overflows, but others were due to high C-factors between the hats of redistributors or collectors.

Increased reflux and boilup test If there is no overflow, the separation should improve. An overflow-initiated flood would usually cause separation to deteriorate. In Case 7.1 in Ref. (201), and in many other cases experienced by the author, up to a certain reflux flow rate, separation improved. When the reflux flow rate was raised beyond that, the distributor or collector overflowed, flooding initiated, and the separation deteriorated. In one caustic absorber (77), raising liquid circulation worked well up to a limit, beyond which overflow occurred, initiating flood which caused caustic carryover from the top of the tower. This was identified by an in situ water test during an outage.

4.6.2 Overflows in Distributors, Redistributors, and Parting Boxes

In distributors and redistributors Overflow-induced flooding (Section 4.6.1) that propagates up the tower is most severe with pan distributors due to their small riser areas. Trough distributors have much larger vapor passage area, which often allows channeling of vapor through some of the vapor passages with the overflowing liquid finding its way down through the others. Sheet or rivulet formation by the overflowing liquid makes it easier to descend. This drains the accumulating liquid and alleviates the flood, at the expense of severe maldistribution and efficiency loss in the bed below. The overflow process may unpredictably alternate between flooding, flooding and dumping, and the channeling modes, generating tower instability.

The troughs in some distributor designs contain slots near the top that direct overflowing liquid into guide tubes, that in turn take this liquid to the bed, out of the path of the vapor. These also minimize maldistribution. The effectiveness of these designs varies with the amount overflowing and the guide tube design.

Figure 4.6a is a photograph taken during a turnaround in situ water test of a redistributor on the verge of overflowing. At that liquid flow rate, the tower worked well, but when raised

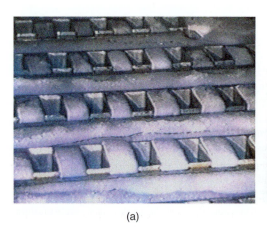 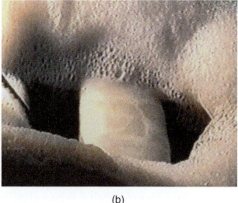

(a)   (b)

Figure 4.6   Packing distributor overflow as photographed during in-situ water test. (a) froth/liquid reaching top of distributor vapor risers; (b) liquid pouring into vapor risers. During operation, vapor flowed through risers and entrained overflowing liquid.

beyond this point, an overflow occurred. Figure 4.6b is a close-up of two risers experiencing overflow in the water test. During normal operation, vapor flowing through the risers prevented liquid descent. Beyond the overflow rate, flooding initiated, propagating through the bed above and causing liquid carryover into the tower overhead product stream.

Diagnosing overflows   The techniques below, as well as gamma scans (Chapters 5 and 6), and to a lesser extent, thermal scans (Chapter 7), have been effective for diagnosing overflows.

An invaluable tool for diagnosing overflows is the head–flow relationship for packing orifice distributors, given by the Torricelli Orifice Equation:

$$Q = 5.67 K_{\mathrm{D}} n_{\mathrm{D}} d_h^2 h^{0.5} \tag{4.2}$$

where Q is the liquid flow rate, gpm; K_{D} is the orifice discharge coefficient, with a recommended value of 0.707 (68, 161, 192, 245); n_{D} is the number of holes; d_{h} is the hole diameter, inches; and h is the net liquid head, inches. The actual liquid height in the distributor, H_{actual}, can be obtained by applying buoyancy and pressure drop corrections to the net liquid head, both of which will tend to raise the liquid height in the distributor:

$$H_{\mathrm{actual}} = h \left[\rho_{\mathrm{L}}/(\rho_{\mathrm{L}} - \rho_{\mathrm{V}})\right]^{0.5} + \Delta P_{\mathrm{vapor\,rise}} \tag{4.3}$$

where ρ_{L} and ρ_{V} are liquid and vapor densities, lb/ft^3, and $\Delta P_{\mathrm{vapor\,rise}}$ is the vapor pressure drop through the distributor risers (pan distributors) or the vapor rise areas (trough distributors), inches of liquid. In most pan distributors, $\Delta P_{\mathrm{vapor\,rise}}$ is of the order of 0.25 in. of liquid (161, 192), and is less in trough distributors. For vacuum and atmospheric systems, the buoyancy correction is negligible, but it becomes significant at higher pressures (>50–100 psia). The vapor rise pressure drop can be estimated from Eq. (4.4)

$$\Delta P_{\mathrm{vapor\,rise}} = 0.187 K_{\mathrm{R}} \left(\rho_{\mathrm{V}}/\rho_{\mathrm{L}}\right) V_{\mathrm{R}}^2 \tag{4.4}$$

where V_{R} is the vapor velocity through the risers or vapor passages, ft/s. K_{R} is the hydraulic constant representing the number of velocity heads lost through the risers. As a conservative estimate, $K_{\mathrm{R}} = 2.5$, based on the head loss through an orifice. In reality, the riser is a short pipe with entrance and exit losses (hydraulic K's of 0.5 and 1, respectively), giving $K_{\mathrm{R}} = 1.5$. The correct value is probably somewhere between, depending on the riser length, with $K_{\mathrm{R}} = 1.5$ more appropriate for long risers (>18″).

Applying Eq. (4.2) for diagnosing overflows Consider a distributor with wall perforations about 1 in. above the floor. For turndown, a minimum net head of about 2 in. above the perforations is required. Turning up the flow to twice the minimum will raise the net head to $2 \times 2^2 = 8$ in. per Eq. (4.2). Since the perforations are already 1 in. above the floor, the liquid height will be 9 in. above the floor, assuming negligible buoyancy and pressure drop corrections in Eq. (4.3). If one wants to turn up the flow to quadruple of the minimum, the required liquid head above the perforations by Eq. (4.2) will need to be $2 \times 4^2 = 32$ in., which is 33 in. ($=32 + 1$) above the floor. To safeguard from overflow, the troughs need to be about 40 in. high. The author had never seen a commercial distributor with even close to 40 in. tall troughs. Typical trough heights in the industry are 8–12 in., mostly closer to 8 in.. Of these, the bottom inch is the elevation of the perforations above the floor, and the top inch is often slotted to permit orderly overflow should it occur. With 8 in. tall troughs, a liquid height of 9 in. above the floor will be 2 in. above the bottom of the overflow slots and will overflow into the vapor passages. With 12″ tall troughs, the liquid level will be 2 in. below the bottom of the overflow slots.

This example illustrates why overflow is such a severe problem. Per the orifice Eq. (4.2), the liquid head rises parabolically with the liquid flow rate and very easily reaches the overflow. Taller troughs or risers will make the distributor much more robust for overflow prevention.

Do not overlook the pressure drop term. In two small-diameter towers (1.25–1.5 ft), the vapor riser areas in pan distributors were about 10% of the tower cross-section areas, and the number of holes was excessive. The liquid head was nearly equal to the vapor pressure drop through the distributors. This is believed to have induced vapor to break through the distributor liquid, causing frothing, turbulence, and the observed poor efficiency (433).

In parting boxes Overflows occur not only in the distributors but also in their parting boxes, which are (Figure 4.1) the boxes that split the incoming liquid to the various toughs of a distributor. Figure 4.7 shows a photograph of a liquid overflow in a water test of a model parting box. The consequences are much the same as those of a distributor overflow.

Figure 4.7 A Liquid overflow in a water test of a model parting box.

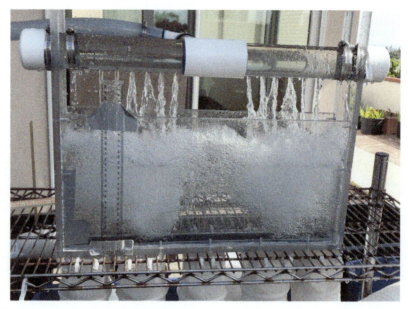

Figure 4.8   A water test of a sparger-fed model parting box shows significant aeration and frothing at the parting box, which raises froth heights to above the liquid head predicted by the orifice equation. Also note the clear liquid zone right under the tee. (From Kister, H. Z., "Practical Distillation Technology," Course Manual. Copyright © 2020. Reprinted with permission.)

In parting boxes, there is often considerable aeration, which drives up the froth level, resulting in an early overflow (247, 248). Water tests with the sparger-fed model parting box, as depicted in Figure 4.8, showed froth heights typically 20–30% taller than the liquid heads calculated using the orifice Eq. (4.2). Using dip tubes that bring liquid from the feed pipe to near the parting box floor eliminates the frothing, but generates a horizontal momentum effect that also causes the liquid head in the parting box to exceed the liquid head calculated using the orifice equation by 20–30% (249). Dip tube designs that break the momentum can reduce the parting box liquid heads to the values calculated using the orifice equation (249).

4.6.3   Overflows in Collectors

Packed tower collectors are common overflow producers. From the collector, liquid descends to a redistributor either by downcomers, downpipes (most common), or by perforations. If the descent is by perforations, the orifice Eq. (4.2) can be used to check for overflow, with any possible aeration taken into account. If the descent is by downpipes or downcomers, the residence time of liquid on the collector between the overflow level and the top of the downpipe should be calculated. If this time is less than 30–60 seconds, the liquid is likely to be aerated and the downpipes need to be sized for self-venting flow, i.e., a flow pattern in which liquid descends while any trapped vapor bubbles disengage back up.

Figure 4.9 shows an excellent correlation for self-venting flow, initially established by Simpson (439) and later by Sewell (425). Above the diagonal, the flow is self-venting. Below the diagonal, the downpipes are likely to choke, causing the liquid to back up the top of the risers, generating an overflow. The overflow liquid is maldistributed and may generate flooding just like other overflows. The correlation is often expressed (439) as maintaining the liquid Froude Number in the downpipes less than 0.31.

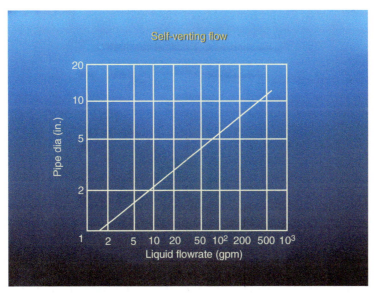

Figure 4.9   A recommended correlation for self-venting flow. (From Sewell, A., The Chemical Engineer, No. 299/300, p. 442, 1975. Reprinted with permission.)

The cases described in Refs. (416) and (417) are classic examples of painful collector overflow experiences. To improve the yield of the valuable jet fuel, trays in the jet fuel–diesel separation section (as well as other sections) of an atmospheric crude tower were replaced by structured packings. Upon restart, the separation worsened instead of improving, and the jet fuel yield was reduced by half. A superb troubleshooting investigation diagnosed the root cause to be collector overflow generating maldistribution in the jet fuel–diesel separation packed bed below. The liquid residence time in the collector was low, and the downpipes transporting its liquid to the distributor below were undersized for self-venting flow. Level measurements and gamma scans showed liquid level reaching the overflow, as well as pulsations indicating frequent intermittent dumps.

Two years later, the same refinery performed a similar revamp in a parallel (second) crude tower of similar design. Incorporating the lesson learned, a chimney tray collector with downpipes sized for self-venting flow was specified to the supplier. Upon restart, jet fuel–diesel separation was not better, with low jet fuel yield again. Tests, simulations, gamma scans, and computer-aided tomography (CAT) scan pointed again to a collector overflow. This time, the overflow was caused by excessive hydraulic gradients on the chimney tray, incurred by chimneys perpendicular to the liquid flow, similar to the problem described in Section 8.10 and Figure 8.11a. Additional obstruction to liquid flow was due to triangular chimney reinforcements that severely hindered flow.

Learning from these experiences, when the first unit was turned around, the collector was replaced with a chimney tray containing smaller chimneys that occupied less of the cross-section area and allowed liquid movement. The downpipes were replaced by down boxes with ample area for self-venting flow. This time, excellent separation and the desired yield were achieved.

The second unit was less fortunate. After its revamp, it performed even worse than before (417). An in situ water test during an outage showed skewed liquid distribution. Visual observations and CFD showed that the long chimneys on the collector and their positions too close to downpipes and down boxes created "shadow" regions (Figure 8.22, Section 8.18, Case 2). The long chimneys obstructed liquid flow to these regions, so their down boxes received much less liquid than other unobstructed down boxes. Success was achieved by replacing with a new chimney tray in which liquid movement to the down boxes was unobstructed by chimneys.

Reference (416) states that overflow from collectors that had low residence times and whose downpipes were undersized for self-venting flow occurred in two other refineries by the same company. All these towers were made to work by good chimney tray design.

Case 7.1 in Ref. (201), as well as Refs. (129, 275), and (482) describe case studies of very poor packed tower performance resulting from collector overflows. The author is familiar with many more.

A summary of the lessons learned for collectors is expressed in Ref. (417) "**If it starts the wrong way, it will keep that way.**" If the collector delivers an uneven feed to the distributor, distribution to the bed is likely to remain poor.

4.6.4   Overflows due to Plugging, Foaming, Impingement

Overflows due to plugging   As stated earlier, typical liquid levels in commercial distributors at the design rates are not far from the overflow. Consider a trough distributor with 12″ tall troughs (which is taller than typical), holding 8 in. of liquid above wall perforations 1 in. above the floor. Should 15% of the holes plug, n_D in Eq. (4.2) will become 0.85 times its previous value, raising the liquid head from 8 in. to $(8/0.85^2=)$ 11 in., i.e., to the overflow (assuming no slots at the top). Even if the distributor operation was at turndown, and the liquid head were only 4 in. above the perforations, it would take only 40% plugging to get to the overflow $(4/0.6^2 = 11\text{ in.})$. So the liquid head again rises parabolically with the number of plugged holes.

The result is that due to the parabolic relationship in the orifice equation, it does not take much plugging to generate an overflow. This also teaches that any significant plugging in a gravity distributor is likely to generate an overflow. For these reasons, overflow troubleshooting must be a priority on packed tower troubleshooting.

Overflows due to impingement   A vapor stream impinging on the liquid surface in a collector, parting box, distributor, or redistributor will generate waves and unevenness on the liquid level, with peaks that exceed the trough walls and bring about premature overflows. Carefully check using sketches to scale (Chapter 8) for vapor-containing feeds, or vapor issuing from chimneys and deflected downward by their hats, blowing on the liquid surface.

Overflows in foaming systems   Foam formation is likely to lead to liquid presence as a foam at taller elevations than those calculated from Eq. (4.2), and to premature overflows. Falling liquid promotes foaming, and liquid fall height should be minimized. Free fall of more than 6 in. should be avoided (40). Section 2.10 and Figure 2.27 describe relevant cases.

4.7   TROUBLESHOOTING MALDISTRIBUTION AT TURNDOWN

Liquid maldistribution tends to lower packing turndown (192, 268, 462)

The orifice Eq. (4.2) is also powerful for addressing distributor turndown. In the example in the previous section, the liquid head at maximum flow was 8 in. above the perforations. At half that flow (a turndown of 2:1), the head declined to $(8/2^2=)$ 2 in. Often, 2 in. is considered the lowest head needed to ensure a reasonable distribution. If instead of a turndown of 2:1 it is desired to go to a 3:1, the liquid head per Eq. (4.2) will be $(8/3^2=)$ 0.9 in. At such a low head, the liquid is likely to be maldistributed. Also, at this low head the distribution is likely to be sensitive to even small degrees of out-of-levelness, waves, turbulence, and uneven frothing. The turndown on most orifice distributors is about 2:1, which is limited by the orifice equation.

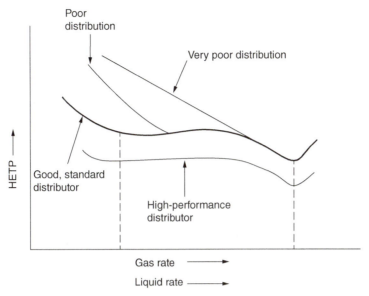

Figure 4.10   Effect of maldistribution on packing efficiency at turndown. (From Kister, H. Z., "Distillation Operation," Copyright © McGraw-Hill, 1990. Used with permission of McGraw-Hill.)

There are certain features that packing suppliers often incorporate to enhance turndown. A common feature is using two perforations, one above the other, as shown in Figure 4.5. Often the perforations are of different sizes, the lower smaller ones active at low liquid rates, and both active at high liquid rates. This is not entirely trouble free. It may experience a "bump" when the liquid level just rises past the upper perforations, especially when these upper perforations are larger (which is often the case), as the liquid head building above the upper perforations will be too small to evenly distribute the liquid issuing from them. In a recent case, we identified this issue as one of the causes of poor packing efficiency experienced in a tower. Also, with two perforations along the wall, the lower ones usually need to be smaller, and therefore more prone to plugging, than with a single perforation.

Figure 4.10 illustrates that packing turndown is drastically reduced by maldistribution. The "standard distributor" curve depicts typical variation of HETP with vapor or liquid load at a constant L/V ratio. The two upper curves depict a progressively lower quality of initial liquid distribution. The unevenness produced by maldistribution is amplified when the liquid flow rates and the liquid heads in the distributors are lower. A curve like the upper curve in Figure 4.10 is a clear indication of poor distribution.

4.8   TROUBLESHOOTING DISTRIBUTOR OUT-OF-LEVELNESS

Out-of-levelness in distributors has been a common issue causing poor efficiency in packed towers, as demonstrated in numerous case studies (137, 201, 450, 483, 508).

The orifice Eq. (4.2) is also powerful for addressing issues related to distributor out-of-levelness. In the example in the previous section, the liquid head at a turndown of 3:1 was about 1 in. Consider a tilt of 0.5 in., such that at the liquid head is 1 inch at the shallow end and 1.5 in. at the deep end. If the liquid flow rate at the shallow end (1 in. head) is 100 gpm, then at the deep end (1.5 inch head) the flow per Eq. (4.2) will be $(100 \times 1.5^{0.5}=)$ 122 gpm, a

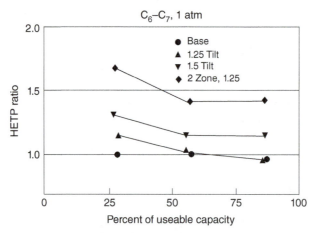

Figure 4.11   Effect of simulated tilt on packing efficiency. (Reprinted with permission from Kunesh, J. G., L. Lahm, and T. Yanagi, Industrial and Engineering Chemistry Research, 26(9), p. 1845, 1987. Copyright (1987) American Chemical Society.)

difference of 22%. Since most towers are designed to operate at 20–30% above minimum reflux (193), a 22% flow variation can easily get the bed section beneath the shallow end into a pinch, approaching or even falling below the minimum reflux. At this low liquid head, even a small amount of out-of-levelness will be disastrous to packing efficiency.

Furthermore, at such low liquid heads, part of the distributor where the liquid heads are low may approach drying up, making the problem even greater.

Consider the same tilt when the liquid head in the distributor is at 8 in. The shallow end will have a liquid level 8 in. above the perforations, and the deep end will have a level of 8.5 in. If the liquid flow rate at the shallow end (8″ head) is 100 gpm, then at the deep end (8.5 in. head), the flow rate per Eq. (4.2) would be $(100 \times [8.5/8]^{0.5} = 103$ gpm, a difference of only 3%. So at high liquid heads in the distributor, out-of-levelness is a much lesser issue than at low liquid heads.

The author was involved with tilt testing on a specialty distributor. The tests found that a 0.5 in. tilt made little difference to packing HETP. This raised a few eyebrows until it was appreciated that the test was performed with a 10″ liquid head in the distributor. At such tall heads — and only at such tall heads — the distributor becomes tolerant to out-of-levelness.

Figure 4.11 shows FRI test data (268) for separating C_6 from C_7 hydrocarbons in a 4 ft tower, packed with 1 in. Pall rings operated under different types of simulated maldistribution. The y-axis is the ratio of measured HETP to the base (excellent distribution) HETP. The two lower curves simulate distributor tilt, expressed as the ratio of liquid flow at the deep end to that at the shallow end (for instance, 1.25 means that the flow at the deep end was 1.25 times that at the shallow end). The x-axis is the percent of usable capacity at a constant L/V operation, so the higher this percent is, the higher the liquid flow rate and the higher the liquid head in the distributor. The diagram echoes the above message, showing a much greater effect of tilt at turndown, when the liquid head is lower.

Closely review the out-of-levelness of the parting box. Parting box tilt will induce preferential higher flows to the troughs beneath the deep end and lower flows to those beneath the shallow end, as reported in one case (409).

Hydraulic gradients, especially in narrow troughs, can also be troublesome. The hydraulic gradient in a trough can be calculated from (40, 192, 326)

$$h_{\mathrm{hg}} = 0.187 \, v_{\mathrm{H}}^{2} \tag{4.5}$$

where

h_{hg} = hydraulic gradient head, inches of liquid
v_H = liquid horizontal velocity in the trough, based on the wetted trough area, ft/s.

   Distributor troughs with floor perforations have less wetted area than those with wall perforations and, therefore, larger hydraulic gradients. It was recommended to keep the maximum transverse liquid velocity in the troughs less than 1 ft/s (40). The maximum value is at the point of liquid supply to the troughs.

4.9   TROUBLESHOOTING DISTRIBUTOR FEEDS

Feed entry ranks among the most common causes of maldistribution and capacity bottlenecks in packed towers. The following guidelines are recommended:

1. Draw a sketch to scale showing the inlet arrangement (Chapter 8). This is the most powerful technique for identifying these problems. Sections 8.13 and 8.18–8.23 contain cases involving packed tower feeds.

2. Check for the presence of vapor in all liquid feeds. Liquid feed distributors cannot handle vapor, even in seemingly small quantities. Note that a small quantity of vapor by weight, say 1%, can mean 90%+ vapor by volume due to the lower vapor density. If any vapor is present, refer to the guidelines in Section 4.13.

3. Beware of the following common situations when unexpected vapor may be present in the feed. Closely check with a flash calculation at column entrance conditions. Critically review simulation results; many flash a feed to the tower pressure without reporting the vapor/liquid split resulting from this step, creating a false impression that the feed is all liquid at the tower entrance.
   - Liquid feed flashed from a higher pressure source to a lower pressure column.
   - Feeding the bottom stream from one column to another.
   - A preheater on the liquid feed stream.
   - Liquid feed to the tower is steam-traced or heat-traced.
   - Liquid feed to the tower contains a highly volatile component.
   - Feed mixing two liquid streams at different temperatures.
   - Alternative feed that is only occasionally used.

4. Use the sketch in item 1 above to closely examine the mixing between the feed and the tower liquid. Case 4 in Section 8.18 (and Figure 8.24) is an excellent example. This guideline is imperative when dealing with extractive distillation. The success of extractive distillation depends on good mixing between the solvent and reflux, and between the feed and solvent. Both must be closely reviewed to ensure good mixing. The cases in Section 8.19, and Figures 8.25–8.27, are excellent examples.

5. Closely check the dimensions in the sketch. Look for any restrictive passages for liquid flow.

6. When feeding via one or more parting boxes, check the following (248, 249):
   - Apply the orifice Eq. (4.2) to calculate the liquid head in the parting box. If the calculated head is more than 70–80% of the parting box height, check with the detailed

guidelines in Refs. (248) and (249) for the possibility of overflow due to frothing or horizontal momentum effects.

- Perform a Gamma scan of the parting box, shooting chords parallel to the length of the box, to measure the liquid level. Request the scanners to shoot every 1 inch along the box elevation to get a good handle on whether the box is overflowing.

- Calculate the liquid velocities in the sparger holes or dip tubes and sparger pipes. Hole/dip tube velocities should be 4–5 ft/s or less. The hole/dip tube velocities should not be less than the sparger hole velocity.

- Closely compare the liquid entry to the parting box to the various alternatives in Refs. (248) and (249). Many common sparger arrangements, with or without dip tubes, give mediocre, even poor split of liquid to the distributor troughs below, especially at lower liquid heads.

- Verify that the parting box liquid ends up in the troughs and does not miss them.

- Test the tower at higher or lower liquid flow rates. If the packed bed produces more stages at higher liquid flow rates, one possibility is poor liquid split to the troughs at turndown. If the packed bed produces more stages at turndown, one possibility is an overflowing parting box. Keep in mind that improvements at lower or higher liquid loads may also be due to many other factors such as distributor and packing performance.

7. When liquid from a collector or parting box descends through several downpipes or down boxes, closely examine (with the aid of a sketch to scale, see Chapter 8) whether there is an even liquid split to the downpipes. In many cases (275, 390, 417), uneven split caused poor efficiency in the bed below. In Case 1 in Section 8.18, downpipes from a collector discharged into regions in the distributor almost totally enclosed within chimney walls, which obstructed flow to the rest of the distributor (Figure 8.21). In Cases 2 and 3, obstruction by chimneys or down boxes generated an uneven liquid split, causing liquid maldistribution and poor separation.

8. Parting boxes that handle highly subcooled (>50–70°F) feeds are prone to damage. Upon an abrupt feed rate step-up, vapor from the tower collapses onto the subcooled liquid, exerting strong forces that can deform troughs and parting boxes. Figure 4.12 is one example.

4.10   EVALUATION OF DISTRIBUTOR IRRIGATION QUALITY

An excellent evaluation method was developed by Moore and Rukovena (326). This method has gained wide acceptance (40, 192, 275, 336, 346, 347). Over the years, the author and his colleagues successfully extended this method to situations not addressed by the original Moore and Rukovena paper, and these extensions are incorporated in the following discussion.

The Moore and Rukovena method calculates a distribution quality index D_Q in percent, given by

$$D_Q = 0.4\,(100 - A) + 0.6\,B - 0.33\,(C - 7.5) \tag{4.6}$$

Typical values of D_Q range from <40% for poor distributors, 40–70% for standard distributors, 75–90% for intermediate distributors, and >90% for high-performance distributors (326). To determine D_Q, each distributor drip point is represented by a circle. The center of each circle is where the liquid from this point strikes the top of the bed. The total area of all the circles is set equal to the column cross-section area, so when all the drip points have the same diameter, the

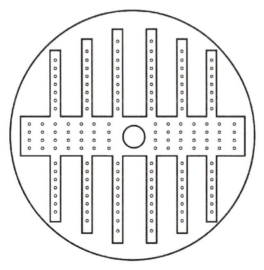

Figure 4.12   Reflux parting box fed by 70°F subcooled liquid via a central pipe. The parting box region under the pipe was repeatedly pushed downward.

area of each circle equals the tower cross-section area divided by the number of drip points. An extension by the author for distributors containing drip points of different diameters still sets the total circle area to equal the tower cross-section area, but proportions the circle area for each hole size to its hole area.

The terms in Eq. (4.6) are evaluated as follows (Figure 4.13a):

A is the percent of the column cross-section area not covered by the circles. It provides a direct measure of the unirrigated area at the top of the bed.

B is evaluated by selecting the most underirrigated or overirrigated continuous area at the top of the bed occupying one twelfth of the tower cross-section area. This is the area where the deviation from the average flow is maximum, and is a measure of the large-scale maldistribution. If this area is underirrigated, B is the ratio of the circle area in this region to the area of the region (1/12th of the tower cross-section area). If overirrigated, B is the ratio of the area of the region to the area of the circles. The lowest value of B anywhere at the top of the bed, expressed as a percent, is used in Eq. (4.6).

C is the total area of overlap of adjacent drip point circles. It is expressed as a percent of the column cross-section area and represents the area that spreads the liquid unproductively.

Figure 4.13b–d illustrates the application of this technique to evaluate irrigation quality for various distributors. Figure 4.14a shows an application (346) in which this method diagnosed poor irrigation ($D_Q = 45\%$) as the root cause of high packing HETP. The distributor preferentially sent liquid to the outer regions at the top and bottom of the diagram. Replacement by a high-performance distributor ($D_Q = 90\%$; Figure 4.14b) improved staging by 40%. Application of the Moore and Rukovena method to trough distributors with side holes and external drip guides directing the liquid into the packings (336) is illustrated in Figure 4.15.

The Moore and Rukovena technique can be applied at three different levels:

- *Basic*, just drawing the circles per rules above. In all the cases the author experienced, this gave a good qualitative assessment of the distribution quality. This can readily be seen by inspecting Figures 4.13–4.15.

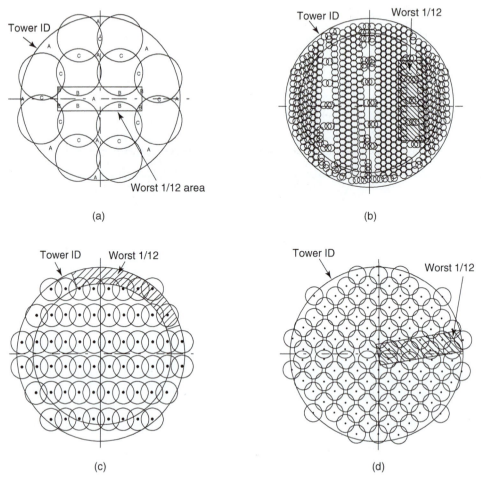

(a)   (b)

(c)   (d)

Figure 4.13   Application of the Moore and Rukovena method for distributor quality rating (a) areas considered in liquid distribution quality rating (A = cross-sectional area not covered by point circles; B = point circle area in 1/12 tower area; C = area of overlap of point circles). (b) standard quality distributor, $D_Q = 60\%$; (c) standard quality distributor, $D_Q = 72\%$; (d) intermediate quality distributor, $D_Q = 84\%$. (From Moore, F., and F. Rukovena, "Liquid and Gas Distribution in Commercial Packed Towers," Paper presented at the 36th Canadian Chemical Engineering Conference, October 5–8, 1986. Reprinted Courtesy of F. Rukovena, Consultant.)

- *Intermediate*, also calculates the D_Q. This gives an excellent quantification, which the author found very reliable.
- *Advanced*, D_Q can be applied together with Figure 4.2 to quantitatively estimate the stages loss to maldistribution. The author's experience with this advanced level is limited.

The author highly recommends this method for evaluating distributor irrigation at the basic and intermediate levels.

With small-diameter towers, B may be difficult to define. In this case, only the basic level can be applied with confidence. With a small number of drip points, double and triple overlaps may incur difficulties (40).

For structured packing, the Moore and Rukovena method may not fully characterize the distribution quality because it does not credit the liquid spreading performed by the top layer of

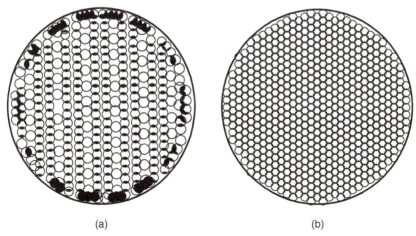

(a)   (b)

Figure 4.14   Application of the Moore and Rukovena method to justify distributor replacement. (a) Original distributor, $D_Q = 45\%$; (b) Replacement distributor, $D_Q = 90\%$. (From Olsson, F. R., Chemical Engineering Progress, p. 57, October 1999. Reprinted Courtesy of Chemical Engineering Progress (CEP). Copyright © American Institute of Chemical Engineers (AIChE).)

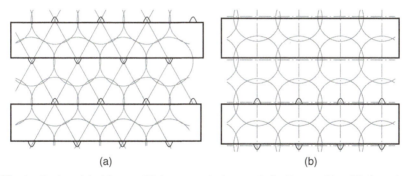

(a)   (b)

Figure 4.15   Application of the Moore and Rukovena method to trough distributors with wall holes and external drip guides directing liquid downwards (a) triangular pitch, $D_Q = 95\%$; (b) square pitch, $D_Q = 90\%$. (From Niggeman, G., A. Rix, and R. Meier, "Distillation of Specialty Chemicals," Chapter 7, pp. 297–336, in A. Gorak and H. Schoenmaker (ed.) "Distillation Operation and Application," Academic Press, 2014. Copyright Elsevier 2014. Reprinted with permission.)

packings, and therefore may penalize distributors that irrigate the packings in a drip line. Spiegel (457) proposed an alternative "wetting index," which accounts for the liquid spread at the top layer of packing, but unfortunately did not provide generic guidelines for the application of this method. The author has had good experience applying the Moore and Rukovena method also to distributors that irrigate structured packings in a drip line. It works at least on the qualitative level, as long as one recognizes that the liquid will spread parallel to the layers. The author concurs with Spiegel's point that the quantitative value of D_Q may penalize such distributors.

Perry et al. (362) presented a useful method, which they conceptually credit to Moore, for troubleshooting irrigation in the tower wall region. In modern designs, drip point patterns are computer-generated. Due to mechanical constraints and support considerations, the computer often locates outer drip points some distance from the tower wall, producing underirrigated zones, which go undetected. The Perry et al. method divides the column into three concentric zones of equal areas, and the number of drip points in each zone is counted by hand. This procedure can readily detect underirrigated zones near the tower wall, as illustrated in Figure 4.16, which is

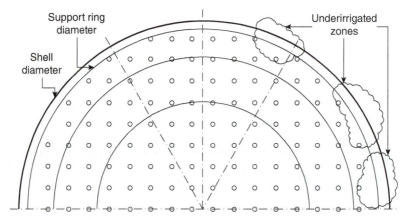

Figure 4.16 Application of the Perry et al. method using three concentric zones of equal area to detect unirrigated areas near the tower wall. The underirrigated areas are marked with clouds. (From Niggeman, G., A. Rix, and R. Meier, "Distillation of Specialty Chemicals," Chapter 7, pp. 297–336, in A. Gorak and H. Schoenmaker (ed.) "Distillation Operation and Application," Academic Press, 2014. Copyright Elsevier 2014. Reprinted with permission.)

its main application (336, 362). For simplicity, Figure 4.16 does not specifically show the vapor risers, but these need to be taken into consideration when applying this method.

When reviewing the irrigation pattern, ensure that the liquid drip points are not too close to the vapor passages to be entrained by the vapor. Also, check that any drain holes do not pass quantities of liquid large enough to generate maldistribution. Finally, confirm that the distributor supports and packing hold-downs do not interfere with liquid distribution. In one case (257, 261), the distributor was supported by $4'' \times 8''$ plates that covered 8% of the packing, leading to poor separation.

4.11 TROUBLESHOOTING FOR VAPOR MALDISTRIBUTION

High-velocity jets entering the tower due to undersized gas inlet and reboiler return nozzles are the prime source of vapor maldistribution. These jets persist through low-pressure-drop devices such as packings. Vapor distributors, even well-designed inlet baffles, have alleviated such problems. Vapor distribution is most troublesome in large-diameter columns. Below are some guidelines for troubleshooting vapor maldistribution. In addition, many of the guidelines in the next section dealing with flashing feeds also apply for vapor distributors.

1. Strigle (462) recommends a vapor distributing device whenever the vapor nozzle F-factor $F_N (= u_N \rho_G^{0.5})$ [u_N is the vapor velocity at the narrowest area of the inlet pipe or nozzle, ft/s, and ρ_G is the gas or vapor density, lb/ft³] exceeds 22 ft/s (lb/ft³)$^{0.5}$, or the kinetic energy of the inlet gas exceeds eight times the pressure drop through the first foot of packing, or the pressure drop through the bed is less than 0.08 in. water/ft of packing. Cases of poor packing performance at high F_N values have been reported (69)

2. An alternative criterion proposed by Moore and Rukovena (326) replaces the $F_N < 22$ rule in item 1 above by a rule to keep $F_N < 52.4 \, \Delta P^{0.5}$, where ΔP is the bed pressure drop in inches of water per foot of packing.

3. When none of the criteria in items 1 and 2 is violated, vapor maldistribution is unlikely to be troublesome, so there is no need for a vapor distributor (326). This applies to tower diameters less than 20 ft.

4. For nozzle F-factors in the range $52.4 \; \Delta P^{0.5} < F_N < 81.2 \; \Delta P^{0.5}$, Moore and Rukovena (326) recommend a simple vapor distributor such as a sparger pipe. At higher F_N, a sophisticated vapor distributor should be used (326).

   Kabakov and Rozen (186) measured severe gas maldistribution in 15-ft diameter CO_2 amine absorbers containing 50 ft of random-packed beds. They did not report values for F_N. The efficiency in these columns was about half that measured in another column that had a specially designed inlet gas sparger. Schafer et al. (415) report a case where adding a multi-vane vapor distributor greatly improved aromatics stripping from water.

   Using CFD, Wehrli et al. (499) found that a very simple device such as the V-baffle (Figure 8.14b) gives much better distribution than a bare nozzle. Both Wherli et al. and Schafer et al found that a more sophisticated sparger or vane distributor was even better.

   With a V-baffle, use a sketch to troubleshoot for possible adverse effects produced by the baffle. Some cases are discussed in in Sections 8.13 and 8.16. It is important to verify that vapor from the V-baffles does not impinge on the liquid below or interact to generate vapor maldistribution, as explained in Section 8.13 and in Case 3 in Section 8.16.

   In one case (271), with $F_N \gg 81.2 \; \Delta P^{0.5}$, the entering vapor jet repeatedly damaged the bottom of an 11-ft tall bed containing sturdy grid packing. The entry was similar to that in Figure 4.17a, but there was no chimney tray between the inlet and the bed. The damaged packing was opposite the vapor entry (Figure 4.17a). High vapor velocity, possibly with entrained liquid, was "literally ripping the bed apart from the bottom (271)." In another case (158), some coking was found in the bottom (slurry) grid bed of an fluid catalytic cracking (FCC) fractionator, most pronounced on the side of the column opposite the vapor feed inlet nozzle (again, similar to Figure 4.17a but no chimney tray). The authors (158) comment that "this type of slurry bed coking is not uncommon in the industry."

5. The best vapor distributor in distillation columns is pressure drop. A high-pressure-drop vapor distributor provides good distribution, while a low-pressure-drop distributor does little to improve vapor distribution, as demonstrated (69) and illustrated in Figure 4.17. Schafer et al. (415) report a case in which adding a high-capacity tray between the stripping steam distributor and the bottom bed, which improved vapor distribution by increasing pressure drop, greatly improved aromatics stripping from water. In another

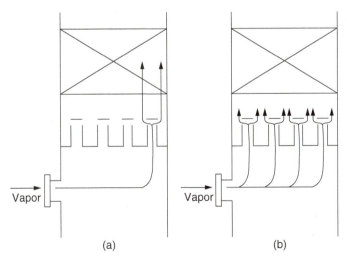

Figure 4.17   Inlet vapor jet traveling through a vapor-distributing device (a) low pressure drop device, vapor maldistributed through the bed; (b) high pressure drop device, vapor distributed evenly to bed.

case (415), replacement of ceramic packings with high-efficiency metal packings lost stripping efficiency because the lower pressure drop of the metal packings was less effective in correcting the vapor maldistribution. Improving vapor distribution solved this problem. In Case 6.1 in Ref. (201), increasing the vapor distributor pressure drop helped eliminate vapor maldistribution.

To evaluate the effectiveness of the vapor distributing device, calculate its pressure drop and compare it to the inlet nozzle velocity head. Pressure drops through vapor distributors typically range from 1 to 8 inches of water (326) and can be calculated using Eq. (4.4). Strigle (462) recommends that the pressure drop through the vapor distributor should be at least equal to the velocity head at the inlet nozzle.

6. In the loading region, i.e., at high gas rates, liquid and vapor maldistribution interact (460). A poor initial liquid distribution may channel vapor and even induce local flooding. Similarly, liquid distribution pattern in the bottom structured packing layers is influenced by strongly maldistributed inlet vapor (350) or by local liquid accumulation (local flooding) induced by high vapor loads (Section 5.5.15). The author had experiences where increasing liquid loads helped alleviate vapor maldistribution in the bed.

Outside the loading region, the influence of the liquid flow on gas maldistribution is small or negligible.

7. The effect of gas maldistribution on packing performance is not well understood. FRI's commercial-scale tests (55) showed little effect of gas maldistribution on both random and structured packing efficiencies. In one case, blocking the central 50% or the chordal 30% of the 4 ft tower cross-section area beneath a 5 ft-tall bed of $250\,m^2/m^3$ structured packing had no effect on packing efficiency, pressure drop, or capacity. The blocking did not permit gas passage but allowed collection of the descending liquid. Simulator tests with similar blocking (350) differed, showing that a 50% chordal blank raised pressure drop, gave a poorer gas pattern, prematurely loaded the packing, and generated maldistribution that penetrated deeply into the bed.

8. Midspan I-beams can be deep, sometimes ranging from 1 to 2 ft. When the bottom of the I-beam is too close to the liquid level, it may interfere with the vapor distribution to the bed. For good vapor distribution, it was recommended that the pressure drop for half the vapor flow passing under the I-beam (between the bottom of the support and the liquid level) should be less than 10% of the pressure drop of the packed bed above (286). Troubleshoot this possibility by drawing a to-scale sketch as shown in Section 8.12, Figure 8.13a.

9. I-beam interference can be troublesome also on chimney trays, when the bottom of the I-beams gets close to the chimney hats. Check this by drawing a to-scale sketch as shown in Section 8.12, Figure 8.13b.

10. Chimney trays with risers ending too close to the bed above may initiate vapor maldistribution. For good distribution of vapor to a packed bed above chimney tray riser hats, Lieberman (285) and Lee (275) advocate a hypothetical angle of 30°, as illustrated in Section 8.12, Figure 8.13c. One case was presented (275) in which a 70° angle was too wide, generating vapor maldistribution and poor packing efficiency in the bed above.

11. Item 10 above extends to vapor inlet distributors discharging too close to the bed above. Hanson et al. (155) reported a case of severe coking in the grid wash bed of a coker main fractionator with a 54″ feed nozzle discharging via a specialized vapor feed distributor. The top of the distributor was only 9 in. beneath the bed. The successful solution relocated the grid bed higher in the tower.

12. CFD has been powerful for analyzing the effects of gas inlet geometry on gas maldistribution in packed beds. Ali et al. (5) showed that the gas velocity profiles from a commercial CFD package compared well with those measured in a 1.4-m simulator containing structured packing with commercial distributors and collectors. Their CFD model guided them to a collector design that minimized gas maldistribution. The application of CFD for evaluating vapor maldistribution is discussed at length in the next section with special reference to troubleshooting refinery vacuum tower feeds maldistribution.

4.12   VAPOR MALDISTRIBUTION IN THE FEED ZONE TO REFINERY VACUUM TOWERS

This is one application in which CFD technology has been extensively applied.

This tower is unique in that the feed to the tower, typically 30–70% vapor by weight, becomes >99.9% vapor by volume under the deep vacuum at the tower inlet. Physically, the feed enters at close to sonic velocity with the liquid dispersed in the vapor as a fine mist. This feed often enters via a tangential nozzle and a vapor horn device (Figure 4.18) that flings the feed against the tower wall. The wall catches the liquid and runs it down the wall, while the vapor expands to the tower diameter. An alternative arrangement uses a radial nozzle and a vane distributor (12, 245, 499).

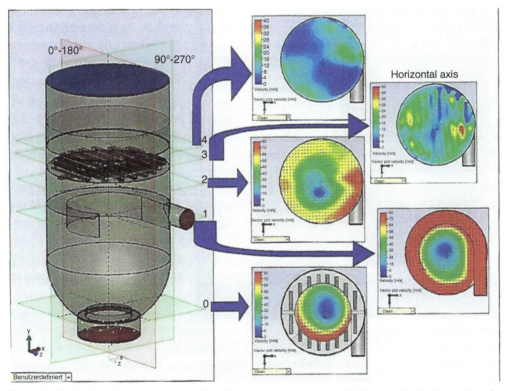

Figure 4.18   Application of computational fluid dynamics to analyze vapor distribution at the feed zone of a refinery vacuum tower, showing the flow profile at different elevations. (From Costanzo, S., S. M. Wong, and M. Pilling, Hydrocarbon Processing, p. 81, September 2010. Reprinted with permission.)

In either case, the vapor from the feed zone rises through a chimney tray (Figure 4.18, also Figure 7.2a) to a packed bed (termed "the wash bed"), which is the most critical bed in the tower. Economic considerations necessitate minimizing reflux to this bed. Maldistribution of vapor or liquid may render the distillate from the top of the bed (heavy vacuum gas oil, or HVGO) to be off-spec or form dry spots in the bed that can severely coke, limiting the capacity and run length of the tower, and often of the entire crude unit. These make vapor distribution to the bed critical.

Several studies were reported on how the vapor inlet geometry affects vapor distribution to the wash bed. Lin et al. (296) tested a 3-ft ID air–water scaled-down model of the flash zone of a large vacuum tower, transfer line, feed nozzles, and slop wax chimney tray (no vapor horn). They observed the vapor preferentially rising at the opposite side of the feed inlet (similar to Figure 4.17a), with some feed liquid deflecting into these chimneys. Vaidyanathan et al. (487) and Torres et al. (482) applied CFD to examine the effect of the geometry of a chimney tray above the inlet on vapor distribution and liquid entrainment. Paladino et al. (355) demonstrated that the presence of liquid in the feed affects the gas velocity profile and must be accounted for in modeling. Paladino et al. (355) and Waintraub et al. (497) used two-fluid models to explore the velocity distributions and entrainment generated by different designs of vapor horn distributor designs. Wherli et al. (499) and Kolmetz et al. (260) presented CFD-produced flow patterns for vacuum tower inlets, including radial bare nozzle, V-baffle, vane inlet device, and tangential entry with a vapor horn. Wehrli et al. (500) produced pilot-scale data simulating a vacuum tower inlet, which can be used for CFD model validation. Remesat and Riha (394) describe at length the application of CFD for vacuum tower feed inlet design.

For a refinery vacuum tower fed via a vapor horn, Figure 4.18 (75) shows CFD-generated horizontal ("x–z") vapor velocity profiles at different elevations between the vapor horn and the wash bed. The arrows, slightly challenging to see, indicate the direction of flow. At plane 0, situated below the horn, Figure 4.18 shows a high swirl velocity of 70–80 m/s, directly beneath the region where the horn terminates. This swirl may generate a vortex at the bottom sump. If there are stripping trays in the smaller-diameter section, this swirl may upset the top tray or the chimney tray that directs the liquid to the trays. Plane 1, which slices the vapor horn, shows very high horizontal vapor velocities at the horn (as could be expected) and much lower velocities in the vapor rise region. Keep in mind that the velocities shown are in the horizontal plane. More important in this region are the vertical ("y") velocities, which determine the potential for excessive entrainment and a premature system limit. It was recommended (394) to keep the peak C-factor (Eq. 2.1) based on the area open to vapor ascent below 1 ft/s.

Plane 2, right above the horn, shows the first transition from the feed zone profile to the flow profile entering the chimney tray and is useful for evaluating the effectiveness of the vapor horn as a vapor distributor. Plane 3 is just below the chimney tray. Comparing Planes 2 and 3 shows that the vapor velocities have significantly calmed down in the vertical space between the top of the horn and the chimney tray. This is well in line with the industry experience (155) that keeping the feed inlet nozzle top at least 5 ft below the chimney tray effectively retards entrainment from the flash zone. Finally, Plane 4, above the chimney tray and just below the wash bed, shows further slowing down of the horizontal vapor velocities. Unevenness on this plane is likely to persist in the wash bed. The reduction in velocities from Plane 3 to Plane 4 indicates that this chimney tray improves vapor distribution.

More important than the horizontal vapor velocities below and above the chimney tray are the vertical velocities. Figure 4.19a shows CFD-obtained vertical ("y") velocity profiles at and above the chimney tray of a different vacuum tower (in this case, the feed enters via a vane distributor, not a vapor horn). The diagram on the right shows high-velocity peaks directly above the vapor outlets from the chimneys as well as around the periphery. These high-velocity peaks would persist into the wash bed and explain the reduced product quality, premature damage

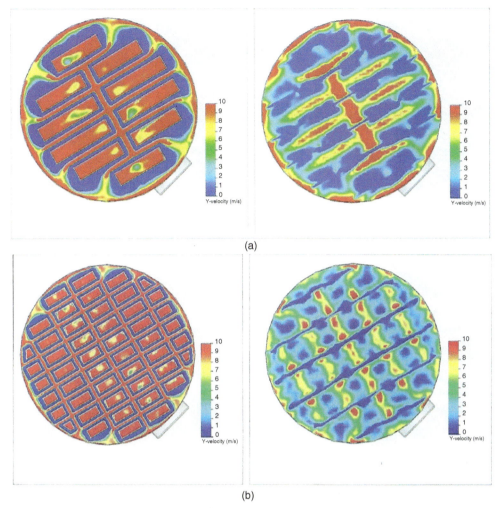

(a)

(b)

Figure 4.19   Vertical ("y") velocity profile provided by a CFD analysis of a vacuum tower feed region (a) existing design; (b) revised proposed design. In both, left is at the chimney tray, right is above the chimney tray just below the wash bed. (From Ray, D., A. Arora, and A. Phanikumar, p. 40, Revamps 2009. Reprinted with permission.)

to the packing, and high slop wax production experienced (391). Similar peaks were reported earlier by Vaidyanathan et al. (487), who attribute them to vapor from two neighboring chimneys merging to form an upward jet that is dependent on the hat design. Torres et al. (482) also observed these peaks. The lower diagrams in Figure 4.19 show that these issues can be mitigated by an enlarged feed nozzle (68″ instead of 52″) and better vane distributor and chimney tray designs.

CFD analysis can be invaluable for interpreting field data. In one case (335), a high pressure drop was measured between the flash zone and the top of the wash bed, suggesting wash bed coking. A CFD analysis found that high pressure drop was due to the flash zone pressure measurement, which included a dynamic velocity head component, and not due to coking.

For the results of the CFD analysis to be valid, an accurate representation of the feed properties and velocity profile is necessary (394). A feed profile is created for the CFD analysis using

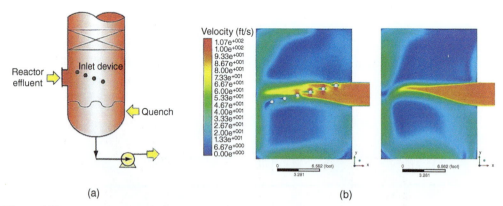

Figure 4.20   CFD simulation of FCC main fractionator inlet (a) a schematic showing the staggered parallel pipe distributor that has been successful in alleviating coking in the packing above the tower side opposite to the inlet; (b) CFD profiles, comparing the flow patterns from a bare nozzle (right hand side) with the flow pattern with the staggered parallel pipe distributor (left hand side). With the distributor, the jet widens near the wall opposite the inlet. (From Niccum, G., and S. White, PTQ, Q1, p. 133, 2014. Reprinted with permission.)

a 3-dimensional mixture of turbulent compressible vapor flow, with liquid droplets moving with the vapor. It is equally important to correctly represent the geometry, and with an existing tower, to validate the CFD model with plant data (394, 506). There are several assumptions involved in CFD modeling, and if these deviate from reality, so will the results and conclusions.

FCC main fractionator feed inlet   With a hot (1000°F) reactive all-vapor feed to this fractionator, only one simple distributor is known to be successful: a series of staggered parallel pipes mounted in front of the inlet (Figure 4.20a). The reason for the success of this distributor in reducing coking on the wall opposite the inlet has been obscure. With the distributor, wall temperatures opposite the inlet seem to be typically 40–60°F hotter than directly above the inlet. Without the distributor, the difference is usually higher, but not by much.

A preliminary CFD study (334) provided a plausible hypothesis. The pipes widen the inlet vapor jet and scatter it on the wall opposite the inlet (Figure 4.20). Without a distributor, the jet remains focused on a narrow region opposite the inlet. The scattering spreads the liquid impacting the back of the tower over a wider area. Additionally, as the impingement region is less defined, the shear stresses are more evenly distributed and shear the coke. This study is preliminary, but shows another valuable development in CFD applications to tower feed modeling.

4.13   TROUBLESHOOTING FLASHING FEEDS ENTRY

Liquid entry is typically designed for an inlet velocity of less than 4–5 ft/s. Figure 4.21 shows two nightmare scenarios of flashing feed entering liquid distributor parting boxes. In Figure 4.21a, water at 95°C entered a parting box of a bitumen stripper at a velocity of 3 ft/s. To improve performance, a consultant proposed reducing tower pressure from slightly above atmospheric to vacuum. Upon drawing vacuum, the feed flashed, and due to the low steam density, the feed velocity increased to 100 ft/s, which sprayed the feed liquid everywhere, precluding any decent distribution.

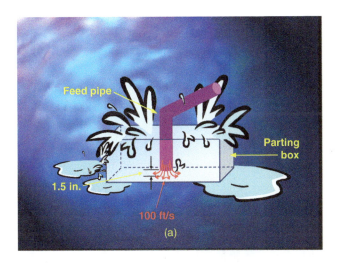

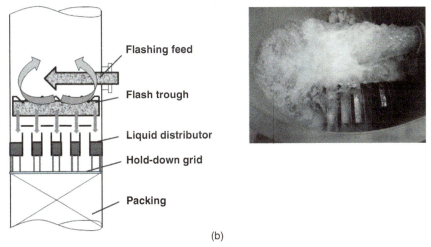

(b)

Figure 4.21   Bad practices of flashing feed entering liquid distributor parting boxes. (Part (a): From Kister, H. Z., W. J. Stupin, and S. Stupin, TCE, p. 44, December 2006/January 2007. Courtesy of IChemE. Part (b): From Schultes, M., "The Raschig Super-Ring is Pushing the FRI Test Column to New Limits," The Kister Distillation Symposium, Topical Conference Proceedings, p. 381, the AIChE National Spring Meeting, San Antonio, TX, April 29–May 2, 2013. Courtesy of Raschig, AG.)

In Figure 4.21b, a flashing feed entered via a side nozzle into a V-notch trough that crossed the tower diameter. The design intention was for the flashing liquid to fall into the parting box and descend via the notches into the troughs of the distributor below, and for the vapor to disengage and rise to the bed above. Tests at the Raschig shop (420) (Raschig were not involved in the original design, only in the troubleshooting) showed that the high momentum of the incoming flashing feed pushed the liquid to the end of the parting box opposite the inlet nozzle, with some of it missing the parting box altogether (photograph in Figure 4.21b). The result was terrible distribution to the packing.

Other cases of flashing feeds entering liquid distributors causing hydraulic and separation problems were reported (71, 239, 240, 433). In one case in which a flashing feed entered a false downcomer, the downcomer was found repeatedly blown apart and lying on the distributor (271).

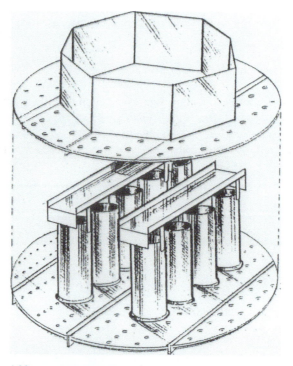

Figure 4.22 A typical gallery distributor design. (Courtesy of KochGlitsch, LP.)

With flashing feeds, the vapor and liquid need to first be adequately separated, then the liquid descend onto a packing liquid distributor. The vapor must not enter the liquid distributor. There is a variety of flash box designs that achieve this, as described in several literature references (199, 245, 462). One of the most popular types is gallery distributors (Figure 4.22). The feed nozzle is located well above the upper plate. The gallery wall facing the inlet nozzle deflects the incoming vapor upward and the incoming liquid downward. The liquid thus separated from the feed joins the liquid raining from the peripheral zones of the bed above, and both descend through the holes in the upper plate. The lower plate is a liquid distributor to the bed below. Vapor from the risers of the distributor below rises through the hollow inside of the gallery.

In many galleries or other flashing feed distributors, liquid from the gallery descends to the distributor below using downpipes instead of floor perforations. To prevent overflow, these downpipes should be adequately sized for self-venting flow (Figure 4.9). The downpipes, either from the gallery or from a collector above, should enter the distributor below without restricting liquid equalization in the distributor, as occurred in Case 1 in Section 8.18.

The impact of the feed on the gallery wall tends to deflect the incoming jet upward. The top of the gallery in this region should be covered to prevent entrainment. Likewise, at the inlet, the feed splits into two streams, with the momentum pushing each stream horizontally along each wall of the gallery. Right across from the inlet, the remnants of the two streams collide. This region too can produce entrainment, which can be prevented by a gallery cover at that location.

When the feed enters too close to the gallery floor, it can entrain the gallery liquid, as described in the case study in Section 8.13. Reference (37) describes severe entrainment from a high-velocity flashing feed that entered via a 10″ inlet. The bottom of the inlet was 10.7″ above a shed deck that held some liquid.

When calculating liquid heads in the gallery using Eq. (4.2), account for liquid raining into the gallery from the bed above, in addition to the feed liquid. Also, there is likely to be some frothing of the liquid, and this needs to be considered in the evaluation but cannot be calculated.

Gallery distributors are prone to damage at the feed inlet point due to severe hydraulic pounding of the gallery wall by the entering feed. There have been cases where sudden increases in the quantity of entering vapor have flattened galleries. Such upset conditions should either be eliminated or catered for by mechanical strengthening.

It is important to ensure that packed tower feed pipes operate outside of the slug flow regime. Cases 6.8 and 6.9 in Ref. (201) describe experiences with packed tower instability, premature flooding, and poor separation resulting from slug flow in the feed.

4.14 TROUBLESHOOTING NOTCHED DISTRIBUTORS

V-notches, Y-notches (Figure 4.23), and rectangular slots are successfully used in high fouling rate services such as slurries, vent scrubbers, quench towers, and where other distributors experience plugging (40, 83, 192, 258, 260). Rectangular notches are also used in some high-quality two-stage distributors. Notched troughs are insensitive to plugging, corrosion, and erosion. They minimize stagnant zones, keep particulates moving, and allow the liquid to sweep the trough clean. The notches typically discharge the liquid directly into the up-flowing vapor.

The recommended equations for flow through notch is from *Perry's Chemical Engineers' Handbook* (5th ed., p. 5–17; some later editions are not clear on the units). For rectangular notches:

$$q = 0.415 \, (L - 0.2 \, h_o) \, h_o^{1.5} \, (2g)^{0.5} \tag{4.7}$$

For narrow rectangular notches, Eq. (4.7) simplifies to

$$q = 0.386 \, L \, h_o^{1.5} \, (2g)^{0.5} \tag{4.8}$$

For triangular notches, not fully covered by liquid:

$$q = 0.31 \, h_o^{2.5} \, (2g)^{0.5} / \tan \Phi \tag{4.9}$$

where

q = volumetric flow rate, ft^3/s
L = width of rectangular notch (Figure 4.24a), ft
h_o = height of liquid above bottom of notch (Figure 4.24a,b), ft
g = gravity acceleration, 32.2 ft/s^2
Φ = angle in degrees, measured up from a horizontal line drawn at the apex (Figure 4.24b)

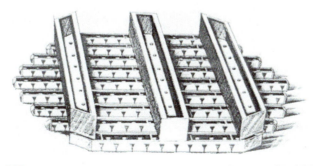

Figure 4.23 A notched trough distributor with Y notches. (Courtesy of KochGlitsch, LP.)

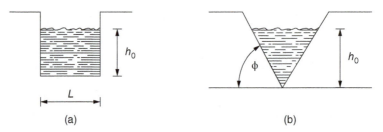

(a)   (b)

Figure 4.24   Terms used for flow through notches (a) rectangular notch, Eqs. (4.7, 4.8) (b) triangular notch, Eq. (4.9).

Per Eqs. (4.7)–(4.9), the flow rate rises to the power of 1.5 and 2.5, respectively, for rectangular and triangular notches, in contrast to flow rates rising to the power of 0.5 with orifice distributors (Eq. 4.2). This means a much lower range of liquid heads. A typical V-notch distributor will have a liquid head of 1–3 in. over its entire operating range (83). If at a head of 1 in. it handles 100 gpm, Eq. (4.9) gives that increasing the head to 3 in. will raise this flow rate to $(100 \times 3^{2.5} =)$ 1600 gpm. This is a huge dependence on liquid head. On the credit side, this gives good turndown. On the debit side, this makes the distribution sensitive to small changes in head and to fabrication irregularities from notch to notch, and makes it difficult to incorporate a large number of drip points, which inherently reduces irrigation quality. This strong dependence on the liquid head also makes V-notch (and to a lesser degree, rectangular notch) distributors far more sensitive to out-of-levelness, hydraulic gradients, and surface turbulence, and these could adversely affect their performance. The hydraulic gradient can be calculated using Eq. (4.5).

At low rates, the issuing liquid may wrap itself around the troughs (Figure 4.25a). At high rates, liquid streams from adjacent troughs may join each other into a single stream, thereby halving the number of distribution points (Figure 4.25b). Both of these were observed by Olsson (346) in distributor water tests. The solution was to add drip guides (Figure 4.25c) that direct the liquid to the desired points.

At high liquid rates, the flow down the length of the troughs is typically very high, and the flow out of each "V" is no longer uniform due to variance in the liquid momentum (Figure 4.26). Furthermore, at high liquid rates, the liquid may submerge the notches and maldistribute from then on.

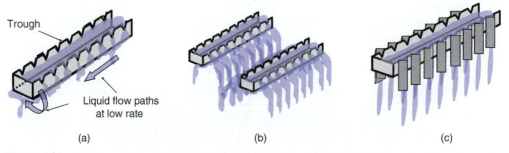

(a)   (b)   (c)

Figure 4.25   Problems experienced with notched trough distributors (a) flow dribbling down the side and wrapping itself around the trough at low rates; (b) liquid streams from adjacent troughs joining into a single stream at high rates; (c) drip guides to direct the liquid into the desired location. (From Olsson, F. R., Chemical Engineering Progress, p. 57, October 1999. Reprinted Courtesy of Chemical Engineering Progress (CEP). Copyright © American Institute of Chemical Engineers (AIChE).)

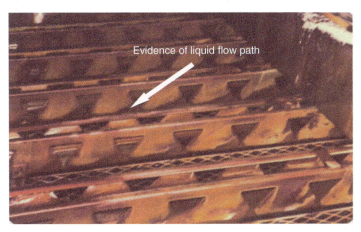

Figure 4.26   Watermarks on a notched trough distributor showing liquid issuing with a sideways momentum. (From Olsson, F. R., Chemical Engineering Progress, p. 57, October 1999. Reprinted Courtesy of Chemical Engineering Progress (CEP). Copyright © American Institute of Chemical Engineers (AIChE).)

Ensure that the liquid level does not reach the top of the notches. Once it does, the notches become triangular orifices, and the mode of operation totally changes, producing maldistribution and/or overflow.

4.15   TROUBLESHOOTING SPRAY NOZZLE DISTRIBUTORS

In spray distributors (Figure 4.27), a header pipe and laterals supply liquid to spray nozzles arranged in a pattern, typically elevated 18″ to 36″ above the packed bed. From the nozzles,

Figure 4.27   Water testing of a spray distributor. (Courtesy of Sulzer Chemtech.)

liquid is sprayed as a wide-angle (typically 120°, but sometimes 90°) full cone. The cone is seldom homogenous. The bed coverage is determined by the number of nozzles, spray angle, and height above the packed bed.

It is a good practice to restrict the spray nozzle pressure drop to 10–15 psi, but up to 20 psi is used. This minimizes the entrainment and also minimizes the likelihood of a spray angle collapse. FRI's experiments with wide-angle (120°) sprays (107) showed the spray cone angle to collapse from 120° to about 45° from the same spray nozzles when the spray nozzle pressure drop was increased between 10 and 20 psi with subcooled liquid. This spray angle collapse is closely related to the collapse of condensing vapor onto subcooled feed liquid (107).

Entrainment Depending on the nozzle pressure drop and vapor superficial velocity in the tower, spray distributors can generate a significant amount of entrainment (up to 25% or more). Pigford and Pyle (364), and later Lin et al. (296, 297) and Trompiz and Fair (484), measured entrainment in pilot-scale simulators using the air–water and air–Isopar systems. All these tests found entrainment to strongly increase with the gas velocity (C-factor, Eq. 2.1) and with the nozzle pressure drop. Some dependence on the nozzle type was observed. An excellent correlation for calculating the entrainment is provided by Trompiz and Fair (484).

Spray footprint diagram The spray circles overlap, so some areas at the top of the packing are overirrigated while others are underirrigated. The objective is to keep this maldistribution small-scale, but this objective is not always met. Spray patterns tend to be empirical. Generally, the distribution quality of spray distributors using established spray patterns is not as good as that of orifice distributors, and can be a lot worse when using nonstandard patterns.

A good way to check the irrigation quality is to prepare a spray footprint diagram. The diameter of the footprint of each spray circle on top of the bed is approximated by

$$D = 0.92 \times 2 H \tan(\Phi/2) \tag{4.10}$$

where

H = height of bottom of the spray nozzle above the bed (Figure 4.28a), inches
D = diameter of circle drawn by the spray footprint on the bed, inches
Φ = spray angle in degrees, measured up from a horizontal line drawn at the apex (Figure 4.28b)

The correction factor of 0.92 is an empirical factor accounting for gravity, suitable for 18″–30″ height between the bottom of the nozzles and the top of the packing.

The footprint of each spray nozzle is drawn as shown in Figure 4.28b. The pattern in Figure 4.28b is a common pattern that minimizes large-scale distribution with small circle overlaps. This pattern has worked successfully in many installations. There are many other patterns, some with much more extensive circle overlaps, that have also worked well. The key is to prepare a diagram like Figure 4.28b and to critically review it to minimize large-scale maldistribution. To optimize the pattern, the locations of some of the nozzles or the height above the bed can be adjusted on the diagram, and if warranted, a new header can be designed.

Spray pressure drop test This should be the very first test when dealing with a spray distributor. Make sure the flow to the sprays is correctly monitored. Check it against pump curves and control valve settings. Then install a simple, calibrated pressure gauge in the liquid line to the sprays, downstream of any control valve, filter, or heat exchanger. The measured spray nozzles pressure drop equals the pressure gauge reading, minus the tower pressure (inferred from the tower pressure transmitter reading), minus the liquid static head in the line to the sprays

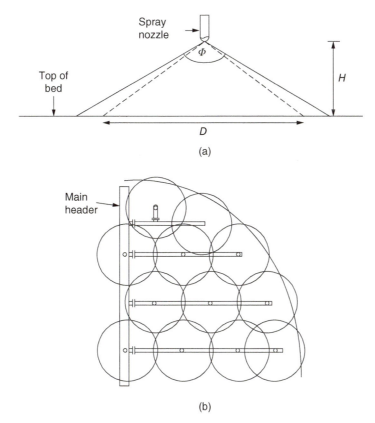

Figure 4.28 Spray footprints diagram. (a) Terms used in Eq. (4.10); (b) sprays footprint diagram.

(calculated from the elevation difference), and minus the friction loss in the liquid line to the sprays (calculated from the liquid flow rate and line size, and is often negligible). Compare the measured spray nozzles pressure drop thus obtained to the pressure drop across the spray nozzles predicted from the spray vendor catalog at the measured flow rate. Do not forget to apply the density correction as prescribed in the spray vendor catalog, as the catalog flows are based on water.

Typically, pressure drop across the sprays at design flow rates is 10–20 psi. For a given flow rate, a measured sprays pressure drop significantly above that calculated from the catalog means plugged spray nozzles; a measured sprays pressure drop well below that calculated from the catalog means broken header or spray nozzles. Measuring the sprays pressure drops at several different liquid flow rates and comparing to the sprays vendor catalog is ideal and more informative. From the comparison of measured versus calculated pressure drop, the number of plugged nozzles can be estimated.

Several case studies where this technique diagnosed the root cause of poor performance of a wash bed in a refinery tower were described in Ref. (130). In one, the pressure drop was zero because a header flange disintegrated due to rigid bolting of the header to the tower wall with no allowance for thermal expansion. In another, the pressure drop was 0.2 psi due to disintegration of a thin-gauge flange. In a third case, a pressure drop of 94 psi was measured where 11 psi was expected, indicating severe plugging. In another case (443), oversized nozzles and header leaks due to missing gaskets led to a pressure drop of 1 psi. The author can add many more to the list, including one instance (209, 211) when the pressure drop was 150 psi; most of the nozzles were plugged solid.

Turndown   Spray nozzles typically have a 2:1 turndown ratio. A high nozzle pressure drop produces fine spray droplets that can be entrained into the vapor. At very low pressure drops (typically <3 psi) the sprays droop, or even come out as "pencils" from the nozzles.

Fouling or plugging can be a concern with spray nozzles, but less so than for gravity distributors. The number of spray nozzles per square foot of bed cross-section area is much smaller than the number of drip points per square foot with a gravity distributor. This permits larger spray nozzle openings compared to gravity distributors, which reduces their plugging tendency. On the debit side, the tortuous path inside the spray nozzle, which generates the spray cone, produces a multitude of low-velocity spots where solids can be trapped and accumulate.

Filters, strainers, and/or screens should be used in the outside liquid pipes to the sprays to minimize the fouling per the guidelines in Section 4.5. Spray headers should be checked for the absence of, or at least the minimization of, stagnant pockets where solids can accumulate or liquid can overheat. Solids accumulating in such pockets often spall off and plug the spray nozzles. Lines from the filters to the tower should be fabricated out of corrosion-resistant metallurgy.

Flow tests   It is recommended that all spray nozzle systems be flow tested inside and/or outside the towers with water (if possible) to check the coverage. Where liquid distribution is critical, such as for wash sections of refinery vacuum towers, the sprays should be tested at every turnaround. If one spray nozzle fails, a significant cross-section area of the column will not be wetted properly. Tests have uncovered plugged nozzles (e.g., Figure 4.29), leaks, liquid shooting right down, loose nozzles, corrosion or erosion damage, and debris. Prior to the tests, the lines leading to the spray headers need to be flushed without going through the nozzles (505) to prevent the introduction of debris. The quality of water supply needs to be monitored. Dirty water aggravates plugging, and high-chloride water causes stress corrosion of stainless steel.

Testing inside the tower is ideal but is sometimes disallowed due to various safety practices and concerns. In this case, the header should be removed and tested outside. Also, the testing may reveal a need for tack or seal welding, which may mandate removal and outside testing.

Chapter 23 in Lieberman's book (287) describes his personal experience with spray testing inside one tower, which highlights some potential safety challenges. These include narrow access that slowed emergency exit, release of pockets of hydrocarbons and H_2S trapped behind

Figure 4.29   Distributor water test finds plugged nozzle. (From Sanchez, J. M., A. Valverde, C. Di Marco, and E. Carosio, Chemical Engineering, p. 44, July 2011. Reprinted with permission.)

plugged nozzles when the nozzles were loosened by hand, and excessive water pressure due to poor communication from inside the tower to outside. The high water pressure created noise (making communication even more difficult), mist that fogged safety glasses, and flushed out trapped hydrocarbons and H_2S into the tower.

When liquid distribution is critical, like in sprays to wash beds in refinery vacuum towers, the flow test should include obtaining a pressure drop versus flow curve. Control valve settings and valve openings should also be recorded. This test provides a useful baseline reference for future troubleshooting when plugging or breakage of the distributor is suspected.

Other than the pressure drop versus flow curve, a visual observation is usually sufficient in spray tests. Should a close evaluation of distribution quality be desired, area samples can be collected to study the flow variation. The visualization also gives insight into the droplet size range and whether collision of sprays from different nozzles creates dry spots or overirrigated spots underneath. Spray nozzle suppliers can often provide drop size distribution data determined from water tests. One watchout is that liquids that have lower surface tensions and densities than water produce smaller drop sizes.

Watch out for drain holes. These are seldom needed in spray distributors. With a high pressure drop across the spray nozzles, drain holes can generate powerful, undesired jets that damage packing.

For high-flow systems, such as pumparounds, a water test may require too much water to be practical. The distribution in pumparound sections is usually far less critical than in the wash section, so the test is often skipped. However, if practical, the test can still be performed by circulating water through the pumparound. Another way of testing nozzles for high-flow systems is by building a rig outside the column that can test one to three nozzles at a time. A less effective test is to run air through the spray header. It can identify nozzles that are plugged or shoot right through.

In some small towers, a single-spray or a few-spray distributor is used. Due to the non-homogeneity of the spray, these may give a very poor distribution. These should be carefully water-tested, and the nozzles should be replaced or adjusted in situ to improve the spray pattern as needed. The tests should be observed via a view port (if available) or an open manhole (505).

4.16 DISTRIBUTOR WATER TESTS

4.16.1 Gravity Distributor Water Tests in the Supplier Shop

Conducting water tests in distributors is essential because what the designer imagines does not always happen. Furthermore, many features in distributors are based on wishful thinking. Does wishful thinking work? Put water in it, and you will find out.

Most major suppliers have well-equipped test rigs capable of testing distributors up to 30 ft in diameter or more. Larger distributors can be tested in sections. The suppliers charge for these tests, but their costs are usually reasonable because it is in their interest to ensure their distributors achieve the design targets. There may also be a schedule impact. These penalties, however, are dwarfed by the costs of poor-quality products or lost production from a column. Often, the water test makes the difference between a working tower and a packing failure or troubleshooting case study. The following guidelines have been recommended (192, 346, 347) for distributor testing:

1. The industry's experience has been (192, 198, 346, 347) that
   - most distributor failures could have been prevented had the distributor been water-tested
   - among water-tested distributors, there have been very few failures, and

   • very few distributors pass water tests without any changes. The author polled experts in the field, and in their experience, the fraction of distributors that pass with no significant change ranges from 0% to 30%.

2. Every distributor should be water-tested, even if they are "alike" (171, 346, 347). Small variations in hole punching or drilling during fabrication lead to large, undesirable variations in flow. A common fabrication error is holes punched in different directions on different troughs or even on each wall of the same trough, so in some troughs the liquid faces the smooth edges of the holes while in others it faces the rough edges that may or may not be deburred. Touching with one's fingers is usually ineffective for identifying such issues. One water test found levels in some troughs about 1/3 deeper than others due to punching in different directions (111, 112). The author is familiar with a fabrication error that made holes in some troughs 5% larger than those in other troughs. These errors are detected as uneven flows in the water tests and can be corrected prior to installation.

3. The actual feed piping and parting boxes often influence the liquid distribution and need to be included in the water test. If this is impractical, request the supplier to build a very close replica for the test. In several cases, liquid from the parting box descended outside some of the troughs, or liquid was poorly metered to the troughs due to incorrect hole diameters. In one case, a sparger with dip tubes dumped more liquid into the parting box inlet area, forming a hydraulic gradient due to the narrow flow areas between the dip tubes and parting box walls. In another case, a parting box developed a similar hydraulic gradient because the liquid entered one end of the box from a downcomer. All these flaws were diagnosed and corrected in time thanks to the water tests.

4. If practical, attempt to include the collectors that bring liquid from the bed above in the test. A key lesson learned from a succession of collector malfunctions that led to poor separation (416, 417, summarized in Section 4.6.3) is "If it starts the wrong way, it will keep that way." If the collector delivers an uneven feed to the distributor, distribution to the bed is likely to remain poor.

5. Send a process engineer, and possibly also a mechanical engineer, to attend the test. Do not just rely on the supplier to test the distributor and provide statistics. It is important to pay close attention and review not only the statistical data but also to actively witness the test and observe the various flow patterns and possible nonuniformities (83). A full vision and understanding of issues are difficult to acquire from a distance.

   Should a large flaw be detected, some tough decisions may be needed, and distributor rebuild may be required. In some cases, the distributor can be modified and retested while on the test stand, saving doubts and later repetition. Inaccessibility may prevent some key measurements, and an alternative plan may need to be devised.

6. Good distributor tests have two components: qualitative and quantitative. In the qualitative part, watch the distributor over the entire flow range and ensure that (346, 347):

   • Each stream is independent and hits the packings at the desired point

   • Each stream is uniform in size

   • There is no splashing around feed pipes and parting boxes

   • There are no turbulent regions in parting box or distributor

   • Liquid levels do not approach the overflow

   • There are no significant hydraulic gradients in the troughs (pay special attention at the minimum rates) or parting boxes

   • Reflection of liquid at the far end of the troughs does not cause turbulence or eddies

- Liquid from all holes or notches of the parting box descends into the troughs and does not pour into the space between troughs

- Liquid streams from distributor holes pass a sufficient distance away from the vapor passages and are not deflected into them, where they can be entrained by the vapor as has been experienced (387).

- There is good liquid mixing in redistributors and feed distributors. This is critical at low liquid rates (171) when totally different streams enter a redistributor (Section 8.18) and in extractive distillation (Section 8.19). Injection of food dye or hot water can check mixing.

7. The quantitative part of the test measures the following:

- Flow of liquid from the feed pipe

- Flow and sampling from individual orifices. Most suppliers run statistics on these samples to produce a "coefficient of variation" (COV), which is the ratio of the standard deviation of the flow rate samples to their mean value. Usually they aim at COV <5% (53, 362) at full rates and <10% at turndown (83, 362). For *V*-notch distributors, this COV constraint is often relaxed to <15% at full rates and <20% at turndown (83).

- Head–flow relationship over the entire range

- Ensuring equal liquid levels in all troughs

- Leakage of any kind. This includes intended leakage due to weep holes and any unintended leakage.

8. Together with the supplier, videotape and fully document the test. This documentation may prove invaluable in case of future issues.

9. Water testing should be used for all distributors (53, 192, 346). Perhaps the only possible exception is orifice pan distributors for high liquid rates, noncritical separations such as those often used in pumparounds, amine absorbers, and amine regenerators.

Figures 4.25, 4.26, and 4.30 are examples of unexpected flow patterns that are extremely difficult, maybe impossible, to predict from first principles or identify in a normal troubleshooting investigation. Running water through the distributor would uncover them.

Difficulties are commonly encountered during distributor shop tests. Dhabalia and Pilling (83) present an interesting account of the challenges in shop testing of a *V*-notch distributor, including turbulence in the pre-distributor, excess velocity in the interior portions of the arm troughs, and a liquid level about 0.5″ higher at the end than in the middle of each trough. The water test identified the causes of these problems and led to their solution.

One water test revealed that above some troughs, the hole area in the parting box was larger than it should have been, yet surprisingly, the liquid levels in the troughs were even. When the excess hole area was plugged, the liquid head in some troughs was 10–20% higher than in others. A closer check found that some of the perforations were punched from the top down (rough side at the bottom) and others from the bottom up (rough side at the top). The cause was traced back to a fabrication shortcut after the perforated sheet was cut to form troughs.

4.16.2   Gravity Distributor Water Tests In situ

Water testing of distributors in situ is invaluable, both for new distributors and those that have been in operation.

For new distributors, people often raise the question whether it is necessary to retest in situ after the distributor has already been tested in the shop. The merit of the in situ testing is to reveal

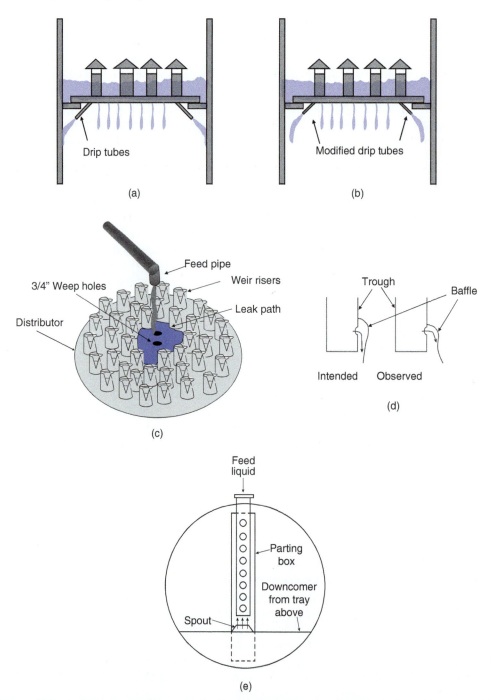

Figure 4.30 Problems uncovered by water testing (a) angled drip tubes were designed to get the liquid under the wide support ring. A water test showed the drip tubes were directing all of the liquid onto the wall; (b) the drip tubes were shortened to correct the flow problem; (c) in-situ water test detects liquid leaking from weep holes and through gaps around the weir risers, not even getting to the notches that are intended to provide distribution; (d) outward deflection of liquid guide baffle guides the liquid to an unintended location; (e) excessive hydraulic gradient observed in parting box. (Parts (a)–(c): From Olsson, F. R., Chemical Engineering Progress, p. 57, October 1999. Reprinted Courtesy of Chemical Engineering Progress (CEP). Copyright © American Institute of Chemical Engineers (AIChE).)

issues that cropped up during installation and remained unnoticed. The piping in the plant differs from the setup in the supplier's shop; the distributor supports in the plant differ from those in the shop; assembly errors could have occurred; debris and dirt could have lodged in some holes and plugged them; new leaks could have developed during assembly; feed pipe flanges may leak and damage could have occurred during transportation and mounting. Each of these can make a difference to the liquid distribution.

There are excellent examples of the merit of in situ testing. Bouck (43) found that at the minimum design liquid rate, the feed pipe cycled between being covered and uncovered by liquid in the distributor, changing the pressure drop across the control valve, which made it difficult to hold the setpoint flow rate. An in situ water test by Schnaibel et al. (417) provided an unexpected finding that liquid flow to some down boxes in a collector was obstructed by long chimneys, so these boxes received much less liquid than unobstructed down boxes. Another in situ test (264) identified maldistribution due to partial plugging of distributor holes by rust and debris as the root cause of poor performance. Diaz et al. (84) discovered a flange leak during an in situ water test, that contributed to the poor packing efficiency in a chemical vacuum batch distillation column. In a case the author is familiar with, a distributor that tested well in the shop showed maldistribution when tested in situ due to a sag in the middle. The sag did not occur in the supplier's shop because it was better supported.

For distributors that have already been in service, the above holds, and in addition, residual dirt, corrosion of holes, damage to delicate parts, bending and sag, loosening of nuts, bolts, or clamps, damage to flanges and gaskets, and unexpected splashing could all adversely affect the distribution.

In situ tests with the packings in place afford the additional merit of showing the liquid distribution pattern as it comes out of the bed. Poor patterns may develop due to an anomaly or plugged or damaged section in the packed bed, or the development of excessive wall flow.

Although in situ water tests usually lack the sophisticated instrumentation and measurement equipment available for supplier shop tests, they usually provide the information needed for troubleshooting. An attempt should be made to run the test at the normal liquid flow rate and test turndown as necessary. The water for the distributor can be provided from the water mains via the distributor feed pipes, pumped from a specially installed water tank, or simply provided by a water hose. Good solid-free and noncorrosive water quality is mandatory for the test, and in the case of stainless steel, also low in chlorides. The quantity of water needed for the tests should be estimated ahead of time, and special pumps should be ordered as needed. The setup used must be approved by the health, safety and environmental personnel. Water exiting the bed can be sampled by a cup that is tied to a rod and is moved around to different points, with its filling time at each point recorded.

It is useful to watch the water emptying out of the distributor after the water supply is stopped. The last streams of water should come out randomly or reasonably evenly. If the last $\frac{1}{4}$–$\frac{1}{2}$ in. of water come out from one area, a levelness issue is implicated.

Figure 4.30c shows an in situ water test that explained a very poor water-ethylene glycol separation (346, 347). All the distributor liquid leaked at the tray floor through $\frac{3}{4}$ in. weep holes and through gaps where the weir risers fitted through the deck. Figure 4.30d shows a baffle deformed in service, deflecting the water issuing from distributor holes in an undesired direction. Figure 4.30e shows a 12-ft long parting box receiving downcomer reflux liquid from the trays above as well as fresh liquid feed. A water test showed a hydraulic gradient in the box with preferential flow and overflow at the downcomer end of the box. Section 9.2.8 describes how an in situ water test showed maldistribution that gave poor separation for decades and was not picked by gamma scans.

A caustic absorption tower with four circulating pumparound beds was water-tested in situ by circulating demineralized water and observing the action through the manholes (77). The tests revealed plugged perforations and a need for mechanical cleaning in one of the distributors. In addition, they revealed overflow at 75% of the design liquid flow rates, which explained the carryover they were experiencing during operation and forced operation at reduced circulation. In another caustic absorption tower, a water test on the collector from a pumparound circuit found no leaks, but circulating water through the circuit with the manhole open found that some of the collector hats overflowed into the risers.

In small towers, it may be possible to test the vapor distribution. An air blower can be connected to the vapor entry nozzle and operated at normal flow. If safe and practical, it may be possible to measure the flow distribution in the tower using an anemometer. Another test advocated by Willard (505) is to run water for several minutes until the packings are fully wetted, then turning the water off while allowing the air to flow. Packings moist at the center and uniformly wet at the walls indicate good vapor distribution.

Chapter 5

Qualitative Gamma Scans Troubleshooting: The Basic Diagnostics Workhorse

"Gamma scans for distillation are analogous to X-rays in medicine. The process engineer is analogous to the doctor, the gamma scan expert is analogous to the X-ray expert. Until you bring the two resources together, you do not have a good diagnosis."

—(Henry Kister, author)

5.1 GAMMA-RAY ABSORPTION

Distillation column gamma scans employ radioactive sources in the 500- to 2500-keV range (66). Compton scattering is the prime process responsible for the attenuation of these rays. The radioactive source usually used is cobalt-60, and less frequently cesium-137. A summary of the principles follows; a detailed description is available elsewhere (66).

When gamma rays pass through a medium from a radioactive source to a detector, some of their radiation is absorbed by the medium. The amount of radiation that is transmitted (not absorbed by the medium) is given by (66, 67, 423)

$$I = I_0 \exp\left(-\mu\, \rho\, x\right) \tag{5.1}$$

where I is the radiation intensity in counts per second, as measured at the detector; I_0 is the radiation intensity of the source, in counts per second; ρ is the density of the medium; x is the thickness of the medium; and μ is the absorption coefficient, which depends on the gamma-ray source and the medium material.

When the gamma-ray energy exceeds 200 keV, μ becomes independent of the chemical composition of the medium, and the absorption becomes a function of the product of the medium density by the medium thickness (66, 67). For a horizontal chord of fixed length (Figure 5.1a), the intensity of radiation received at the detector is therefore a function of the density of the medium. If the gamma ray passes through metal (very high density) or liquid (high density), the intensity received by the detector is relatively low, but if the ray passes through a vapor space (low density), the detector reading is high.

The source and detector are lined up on the same horizontal plane, positioned some distance above the top tray (or the upper tray of the desired scan), and a reading is taken (a gamma

Distillation Diagnostics: An Engineer's Guidebook, First Edition. Henry Z. Kister.
© 2025 American Institute of Chemical Engineers, Inc. Published 2025 by John Wiley & Sons, Inc.

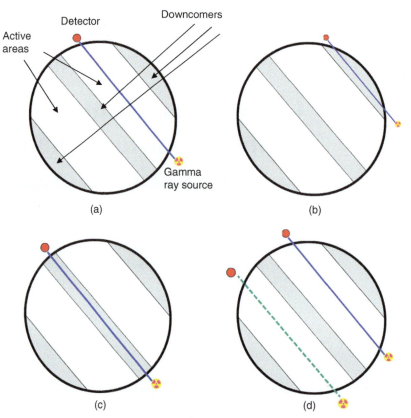

Figure 5.1 Good gamma scanning practices, shown in reference to 2-pass trays. (a) Active area scan and terms of reference; (b) side downcomer scan; (c) center downcomer scan; (d) active area scan looking for maldistribution.

ray is "shot") either across the tray active area or across the downcomer (as desired). Both the source and detector are then simultaneously lowered to the next vertical position, where the next reading (gamma-ray shot) is taken. The source and detector are thus simultaneously moved vertically straight down the column. Metal guides attached to the column and mechanized systems that move the detector and source synchronously down the guides are used for this purpose. Radiation intensity reaching the detector is recorded at each vertical position. High recorded intensities indicate vapor spaces along the ray's path; low recorded intensities are interpreted as passage through liquid or solid. The vertical intensity profile thus obtained provides information on column behavior and identifies the location and nature of column irregularities.

5.2 QUALITATIVE GAMMA SCANS

This is the most common and least expensive gamma scanning technique, accounting for over 80% of the gamma scan applications. In tray towers, shots are typically taken every 2 in. of vertical height. The shooting continues to the bottom of the tower or to the bottom of the section of interest. After completion of the scan, additional chords can be shot. Often, a second chord is shot through the downcomers (Figure 5.1b or c) or through another pass of the tray to look for channeling (Figure 5.1d).

Figure 5.2a (66, 67, 283, 503) and 5.2b (283, 503) include fault-condition diagrams illustrating how different types of irregularities show up on a gamma scan.

- A typical normal operation tray scan will show a high-density (or low-detector-reading) region just above the floor of each tray (due to the presence of liquid), followed by a low-density (or high-detector-reading) region in the vapor space between the trays.

- A typical flood appears as a high-density region in the vapor space between trays, with the high density resulting from the accumulation of liquid.

- Foaming typically appears as a region of uniform intermediate density in the vapor space between trays. In non-foaming scans, the density is high near the tray floor and decreases with elevation above the tray. With foaming, the dispersion is uniform like in a bubble bath or in the head on the glass of beer.

- Weeping typically appears as a region of low, uniform density in the vapor space between the trays. Weeping looks like rain, with only a relatively small amount of liquid in the vapor space. Like rain, this amount is uniform throughout the vapor space. Also, under weeping conditions, the peaks at the tray floor are usually low (low froth heights), and the vapor space above the top tray is usually clear.

- Missing, damaged, or otherwise dry trays typically appear as a low-density region where a tray is expected. Note that gamma scans are shot every 2 in., while the tray is only 0.1 in. thick, so a dry tray may look similar to a missing or damaged tray. Gamma scans often do not see the trays; they see the liquid sitting on the tray.

- In a packed column, a packed bed appears as a medium-density region (medium detector reading), a collapsed bed as a low-density region (high detector reading), and plugging or flooding as a high-density region.

- In a packed bed, a gradient indicates irregularity. Denser at the bottom suggests plugging or flooding at the bottom not propagating upward, or missing packings near the top. Denser at the top suggests plugging or flooding near the top.

5.2.1 What Do Qualitative Gamma Scans Show in Tray Towers

Figure 5.3 (45) illustrates a typical evaluation of tray active areas. The y-axis shows elevations in reference to a certain point in the tower. Typically, the top tangent line is marked as zero height, and the y-axis shows distances below it. The x-axis shows the radiation picked up by the detector in counts per unit time, plotted on a log scale. In some scans, Eq. (5.1) is applied to convert the counts to a relative density scale. The absolute values of the densities thus obtained are likely to be inaccurate unless reference scans through pure liquid at the same chord are available.

Marked on the plot are tray locations and sources of external interference (e.g., welds, supports, nozzles, and insulation rings). The actual scan data, i.e., the connected dots from the measurements taken every 2 in., form the solid plot. The "clear vapor" line on the right side of the plot is a reference line giving the transmission of gamma rays through clear vapor. To reliably determine this line, the scan needs to pass through a region where clear vapor exists (typically above the top tray, above the reboiler return inlet, or above a chimney tray, in the absence of flooding or weeping in this region). The "froth height intensity" line provides a simple, approximate empirical measure of froth heights by averaging the radiation counts of vapor and liquid. In Figure 5.3, the vapor and liquid radiation counts are 3800 and 200, respectively, so the froth line is at their average (about 2000). The froth height is approximated by measuring the distance from deck to spray, as shown for Tray 2 in Figure 5.3. For this tray, the froth height is 9 in. (225 mm). A more accurate method for determining froth heights (160) is discussed in Section 6.1.1.

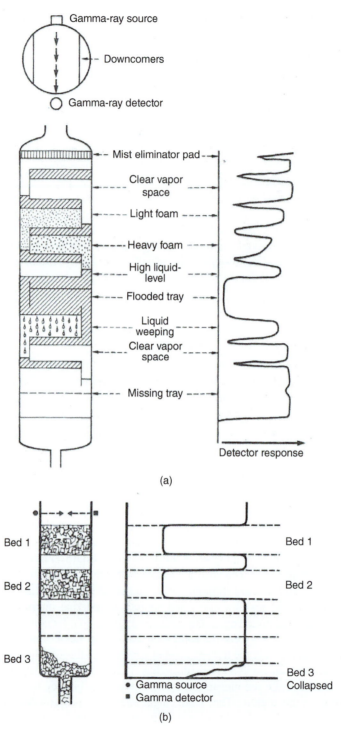

Figure 5.2 Illustrative gamma scans, depicting various types of column irregularities. (a) Tray column; (b) packed column. (Part (a): From Charleston, J. S., and M. Polarski, Chemical Engineering, January 24, 1983. Reprinted with permission. Part (b): From Leslie, V. J., and D. Ferguson, Plant/Operation Progress, 4(3), p. 145, July 1985. Reprinted with permission.)

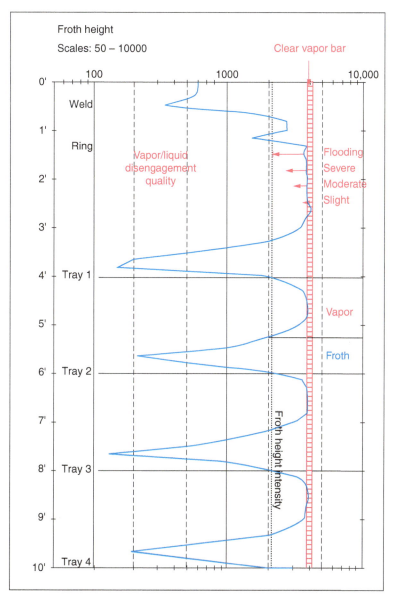

Figure 5.3   A gamma scan illustrating typical evaluation of active areas of trays based on measured froth heights. (From Bowman, J., Chemical Engineering Progress, p. 25, February 1991. Reprinted Courtesy of Chemical Engineering Progress (CEP). Copyright © American Institute of Chemical Engineers (AIChE).)

Gamma scan contractors often use their proprietary methods for froth height determination. Froth heights are expressed either as linear values or as percentages of the tray spacing.

In Figure 5.3, all the vapor space peaks reach clear vapor, indicating little or no entrainment. When some of the spray approaches the tray above, the vapor space peaks do not reach the clear vapor line. This condition is reported as "entrainment." The degree of entrainment can be qualitatively inferred from the proximity of the top of the peak to the clear vapor line. The arrows above tray 1 show the commonly used terminology, ranging from "slight" to "heavy" entrainment and finally "flooding." This is further discussed in Section 5.2.2.

Qualitative gamma scans can readily detect gross abnormalities such as flooding, missing trays, collapsed trays, excessive liquid level in the bottom sump, or heavy foaming. This technique can also help diagnose a flood mechanism and shed light on more subtle abnormalities such as high or low tray loadings, foaming, excessive entrainment, excessive weeping, blockage, and multipass liquid maldistribution. Gamma scans performed on a routine basis can also be used to monitor deterioration in column performance due to fouling, corrosion, damage, and other factors.

5.2.2 Entrainment versus Weeping

Both entrainment and weeping show liquid in the vapor space, yet their appearance differs. Figure 5.4 (485) shows illustrative active area scans that highlight the differences.

Generally, entrainment vapor space peaks (Figure 5.4a) show dense material near the tray floor, becoming less dense as one ascends the vapor space, terminating in a single point before the next tray is encountered. This appearance characterizes the shape of a spray: there are more, larger drops and froth near the tray floor, which show on the scan as high-density material. The entrained drops that reach higher elevations become progressively smaller and fewer. This in turn progressively reduces the concentration of liquid in the vapor space, which shows progressively lower densities as one ascends from the tray floor.

Figure 5.4a illustrates peaks typical of different degrees of entrainment. The uppermost tray, just above the 4 ft mark, shows 6–8 in. froth height with no significant entrainment, as the peak gets close to the clear vapor line. The second tray, just above the 6 ft mark, is holding 10–12 in. of froth and has a peak 85–90% below the clear vapor line, which classifies it as slight entrainment (485). Notice the point of inflection in the slope of the froth, suggesting a spray zone developing above the tray froth. The third tray, just above the 8 ft mark, is holding 16–18 in. of

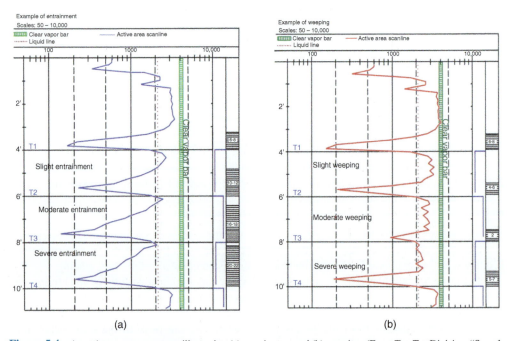

Figure 5.4 An active area gamma scan illustrating (a) entrainment and (b) weeping. (From Tru-Tec Division, "Sample Scan Brochure," 1998. Available from Tracerco, Pasadena, Texas. Reproduced Courtesy of Tracerco.)

froth or spray and has a lower single-point vapor space peak. When the vapor space peak is at about 70–80% of the clear vapor line, it is often classified as "moderate entrainment." The fourth tray, just above the 10 ft mark, has 20–22 in. froth height and is close to the jet flood point. The vapor space peak is at about 50–60% of the clear vapor, and the entrainment is classified as "severe."

Figure 5.5 (510) provides a pictorial illustration of how active area gamma scan peaks vary as internal rates are increased in a tower. Figure 5.5a shows operation at the design rates, which for this tower were well below flood, with froth heights less than half the tray spacing. Figure 5.5b and c shows how the peaks increase as rates were progressively raised to 120% and 130% of design, respectively. At 130% of design, the froth height on the feed tray approached the tray above, with some entrainment to the tray above, as evidenced by the absence of clear vapor above this tray. The trays below the feed still had some margin from flood.

Note that the entrainment classification from the gamma scans does not coincide with the amount of liquid entrained or carried over from one tray to the next. The entrainment reported by gamma scanners is a measure of how much liquid is present at the vapor space peak. Gamma scans do not measure how much of this reaches the tray above. One case history was presented (377), and the author experienced many similar cases, in which a column performed well despite severe entrainment from many of its trays.

Figure 5.4b illustrates different degrees of weeping. As distinct from the point vapor space peaks that entrainment gives, weeping gives flat peaks. Weeping is liquid raining from the tray above, and it will have the same uniform liquid concentration (and therefore the same density in the scan) at any elevation in the vapor space. Another distinguishing feature is that weeping usually occurs at lower vapor loads, where the froth heights are low, while entrainment is most severe at high vapor loads. Likewise, if scans are performed at different vapor loads, the entrainment will increase, while weeping will diminish, with higher vapor loads. Finally, look closely at the top tray. Entrainment will show up above the top tray, weeping will not (as there is no tray above from which the weep comes).

In Figure 5.4b, the uppermost tray, just above the 4 ft mark, is holding 6–8 in. of liquid and has a vapor peak underneath of about 80–90% of the clear vapor. This is classified as "slight weeping." The second tray, just above the 6 ft mark, is holding 4–6 in. of liquid and the vapor peak underneath is at about 70–80% of the clear vapor. This is classified as "moderate weeping." The third tray, just above the 8 ft mark, shows very little (about 2 in. or less) liquid on the tray, characteristic of dry or damaged tray (485), and the vapor space peak underneath is at 50–60% of the clear vapor. This is classified as "severe weeping."

If the scan indicates moderate or severe weeping, it is worthwhile to check the amount of weeping that should be expected by calculation. Good procedures for weep calculation are available in the literature (74, 179, 193, 245, 300). If the calculation predicts only a small amount of weep while the scan shows severe weeping, then there is a good possibility of corrosion, damage, manways left open, or some unexpected hydraulic mechanism (such as channeling or leakage at the support rings) that may cause the weeping. See Section 3.2.2 for troubleshooting for tray weeping.

5.2.3 What Do Qualitative Gamma Scans Show in Packed Towers

In packed towers, a "grid" of four equal chords is often shot, one chord after the other (Figure 5.6a). For each chord, the source and detector are moved simultaneously down the bed, taking shots every 2–4 in. This "grid" gamma scan looks for maldistribution and channeling, which are by far the main causes of packed tower efficiency loss (Section 4.1). When liquid distribution is good, the four chords give the same detector readings, so their plots fall right on top of one another (Bed 1 in Figure 5.7). Differences between the chords are interpreted as bed

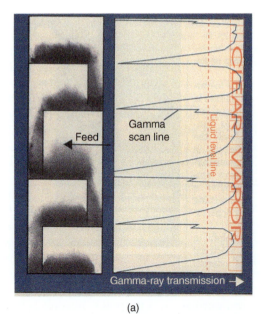

(a)

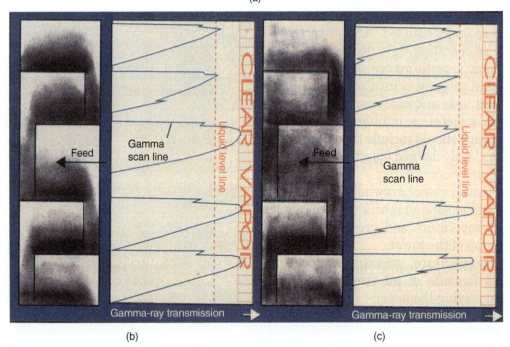

(b) (c)

Figure 5.5 Active area gamma scans as the tower rates are raised and the froth heights increase. (a) Lowest rates, for which the tower was designed; (b) 120% of the design rates, froth heights increase, and the dispersion approaches the tray above; (c) 130% of the design rates, dispersion reaches the tray above, getting close to flood at the feed tray. (From Xu, S. X., C. Winfield, and J. D. Bowman, Chemical Engineering, p. 100, August 1998. Reprinted with permission.)

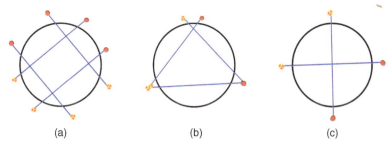

(a)   (b)   (c)

Figure 5.6   Gamma scanning chords used in packed towers. (a) Grid scan, recommended for diameters 4 ft or larger; (b) triangle scan, used for diameters 2–4 ft (c) + scan, used for diameters <2 ft.

maldistribution. For instance, one of the chords in Bed 2 of Figure 5.7 shows more radiation transmission than the other three, indicating less liquid flowing through the bed along this chord. The higher transmission initiates immediately under the distributor, indicating less liquid issuing from the distributor along this chord, and therefore a distributor problem. In this case, the problem was out-of-levelness. Gorrochotegui and Portillo (137) presented a very similar out-of-levelness scan, with two chords showing more radiation transmission than the other two.

The packing chords are usually oriented such that two intersect the other two at an angle of 90°. This arrangement is considered optimal for the verification of the presence and integrity of liquid distributors and collectors.

Judicious setting of the chords can also provide a measurement of liquid height and frothiness in collectors, parting boxes, and distributors, and identify trough or pan overflows or level unevenness causing liquid maldistribution or premature floods. Overflows and plugging (which too will cause overflows) were identified as major troubleshooting issues in packed towers (Sections 4.5 and 4.6). FRI's tests (382) show that commercial gamma scans can provide excellent measurements of liquid levels in the distributor troughs of a test tower.

In addition, grid scans can normally detect the position of the bed, the disappearance of some of the packings (e.g., by being corroded away), significant blockages, and local flooded regions.

Figure 5.8 (64) shows a single chord scan in the FRI 4 ft ID tower containing fourth-generation random packings with the C_6–C_7 test system at 24 psia at total reflux under excellent distribution conditions as verified by FRI. The various curves are for different loadings through the bed, ranging from a dry bed to 102% of the maximum useful capacity (MUC). At 100% of the MUC the HETP was good, and at 102% it doubled. Figure 5.8 shows that up to 79% of MUC, the bed appeared only slightly denser than in the dry scan. The density significantly increased at 100% of MUC. It follows that the relatively high density in the scan implies the proximity of flooding.

Figure 5.9 shows the corresponding distributor liquid level, showing clear increases as the liquid loads were raised. The horizontal lines show how the liquid levels were determined. The liquid levels measured in the distributor were in excellent agreement with the bubbler measurements (64) conducted by Fractionation Research Inc. (FRI).

Vidrine and Hewitt (493) discussed some limitations that constrain the accuracy of the liquid maldistribution diagnostics. For instance, with an 8 ft diameter tower, packing density of 15 lb/ft³, and low liquid wetting, the liquid accounts for only 1.5 lb/ft³. In this case, a 50% maldistribution results in a radiation variation similar to that produced by a 1.25″ shift in both source and detector positions (e.g., due to a platform disturbance or by wind). In addition, the statistical error in radiation counting typically alters the calculated density by about twice the square root of the count rate, which in the cited example is equivalent to a 50% liquid maldistribution. Finally, Vidrine and Hewitt point out that the error rapidly escalates with reduced source strength and

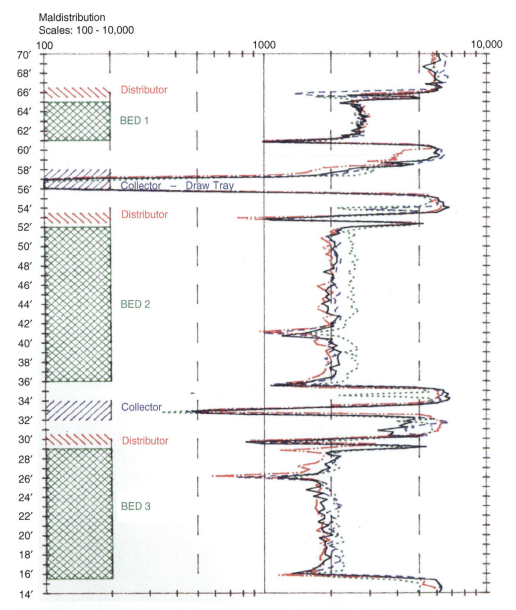

Figure 5.7 A packed tower gamma scan, illustrating excellent liquid distribution in the top bed, gross liquid maldistribution initiating at the distributor in the middle bed, and mediocre liquid distribution in the bottom bed. (From Tracerco, "Process Diagnostics – TRU-SCAN® Case Studies," 2010. Available from Tracerco, Pasadena, Texas. Reproduced Courtesy of Tracerco.)

wider towers. In large-diameter towers (>15–20 ft diameter) the source needed for obtaining a reliable liquid maldistribution diagnostics may be too strong to be practical.

Whenever possible, grid scans should be preferred. In smaller columns (<4 ft diameter), a grid scan may not be practical, and is often substituted (376) by a triangular scan (2–3.5 ft diameter, Figure 5.6b) or by a "+" scan, which means two chords perpendicular to each other (<2 ft diameter, Figure 5.6c). Refer to Section 9.2.8 for a discussion on an issue with + scans.

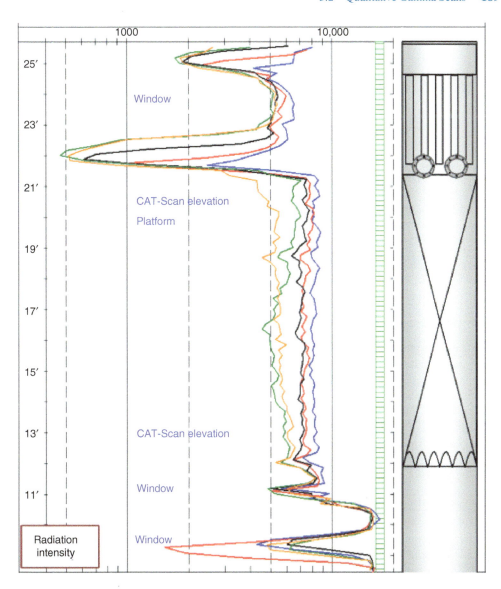

Scan	% Max. useful capacity	Data curve
1	Dry	Blue
2	45%	Red
3	79%	Black
4	100%	Green
5	102%	Orange

Figure 5.8   A single-chord gamma scan of a bed of fourth-generation metal random packing under excellent liquid distribution in a 4 ft ID test column. The curves show scans for total reflux operation ranging from a dry bed to a 102% of the maximum useful capacity. (From Chambers, S., L. Pless, R. Carlson, and M. Schultes, "Gamma Scan and CAT-Scan Data from a Distillation Column at Various Loadings," The Kister Distillation Symposium, Topical Conference Proceedings, p. 119, the AIChE National Spring Meeting, San Antonio, TX, April 29–May 2, 2013. Copyright © FRI. Reprinted Courtesy of FRI.)

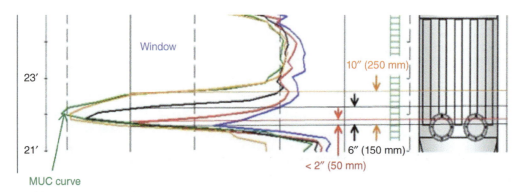

Figure 5.9 A single-chord gamma scan of the distributor in the 4 ft ID test column of Figure 5.8. The curves show scans for total reflux operation ranging from a dry bed to a 102% of the maximum useful capacity. Legend is the same as for Figure 5.8. (From Chambers, S., L. Pless, R. Carlson, and M. Schultes, "Gamma Scan and CAT-Scan Data from a Distillation Column at Various Loadings," The Kister Distillation Symposium, Topical Conference Proceedings, p. 119, the AIChE National Spring Meeting, San Antonio, TX, April 29–May 2, 2013. Copyright © FRI. Reprinted Courtesy of FRI.)

5.2.4 Stationary Monitoring ("Time Studies")

In this application, a number of strategically placed stationary gamma-ray sources are placed along the column shell at elevations coinciding with vapor spaces inside the column. A stationary detector is placed right across from each source at the same elevation. Starting in the unflooded condition, all these detectors will read high as the gamma rays travel through vapor. The rates (reflux, boilup, feed, or whatever variable is studied) are raised until the amount of radiation transmitted sharply falls in one of the vapor spaces, indicating liquid accumulation and therefore flooding. The location where the radiation transmitted initially drops off is where flooding initiates, and the rates at which it occurs are the flood rates.

The author has had extensive excellent experience with this technique, and many other successful applications are described in the literature. In one case study (509), regular gamma scans at high rates identified flooding in two regions, one region near the top and another near the bottom. It was required to determine which floods occurred first. Stationary source and detector pairs were placed between the top two trays, above the bottom tray, and below the feed point. As rates were gradually raised, these points were continuously monitored. This test conclusively showed that the top point flooded first. In another case (146), a stationary monitor was set up 15 in. above the feed tray, this time using neutron backscatter (Section 6.3) instead of gamma scans. When feed rates were raised, the detector saw liquid accumulation well before the pressure drop in the tower started rising, verifying flood initiation at the feed tray.

Stationary monitoring is useful for determining the initial point of flooding, for calibrating liquid level instruments, for detecting entrainment and overflow from packing distributors, and for detecting floods due to high bottoms level. Stationary monitoring is also valuable for accurately determining the flood rates, permitting engineers and operators devise a strategy for pushing the tower to its limits without flooding (45).

One limitation is that the vertical positions of the stationary sources cannot be too close to each other (typically about 6 ft) to avoid radiation interference. This can be overcome by repeating the test after moving the stationary monitoring a short distance up or down. In one case, stationary monitoring detected no flooding on the bottom tray, even though flooding occurred a few trays above. The test was repeated after moving the bottom monitoring one tray up. This time flooding was observed. This suggested flooding by downcomer backup, as the hydraulic loads were slightly higher on the bottom tray, so the only conceivable explanation for no flood at the bottom tray and flooding on the next was that the bottom downcomer was much longer than those above and could tolerate a higher backup.

5.2.5 What is in the Tower Inlet/Outlet Pipes?

Gamma scans can detect plugging, presence of vapor in liquid lines, and presence of entrained liquid in vapor lines. This technique can give reliable quantitative numbers when performed properly.

The basis of the method is Eq. (5.1). For best results, a piece of pipe of the same diameter, thickness, and material as the pipe to be checked is cut in the workshop. A flat plate is welded or flanged to one end, forming a cylinder closed at the bottom and open at the top. The empty cylinder is then gamma scanned. The cylinder is then filled with water and scanned again. From these, the constants in Eq. (5.1) are evaluated. Next, the pipe to be checked is scanned at one or several spots. The density of the material inside the pipe can be interpolated using Eq. (5.1). It is a good practice to scan in a + formation (in a horizontal pipe, top to bottom and left to right).

As an additional check for liquid pipes, it may be desirable to scan the pipe to be checked, or a similar pipe carrying a similar process liquid, at a location where plugging and vapor bubbles are unlikely. This check should yield the density of the process liquid, which should equal that from the simulation.

Our experience with this type of scanning in liquid pipes, both for determining plugging and vapor bubbles, qualitatively and quantitatively, has been excellent. With scanning vapor pipes for liquid entrainment, our experience has been good qualitatively, but the quantitative results did not always match our expectations.

Ferguson (Case 6.8, Ref. 201) reports a clever application of this technique to check for slug flow in a tower feed pipe. They positioned a source and detector across the upward elbow of the feed pipe at the bottom of the vertical section and monitored the intensity of radiation passing through it over time. Rapid cycling of this intensity indicated that liquid was collecting and disappearing, as in slug flow. Garcia et al. (118) reported successfully applying this technique to verify vapor presence and measure density in a liquid product rundown line.

5.3 GAMMA SCANS PITFALLS AND WATCHOUTS

Pitfalls must be avoided, as they can degrade the invaluable information produced by gamma scans into inconclusive, even misleading, diagnostics. The guidelines below can help avoid common pitfalls that adversely affect the validity and quality of information supplied by gamma scans:

1. It is critical to combine the information provided by the scans with information from other troubleshooting techniques such as hydraulic calculations, flood tests, temperature surveys, and vapor/liquid sensitivity tests. It is best to perform the calculations and field tests well before the gamma scans, as information from calculations and tests is invaluable for guiding the gamma scan plan and getting the most out of the gamma scans. Questions like "Do the gamma scans show flood where the hydraulic analysis predicts flood?" or "Are the liquid or vapor loads so small that the trays active areas may dry up?" or "Has there been foaming experience in the service where the gamma scans interpreted the trays to foam?" can go a long way to eliminate "lying gamma scans".

   Distillation troubleshooting is analogous to medical diagnostics. The best way to work with gamma scanners is like a good doctor working with an X-ray specialist. Remember, it is the doctor, not the X-ray specialist, who does the diagnosis on the patient. While the X-ray information is very important, often the doctor cannot diagnose the ailment without it, it is only one piece of the puzzle that needs to fit in with the other pieces. And if it does not, you do not have a diagnosis. Just like a good doctor in this situation, you will need to perform more tests and checks until you can reconcile what you know about the tower with what the gamma scanners tell you.

Sections 5.5.9, 5.5.20, as well as Ref. 18, describe case studies where gamma scans used as the sole basis for troubleshooting led or could have led to misdiagnosis.

2. It is important to provide the scanners with a good set of drawings of the column and internals, which include accurate information on tray or packing location, orientation, and column shell thickness. It is also beneficial to give the scanners a summary of the column history with emphasis on the problems it is experiencing, including specifics such as rates or times at which the problem occurs. This information is essential for good planning and interpretation of the scans.

3. Interpretation of gamma scans may be difficult and requires good knowledge of both scanning technology and column operation. Often, a gamma scan contractor is brought into the plant, shoots gamma scans, and then writes a report. The author has seen many such reports containing interpretations that were way off the mark and misled the team. To avoid this, a person or persons who are familiar with the tower design and operating history need to closely communicate with the contractor before scanning, during scanning, and when the scans are being interpreted.

4. It is essential to have stable plant operation while scanning. Variations due to instability may be misinterpreted as column abnormalities. Also, the active area scans and the downcomer scans, or the four chords in a packed tower grid scan, are shot at different times, and a change in plant conditions may be misinterpreted as an issue with tray, downcomer, or packing hydraulics.

5. Avoid shooting gamma scans in extreme weather conditions. These often promote column instability while slowing down the gamma scans. Both may lead to variations that can be misinterpreted. Also, when scanning narrow downcomers or distributor troughs in windy conditions, it is often necessary to use hard lines or guide wires to make sure the scan path remains constant.

6. When troubleshooting for flood or other capacity limits, always perform a flooded and an unflooded scan. The flooded scan can give a good indication of what is happening and the location of the flood, but it does not offer much information on the cause. The unflooded scan should be shot at loads just below the flood point, and is invaluable in providing information on what is likely and what is unlikely to be the cause.

7. Concisely document the key tower variables (flows, pressure drops, pressures, levels) for each scan. Avoid terms like "high feed rate" and "low feed rate." This is imperative with towers that are repeatedly scanned while varying some process conditions. Request the scanners to include the key conditions in their report. Many a scan had become worthless as no one remembered under what conditions they were shot.

8. Active area scan chords (Figure 5.1a or 5.1d) should be placed strategically, at least 6 in. away from the downcomer walls to positively avoid reflections. If it is important to assess the vapor space below the bottom tray, the scan chords should also be strategically placed away from the seal pan.

9. Some scanners prefer to shoot downcomer scans perpendicular or angled to the outlet weir (e.g., Figure 5.10) instead of parallel to the weir, as shown in Figure 5.1b and c.

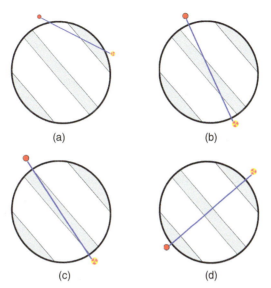

Figure 5.10   Bad downcomer gamma scanning practices, not recommended, shown in reference to 2-pass trays. (a) Side downcomer scan; (b) center downcomer scan; (c) center downcomer diagonal scan; (d) side and center downcomer scan, terrible practice.

The author has had very few satisfactory experiences with downcomer scanning angled or perpendicular to the weirs. Such scans are very difficult, often impossible, to correctly interpret due to the coupling between the downcomer and tray dispersions. This coupling renders such scans a breeding ground for misleading diagnostics. The problem is aggravated when the downcomer is sloped, because the relative portion of the chord passing through the downcomer decreases as one descends. Furthermore, at the top of downcomers, angled gamma rays pass through support beams, which may be quite thick; their presence shows up in the same manner as extra liquid. A chord passing diagonally through the downcomer from one wall to the opposite wall (Figure 5.10c) may pass through the thick bolting bars and should also be avoided. In one 20 ft diameter tower, the downcomer was scanned per Figure 5.10c. The initial scan report concluded a downcomer flood, but the final report backed off, stating that no definitive analysis could be made due to suspected interference from "heavy construction elements." For these reasons, the author recommends always performing downcomer scans parallel to the weir (Figure 5.1b or 5.1c) and not perpendicular or angled to it (Figure 5.10).

10. Side downcomer scan chords (Figure 5.1b) may be shallow, with scattering and reflection lowering accuracy. When the column shell is thick (e.g., high-pressure services), the metal can absorb much more radiation than the fluid. Consider a scan along a 12 in. wide chord in a side downcomer containing liquid of a specific gravity of 0.5 with 50 percent aeration. The mixture's specific gravity will be 0.25, much less than the specific gravity of steel, which is about 8. If the column shell is 1 in. thick, it will absorb approximately five times more of the radiation than the process fluid. For this reason, in high-pressure columns, it is best to scan center downcomers (Figure 5.1c) rather than side downcomers (Figure 5.1b), and caution is required when interpreting side downcomer scans with short chords.

11. Side downcomer scans should be run at a high threshold setting on the radiation counting equipment in order to minimize scattered radiation. A high threshold setting electronically discriminates low energy or scattered radiation, resulting in higher-resolution data.

12. With the gamma scans shot every 2 in., and the tray thickness being about 0.1 in., it is difficult for gamma scans to distinguish dry trays from missing trays. Gamma scans do not see trays; they see the liquid sitting on the trays. This can be a major issue in low-liquid-load or low-vapor-load trays. It is important to check using the hydraulic analysis whether there are low liquid and/or vapor loads, and if so, to repeat the scan at higher values. If the rescan at the higher vapor or liquid loads still sees no liquid on the tray, then the missing tray diagnosis is supported. Sections 5.5.9 and 5.5.20 give excellent illustrations.

13. Inadequate feed and reflux distribution into trays, as well as poorly designed side draws and chimney trays inside the tower, are common sources of bottlenecks. Adequate scanning for partial flooding and maldistribution in the vicinity of the relevant points should be included.

14. With packed towers, it is valuable to measure liquid levels in collectors and distributors when practical. Overflowing collectors and distributor troughs is one of the most frequently encountered malfunctions reported in packed towers (Section 4.6), but may require extra scans to avoid interference from the trough metal. In narrow troughs, level measurements may be difficult, so seeking the scanner's advice is crucial.

    Tray towers are not immune to problems with overflowing collectors, and it is valuable to measure liquid levels on them, as described in Section 5.5.19.

15. The desired scan intervals need to be set. Scanners usually shoot gamma rays every 2 in. as they descend the tower. This may be too large an interval when scanning packing liquid distributors or chimney trays to determine liquid level, regions under spray distributors to identify plugged or broken spray nozzles, froth heights at short tray spacings, or short downcomers. The scanning can be easily performed at smaller intervals (every 1 in.), but this raises the costs. Only the critical region needs to be shot at these small intervals.

16. Once the desired scan chords are set, the scanners should be consulted. Practical difficulties are frequent due to nozzle or pipe interference, restricted access, difficulty of passing the source or detector through platforms, and other issues. A review by the scanners can foresee difficulties and formulate solutions. These solutions may involve drilling holes in some platforms, moving a chord a few inches, or skipping a few unimportant trays. At other times, it may favor a switch from one side of the tower to the other. Additional scaffolding or other equipment may be needed, and should be set up ahead of the test. At times, the practical difficulties make it impossible to shoot some desired chords, and an alternative plan would need to be devised.

    There is much that the scanners do to contribute by good choice of the sources used and the degree of collimation. Gamma radiation does not propagate as a perfectly collimated beam. It tends to scatter, at times producing "fuzzy" pictures. In most of the author's investigations, the scanners' reviews made beneficial changes to our plans.

17. Certain tower features may interfere with the scans. Wind deflectors (curved, thick protrusions on the outside wall, designed to reduce the wind bending moment on the tower), tray support trusses, and internal baffles may make it difficult to get even a good scan of the active areas, and rule out the scanning of packed towers. In one case (146), the use of surrogate support rings extending 5.5 in. below Superfrac® trays at 16 in. spacing, as well as spreader plates inside downcomers, hampered the correct interpretation of the scan. The scan plan should devise a strategy to get around such obstacles as much as practicable.

18. Design features incorporated in high-capacity trays, such as forward push, truncated downcomers, curved downcomers, and complicated supports, make the interpretation of high-capacity tray scans difficult, especially at close tray spacing (146, 427). Judicious

multi-chordal gamma scans with quantitative analysis (Section 6.1.6) are the best tool for diagnosing subtle issues with these trays.

19. Radioactivity needs to be considered. Towers containing radioactive materials do not produce good gamma scan results. Gamma scans can interfere with nucleonic level transmitters. Level controls using such devices need to be switched off and the operators should be alerted to their erratic readings during the tests.

20. The process/operation investigation leader should be available throughout the scanning tests, ideally on the site. Practical issues always arise during the tests, requiring important on-the-spot decisions. Some chords yield unexpected results; weather or plant conditions suddenly change, all contribute to the need to change plans. Any such decisions should be made jointly between the process/operations team leader and the scan team leader. A poor decision may result in a costly repetition of several measurements or an impaired diagnosis.

5.4   GAMMA SCAN SHORTCUTS: COST VERSUS BENEFITS

Adding scans increases costs. However, the main cost of gamma scans is the transportation and travel time of the scanners and their equipment to the site, getting through the required safety and work permits, and marking up the column and setting up the equipment. While the scanners are already on site, the cost of performing additional scans is relatively low. Still, this is an expense that should be saved if the scans are unnecessary. This additional cost needs to be considered against the potential benefits.

5.4.1   Dry (Empty Column) Scans: Yes or No?

These scans are conducted at the turnaround or during an outage in which the liquid is drained from the trays or packings. The benefit of dry scans is that they provide a reference where there is no liquid in the tower. These scans are most beneficial in the following situations:

- Packed beds operating at low liquid rates. Liquid maldistribution can be very difficult to diagnose at low liquid rates due to the minimal liquid holdup (Section 5.2.3). The radiation absorption is more attributed to the packing metal than to the liquid. A dry scan is invaluable to decouple the metal radiation absorption from that of the liquid.

- Towers containing ceramic or carbon packings or trays. The ceramics usually absorb more radiation than the liquid, and without a dry scan it may be very difficult, even impossible, to draw conclusions on distribution or froth heights.

- Where there are strong interferences from column internals (like truss beams), external or internal piping, platforms, supports, flanges, and lagging.

- In foaming services with packings. It may be difficult to diagnose foaming without knowing how much of the radiation absorption is attributed to the packing. Here too the dry scan enables decoupling the metal radiation absorption from that of the foam.

- Packed beds in large towers. Similarly, packed beds containing grid packings (which are fabricated of thicker metal). The dry scan again helps decouple the metal absorption from the liquid.

- When it is important to determine the liquid level in distributor troughs. Often, trough scans see more metal than liquid, and decoupling between the two can help.

In most other situations, there is less incentive to perform a dry scan.

5.4.2   Initial Operation Scans (Often Referred to as Baseline Scans): Yes or No?

These scans are conducted a short time after the tower returns to full production rates following a turnaround in which the tower was cleaned. The initial operation scans provide a reference when the internals are clean and undamaged. In the following situations these scans are invaluable and therefore frequently performed:

- When fouling, corrosion, or damage have been experienced or where there is a high potential for any of these to occur. This scan provides an invaluable reference to compare when tower performance drops off due to one of these issues, and can help identify the issue and guide preventive actions.

- When intermittent foaming is expected. The initial operation scan needs to be shot in the absence of foaming, and can later be used to diagnose foams and distinguish them from other process issues.

- Where nozzles, supports, electrical conduits, platforms, or other sources of external interference can obscure the results.

5.4.3   Changed Conditions Monitoring Scans

To track changing conditions inside the tower (fouling, damage, corrosion), it is useful to shoot gamma scans of the tower or the affected region at regular intervals. If damage due to upsets is suspected, a new scan can be performed following every suspected damage episode. These scans are invaluable for

- Optimizing tower operation for the changed conditions
- Avoiding an unnecessary early shutdown to repair the damage or clean
- Keeping the tower operating as long as possible in the changed conditions
- Ordering the correct replacement equipment.

The gamma scans cost usually dwarfs the savings from achieving any of the above objectives. The information provided by the gamma scans monitoring is invaluable for optimizing the process and implementing the correct, optimal temporary and/or permanent solution.

5.4.4   Performing Downcomer Scans: Yes or No?

The author's experience is that at least half the time downcomer scans can contribute enough to the diagnostics to justify their costs. Typically, we see downcomer scans performed about 10% of the time. Failure to cure a problem due to misdiagnosis can bear tremendous costs, and good downcomer scans can go a long way to avoid such failure. Downcomer scans are generally not performed often enough in the industry.

In order to make an educated decision as to whether to proceed with a downcomer scan, consider the following:

- Check whether excessive backup, high downcomer entrance velocities, plugging, damage, foaming, and seal breaks (especially in truncated downcomers) are on the troubleshooting list. Pay special attention to issues that cannot be predicted by calculation. If any of these are high-possibility items there is a strong incentive to perform the downcomer scans.

- Perform hydraulic calculations to evaluate the proximity of the operating conditions to downcomer bottlenecks. For members of FRI, their software is excellent. For all, good results can be obtained using the correlations in *Perry's Handbook* (245), other texts (e.g., 34, 193, 299, 451, 492), vendor software, or the software available on major simulators with appropriate selection of the suitable correlations.

- Perform judicious tests to check for various theories. If downcomer scans can shed light on any of the theories, there is a strong incentive to proceed with the scans.

- If previous active area gamma scans are available, review them closely. Active area gamma scans showing low froth heights and clear vapor spaces indicate that the trays are not approaching a limit, so if there is a limit in the tower at slightly higher rates, there is a good chance it will be in a downcomer (or downcomers).

5.4.5   Can We Learn More about the Bottleneck?

The "bottleneck" referred to may be a hardware bottleneck, or it may be a bottleneck induced by fouling, damage, corrosion, or foaming. Brown Burns et al. (52) demonstrated the value of additional scans for diagnosing the nature of a bottleneck. The fractionator under investigation experienced bottlenecks in three different regions, all of which were associated with fouling or corrosion. Once the initial scan identified the bottleneck regions, Brown Burns et al. judiciously varied the vapor and liquid loads throughout each bottlenecked section of the tower. In some sections, the liquid loads were slightly changed, while in others, they were quite drastically changed. The result was a much better diagnosis, leading to measures that optimized operation and guided turnaround modifications. Scanning costs can be minimized if these additional scans are performed right after the original scan.

5.4.6   Would Our Proposed Modification Solve the Problem?

In many investigations, there is a proposed expensive solution on the table, but there is uncertainty about its effectiveness. There may even be concerns that a proposed modification will make the tower worse. An opportunity may arise to inexpensively change only a few trays, and once back in operation, gamma scan the tower to evaluate. A team led by the author found this approach invaluable when troubleshooting a 300-tray tower that did not work too well. Initially, only 10 trays were replaced. After gamma scans demonstrated much better performance for these 10, the rest were confidently changed. A similar approach was applied by Harrison (160) and Fulham and Hulbert (114).

5.5   SOME APPLICATIONS OF QUALITATIVE GAMMA SCANS

In this section, we present several case studies that address some frequently asked questions, and illustrate some key considerations when interpreting qualitative gamma scans.

5.5.1   Distinguishing Fact from Interpretation

The gamma scan report contains both facts and interpretations, and often the two are not clearly distinguished. Failure to distinguish fact from interpretation has been the cause of many incorrect diagnoses and failed fixes.

Facts

Any theory developed to diagnose column problems must be consistent with the facts. If a theory is inconsistent with the facts, it should be discarded. *Measurements* reported by the gamma scans are facts, within the accuracy limits of the measuring device. They are not any different than a temperature, pressure, or flow measurements and should be treated the same way. Below are listed *measurements* usually supplied by the gamma scans

- *Froth heights.*

    For instance, refer to Figure 5.3. For these peaks, the froth heights are reported as 9″. Measurement of the froth height is not highly accurate, typically varying by plus or minus 20–30%. So, if 9 in. are measured , it could be as high as 12″ or as low as 7″, but it is highly unlikely to be 20″. That is a fact.

- *Presence/absence of clear vapor.*

    Again, refer to Figure 5.3. In the vapor spaces, the peaks between the trays reached the clear vapor line. This means there is no liquid there. A gamma ray was shot and reached the detector at full strength, signifying it traveled through vapor only. Had there been liquid in the path, it would not have reached clear vapor.

- *Qualitative evaluation of the degree of material density.*

    A large absorption of gamma rays means dense material. This can be liquid or solid. Distinguishing solids from liquids will be an interpretation, not a fact. All that the scans can factually determine is the presence of something dense in the path.

- *Flooding.*

    Vapor spaces consistently filled with dense material indicates liquid accumulation, i.e., flooding. Typical accuracy is plus or minus 10%. So the dP may not have risen yet, but the scans can already detect the initiation of the liquid accumulation.

- *Packing maldistribution.*

    In Bed 2 in Figure 5.7, one chord clearly and consistently shows higher transmission than the others. This means it is seeing less liquid. In Bed 1, all the chords fall together, which means good distribution. While maldistribution is a fact, before concluding that distribution is good, one needs to consider the possibility that the liquid rate is too low to give a good indication of maldistribution, as discussed by the quotes from Vidrine and Hewitt (493) in Section 5.2.3. One also needs to consider that the scans in Figure 5.6 are unlikely to show a condition in which there is more or less liquid pouring into the column center zone.

- *Missing packings.*

    Gamma scans were shot through a packed bed and should see metal. If they reach the detector at full or almost full strength (like Bed 3 in Figure 5.2b), it indicates that the packings are not there.

- *Trays not holding liquid.*

    Gamma scans were shot through trays should see liquid sitting on the trays. If they reach the detector at full or almost full strength, the trays do not hold liquid. This means that the trays are either not there, or they are damaged, or they are dry (due to too little liquid flow or too low vapor flow). The scans cannot distinguish between these issues, all they can tell is that the trays are not holding liquid for some reason.

Interpretations   Everything else is an interpretation. Interpretations range all the way from the very conclusive interpretation, nearly a fact, to the very inconclusive interpretation bordering on fantasies. The challenge for process and operation engineers is to evaluate how conclusive each interpretation is. Plugging, missing trays, foaming, call of jet versus downcomer flooding, are examples of interpretation. In the forthcoming examples, we will carefully distinguish the facts from the interpretations and evaluate the conclusiveness of the interpretations.

5.5.2   Is the Flood a Jet Flood?

Following a major retrofit, one column (483) experienced poor separation, a higher than normal pressure drop, and liquid carryover at the design rates. Theories included poor design or damage during recommissioning.

Two gamma scans were conducted. Figure 5.11 (483) shows the top seven trays. The green dashed scan was at design rates, the blue heavy scan at rates slightly reduced to the point of getting the column out of trouble.

The heavy scan shows all trays holding liquid, with froth heights about 8–10 in. and clear vapor spaces above (all peaks reaching clear vapor). The dashed scan showed little difference from the heavy scan on trays 5–7. From tray 4 up, it showed considerably taller froths, growing progressively taller from tray 4 to tray 1. There was no clear vapor above the top four trays, with progressively more liquid in the vapor spaces as one ascends from tray 4 to tray 1. The appearance on trays 2 and 3 will often be called "*flood*" by the scanners. The flood criterion varies with the scanners but is always based on the froth heights and the liquid content in the vapor spaces. Flood is the accumulation of liquid, and this happens in the vapor spaces and is the reason for the high liquid content of the vapor spaces above trays 1–4 in the dashed scan. So this scan sees flooding in the top 3–4 trays.

Everything up to this point has been facts.

The reply to the question of jet flood or another type comes from a very conclusive interpretation. In Section 2.2, jet flood was defined as follows: "As the vapor velocity is increased, froth or spray height on trays rises. When the froth or spray approaches the tray above, some of the froth or spray is aspirated into the tray above as entrainment. Upon further increase in vapor velocity, massive entrainment of the froth or spray begins, causing liquid to accumulate on the tray above and flood it." The unflooded scan (heavy curve) was shot just below the flood point, i.e., when the flood was approaching. The froth heights in this scan were 8–10 in., well below the tray spacing of 30 in. (the tray spacing can be inferred from the scan as the distance between successive trays), and the vapor spaces were clear, without any signs of liquid aspiration into the tray above. Therefore, it can be concluded with confidence that the trays were nowhere near jet flood, and the flood was caused by another mechanism.

The scan findings denied the theories of poor design and damage. The investigation looked elsewhere and found that the flood was caused by an improperly installed orifice plate in the reflux line, causing the meter to grossly understate the reflux rate. As a result, the plant ran the tower with too much reflux, which caused a downcomer or a reflux entry point flood. Adjusting the reflux established good operation and circumvented the need for a shutdown.

5.5.3   Diagnosing the Correct Flood Mechanism and Arriving at the Effective Fix

Kister, H. Z., and C. Winfield, Ref. 227. Reprinted Courtesy of Chemical Engineering Progress (CEP). Copyright © American Institute of Chemical Engineers (AIChE)

A large stabilizer equipped with 4-pass valve trays was operating right at its maximum throughput limit and bottlenecked the capacity of an entire gas plant (227). Any increase in rates would generate flooding in the bottom section of the stabilizer, as was seen by instability and high pressure drop. With a turnaround approaching, it was desired to debottleneck the stabilizer to raise the capacity of the plant by 25%. The economics were such that any modifications external to the tower (e.g., adding a preheater) were out of the question. Downtime was at a premium, so the revamp was a one-shot deal: a failure would mean five years or more at the failed conditions, with a huge economic penalty.

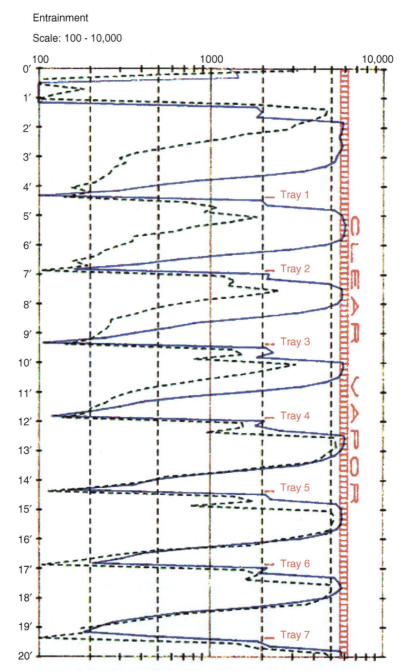

Figure 5.11 An active area gamma scan, illustrating normal operation in the unflooded scan (blue, heavy) and flooding near the top of the tower (green, dashed). (From Tracerco, "Process Diagnostics - TRU-SCAN® Case Studies," 2010. Available from Tracerco, Pasadena, Texas. Reproduced Courtesy of Tracerco.)

Table 5.1   Hydraulic calculations.

Tray number	Bottom	4–6 Trays from bottom	Just below feed
% Jet flood	52	52	17
% Froth in downcomer	61	60	41
% Downcomer choke	98	99	38

There was a debate on the flood mechanism, with some calculation methods suggesting a jet flood, while others suggested a downcomer choke. Correct determination of the limiting flood mechanism was central to achieving an effective debottleneck of the tower. It therefore became necessary to positively identify the flood mechanism.

Results of our hydraulic calculation at the rates where flood was observed are shown in Table 5.1. The concept behind Table 5.1 is similar to that of Table 2.1 in Section 2.3. We advocate always performing a hydraulic calculation such as in Table 2.1 or Table 5.1 prior to gamma scans (see Section 5.3, item 1). Table 5.1 shows results from our calculations, which were challenged by other team members.

To positively determine the flood mechanism, three scan chords were shot: one through the active areas, another through the center downcomers, and the third through the off-center downcomers. All scans were shot parallel to the downcomer walls, as recommended in item 9 of Section 5.3. Side downcomers were not scanned due to concerns about interference from thick tower walls, as described in Section 5.3, item 10. Each of the downcomer scans was shot twice: a "flooded" scan at the loads where flood was observed to initiate and an "unflooded" scan at loads about 3% below that point, with the tower operating at stable conditions. The active area scan was shot only in an unflooded condition.

Figure 5.12 shows an overlay of the unflooded scans of the center downcomers (red scan) and the active areas (blue scan) near the bottom of the tower. The active area froth heights were 10–12″ out of 24″ tray spacing, and the vapor peaks approached the clear vapor line. This means that the active areas were nowhere near jet flood, similar to the heavy blue scan in Figure 5.11. When a tray approaches jet flood, the spray heights get close to the tray spacing, and one cannot see clear vapor. Another interesting observation is that the peaks above the downcomers from trays 2 and 4 are very similar to those above the active areas, which validates the scan quality. The downcomers from trays 2 and 4 were full of froth nearly to the top and were getting close to a limit. This strongly supports a downcomer flood mechanism.

Figure 5.13 shows an overlay of the center downcomer scans for the flooded (red scan) and unflooded (blue scan) conditions. There is a clear change from the unflooded to the flooded scan: in the flooded scan, the downcomers from trays 6 and 8 were full of froth and have begun to flood. The downcomers from trays 12, 14, and 16 look normal and were not flooded in either scan. In the flooded scan, the froth in the downcomers from trays 6 and 8 began to back up into the vapor space areas above them, and the peaks above these downcomers no longer approached the clear vapor line. This is all a clear indication of flood initiating in the downcomers and propagating into the active areas. Overall, the gamma scans conclusively showed that the tower was flooded by downcomer flooding in the lower 8 trays.

While downcomer scans can distinguish jet flood from downcomer flood, they cannot distinguish downcomer backup from downcomer choke flood. Both these downcomer mechanisms show up as full downcomers. The distinction can be drawn by calculation (Table 5.1), which in the stabilizer case favored downcomer choke. Also, as stated in Section 2.2, downcomer choke is common in high-pressure distillation, such as in a stabilizer. Further, in the absence of

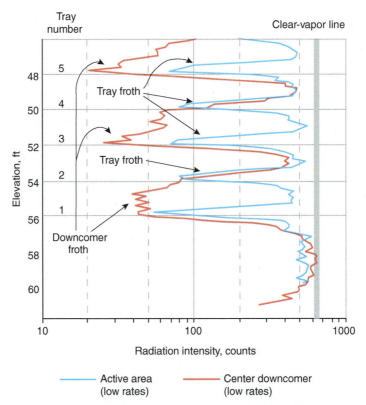

Figure 5.12   Stabilizer unflooded active area scan and center downcomer scan overlay, with the tower operating at stable conditions at rates 3% below flood. (Reprinted Courtesy of Quantum Technical Services, LLC.)

fouling and foaming, downcomer backup is uncommon in well-designed trays of 24″ spacing and higher. All these strongly supported downcomer choke.

Based on this diagnosis, the downcomer top areas were largely increased. To maximize throughput and minimize the loss in the active areas, the downcomers were heavily sloped, with the downcomers' top areas being three times their bottom areas. The debottleneck solution exceeded expectations, with a capacity gain of 29%.

5.5.4   Does the Tower Flood Occur at the Expected Location?

For several years, the gas plant of a fluid catalytic cracker (FCC) unit was bottlenecked at a throughput of 57,000 BPSD by the debutanizer column (238). Raising charge rates or reflux would lead to flooding, accompanied by high pressure drop, instability, and poor separation. With an FCC turnaround coming up, a task force was gathered to diagnose the bottleneck and offer a solution.

As in Case 5.5.3, the first step was performing a hydraulic analysis (Table 5.2). The concept behind Table 5.2 is similar to that of Tables 2.1 and 5.1. Table 5.2 shows that the trays operated nowhere near jet flood and that the froth heights in the downcomers were low. There was no history of fouling or foaming in that tower, and the bottleneck was experienced at the same charge rates at any time, confirming unlikely foaming or fouling. The only mechanism nearing a limit was downcomer choke. The downcomer entrance velocity exceeded the good practice

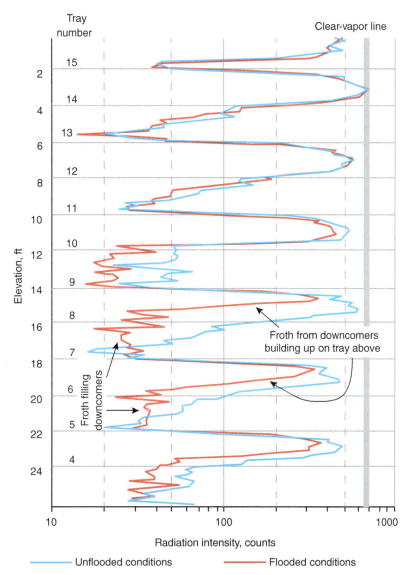

Figure 5.13   Overlay of center downcomer scans under flooded and unflooded conditions. (Reprinted Courtesy of Quantum Technical Services, LLC.)

Table 5.2   Hydraulic calculations.

Tray number	Above feed	Below feed	Bottom trays
% Jet flood	55	39	56
% Froth in downcomer	56	56	69
% Downcomer choke interpolation	71	55	72
% Downcomer choke correlation	90	79	103
Downcomer entrance velocity (ft/s)	0.31	0.44	0.52

guideline (193, 245) of 0.45 ft/s maximum, and one correlation suggested the bottom trays were at flood. As downcomer choke correlations are notoriously unreliable (193, 245), we checked by interpolating measured test data by FRI for a similar system. The interpolation showed that these trays operated at only 72% of downcomer choke. Table 5.2 therefore discounts jet flood and downcomer backup flood, but marks for further study the possibility of downcomer choke flood. If downcomer choke indeed bottlenecked this debutzanizer, the bottleneck would be experienced at the very bottom per Table 5.2.

There was another theory on the table. Liquid from the bottom tray was collected by a chimney tray collector, from which it flowed to a once-through thermosiphon reboiler. The line to the reboiler was undersized for self-venting flow, so it was theorized that liquid on this chimney tray backed to the top of the chimneys and was then entrained and blown onto the bottom tray by the rising vapor, initiating flooding.

The refinery had gamma scans that were shot six years earlier. Gamma scans do not age as long as the process conditions remain identical, which was the case. The six-year old scans (Figure 5.14) were therefore satisfactory for the investigation eliminating the need to rescan. The red heavy scan is the flooded scan, while the blue dashed scan is the unflooded scan. In the bottom section, the flooded scan was shot at vapor and liquid loads 3–5% higher than the unflooded scan.

The trays closest to downcomer choke were the bottom trays (Table 5.2), so if the flood was caused by downcomer choke, the flood would initiate at the tower bottom. Likewise, if the flood was due to entrainment from the chimney tray, the flood would initiate at the bottom trays. However, the gamma scans showed no flood on the bottom trays. Furthermore, although the scans showed that the liquid on the chimney tray was at the top or just below the top of the chimneys, they showed no liquid or entrainment in the vapor space above the chimney tray. The scans therefore invalidated both the downcomer choke theory and the chimney tray entrainment theory.

The scan (Figure 5.14) shows flooding in the tower, but this flood initiated near the feed, where it was not anticipated by either theory. It therefore steered the diagnosis away from downcomer choke and chimney tray entrainment to focus on the feed entry. A calculation showed that the flashing feed came out of the feed pipe slots at a huge downward velocity of 33 ft/s (or rho vee squared of 6000 lb/ft s^2) pointing straight at the valve tray floor. This entry arrangement (Figure 5.15a) is likely to have caused excessive turbulence on the tray, re-entrainment of the feed, and penetration of the flashing mixture through the valves on the tray floor into the downcomer below, as depicted in Figure 5.15b. These actions are likely to produce a premature flood on the trays below the feed, and this flood would propagate up the tower like the scans show.

Based on this analysis, the feed arrangement was modified. Although the other bottlenecks (downcomer choke in the bottom and entrainment from the chimney tray) have not yet been reached, they appeared to be approached, so a good debottleneck needed to cater for them as well. The hardware changes are described elsewhere (238). The end result was the elimination of the flood and an increase in the charge rate from 57,000 BPSD to 72,000 BPSD with no limit in the debutanizer even at 72,000 BPSD.

5.5.5 Foaming or Not Foaming?

The active area scan in Figure 5.16 (485) shows a foam-flooded amine absorber. What gives away foaming is the uniformity of the dispersion. For all trays in Figure 5.16, there are very small differences between the transmission reading near the tray floor and the transmission reading further up in the vapor space, all the way to the tray above. This means a very uniform froth density on the tray. Foams are uniform, like shooting scans through a bubble bath or through the head on a glass of beer (these are foamy dispersions). Contrast this appearance with a non-foaming

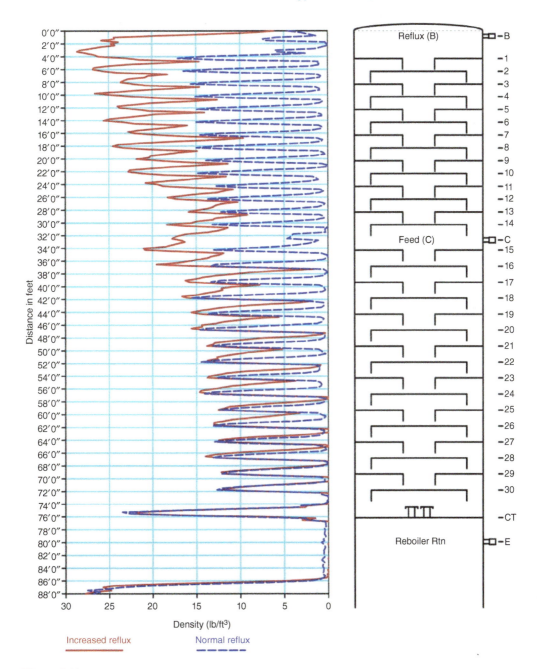

Figure 5.14   Debutanizer tray active area gamma scans. The blue (dashed) scan is at unflooded conditions, while the red (heavy) scan shows flooding initiating near the feed. (From Kister, H. Z., D. E. Grich, and R. Yeley, PTQ Revamps and Operation, 2003. Reprinted Courtesy of PTQ.)

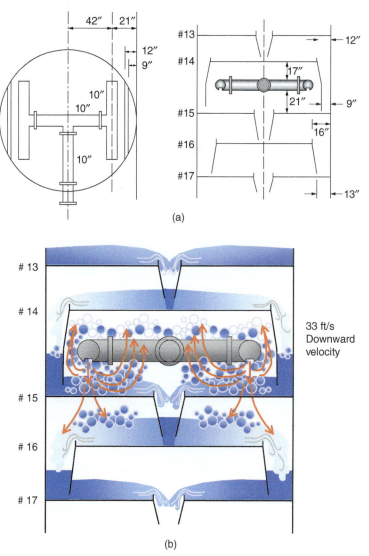

Figure 5.15 Debutanizer feed entry arrangement that caused tower and gas plant bottleneck. (a) Geometrical arrangement; (b) flashing feed issuing downwards at the very high velocity of 33 ft/s onto the tray floor, causing entrainment and downcomer choke at the feed point that led to a premature flood. (Part (a): From Kister, H. Z., D. E. Grich, and R. Yeley, PTQ Revamps and Operation, 2003. Reprinted Courtesy of PTQ. Part (b): From Kister, H. Z., "Practical Distillation Technology," Course Manual. Copyright © 2013. Reprinted with permission.)

flood, such as the flood on trays 2 and 3 in Figure 5.11. In the flooded scan in Figure 5.11, the transmission at the tray floor is much lower than further up in the vapor space, indicating a density gradient, with dense froth at the tray floor and much less dense froth further up. This is typical of non-foaming floods.

Another thing that supports foaming in this case is the service. Foaming is a service-specific phenomenon (198, 201). Some services have high foaming tendencies, others do not. As has been pointed out (Section 2.10), acid gas absorbers and regenerators are foaming systems. On the other hand, debutanizers do not foam (unless some foam-generating impurity like caustic

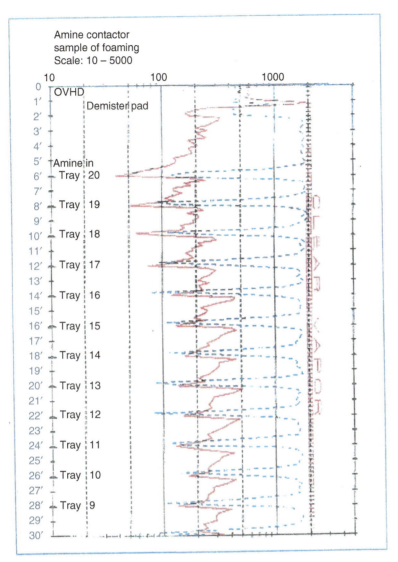

Figure 5.16 An active area gamma scan illustrating foaming in an amine contactor. The red (heavy) line shows the trays foam-flooded. The blue (dashed) line is an unflooded scan of the same active area, showing that the active areas are not heavily loaded. (From Tru-Tec Division, "Sample Scan Brochure," 1998. Available from Tracerco, Pasadena, Texas. Reproduced Courtesy of Tracerco.)

gets in, an infrequent occurrence). So an appearance like Figure 5.16 in an amine absorber is a very conclusive foam diagnostic. On the other hand, if there are three trays in a debutanizer with a uniform froth appearance, it will be a very inconclusive foaming diagnostic. The author was involved in three troubleshooting investigations in which gamma scanners interpreted debutanizer scans to indicate foam. Three times they were wrong. In all three, there was a uniform dispersion for 2–4 trays, which misled the interpretation.

In Figure 5.16, note the unflooded scan (blue, dashed). Froth heights are very low, suggesting the trays are nowhere near jet flood. In fact, the flat peaks showing liquid in the vapor spaces not far from the clear vapor line suggest weeping (see Figure 5.4b). Foams generally do not bottleneck trays. They usually bottleneck downcomers, as described in Section 2.10. Just before the flood, the active area looks clear. A scan through the downcomers just below the flood point would have shown the downcomers close to full.

Combining downcomer scans with active area scans is a powerful method to evaluate foaming as well as the effectiveness of antifoam addition. As foams usually bottleneck downcomers rather than trays, it is of utmost importance to scan downcomers when troubleshooting for foam or evaluating the effectiveness of antifoam addition. This is shown in the next case study (227). There are other experiences reported (1, 469) of foaming towers in which active area scans below the flood did not show foaming signs.

A chemical tower was experiencing reduced throughput with very poor separation. Gamma scans of the trays active areas and even numbered downcomers revealed that foam was present throughout the height of the column. The foam was recognized by the uniform dispersion on the flooded trays as seen in Figure 5.17a (red); this scan is very similar to Figure 5.16, with absorption of gamma radiation near the tray floors much the same as near the top of the vapor spaces. The scan of the even-numbered downcomers in Figure 5.17a (blue) shows much the same, with uniform dispersion through the scanned downcomers. The foamy uniform dispersion in each downcomer extended all the way to the tray above. Both the active area and downcomer scans provided conclusive evidence for foaming.

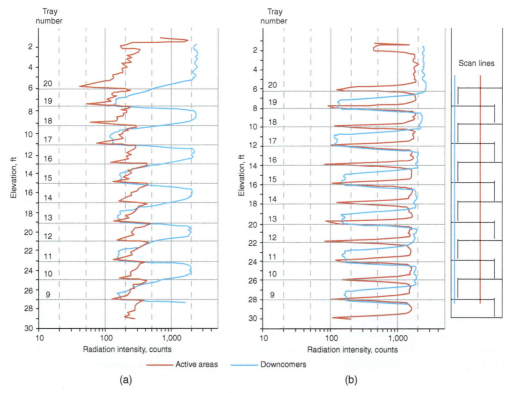

Figure 5.17   (a) Gamma scans show foaming on active areas and in downcomers. (b) Gamma scans show effectiveness of antifoam on active areas and in downcomers. (From Kister, H. Z., and C. Winfield, Chemical Engineering Progress, p. 28, January 2017. Reprinted Courtesy of Chemical Engineering Progress (CEP). Copyright © American Institute of Chemical Engineers (AIChE).)

An antifoam injection was initiated. The success of antifoam treatment depends on the agent used and its concentration. Antifoams are surface-active agents, and too much antifoam can be worse than too little (Section 2.10). Several trials with different agents and concentrations are sometimes required. Gamma scans through the active areas and downcomers are powerful for evaluating the effectiveness of an antifoam injection.

Figure 5.17b shows the active areas (red) and downcomers (blue) gamma scans used to evaluate the effectiveness of an antifoam injection into the tower. The scans no longer showed foam on the trays. Once the foam was eliminated, the active areas of the trays were not highly loaded, with froth heights well below half the tray spacing and clear vapor spaces above (similar to the dashed scan in Figure 5.16). The downcomer scans still show some foam in the downcomers, but the foam heights had subsided and there was good clearing near the very top of the downcomers. Although the foam was gone from the trays and no longer bottlenecked the tower, there was still a good height of foam in the downcomers that could build up to the tray during changes or upsets. So the antifoam effectively alleviated the foam in the downcomers, and removed the foam from the active areas.

5.5.6 Insights into Fouling Patterns/Shortcuts Lead to Misinterpretations

Kister, H. Z., and C. Winfield, Ref. 227. Reprinted Courtesy of Chemical Engineering Progress (CEP). Copyright © American Institute of Chemical Engineers (AIChE)

Downcomer gamma scans can reveal what calculations would not, but only if performed and interpreted correctly, as illustrated in the following case (227). A petrochemical tower in a fouling service, equipped with 2-pass valve trays at 36″ spacing, experienced flooding mid-way through the run. The flooding showed up as high pressure drop and loss of separation. To plan modifications for the next turnaround and to devise an operation strategy for the rest of the run, it was necessary to understand the nature and location of the flood.

Both active area and downcomer scans were shot. The flooded active area scans showed large froth accumulation on the top tray. On the lower trays, froth heights were low, and clear vapor was reached less than 12″ above the trays. Since the flooding occurred on the top tray, it was essential to scan the downcomers from that tray. The challenge was that these were side-downcomers, and practical difficulties precluded scanning them parallel to the weir.

Fortunately, this was one of the few cases in which one could get away with a chord that intersects the downcomer wall at a small angle (about 30°). Usually, when a chord crosses the downcomer wall into the active area, it is impossible to decouple the downcomer dispersion from the tray dispersion (Section 5.3, item 9). But here the tray dispersion terminated well below the tray above, so from 12 in. above the tray and up to the next tray there were 24 in. in which the scan saw only the downcomer dispersion.

Figure 5.18 shows the scans for the two side downcomers. The two downcomers from tray 1 showed widely different dispersions. The blue curve shows a dense dispersion on the top half, with a gradually decreasing froth density as one descends the downcomer. At the downcomer bottom, the froth density was about the same as the tray froth density above tray 3. The red curve showed a low-density dispersion throughout the downcomer, very similar to the tray froth above tray 3 below. This suggests that this downcomer passed vapor, just like a tray does.

Returning to the blue chord, the initial interpretation of the density gradient was of a seal loss in this downcomer impeding the descent of liquid from the tray above. A close check of the scan chord against the drawing revealed that this was a misinterpretation. The side downcomers were sloped (Figure 5.19), and below the top 18 in., the scan chords passed through progressively shorter lengths of downcomer as one descended. Near the bottom of the downcomers, the

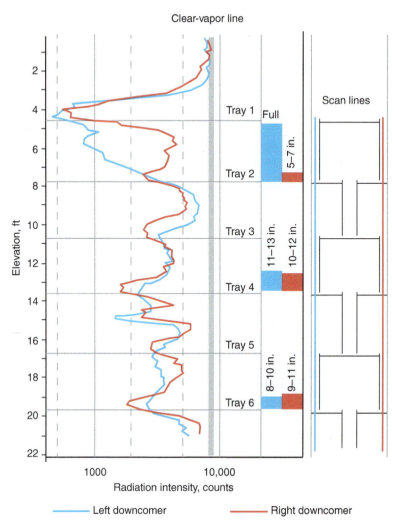

Figure 5.18 Scans of two side downcomers reveal very different behavior for the side downcomers from tray 1. (From Kister, H. Z., and C. Winfield, Chemical Engineering Progress, p. 28, January 2017. Reprinted Courtesy of Chemical Engineering Progress (CEP). Copyright © American Institute of Chemical Engineers (AIChE).)

chords missed the downcomers altogether, catching only the tray dispersion (Figure 5.19). The switch from dense to highly aerated dispersion did not indicate loss of seal; rather, it indicated the downcomer chord becoming progressively shorter due to the slope. It is therefore likely that the downcomer through which the blue chord passed was all full with the dense liquid seen in its upper part.

Putting the information together, tray 1 appeared to be plugged, incurring a very high pressure drop that in turn led to a downcomer backup flood. The plugging was nonuniform, with most of the liquid descending via the side downcomer that showed high dispersion density (blue scan). Due to the high pressure drop across the top tray and the liquid maldistribution, it appears that the vapor managed to break the seal of the opposite downcomer, causing its dispersion to appear like part of the active area (red chord). This vapor breakthrough would channel more liquid to the downcomer that had the high dispersion density.

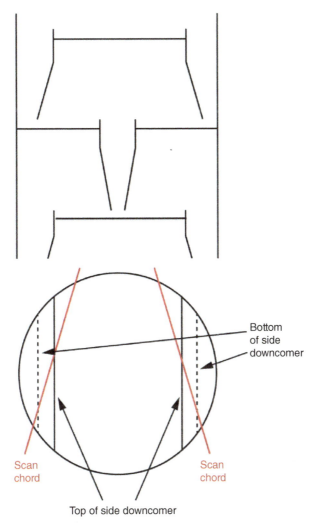

Figure 5.19 Scan chords passing through progressively shorter lengths while descending sloped side downcomers. (From Kister, H. Z., and C. Winfield, Chemical Engineering Progress, p. 28, January 2017. Reprinted Courtesy of Chemical Engineering Progress (CEP). Copyright © American Institute of Chemical Engineers (AIChE).)

This interpretation supported the theory that reflux maldistribution in the tower contributed to the plugging problem. A turnaround inspection later confirmed the nonuniform nature of the plugging, and there were additional water marks pointing to the maldistribution.

5.5.7 Multipass Trays Maldistribution

An existing tower experienced a premature flood near the bottom (241). The tower bottom section contained 4-pass fixed valve trays, in an equal bubbling area arrangement. The trays had split downcomers, with a gap for supporting I-beams between, which also provided full vapor equalization and partial liquid equalization. To balance the liquid flow to the passes, the side weirs were swept back, and the side outlet weirs were shorter (1.5″) than the center weirs (3″).

Maldistribution between the passes was one of the theories to be tested. Chords were shot through the outer (red) and inner (blue) active areas (Figure 5.20a) at steady, unflooded operation. The scans (Figure 5.20b) show denser froth on the outer (side) panels than on the inside (center) panels. At face value, the difference between the peaks looks rather small on Figure 5.20b, but one must keep in mind that the chord through the inside panels was 30% longer than the chord through the outside panels, so it sees 30% more material. Using the same calibration, the inside chord should have produced a lot more absorption, not less. A subsequent analysis by the scanners, which compared tray froth densities on the inner scan paths and the outer scan paths taking into account the respective scan path distances, determined that the outer panel froth was 22% denser. This gamma scan provides conclusive evidence that the L/V ratio on the outside panels was considerably higher than that on the inside panels.

In another tower containing 4-pass trays (327), nine scan lines were shot, giving full mapping of all active areas and downcomers. This ideal mapping pattern showed uneven loadings of active areas, and high and uneven loadings of downcomers. There were even large differences between the scans of the two opposite side panels in one section of the tower. These scans identified maldistribution between passes and obstruction of the downcomer entrance area by sumps and I-beams (see Section 8.5) as the major factors leading to premature flooding and poor separation in the tower.

5.5.8 Fouling-Induced Maldistribution in 2-Pass Trays (180)

Fouling that preferentially plugs one panel more than other panels can produce maldistribution between the panels in 2-pass trays. Such maldistribution is not always addressed in scan reports, but it can be visible in the scans. When observed, it provides a valuable diagnostic. Figure 5.21 illustrates this appearance.

In a refinery atmospheric crude tower, trays 10–13 were a heavily liquid-loaded pumparound section. The pumparound liquid was drawn from a sump located beneath the downcomer from tray 10. Some of this liquid was drawn as the diesel product; the rest was cooled externally and returned to tray 13. Trays 6–9 were a lightly loaded fractionation (wash) section.

In the normal ("unflooded") scan, the only flooded tray was tray 10. Both the trays above and below were not heavily loaded. The flooded scan shows the propagation of the flood up the tower. Both scans conclusively verify flood initiation on tray 10.

The taller and denser froths observed on the southwest (SW) side of trays 6–9 in the flooded scan persisted on tray 9 in the "normal operation" scan. They became much less pronounced on trays 6–8. In the flooded scan, the northeast (NE) side of trays 6 and 7 approached drying.

This maldistribution observation tied in with the temperature indicator on the NE reading 15°F hotter than the temperature indicator on the SW, both under "normal" and "flooded" conditions. In previous runs, when tray 10 was not fouled, the two temperature indicators read the same. Together, the scans and temperature difference verified maldistribution between the SW and NE panels, with more liquid descending on the SW.

Downcomer scans (180) established flood initiation in the center downcomer from tray 10. Vapor/liquid sensitivity tests (Section 2.7.1) determined the flood to be sensitive to both vapor and liquid, leading to the conclusion that the flood was a downcomer backup flood. Putting the pieces together established the plugging of tray 10 as the root cause of the downcomer backup flood, with the plugging occurring unevenly on the SW and NE panels. This finding denied alternative theories and guided the troubleshooting team to the best path forward.

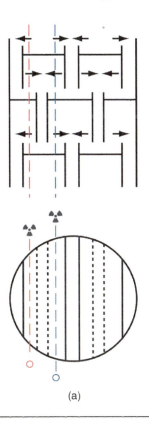

(a)

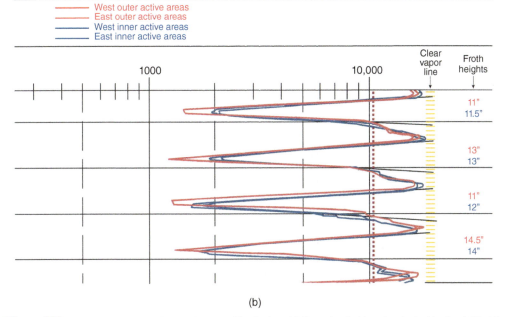

(b)

Figure 5.20   Gamma scans exploring multipass maldistribution. (a) Scan chords. Note that the inside chord (blue) is longer than the outside chord (red). (b) Gamma scans comparing the inside and outside chords. (From Kister, H. Z., R. Dionne, W. J. Stupin, and M. R. Olsson, Chemical Engineering Progress, April 2010. Reprinted Courtesy of Chemical Engineering Progress (CEP). Copyright © American Institute of Chemical Engineers (AIChE).)

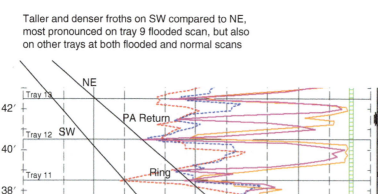

Figure 5.21   Active area gamma scans under "flooded" (dashed scans) and "normal" (continuous scans) operation showing fouling-induced maldistribution in 2-pass trays (SW = Southwest, NE = Northeast).

5.5.9 Weeping/Dense Liquid/Missing Trays

A depropanizer in a refinery alkylation unit experienced flooding and poor operation mid-way through the run. The gamma scan at normal operating conditions is the heavy scan on Figure 5.22 (483). For trays 1–8, the scan shows liquid in the vapor spaces (fact). The liquid in the vapor spaces of trays 1–8 is weeping (conclusive interpretation). The following three pieces of evidence put together make the weeping diagnosis conclusive:

- The peaks in the vapor spaces are flat and close to the clear vapor. This means uniform low density, like shooting gamma scans through rain, which gives a uniform low-density appearance (same as in Figure 5.4b and the dashed blue curve in Figure 5.16, both of which indicate weeping). Entrainment usually shows up as a density gradient, with a denser appearance near the tray floor and less dense near the tray above, similar to the gradients in Figure 5.4a.

- Froth heights on the trays are very low, and Tracerco estimates them at 0–4″ (483).

- The vapor space above tray 1 is the only vapor space to reach the clear vapor. As there is no tray above, there is no weeping in this vapor space.

The scan shows dense liquid between trays 11 and 29. This is not an ordinary flood; an ordinary flood is the flood shown in Figure 5.11. This is a column of very dense liquid, or as one of my class participants put it, a "solid liquid."

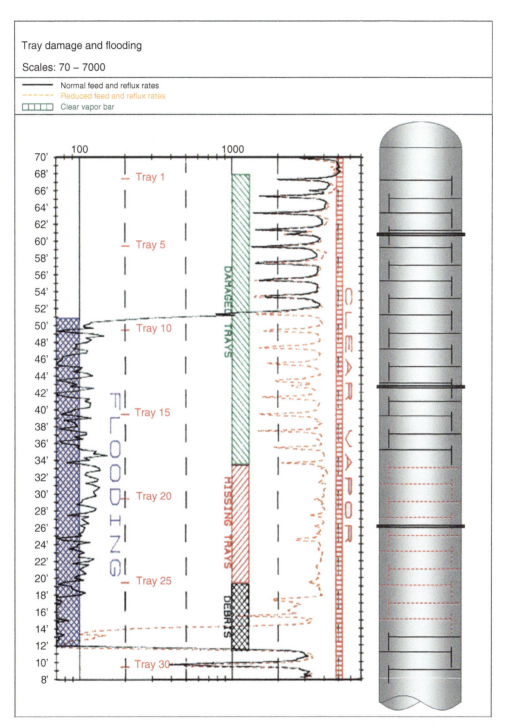

Figure 5.22   An active area gamma scan, illustrating missing trays, weeping and dense liquid accumulation. The heavy curve is "normal operation" scan, with dense liquid accumulation. The dashed curve is the unflooded scan showing missing trays. Both scans show weeping in the upper trays. (From Tracerco, "Process Diagnostics - TRU-SCAN® Case Studies," 2010. Available from Tracerco, Pasadena, Texas. Reproduced Courtesy of Tracerco.)

As stated earlier, the flooded scan shows what is happening. The unflooded scan shows why. To gain better understanding, the rates were reduced and the unflooded scan (dashed curve in Figure 5.22) was shot. In this scan, trays 1–8 did not change much. The liquid column receded to trays 28–29, exposing trays 9–27. Trays 9–18 looked much the same as 1–8. Trays 19–25 were missing. It appeared that debris from the missing trays collected on tray 29, blocking the liquid descent and causing the dense liquid to pile up.

When the tower was opened, trays 1–18 were found to have suffered severe corrosion damage, which accounted for the weeping. Trays 19–28 were blown out and were found on top of tray 29, blocking liquid passage and causing the dense liquid column to build up.

Diagnosing missing trays is not always straight forward. One refinery crude atmospheric tower was experiencing poor removal of diesel (valuable product) from the bottom residue. A gamma scan (209) concluded that the stripping trays were missing. Tray dislodging is a common experience in this service. The appearance was much the same as that of trays 19–25 in Figure 5.22. Based on this, the refiner shut down the crude unit and opened the tower, just to find all the trays there in good condition with no signs of damage, corrosion, or fouling.

A postmortem showed that the stripping steam rates (and therefore the vapor loads) in the stripping section of this tower were low. The C-factor calculated from the tower simulation for the active area (Eq. (2.1)) was 0.03 ft/s, compared to a typical flood C-factor in this application of around 0.4 ft/s (Rule 1, Section 2.16), so the trays operated at about 7–8% of flood. At this low C-factor, trays do not hold much liquid (unless specially designed to do so, which they were not). The scans were shot every 2 in.; the trays' thickness was 0.075 in., so it was easy for the scans to miss the trays altogether. On gamma scans, dry or empty trays can look the same as missing trays.

This hydraulic calculation should have been performed before the scan, not after it (Section 5.3, item 1). Had the low vapor load been appreciated prior to the scan, it would have been easy to triple the steam rate for a second test. In this second test, the C-factor would have been 0.09 ft/s, about 20% of flood, and the gamma scans would have easily seen froth on the trays. An expensive and unnecessary shutdown would have been circumvented.

In other experiences (one described in Section 5.5.20), a scan concluded there was tray damage. Before shutting down for repair, the author was requested to review the scanners' conclusions. Rough hydraulic estimates based on plant meters showed very low liquid rates through these trays, again raising the question of whether the trays were just dry or damaged. Here, the test was to double the liquid rate and rescan. The rescan was similar to the original scan, confirming damage rather than just drying out.

5.5.10 Plugging, Flood, and Fouling Monitoring in a Packed Bed (90, 372)

A refiner experienced a high pressure drop and an off-spec gasoil side product in their atmospheric crude tower. Gasoil is the first side draw above the tower feed. Between the draw and the feed point, a bed of structured packings (the "wash bed") used a small reflux ("wash") stream to remove heavies from the vapor. Economics favor running this reflux at the minimum that will keep the gasoil on-spec. Any excess reflux ends up in the far less valuable tower bottom ("atmospheric resid") product with a severe economic penalty.

A grid scan (Figure 5.23a) showed a large buildup of dense material throughout the bed. The dense material rose past the 150 mm vapor space between the distributor and the bed. The distributor (located right above the bed) was full of massive, dense liquid, probably entrained from the bed. This indicated severe flooding in the bed. The very high density can only be explained by severe plugging. The wash bed was plugged solid. All these are facts based on the scan data, not on any interpretation.

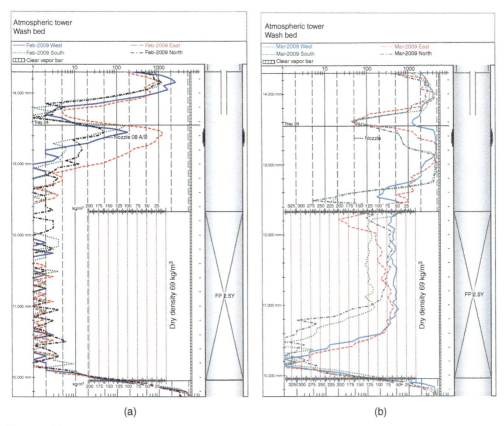

(a)   (b)

Figure 5.23   Gamma scans of atmospheric tower wash bed. (a) Plugged and severely flooded after years in service; (b) following a chemical wash, attempting to clear the fouling. (Reproduced Courtesy of Tracerco.)

In an attempt to circumvent a total shutdown, the refiner temporarily interrupted the feed and performed a chemical wash on the bed. Figure 5.23b shows a scan of the bed following the wash. There was a major improvement in the upper part of the bed. There was little improvement in the bottom part of the bed. It appears that there was residual fouling in the bottom part of the bed that the chemical wash was unable to remove. This part remained plugged and flooded. Note the difference between the chords in the upper parts of the bed. It appears (interpretation) that the chemical wash was more effective along some chords than along others. The tower was shut down shortly after. The packing near the bottom of the bed was heavily fouled, and to remove some of the bed, the packing needed to be literally torn apart. Packings at the top of the bed (inspected a short time after the chemical wash) were fairly clean.

The bed was replaced with in-kind. Anticipating fouling, a monitoring program by performing grid scans about once per year was started. Figure 5.24a shows the grid scan immediately after the tower returned to service following the packing replacement (year 2009). There is no sign of fouling. The liquid distribution looks excellent. Figure 5.24b shows yearly shots of one chord of the grid scans over the next five years, taken at comparable hydraulic loads. Per good practice, all four chords of the grid scan were shot each time. Figure 5.24b shows gradual buildup of fouling material at the bottom of the bed over time. The dark blue curve on the right is the same as the blue curve in Figure 5.24a and shows the scan immediately after returning to service. The red curve shows that as little as a year later fouling started at the bottom 200 mm of the bed, but the rest

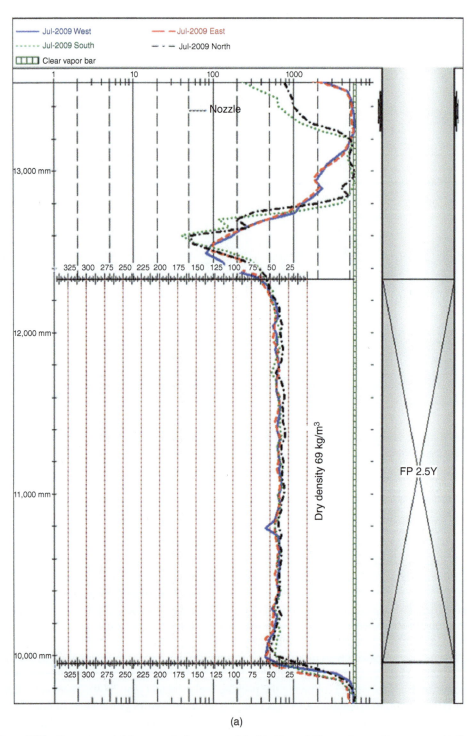

(a)

Figure 5.24 Gamma scans of the atmospheric tower wash bed in Figure 5.23 over the next five-year run. (a) Shortly after returning to service following bed replacement; (b) yearly shot of one chord of the grid scan over the next five years; (c) comparison of the full grid in January 2014 (before replacement) and in July 2014 (after replacement with similar packing). (Reproduced Courtesy of Tracerco.)

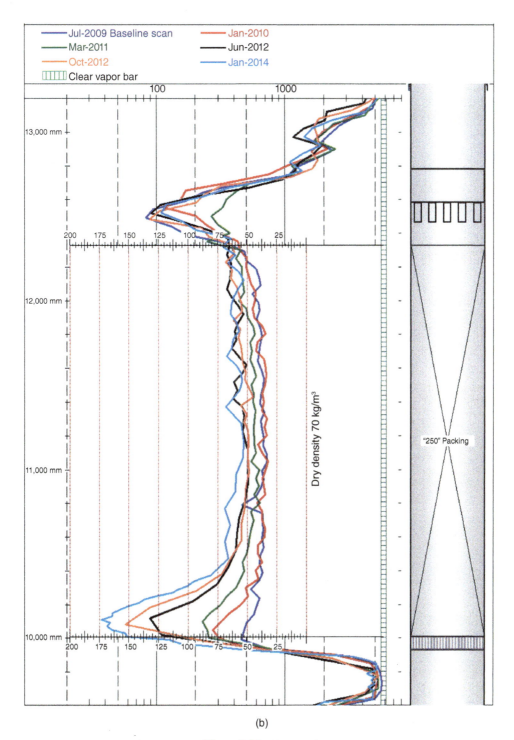

(b)

Figure 5.24   (*Continued*)

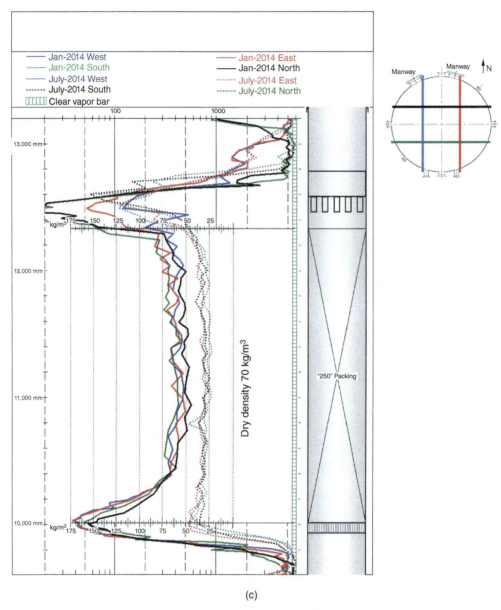

(c)

Figure 5.24 (*Continued*)

of the bed remained unaffected. A year later (2011, green), the fouling at the bottom expanded. In the 2011 scan, a higher absorption can be seen throughout the bed. The fouling at the bottom of the bed, and the absorption of radiation in the upper part of the bed, intensified over the next three years. In these three years, there appears to be flooding in the bottom 200–400 mm of packing.

The higher radiation absorptions in the upper part of the bed merit discussion (all interpretation, but quite conclusive). One possibility is fouling in the upper part as well. A counter-argument is that this absorption is uniform with bed height, while fouling is likely to be more intense near the bottom, which would have produced a density gradient, not a uniform profile with height.

A stronger argument is based on Figure 5.23b. Following the chemical wash, in which some of the fouling matter was removed, the south and east chords showed considerable radiation absorptions in the upper part of the bed, which were quite uniform with height (especially the south chord), similar to those in Figure 5.24b. When the tower was opened, the packings in the upper part of the bed were fairly clean; nothing that can explain the observed high absorptions there. Historically, the fouling was always observed in the bottom of the packings.

The possibility of higher liquid loads in the bed can also be discounted, as the reflux rates used in the wash bed were very low and could not account for the absorptions in the upper part of the bed, nor for the south curve in Figure 5.23b. This leaves one conceivable mechanism: flooding due to plugging in the bottom of the bed generated large liquid entrainment, which propagated throughout the bed and eventually descended. As the flooding at the bottom gets worse, so does the entrainment, liquid recycle in the bed, and the liquid retention in the upper part of the bed. In Figure 5.23b, following the chemical wash, the liquid entrainment/recycling channeled preferentially along the south and east chords.

Figure 5.24c compares the grid scans before and after bed replacement with a similar packing in 2014. It shows that before bed replacement there was liquid accumulated above the bed and backing up into the distributor in the south and north chord (black and green curves).

The monitoring was valuable for the refiner for providing a better understanding of the mechanisms involved and for planning the next turnaround.

5.5.11 More Plugging and Flooding in a Packed Bed (485)

A chemical column containing structured packings experienced fluctuating pressure drops and off-spec overhead product. Figure 5.25 shows a scan of this column. The lower bed shows good liquid distribution and normal operation.

The upper bed shows high density and flooding. The absorptions are irregular. The dashed black chord shows much higher density than the others. The dotted purple chord returns to normal bed density at an elevation of about 43 ft. There are irregularities among the chords at elevations of 42–44 ft. All observations so far are facts. The irregular density profile, with some chords showing more radiation absorption than others, suggests plugging rather than regular flooding (inconclusive interpretation). The fact that the flooding begins at the bottom of the upper bed and tends to clear, at least to some degree, near its top suggests that the plugging begins and is most severe at the bottom of the bed (conclusive interpretation).

Upon entering the tower, the plant found a layer of polymerized material had formed on the support plate of the upper bed. This polymerized material was responsible for the flooding. The plugging backed up several feet of liquid in the upper bed, giving the observed absorptions.

Another case (480) Grid packing in a plugging chemical service was replaced by an alternative grid packing in an endeavor to increase run length, but with the new grid the run was drastically shorter. Comparing gamma scans showed that with the old grid solids filled the bed from top to bottom before there was an operational issue. With the new grid, the solids accumulation point shifted deeper in the bed, with the upper sections of the bed not plugged. Upon inspection, it was noticed that large chunks of solids had come off the packing and were clogging the flow channels, leading to a premature flood.

5.5.12 Flood due to Crushed or Damaged Random Packings (509)

Figure 5.26a is a scan of a tower immediately after returning to full-rate operation following a turnaround where the packings were replaced with in-kind. At this time, it was meeting all

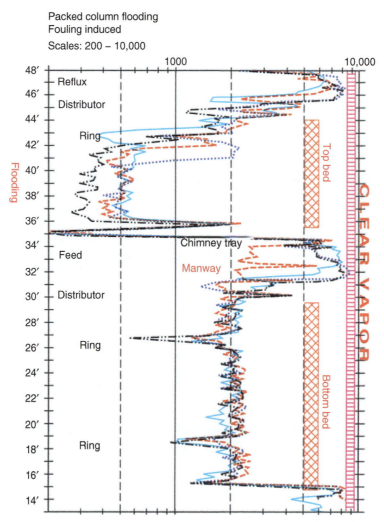

Figure 5.25   Plugging-induced flooding at the top packed bed of a chemical tower. (From Tru-Tec Division, "Sample Scan Brochure," 1998. Available from Tracerco, Pasadena, Texas. Reproduced Courtesy of Tracerco.)

expectations on product specifications, capacity, and pressure drop. An unusual feature is a density gradient through the bed. The density gradient may indicate higher liquid loads or compression of the packings as one descends the bed. The bottom of the bed appears closer to flood than the upper portions.

Figure 5.26b is a scan shot just prior to the turnaround, when the tower was experiencing problems. The density gradient intensified. The bottom of the bed was flooded. Liquid maldistribution is apparent in the flooded bottom region, probably connected with the flood. From 9 ft elevation down to the bottom of the bed, the density of the bed was higher than at equivalent elevations in Figure 5.26a. The packed height was about 2 ft shorter than it should have been, indicating (conclusive interpretation) crushed or damaged packings in the bottom of the bed as the cause of flooding.

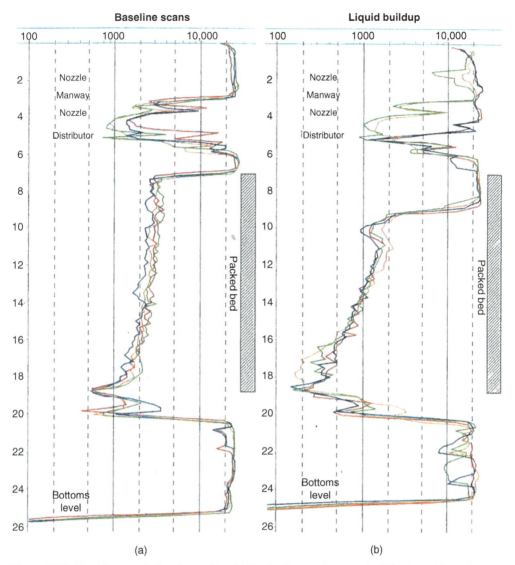

Figure 5.26 Flood due to crushed random packing. (a) Shortly after returning to service following packing replacement with in-kind; (b) just before the turnaround showing flooding at the bottom due to crushed or damaged packings. (From Xu, S. X., and L. Pless, Chemical Engineering, p. 60, June 2002. Reprinted Courtesy of Chemical Engineering.)

5.5.13 Missing Random Packings (373)

A 2 ft diameter hydrogen fluoride stripper was experiencing poor separation. Due to the small diameter, a + scan (Figure 5.6c, Section 5.2.3) was performed (Figure 5.27). In the top two beds, the uniform density profiles and the two chords traveling together indicate that these beds were operating normally with good liquid distribution. The low-density readings in the upper 8.5 ft of the bottom bed, where radiation transmission reached the clear vapor line, indicates that the packings were not there. At the bottom of the bottom bed, there was a small region of high density,

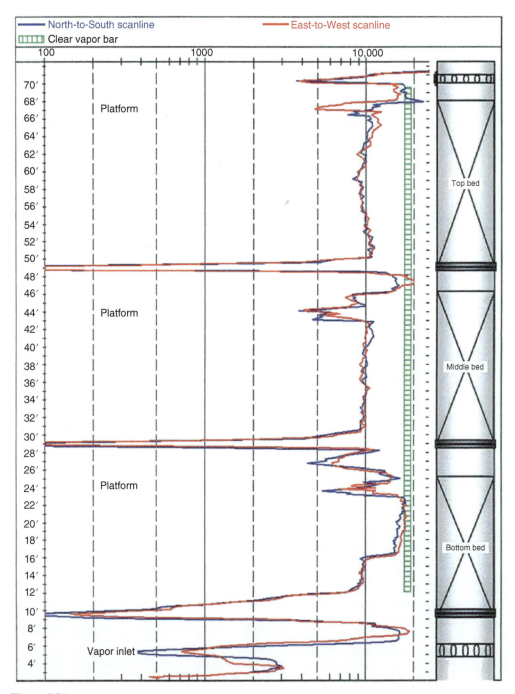

Figure 5.27   Gamma scans indicate the top and middle beds appeared to be in place but the bottom bed was missing approximately 8.5 ft of packings. (Reproduced Courtesy of Tracerco.)

indicating flooding or dense material. These are all facts. When the column was opened, a piece of the support grating was found bent and holding a pocket of packings in place. A lot of the packings in the bottom bed migrated out through the damaged support.

5.5.14 Displaced and Damaged Structured Packings (89)

During startup, a high liquid level was observed in the bottom of a refinery vacuum tower. Once operation was stabilized, the tower experienced problems meeting the color and metal specs on the lowest side product, the heavy vacuum gasoil (HVGO). To achieve a product of acceptable specs, the unit was forced to reduce the HVGO yield, losing considerable HVGO to the far less valuable bottom residue product. The tower is shown in Figure 7.2.

Figure 5.28a shows a scan of the "slop wax" (or "wash") packed bed, located between the tower feed and the HVGO side draw. This bed uses a very small liquid reflux to remove heavy color bodies and heavy and entrained metallic compounds from the HVGO side product. The bed contained grid packing at the bottom, with structured packings mounted above them. Throughout the bed, the east chord (black curve) shows high density. The south chord (red curve) shows clear vapor at the top two feet of bed, suggesting some packings are missing. On the north chord (blue curve), and to a lesser extent on the east chord, there appear to be some dense absorptions above the bed. As the distributor was a spray distributor and the liquid rate was quite low, these unequal heights at the top of the bed could not be explained by liquid maldistribution. They most likely suggest packing displacement.

The chimney tray under the wash bed (the "slop wax" chimney tray) appears lightly loaded and operating dry (compare to the HVGO chimney tray in the next paragraph). The absorptions are probably due to the chimney metal. This tray is usually designed for dry operation, as any liquid sitting on the tray at high tower temperatures is likely to turn into coke. The level indicator on this tray showed dry operation. So the light density for this tray in the scan should not be interpreted as damage. As with normal trays (Section 5.5.9), it is difficult to distinguish a dry tray from a damaged tray.

Above the wash bed there was an HVGO chimney tray, which collected the liquid from the HVGO bed above. Some of this liquid became the HVGO product, some was refluxed to the slop wax bed below, and the rest was cooled externally and sprayed at the top of the HVGO bed above (Figure 7.2) to condense the HVGO by direct contact heat transfer. Figure 5.28b shows the HVGO chimney tray in place and holding liquid.

Figure 5.28b shows displaced packings in the HVGO bed. On the north (blue curve), the top two feet of the bed show clear vapor. On the west (green curve), there is a shorter distance of clear vapor. On the east (black curve), there is about 1 ft of packing displaced above the bed.

When the tower was opened, packings were found displaced both in the wash and HVGO bed, as shown by the scans.

5.5.15 Flooding Induces Channeling in Deep Vacuum Packed Tower (232)

This is the same chemical vacuum tower discussed in Section 2.9.1. The bottom section of the tower had two beds containing high-efficiency structured packings, with a vapor side draw between them. The tower was operated at about 60% of flood and experienced a high heavy impurity in the side product. In a test, the reboiler duty and reflux rates were raised to the anticipated flood point to explore for possible hydraulic issues or fouling.

The pressure drop versus C-factor curve (Figure 2.24a) showed no point of inflection, which is often used to identify flood, even when the expected flood point C-factor (Eq. 2.1) of 0.37 ft/s at the bottom of the bottom bed was exceeded. Raising vapor and liquid loads reduced the temperature spreads in the bottom bed (Figure 2.24b), as often experienced with wire mesh packings,

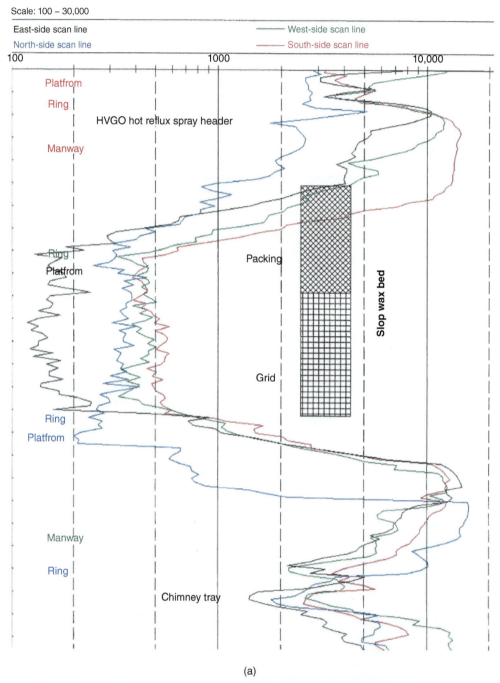

(a)

Figure 5.28 Packing displacement in refinery vacuum tower. (a) Wash bed. (b) HVGO bed. (From Duarte Pinto, R., M. Perez, and H. Z. Kister, Hydrocarbon Processing, April 2003. Reprinted Courtesy of Hydrocarbon Processing.)

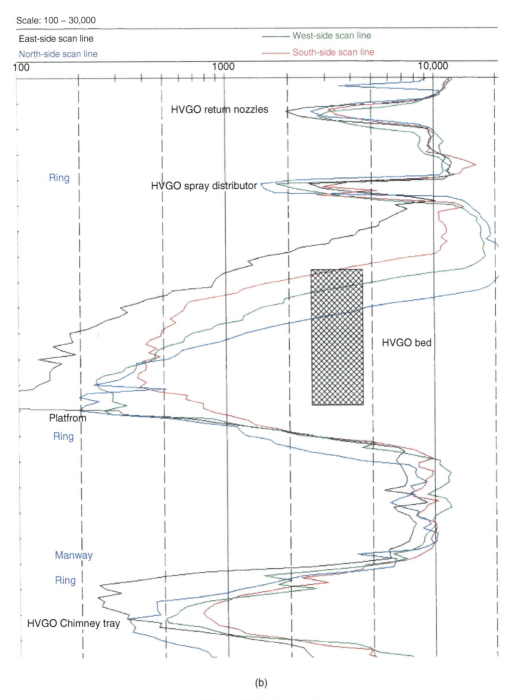

Scale: 100 – 30,000

East-side scan line   West-side scan line
North-side scan line   South-side scan line

(b)

Figure 5.28   (*Continued*)

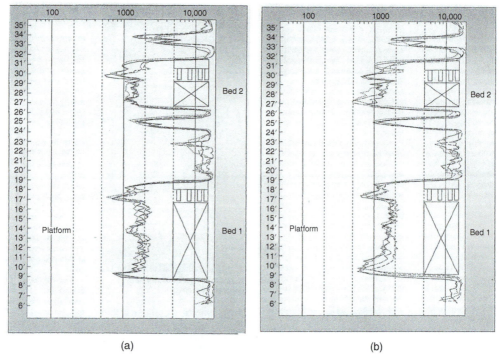

Figure 5.29   Gamma scans of vacuum tower. (a) C-factor of 0.3 ft/s, good distribution in bottom bed. (b) C-factor of 0.36 ft/s, showing liquid accumulation and maldistribution in the bottom 2 ft of the bottom bed. (From Kister, H. Z., R. Rhoad, and K. A. Hoyt, Chemical Engineering Progress, September 1996. Reprinted Courtesy of Chemical Engineering Progress (CEP). Copyright © American Institute of Chemical Engineers (AIChE).)

like the packings in this bed (193). At a C-factor of about 0.36 ft/s, the middle temperature of the bottom bed abruptly shot up, becoming the same as the bottom temperature, indicating loss of separation between the two points. The packing efficiency sharply dropped, which indicated flooding. The unsolved mystery was why the flood was not accompanied by a rise in pressure drop. The gamma scans explained this behavior.

Figure 5.29a and b shows the gamma scans at C-factors of 0.30 and 0.36 ft/s, respectively. Both scans show good liquid distribution in the bottom bed down to the bottom 2 ft. In the 0.30 ft/s C-factor scan, the good liquid distribution persisted all the way to the packing support. However, in the 0.36 ft/s scan, maldistribution appears to have set in the bottom 2 ft or so of bed, with liquid accumulating along two of the four chords. This accumulation is a symptom of flooding. Figure 5.29b suggests that the liquid accumulation generated channeling; the liquid accumulated only along two chords. Presumably, the vapor kept rising freely through the regions in which the liquid did not accumulate. This means that the vapor did not need to penetrate through a static head of liquid, which explains why no steep rise in pressure drop was apparent even though the packing flooded (well-supported interpretation).

5.5.16   Distributor Overflow and Maldistribution (377)

A 9 ft diameter stripping tower in a synthesis gas plant contained two random packed beds with three trays above. The tower has a history of fouling, with several episodes detailed with gamma scans in Ref. 377. The episode below focuses on a scan (Figure 5.30) shot in September

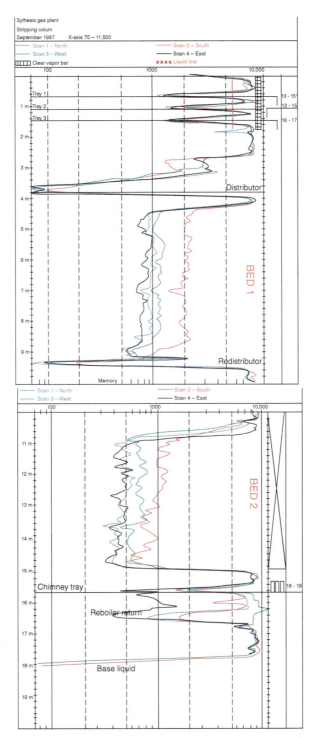

Figure 5.30   Grid scan of stripping tower showing significant liquid maldistribution in both packed beds and that the liquid distributor and redistributor were liquid-full and overflowing. (From Pless, L. and B. Asseln, PTQ, p. 115, Spring 2002. Reprinted Courtesy of PTQ.)

1997 (377), a short time after the tower returned to full-rate operation following a turnaround in which the fouling material was removed.

Figure 5.30 shows both beds operating with significant liquid maldistribution with the south side in both (red scan) receiving less liquid. The problem appeared to initiate at the top of each bed, suggesting distributor problems.

The liquid level in the distributor above the top bed was uneven. The liquid heights of the south (red scan) and west (green scan) chords were 50 and 20 cm, respectively, taller than those on the north (blue scan) and east (black scan). The levels of liquid measured on the south and west were high enough to have the distributors liquid overflowing into the vapor passages. Likewise, the redistributor between the two beds was liquid-full and overflowing. Overflowing liquid is never properly distributed (Section 4.6.1) and would account for the maldistribution in the beds in Figure 5.30.

Another observation is that the packed height in the lower bed was about 40 cm below the bed limiter. The top of the packing of the upper bed was not even, varying from 75 cm to 1 m below the upper bed limiter. The lower bed heights could be due to either under-filling of the beds or settling of the packings (which is not unusual). There is no indication of crushed packings that would have shown up as a density gradient like it did in Case 5.5.12.

Despite the issues reported, the tower operated stably and there was no flooding.

5.5.17 Overflow, Entrainment, and Maldistribution from Flashing Feed Distributor

An ammonia plant absorber–stripper system experienced excessive slip of carbon dioxide from the top of the absorber. The absorber and stripper were studied using hydraulic calculations, temperature surveys, and gamma scans. One of the major bottlenecks found was the flashing feed distributor that brought the rich solvent to the top of Bed 2 in the stripper.

The hydraulic calculations showed that the random packings near the top of Bed 2 should have been operating at 88% of flood. The temperature survey showed severe maldistribution throughout Bed 2.

Figure 5.31 is the 2-chord gamma scan for Bed 2. It was intended to perform a full grid scan, but strong winds and issues with scaffolding prevented shooting two of the chords. The x-axis is expressed in relative density units, with the clear vapor line taken as zero density. The density for other locations can be inferred from Eq. (5.1) in Section 5.1. Figure 5.31 shows that the density at the top six feet of the bed was 50–100% higher than the density in most of the bed, suggesting flooding. There is a definite maldistribution in the bottom 15 ft of the bed. There is also a high-density region at the bottom three feet of the bed, also suggesting flood. The flooding in the bottom of Bed 2 may be due to compression or some fouling of the packing (the packing type in this tower was prone to either). There is severe overflow and flooding on the top of the distributor, with froth height well above 60 in. A large amount of liquid is present even well above the feed pipe.

A close review revealed that the flashing rich solution entry was highly nonstandard. The flashing feed entered the tower via a sparger pipe with downward-pointing holes, exiting the holes with a downward velocity of 10 ft/s. At that velocity, it would have taken half a second to reach the liquid level in the distributor below with substantial impingement. This, in turn would froth up the distributor liquid, similar to Figure 4.8, with froth overflowing over the top of the distributor. Some of the vapor jets would penetrate right through the distributor holes into the top of Bed 2. The recycle vapor would then disengage and turn upward, increasing the vapor load, and would flood the top of the bed. Both the vapor penetration and the distributor frothing would account for the maldistribution.

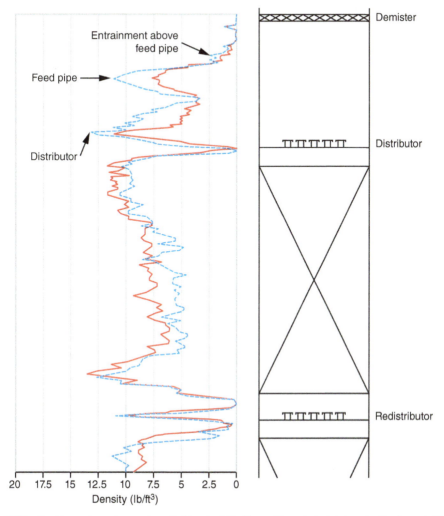

Figure 5.31   Frothing and entrainment from flashing-feed distributor. The main bed, below the distributor, is "Bed 2".

5.5.18   Tower Overfill due to Excessive Pressure Drop in Kettle Reboiler Piping

A tower experienced flooding, thought to be due to a tray limitation. Retraying the tower was being considered. Prior to the retray, the tower was gamma scanned.

The scan showed flooding (Figure 5.32) but provided a different story about the cause. The scan shows no clear vapor space below the bottom tray. It also showed the bottom sump liquid level reaching the reboiler return inlet nozzle. The tower had a kettle reboiler, so the liquid level in the tower bottom sump was not controlled but was set by the pressure drop in the kettle reboiler circuit. When this liquid level exceeds the reboiler return elevation, the reboiler return vapor will entrain massive quantities of this liquid into the trays, initiating flood (Section 2.13).

Therefore, the tower flooding was caused not by a tray bottleneck, but by excessive pressure drop in the reboiler circuit. An excessive pressure drop in kettle reboiler circuits is a common cause of high bottom levels that initiate flooding in towers (Section 2.13).

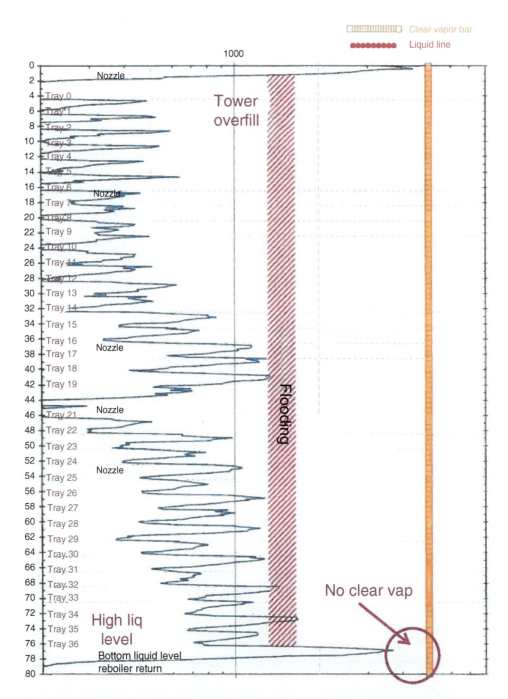

Figure 5.32   Gamma scan showing high bottoms liquid level generating flood throughout the tower.

5.5.19 Is a Collector (or Chimney Tray) Overflowing? Does this Initiate Flooding?

When a packing distributor, redistributor, or a chimney tray overflows, liquid enters the vapor passages. The vapor rise area is typically 50%, 25%, and 25% for trough distributors, pan distributors, and chimney trays, respectively. With C-factors (Eq. (2.1)) of high-capacity trays and packings often approaching 0.4 ft/s (Section 2.16), a pan distributor or chimney tray overflow is likely to induce system limit flooding. This is further discussed in Section 8.11, with typical examples in Cases 1 and 2 of this section.

Gamma scans are powerful for identifying such bottlenecks. A liquid level reaching or exceeding the top of the chimneys indicates a likely overflow. The gamma scans can also detect whether flood initiates at that point, or whether there is liquid descending to the section below via the chimneys. In three different columns (118, 330, 489), gamma scans showed liquid level at the top of the chimneys of a chimney tray. In two of these columns (330, 489), the liquid going to a kettle reboiler overflowed the chimneys. In one of them (330), the liquid descended through the chimneys, bypassing the reboiler and inducing lights into the bottom product; in the other (489), the vapor velocity through the chimneys was too high to allow the liquid descent so the liquid was entrained, flooding the trays above. In the third column (118), the liquid backed up, some descending through the chimneys and contaminating the bottom product, the rest entrained initiating a premature flood of the trays above.

When scanning for overflow, it is essential to judiciously set the scan chords. Poor scan chord selection may lead to inconclusive results. The scan chords should be able to clearly distinguish the liquid level from the chimney metal. Generally, a chord that passes through liquid only is best, while chords that intersect too much metal should be avoided.

Figure 5.33 is an enlargement of the gamma scan in Figure 5.14 at elevations from 68 to 88 ft. The troubleshooting investigation (238) in the tower in Section 5.5.4 needed to test a theory that attributed the tower flood to overflow of the chimney tray collecting the bottom liquid and feeding it to a once-through thermosiphon reboiler below. The theorized high liquid level was attributed to insufficient residence time to degas the liquid with the reboiler feed line undersized for self-venting flow (Section 2.15).

Figure 5.33 confirms that the liquid level on the chimney tray reached the top of the chimneys (18 in.) both at the normal and increased reflux scans. However, trays 29 and 30 showed clear vapor spaces and no flooding. The scans thus confirmed a "flood bottleneck waiting to happen, but not yet reached." Figure 5.33 shows likely weeping from the chimney tray in the normal reflux scan (blue), but no weeping in the high-reflux scan (red). During operation, there were no problems with reboiler operation or with bottoms purity, so no one expected an issue. However, while performing the modifications discussed in Section 5.5.4 during the next turnaround, the chimney heights were increased from 18″ to 36″. Upon return to service, the concentration of lights in the bottom was much lower, which could only be explained by elimination of the overflow.

5.5.20 Damaged or Dry Trays?

Co-authored with Betzalel Blum, Bazan, Haifa, Israel

Following a process upset, a refinery crude tower experienced poor separation between the diesel and the bottom residue. To keep the diesel 95% point (heavies content) on spec, the refinery lost some diesel in the bottom residue and needed to cut charge rates. Tray damage at the upset was suspected. An outage for repair was considered, but a positive diagnosis was required before taking such a harsh and expensive step. The tower was scanned with a focus on the wash section, the section that achieves the diesel rectification and gets it on spec, trays 35 to 41 in the tower.

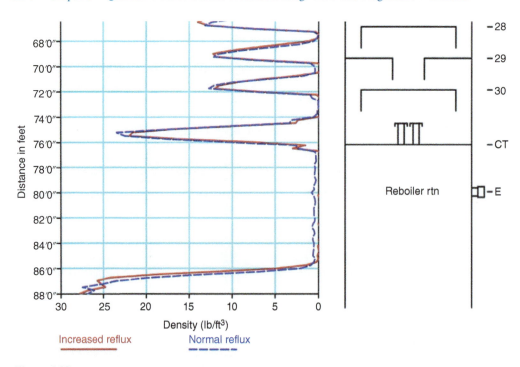

Figure 5.33 Gamma scan showing liquid level reaching or almost reaching the top of chimneys on a chimney tray, but not generating flood. (From Kister, H. Z., D. E. Grich, and R. Yeley, PTQ Revamps and Operation, 2003. Reprinted Courtesy of PTQ.)

Each of the 2-pass trays active areas, as well as the center downcomers, were gamma scanned. The active area scan chords were parallel to and equidistant from the centerline. The south active areas looked dry (red curve in Figure 5.34a); the north active areas (blue curve) appeared more normal, but still on the dry side. The south side of trays 36 and 37 held very little liquid, suggesting damage. The south scan showed some gamma-ray absorptions in the vapor spaces, suggesting the presence of some dense material such as dislodged tray parts. The downcomer scans (not shown) showed about 4 in. of liquid in the center downcomers from trays 35 and 37. So there were more arguments supporting tray damage than tray dryout, but the arguments were inconclusive to justify a unit shutdown.

The two key parameters for tray hydraulics are the C-factor (Eq. (2.1)) and the weir load (Section 2.11), which express the vapor and liquid loads, respectively. At the tray spacing of 30 in. and the low weir loads typical of the wash section, the trays will be near jet flood when the C-factor reaches about 0.4 ft/s (Section 2.16). From the tower simulation, the C-factor was less than 0.23 ft/s, which gives less than 60% of jet flood. With the very small liquid loads and the large tray spacing in that section, downcomer flood would not even be close. So these trays should be operating at a comfortable margin away from flood.

Weir loads are usually extremely low in wash sections (<1–2 gpm/in. of outlet weir length). Based on the plant meters, the weir loads during the scan were even lower, of the order of 0.2–0.6 gpm/in. of outlet weir length. At this liquid load range, the trays will be essentially dry and will look dry on the gamma scans. A dry tray does not look much different from a missing tray (Section 5.3, item 12, and Section 5.5.9). Gamma scans do not see trays; they see liquid sitting on the trays. This is key for interpreting the current gamma scans. This led to the conclusion that the scan interpretation so far was inconclusive.

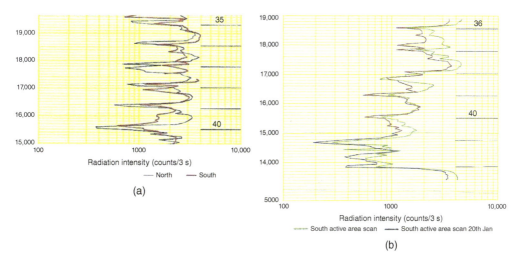

Figure 5.34   Gamma scans of wash section. (a) At normal operating liquid rates, showing difference between south and north active areas; (b) south active area scan at double the liquid rates (red pen) compared to the original liquid rates (green pen).

To distinguish dry from damaged trays, it was necessary to increase the weir loads by a great amount. This was achieved by cutting the diesel draw rate and running what used to be product as reflux to the wash section. The liquid rate was thus doubled, to 0.4–1.2 gpm/in. At weir loads greater than about 1 gpm/in. of outlet weir length (most of the trays), the trays should no longer look dry and should be much easier to distinguish from damaged trays. We then repeated the scans of the south active area and the center downcomers.

At the higher liquid loads, the south active area scans (Figure 5.34b) showed more liquid on trays 37, 39, 40, and 41 (red curve). Not much change on trays 36 and 38. Both trays 36 and 38 had long outlet weirs and therefore low weir loads. There appears to be some flooding initiating above trays 37, 39, and 41 (the trays with the side downcomers). At the low C-factors, there is no explanation to flooding on these trays other than damage. As expected at the higher liquid loads, the liquid heights increased in the center downcomers.

The observation of little change on trays 36 and 38 despite doubling the weir loads, as well as the unexpected flooding, provided strong evidence supporting tray damage. The tower was shut down, and the damage was found and repaired.

Chapter 6

Advanced Radioactive Techniques for Distillation Troubleshooting

"The devil is in the details . . . and so is salvation"

—(origin unknown)

6.1 QUANTITATIVE MULTI-CHORDAL TRAY GAMMA SCANS ANALYSIS

Qualitative gamma scanning (Chapter 5) has been successfully applied to troubleshoot tens of thousands of tray towers for over five decades. Scan data have traditionally been interpreted by visual examination of the scans to detect flood, damage, fouling, foaming, entrainment, froth (or spray) heights, weep, and liquid holdup. In the majority of troubleshooting investigations, where tray maldistribution is not an issue, qualitative observations suffice.

Liquid and/or vapor maldistribution is sometimes encountered in towers containing either conventional or high-capacity trays, especially those with larger diameters. Capacity-enhancing features include push valves that move the liquid forward or sideways, truncated downcomers that boost the tray active area, and multiple downcomers that lower the effective weir loads. These features, however beneficial, bring with them unique challenges. Excessive forward push may generate excessive froth gradients. Truncated downcomers lack a static liquid seal, possibly permitting vapor rise up the downcomers ("blowby"). Multiple downcomers generate nonsymmetrical active areas that may promote channeling and maldistribution. Each of these issues lowers tray efficiencies, increasing the reflux, reboil, and energy requirement, and bottlenecking tower capacity.

Vapor and liquid maldistribution, excessive froth gradients, and vapor rise into downcomers are difficult, often impossible, to diagnose using conventional troubleshooting techniques such as vendor or technology-provider software, ΔP measurement, and conventional single-chord qualitative gamma scans. Judicious multi-chordal gamma scan with quantitative analysis is the best tool for diagnosing these issues. This technique and its development are described in detail in Refs. 204 and 205.

With this technique, several parallel chords, typically 3–6, are shot along the flow path (Figure 6.1). Froth heights, froth densities, liquid heads, and entrainment index data

Distillation Diagnostics: An Engineer's Guidebook, First Edition. Henry Z. Kister.

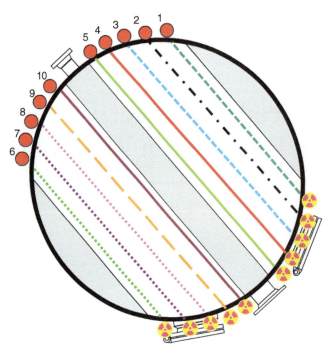

Figure 6.1   Parallel chords used for multi-chordal quantitative gamma scan analysis of a tower containing two-pass trays.

are determined for each tray at each chord using high-accuracy methods as described in Refs. 160 and 204 and outlined below. From the data, the profiles of these variables along the flow path are mapped and are plotted using a "Kistergram," which shows the variable to scale on a tray diagram. From these plots, channeling patterns can be inferred as described in the following.

As this analysis often seeks to identify small differences between chords, accurate positioning of the source and detector is required, and is taken care of by the scanning contractors. Special care is required to accurately position and fasten the metal guides that move the source and detector down the column and to protect them from getting stuck on supports and from flapping in the wind.

Quantitative multi-chordal gamma scan analysis is not a "cure all." It typically costs 5–10 times more than a regular qualitative gamma scan. About half of this extra cost is in the additional scanning chords, the other half is in verifying data validity and consistency, analyzing the scans, obtaining froth heights, froth densities, entrainment, and liquid heads, and putting these on the appropriate graphics. Verification of data validity and consistency is a must, as one set of bad data is sufficient to lead to an incorrect diagnosis. Unnecessary scans are costly, while missed details, shortcuts, and misapplication can (and have) led to incorrect diagnostics and ineffective solutions. We therefore recommend embarking on a quantitative multi-chordal analysis only when there is a real need for it and when there is commitment and resources to do it correctly.

6.1.1   Harrison's Method for Froth Height and Flood Determination

Froth (or spray) height values reported in standard gamma scan reports are excellent for qualitative analysis, but in our experience, they are seldom accurate enough for quantitative analysis, where the channeling patterns are inferred from small differences, and therefore high accuracy is required. Accurate determination requires a laborious method, usually unjustified in low-cost

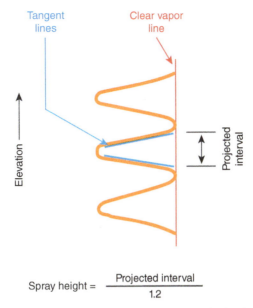

$$\text{Spray height} = \frac{\text{Projected interval}}{1.2}$$

Figure 6.2 Harrison's analysis method for spray (or froth) height determination from commercial gamma scans. (Based on Harrison, M. E., Chemical Engineering Progress, 86(3), p. 37–44, March 1990. Diagram reprinted from Kister, H. Z., Chemical Engineering Progress, p. 33, February 2013. Reprinted Courtesy of Chemical Engineering Progress (CEP). Copyright © American Institute of Chemical Engineers (AIChE).)

qualitative scanning. Harrison (160) developed a highly reliable method for froth height determination, which has become the standard for quantitative analysis. His procedure is as follows:

1. The clear vapor line is determined per Section 5.2.1. This vertical line denotes the amount of radiation transmitted through a dry, obstruction-free vapor space region near the relevant tray section, and is used as a reference. This region may be the vapor space above the top tray, where tray spacing is larger, or the region below the bottom tray, or the vapor space above a chimney tray. If entrainment is a possibility at the relevant region at high loads, a low-load or dry-tower scan is needed to correctly establish the clear vapor line.

2. For the gamma scan peak on each tray, draw a tangent to each side of the peak, and extrapolate each tangent to the clear vapor line. This is illustrated in Figure 6.2.

3. Determine the length of the interval between the points of intersection of the tangents on the clear vapor line (Figure 6.2). The froth or spray height equals the interval length divided by 1.2. Harrison's paper gives a detailed justification of the factor of 1.2.

4. Froth (or spray) heights well below the tray spacing indicate non-flooded trays. Froth (or spray) heights exceeding the tray spacing indicate flooding. Froth (or spray) heights close to the tray spacing indicate a possible flood or flood initiation. Spray (or froth) height on a tray varies dynamically (106) and is not a highly precise measurement, so precision of ±20% is good.

Figure 6.3 illustrates an application of this method to an actual gamma scan of the active areas on one side of two-pass trays at 24 in. spacing in gas processing service. In this tower, flooding occurred above the vapor feed, as also indicated by ΔP measurements. This scan shows both flooded and unflooded trays and depicts a variety of peak shapes often encountered, most of which deviate from the total linearity of Figure 6.2.

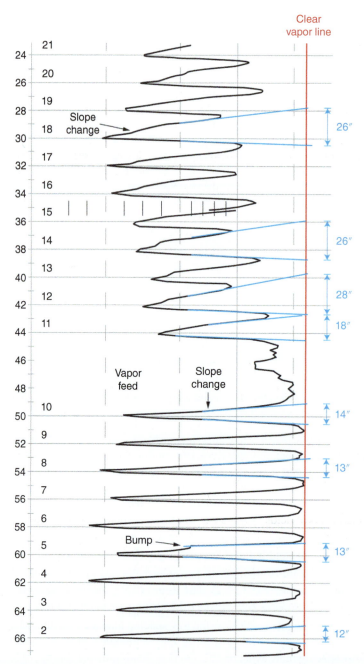

Figure 6.3 Application of Harrison's method for spray (or froth) height determination with a variety of peak shapes. (From Kister, H. Z., Chemical Engineering Progress, p. 33, February 2013. Reprinted Courtesy of Chemical Engineering Progress (CEP). Copyright © American Institute of Chemical Engineers (AIChE).)

Figure 6.3 shows the way the author believes the tangents should be drawn, based on his vast experience with this method. The sides of the peaks on trays 2 and 8 approach total linearity, with only minor deviations near the clear vapor. Here, drawing the tangents is straightforward. The sides of the peaks on all the other trays in the bottom section except for 5 and 10 are linear. The ascending side of the peak on tray 10 shows a larger deviation from linearity, and the tangent is drawn to match the bulk of the peak. Tray 5 shows a "bump" in the ascending side of the peak. As most of the peak is above the "bump," the tangent is drawn to the side of the peak above the bump. For the trays above the feed, the ascending sides of the peaks show larger deviations from linearity, many with a change of slope near the tray floor (trays 12–14, 16–18, and 20). Similar to dealing with the bump on tray 5, the tangents are drawn above the change of slope on the ascending side of the peaks.

Below the feed, Figure 6.3 shows very consistent froth heights, all within the range of 12–14 in. (tray spacing is 24 in.). This is typical of a section where the hydraulic loads and tray design are uniform. The superb consistency of the froth height determination provides validation to Harrison's method. Below the feed, the trays are operated well below flood. On trays 12–21, froth heights are consistently above 24 in., indicating flood.

Alternative to Harrison's Method   Pless (371, 380) recently proposed a method in which the froth height is derived from the total area between the upper side of a peak and the horizontal line at the tray floor. No details were provided about how to derive froth heights from the areas. This calculation is performed by the scanning contractor using a new detector technology and software interpretation. Like in Harrison's method, the froth height to tray spacing ratio is calculated, and expressed as "% tray space" (371, 380). It has been stated (371) that this % tray space correlates well with the percent of flood. The author's experience has been that this method is a marked improvement on the qualitative method for froth heights (Section 5.2.1), but its accuracy still falls short of that of the Harrison method.

Downcomer Froth Heights   While Harrison's method applies to froth heights in tray active areas, it does not extend to downcomer froth heights. The author does not have a satisfactory quantitative method to determine the downcomer froth height. All our interpretations of downcomer scans remain qualitative. Pless and Nieuwoudt (380) state that the Pless quantitative method for froth height (371, 380) yields downcomer froth heights that correlate well with downcomer backup and downcomer choke predictions. This conclusion is based on experimentation at a Koch-Glitsch test tower with Superfrac® trays at 16 in. spacing, with predictions based on the Koch-Glitsch proprietary correlations, and it is unknown how it extends to other situations. The author's experience has been that the quantitative downcomer froth height values provided by gamma scanners have been reasonable, but their correlation with percent downcomer flood has ranged from good to poor.

6.1.2   Application of Harrison's Method Prevents Unnecessary Shutdown

A 9 ft ID chemical tower containing single-pass sieve trays operated at 20–30% above its original design capacity. In an effort to squeeze an additional capacity increase of 3–4%, the hole areas of the 45 bottom trays were increased from 8.5% to 13%. Additionally, radiuses were added at the bottom of the downcomers to reduce downcomer backups. Nothing else was changed. Instead of gaining capacity, the tower lost 5% in capacity.

Gamma scans of the bottom 30 trays, all below the feed and of the same design, showed a steep drop in radiation transmission above tray 20 (Figure 6.4). Based on this appearance, the scanners report concluded that flooding initiated at tray 20 and propagated upward. Both the

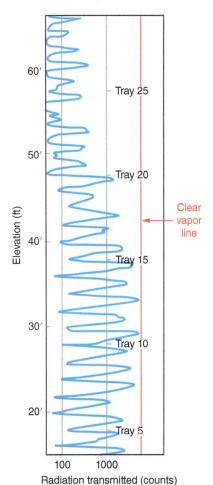

Figure 6.4 Qualitative gamma scan of the bottom 30 trays in the tower discussed in Section 6.1.2, suggesting flood initiating at tray 20. (From Kister, H. Z., Chemical Engineering Progress, p. 33, February 2013. Reprinted Courtesy of Chemical Engineering Progress (CEP). Copyright © American Institute of Chemical Engineers (AIChE).)

vapor and liquid hydraulic loadings slightly declined as one ascended the tower, so the loadings on tray 20 were about 3–5% less than near the bottom. A flood due to a design issue would have initiated on the highly loaded bottom trays, leaving an installation issue as the only conceivable explanation for a flood initiating 20 trays above. To repair the suspected installation error, the plant was making plans to shut the tower down.

To audit the need for a costly shutdown, Harrison's method was applied (Figure 6.5). Harrison (160) defines the "spray ratio" as the ratio of spray height to tray spacing (to improve clarity, this definition is slightly altered here from that in Harrison's original article, while retaining its principles and application). Spray ratios above 1.0 indicate flooding, while spray ratios below 1.0 indicate no flooding. Figure 6.5 confirmed that the spray ratios above tray 20 exceeded 1, but trays 17–19, and to a lesser degree also 15, and even 6 and 7, also were at or near flood. A flood due to an installation issue at tray 20 should have propagated from tray 20 up, and not hovered around some lower trays. This strongly argued against the installation issue theory.

The observation of flood taking place on some lower trays can also be inferred from a close examination of Figure 6.4. The qualitative scan shows flooding on trays 18, 16, and 6, but its spotlight is on the strong intensification of the flooding, prompting a diagnosis of flood initiation on tray 20. In contrast, the spotlight in the quantitative analysis of Figure 6.5 is on where the flood starts rather than where the flood intensifies.

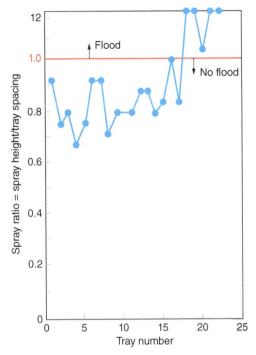

Figure 6.5 Spray ratios derived from the spray heights determined by Harrison's method for the tower discussed in Section 6.1.2, showing flood initiation below tray 20. (From Kister, H. Z., Chemical Engineering Progress, p. 33, February 2013. Reprinted Courtesy of Chemical Engineering Progress (CEP). Copyright © American Institute of Chemical Engineers (AIChE).)

The cause of the capacity loss was later diagnosed to be the onset of vapor cross flow channeling (VCFC; Section 2.12), diagnosed by the method described in Section 6.1.4 (214).

6.1.3   Entrainment Index

Qualitative scans usually infer entrainment from the proximity of the vapor peak between the trays to the clear vapor line. A vapor peak reaching the clear vapor line is usually reported as "no entrainment." Terms such as "slight" (or "light"), "moderate," and "severe" signify increasing distance between the vapor peak and the clear vapor line, as described in Section 5.2.2. In Figure 6.3, as an example, the entrainment on trays 3–9 was described as "none," that on tray 2 as "light," that on trays 11 and 19 as "moderate to heavy," and trays 12–18 were described as "flooded."

To quantify, the entrainment index EI is defined by Eq. (6.1):

$$EI = (\text{transmission, clear vapor})/(\text{transmission, max for the tray}) - 1.0 \qquad (6.1)$$

Our experience has been that generally, an entrainment index indicates the following:

EI	Entrainment
>2	Flood
1.5	Heavy
1.0	Moderate
0.5	Light
0	None

6.1.4   Kistergrams and Their Application

A graphic capable of transforming the large volume of data produced in multi-chordal scans into a clear visual of tray hydraulics was dubbed "Kistergram" by gamma scanning maestro David Fradgley (110). Simply, this graphic constructs "the forest from the trees." A Kistergram is a to-scale elevation sketch of the trays with superimposed quantitative data derived from the scans. Data plotted on Kistergrams include froth (or spray) heights, froth densities, entrainment, and clear liquid heads. A separate sketch is drawn for each variable. Not all these variables need plotting; only those that are needed for a good diagnosis. In the case study below, the froth height and entrainment Kistergrams were sufficient to provide a good diagnosis, so there was no need in the froth density and clear liquid height graphics.

Drawing Kistergrams   Superimposing data on the elevation diagrams to generate Kistergrams uses the following conventions (204):

1. All scales are linear.
2. The *x*-axis shows the actual position of the scan chord on the trays.
3. Froth (or spray) height data are plotted using the same scale as the diagram. When the determined froth height equals or exceeds the tray spacing, it is plotted right on the tray above. Froth height of zero inches is plotted as a point on the tray floor.
4. When the entrainment index is 2 (flood), it is plotted as a point right at the tray above. An entrainment index of zero is plotted as a point on the tray floor.
5. The froth density scale is set to suit the investigation. Most commonly, froth density data are plotted as they are; a froth density of 1.0 (all liquid) is plotted as a point right at the tray above (full tray spacing) while a froth density of 0 is plotted as a point at the tray floor. However, when all froth densities are low, it may be more informative to use a scale where a froth density of 0.5 shows as a point on the tray above (the full tray spacing). While the scale may vary from tower to tower, it is important to use a consistent scale in each tower.
6. Clear liquid heads are plotted so that a liquid head of half the tray spacing is plotted as a point on the tray above, while zero liquid head is plotted as a point on the tray floor.

Case Study: Diagnosing Channeling   When an olefins plant was operated at rates exceeding 105% of design with propane feed, there was excessive liquid carryover from the top of the large-diameter water quench tower. This high carryover was not anticipated from either the design or tray supplier calculations. The tower contained two-pass fixed valve trays with large slot areas (18.5% of the active area). Theories included jet flood, liquid inlet maldistribution, and VCFC. With the trays meeting all the criteria listed in Section 2.12, VCFC was the prime suspect. Theory testing was performed by multi-chordal gamma scans with quantitative analysis.

Figure 6.6a and b contains Kistergrams of froth heights and entrainment profiles, respectively, along the flow paths as measured by the multi-chordal gamma scans shot during a water carryover episode. Froth heights or entrainment reaching the tray above indicate flood or proximity to flood.

Figure 6.6a indicates minimal flooding in the tower. The spray heights on trays 8 and 10 (hydraulically, these were the most loaded trays in this tower) reached 26–27 in. near the center downcomers, a little short of the 30 in. tray spacing. Jet flooding was approached at these locations. All other locations were not near jet flood, as spray heights were 21 in. or less. The entrainment graphic (Figure 6.6b) showed a similar pattern, but identified flooding on trays 8 and 10 near the center downcomers and in the middle of the flow paths, with heavy entrainment in the

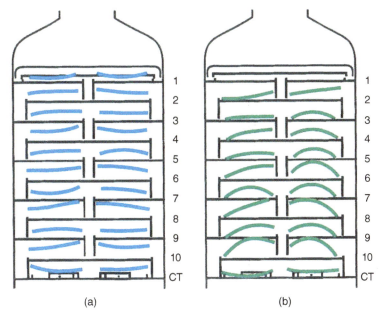

(a)   (b)

Figure 6.6   Kistergrams showing froth heights and entrainment profiles along the flow paths measured by multi-chordal gamma scans for the tower in Section 6.1.4 (a) spray heights and (b) entrainment index. (From Kister, H. Z., K. F. Larson, J. M. Burke, R. J. Callejas, and F. Dunbar, "Troubleshooting a Water Quench Tower", 7th Annual Ethylene Producers Conference, Houston, Texas, AIChE, March 20, 1995. Reprinted with permission. Copyright © American Institute of Chemical Engineers (AIChE).)

same regions on trays 6, 7 and 9. In most other locations, entrainment was low to moderate. For all trays, including 6–10, Figure 6.6b shows very little entrainment near the side downcomers.

Consistent significant differences in spray height and entrainment rates across a tray flow path indicate channeling. Locations of tall spray and high entrainment indicate high local gas velocities. Figure 6.6a and b showed that near the center downcomers and in the middle of the flow paths, gas velocities were much higher than those near the side downcomers. The channeling was most severe at the bottom tray and subsided as one ascended the tower.

The observed spray height and entrainment gradients invalidated the VCFC theory. VCFC is induced by the tray hydraulic gradient, so spray height and entrainment rates are highest in the tray middle and outlet regions, as in the side panels of the case study in Ref. 252. Here, the highly loaded even-numbered trays showed the highest spray heights and entrainment in their inlet regions.

As tall froths and high entrainment are induced by high gas loads, the gas preferably channeled through the center of the tower. The chimney tray (tray CT beneath tray 10) had froth heights exceeding the chimneys (these too are drawn to-scale), especially near the sides, suggesting liquid overflowing the chimneys near the sides, which would channel the vapor preferentially toward the middle.

It may be surprising to talk about chimney tray froth heights rather than chimney tray liquid heights. The gamma scan chords through the chimney tray showed that the chimney tray was as aerated as the trays above, i.e., that it contained froth, not liquid (204, 231).

Combining the findings, the quantitative analysis diagnosed channeling on the trays but denied VCFC. Instead, the channeling root cause was liquid overflow into the chimneys near the side downcomers, with gas channeling preferentially near the center of the chimney tray. Due to the high slot areas of the trays above (18.5% of the tray active area), there was little dry pressure drop to counter the gas channeling. The gas continued to preferentially rise near the center down-comers and the middle of the flow paths. In these regions, froths were tall and entrainment was high. Some entrainment reached the tower overhead, causing the observed water carryover.

Based on this diagnosis, the chimney tray was redesigned to eliminate frothing and hydraulic gradients. The trays open areas were reduced from 18.5% to 13–14% of the tray active areas by blanking about ¼ of the valves. These modifications eliminated the carryover and reinstated trouble-free operation.

6.1.5 Froth (or Spray) Density and Liquid Head Determination

For a given chord length, the transmitted radiation is a direct function of the density of the medium (Eq. 5.1). Qualitative scans at times express the radiation measured by the detector as liquid densities (e.g., Figure 5.31). The highest measured radiation (at the clear vapor line) is assigned a density of zero, with densities for other locations derived from Eq. (5.1). This approximate method is adequate for qualitative analysis but is less satisfactory for quantitative analysis. Starting from the vapor end (zero density), the error in the density escalates as the liquid becomes denser. When determining the liquid head on the tray, a prime variable in many channeling studies, it is the denser liquid which is of greatest interest.

Applying Eq. (5.1) both at the vapor end ($\rho = 0$) and at the liquid end ($\rho = \rho_L$, where ρ_L is the liquid density) to give the value of ($\mu\, x$) improves accuracy, making the froth density determination suitable for quantitative analysis. An adjustment can be made for metal thickness, but it is usually unnecessary. The merit of this method is that it interpolates between both ends.

This method requires a good measurement of transmitted radiation density at the bottom sump. Scan chords need to pass through the bottom sump. In large-diameter towers, the radiation transmitted through the bottom sump liquid may be too little to permit a reliable reference point. Gamma-ray reflection from large reboilers may be problematic, giving excessive radiation counts at the detector. The liquid reference needs to be carefully reviewed to ensure it is suitable for froth density determination.

Tower internals, such as chimney trays, accumulators, or internal sumps, can solve the dilemma. Chords through these travel partly through vapor and partly through liquid, and the length of the chord that passes through liquid can be determined and used as "x" in Eq. (5.1). As the gamma rays passing through these devices partly pass through vapor, the transmission absorption is close to that of the tray liquid, providing a reliable interpolation reference point for froth densities. It is important to judiciously scan these internals when performing a quantitative analysis.

The fractional froth density Φ_{froth}, with liquid assigned a froth density Φ_{froth} of 1.0 and vapor 0, is a convenient term for expressing the froth density. With I (the radiation intensity in counts per second, as measured at the detector), I_0 (the radiation intensity of the source, in counts per second); ρ (1.0, froth density through the liquid), and x (the length of the gamma-ray passage through liquid, ft) known, μ, the absorption coefficient, can be determined from Eq. (5.1). Once μ is known, the froth density $\Phi_{froth,\,elev}$ at a given elevation $elev_{froth}$ (in.) can be determined from the radiation transmission reading by applying Eq. (5.1). Several heights above the tray floor $elev_{froth}$ are arbitrarily selected, usually at regular intervals. In Figure 6.7, the intervals were 4″. For the top elevation of each interval, $elev_{froth}$, the measured radiation intensity is substituted in Eq. (5.1) to give the froth density, $\Phi_{froth,\,elev,}$ at that elevation. For instance, for the mid out curve (green) in

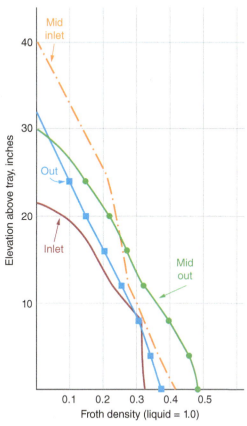

Figure 6.7 Froth density determination. Froth density profiles obtained from four parallel scans along the flow path of a 1-pass tray. The area under each curve divided by the froth height gives the average froth density at that position. (From Kister, H. Z., Chemical Engineering Progress, p. 33, February 2013. Reprinted Courtesy of Chemical Engineering Progress (CEP). Copyright © American Institute of Chemical Engineers (AIChE).)

Figure 6.7, at elevations $0''$, $4''$, and $8''$, the froth densities determined from the detector readings using Eq. (5.1) were 0.49, 0.45, and 0.40, respectively. The average froth density, Φ_{froth}, is the integrated froth density for the peak over the total froth height, h_{froth} (inches), i.e.

$$\Phi_{\text{froth}} = \int (\Delta\text{elev}_{\text{froth}} \, \Phi_{\text{froth,elev}})/h_{\text{froth}} \tag{6.2}$$

The froth height h_{froth} is determined for that peak using Harrison's method, Section 6.1.1. The integration is performed numerically for each peak.

The area under the curve is the integration of ($\Delta\text{elev}_{\text{froth}}$ times $\Phi_{\text{froth, elev}}$) and gives the product of the froth height and froth density $h_{\text{froth}} \, \Phi_{\text{froth}}$. Dividing this area by the froth height h_{froth} gives the integrated froth density Φ_{froth}. This area also gives the tray liquid head (inches) h_{C} at the chord, because

$$h_{\text{C}} = h_{\text{froth}} \, \Phi_{\text{froth}} \tag{6.3}$$

Figure 6.7 shows four different froth density profiles, each for a different chord on the same tray.

Example 6.1   Calculate the froth density and clear liquid height for the "mid out" location (green curve) in Figure 6.7. The froth height for this location determined from Harrison's method was 28 in.

Solution:

First determine the intervals. In this case, the intervals are $0''$, $4''$, $8''$, $12''$, $16''$, $20''$, and $24''$ above the tray floor. The corresponding values of $\Phi_{\text{froth, elev}}$ at these elevations were 0.49, 0.45, 0.40, 0.32, 0.27, 0.21, and 0.13, respectively (Figure 6.7).

Integrating the froth density across each interval will give the clear liquid height, i.e.:

$$
\begin{aligned}
h_C &= \int (\Delta\text{elev}_{\text{froth}} \; \Phi_{\text{froth,elev}}) \\
&= 4 \times (0.49 + 0.45)/2 + 4 \times (0.45 + 0.4)/2 + 4 \times (0.4 + 0.32)/2 + 4 \\
&\quad \times (0.32 + 0.27)/2 + 4 \times (0.27 + 0.21)/2 + 4 \times (0.21 + 0.13)/2 + 6 \\
&\quad \times (0.13 + 0)/2 = 8.23 \text{ in.}
\end{aligned}
$$

$$
\Phi_{\text{froth}} = \int (\Delta\text{elev}_{\text{froth}} \; \Phi_{\text{froth,elev}})/h_{\text{froth}} = 8.23/28 = 0.29
$$

The clear liquid height is 8.23 in. and the froth density 0.29 in the "mid out" location.

The "mid out" location showed the highest froth density, about 0.29 (liquid = 1.0). Although froth heights for the other three chords differed, their froth densities were similar, around 0.22–0.23. In this case, the constant μ in Eq. (5.1) was derived from a scan through a chimney tray, which gave a reliable interpretation. Froth density profile plots from research gamma scans, as reported by a number of scholars (368, 520), have very similar shapes to the curves in Figure 6.7.

6.1.6   Quantitative Analysis for High-Capacity Trays with Truncated Downcomers

A variety of capacity-enhancing features were described in Section 6.1. Potential troubleshooting issues, in addition to those that are encountered in conventional trays, include steep froth gradients due to excessive forward push, vapor penetration into truncated downcomers ("blowby"), and channeling and maldistribution in nonsymmetrical multiple downcomers trays. Each of these issues may lower tray efficiencies and bottleneck tower capacity. Conventional troubleshooting techniques such as vendor and technology provider software, ΔP measurement, and conventional single-chord qualitative gamma scans are usually at a loss for diagnosing such issues. Judicious multi-chordal gamma scans with quantitative analysis are highly effective for diagnosing the above issues.

Case Study: Blowby Diagnosis   This case study deals with gamma-scanning Enhanced Capacity Multi-Downcomer™ (ECMD™) trays. These trays feature multiple truncated downcomers with perforated or slotted bottoms, extending just over half the distance to the tray below, with each tray rotated 90° to the tray below. Figure 6.12 (later) shows a typical layout. Scanning these trays at small (12 in.) tray spacing in an 8 ft ID test simulator column, Shakur et al. (427) concluded, "Gamma scanning cannot be utilized in closely-spaced MD™ trays to detect deck weeping or downcomer blowby (and to) precisely determine froth heights and liquid levels in downcomers."

Indeed, these challenges extend beyond the capabilities of qualitative gamma scans, but are readily diagnosed by proper multi-chordal, quantitative analysis. Froth heights, froth densities, blowby, tray dry-out, and maldistribution on similar trays at as little as 7 in. spacing have been reliably measured using this technique. Problems were diagnosed, and corrective actions based on the diagnoses yielded a complete solution to the problems.

A close examination of the data reported by Shakur et al. by Kister (205) highlights the criticality of adequate scan lines and correct data analysis. Figure 6.8 shows two of their three

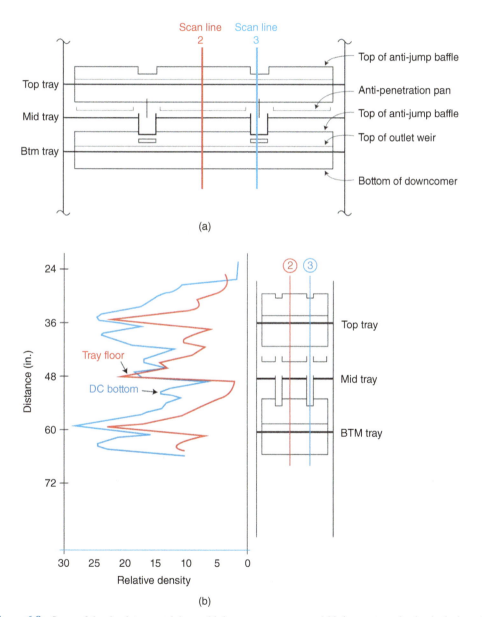

Figure 6.8   Scans of the simulator containing multi-downcomer trays rotated 90 degrees to each other in the investigation by Shakur et al. (427). (a) Scan lines, (b) overlay of the downcomer scan (Scan line 3) and active area scan (Scan line 2) reported by Shakur et al. (427) for "Design," suggesting blowby in the truncated downcomer. (From Kister, H. Z., Chemical Engineering Progress, p. 45, April 2013. Reprinted Courtesy of Chemical Engineering Progress (CEP). Copyright © American Institute of Chemical Engineers (AIChE).)

scan lines: scan line 2 passed through the active areas, while scan line 3 through the downcomers. The third line, scan line 1 (not shown here), passed through the active areas at right angle to the view in Figure 6.8. The scans relevant to this discussion were labeled "Design" and "Blowby." Blowby refers to vapor breaking the seal in truncated downcomers, which can be detrimental to tray efficiency and, in the worst-case scenario, can induce premature flooding (Section 2.11). Therefore, it was a major focus in this investigation.

The only chord that can positively identify blowby is scan line 3, through the middle tray downcomer and parallel to the downcomer walls. Downcomers normally contain fairly clear liquid near their bottom, while trays hold a less dense vapor–liquid mix, such as froth or spray. Tray froth denser than the downcomer froth conclusively indicates blowby, i.e., a significant amount of vapor ascending up this downcomer. To detect blowby, scan line 3 is required, and the downcomer froth density is compared to the tray froth density from scan line 2. The "Blowby" scan presented by Shakur et al. only contained scan line 2 (no scan line 3), and on this basis they incorrectly concluded that blowby cannot be detected on gamma scans. The valid conclusion should have been that blowby requires a downcomer scan such as scan line 3 and cannot be detected from active area scans such as scan line 2.

Unwittingly, a close examination of the scans by Shakur et al. reveals a likely blowby, but not in the condition they anticipated. Figure 6.8b overlays their scan lines 2 and 3 under "Design" conditions. The liquid at the bottom of the downcomer from the mid tray appears more aerated (less dense) than the liquid on the mid tray above! There should not be anything obstructing the view in either scan line. So it is likely that blowby occurred in their "Design" case.

There is one additional source of evidence supporting blowby at the "Design" conditions. At an elevation of about 41 in., both scans pass through the bottom of the downcomers from the top tray, which are perpendicular to the scan lines. At this elevation, the froth density of scan line 3 (about $22 \, lb/ft^3$ on the density scale) is far greater, more than double, that of scan line 2 (about $10 \, lb/ft^3$ on the density scale). Scan line 2 passes right through the downcomer holes, while scan line 3 is nowhere near these holes. The observed density difference indicates vapor passage into the downcomer through the downcomer holes, strongly supporting blowby.

Figure 6.9 shows the downcomers in another tower (different service). These downcomers operated normally, with no blowby. Comparing the downcomer peaks to the active area peaks, the downcomer peaks clearly contain denser liquid, just like they should.

Case Study: Empty Downcomers Figure 6.10 shows a downcomer scan of another tower equipped with trays containing multiple truncated downcomers rotated at 90 degrees to each other. This tower experienced low tray efficiencies. The scan shows empty downcomers from trays 12 to 14 (almost clear vapor). The downcomers from trays 16, 18 and 20 contained dense liquid and operated normally. Good scanning can diagnose dry truncated downcomers.

Case Study: Dry Trays Figure 6.11 shows the active area scan in the upper section of a tower that experienced low tray efficiency. The scan passed through the downcomer holes. The scan shows very little liquid on the even-numbered trays. The little liquid was highly aerated. In that tower, the design liquid loads were not small, so the only conceivable explanation was liquid maldistribution that deprived the central regions of some of the trays of liquid. On this basis, the trays were redesigned and their performance was greatly improved, exceeding design expectations.

A Word of Caution The diagnosis of dry versus normally operating downcomers requires observation of consistent behavior in multiple chords. A scanning anomaly may affect one chord. Each of the scans shown in Figures 6.9–6.11 are representatives of multiple consistent scans.

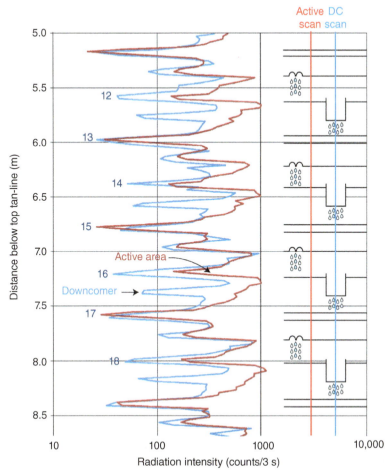

Figure 6.9   Overlay of downcomer and active area scans of multiple downcomer trays rotated 90 degrees to each other with truncated downcomers. Scans show normal downcomer operation with little or no blowby in the truncated downcomers. (From Kister, H. Z. Chemical Engineering Progress, p. 45, April 2013. Reprinted Courtesy of Chemical Engineering Progress (CEP). Copyright © American Institute of Chemical Engineers (AIChE).)

6.1.7   Scan Chord Selection

Too few chords will provide an incomplete picture and an inconclusive, even misleading, diagnosis. Too many chords rapidly escalate the costs with limited added value. The number of trays to be scanned requires similar balance. Missing a chimney tray will deprive the analysis of an invaluable reference. On the other hand, scanning even 60 trays per chord is usually unnecessary and runs up the cost. While each case should be considered on its own merits, the following guidelines can be proposed based on the author's experience:

1. It is beneficial to scan a preliminary top-to-bottom active area chord that can later be incorporated in the quantitative analysis. This chord should be parallel to and at least 6 in. away from the downcomer walls, not at an angle to them, and should descend into the bottom sump. If flooding is suspected, it is best to scan both the flooded and unflooded conditions. If downcomer issues are possible, include a downcomer unflooded scan, with the scan chord following guideline 9 in Section 5.3. If channeling is suspected, consider

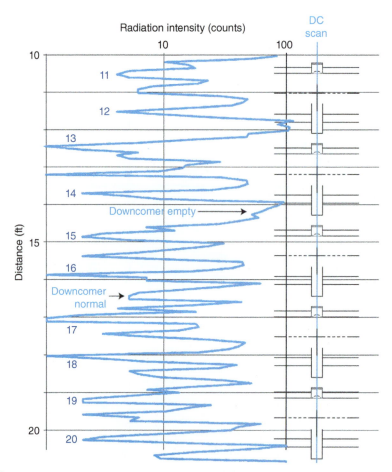

Figure 6.10   Downcomer scans of multiple downcomer trays with truncated downcomers rotated 90 degrees to each other. Scans show normal downcomer operation on even-numbered trays 16–20 and empty downcomer operation on even-numbered trays 12 and 14. (From Kister, H. Z., Chemical Engineering Progress, p. 45, April 2013. Reprinted Courtesy of Chemical Engineering Progress (CEP). Copyright © American Institute of Chemical Engineers (AIChE).)

shooting a second active area chord. If maldistribution is suspected in four-pass trays, scan both the inside and outside active areas of one tower half.

2. Based on hydraulic evaluation (by tray rating programs), operating experience, and other measurements such as ΔP and temperatures, as well as the scan(s) in step 1 above, determine whether quantitative analysis and/or extra scans are needed, and if so, where.

3. Correct selection of the chords and trays to be scanned is critical, as guided by steps 1 and 2 above. For instance, if the problem appears to be confined to one tower section, it is enough to scan only this section. If large sections of tower show similar behavior and have similar hydraulics, scanning of only a few of them, with spot checks of the others, may suffice. In case of flooding, another choice is whether multi-chord scanning is required in the flooded or unflooded condition, or both. The author generally found the unflooded scans to be more informative for identifying the root cause.

4. Consistency and repeatability checks are essential, e.g., by scanning a symmetrical chord on the other side of the tower, or by repeating selected chords on different dates

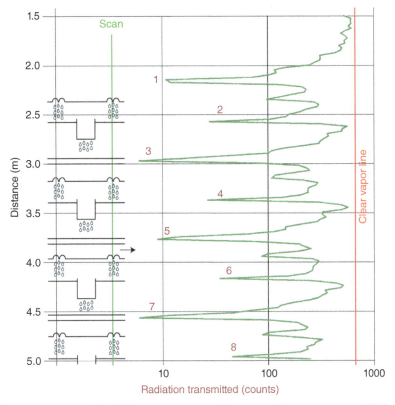

Figure 6.11   Active area scan of multiple downcomer trays with truncated downcomers rotated 90 degrees to each other. Scan shows very little, highly aerated liquid on the even-numbered trays 2–8. (From Kister, H. Z., Chemical Engineering Progress, p. 45, April 2013. Reprinted Courtesy of Chemical Engineering Progress (CEP). Copyright © American Institute of Chemical Engineers (AIChE).)

(but at the same hydraulic loads). Good repeatability of scans shot at different times (but at the same loads) validates scan reliability. A single tray does not make a scan. Often, a scanning anomaly affects the scan of some trays. If a group of trays with similar hydraulics produces a consistent trend, it is more likely to reflect actual hydraulic behavior.

5. Reference points such as bottom sumps, chimney trays, liquid accumulators, and regions of clear vapor are invaluable for quantitative analysis. Aim to include all the relevant reference points in the scans.

6. The flow path from the active area inlet to the outlet can be quite short. Forward push changes the froth height along the length. To evaluate the froth gradient, one needs to scan at a number of well-selected chords along the flow path length. Consideration should also be given to whether liquid from the tray above is likely to descend along the chord.

7. Scans through the downcomers, parallel to the downcomer walls (Section 5.3, guidelines 9 and 10, also Section 6.1.6) are essential for truncated downcomers, as these look out for vapor passage (blowby). Center or off-center downcomers scans are more informative, as side downcomers chords are often shallow. Anti-jump baffles mounted above the downcomers should be reviewed for possible interference. Scanning chords in narrow

downcomers in large-diameter towers may require special techniques, so the scanners should be consulted.

8. With many downcomers, outside chords can be quite shallow and difficult to scan. When scanned, a reliable reference point (e.g., chimney tray, bottom sump) where the density is known should be included.

9. Prepare to-scale tower elevation and plan diagrams, clearly showing the dimensions of each chord and where they are to be shot relative to downcomers. Include supports and obstructions in the diagrams. Once the desired scan chords are laid out, the scanning contractor should be consulted per guideline 16 in Section 5.3.

10. Stable operation during the scans, as well as repeatability and consistency checks, are imperative with multi-chord scans. Adhere to guidelines 4 and 5, Section 5.3.

11. Froth density determination (Section 6.1.5) is the most laborious and time-consuming part of quantitative analysis due to the integration. Any shortcuts here can significantly reduce costs. When analyzing data, question whether this determination is necessary, and if so, for which trays and which runs.

12. Closely adhere to the guidelines in Section 5.3, they also apply to quantitative multi-chord scans. Due to the small differences between chords, the high accuracy required, and the length of time required for the measurements, many of these guidelines are far more critical for quantitative than for qualitative gamma scans.

Case Study: Application of Guidelines 3–9 above for chord selection   Figure 6.12 is a to-scale scan plan for high-capacity sieve trays with truncated multiple downcomers. On these trays, the downcomers do not extend all the way to the tray below, but terminate about half way down the vapor space, with liquid issuing through holes in the downcomer floors. Successive trays are rotated 90 degrees to each other. Dimensions (that should be included on the sketch) have been omitted for confidentiality reasons.

As these trays were spaced only 16 in. apart, the gamma scans needed to be shot at 1 in. vertical intervals. To keep costs down, any unnecessary or marginally useful scans were excluded. The strategy was to focus on a selected region, with limited comparison to others. The region selected was near the center of the tray to avoid shallow chord issues.

Six scan chords were shot. From left to right in Figure 6.12, the "holes" chord scanned the active areas right under the downcomer holes, and the "active" chord scanned the active areas where there is no rain from the downcomer holes. This permitted distinguishing the rain from the holes from the tray dispersion, and evaluating a possible hydraulic gradient. The next chord scanned the downcomers. The "center holes" chord scanned the inside active areas, again right under the downcomer holes, allowing comparison with the "holes" chord on the left. The east–west scan chords on the plan diagram (Figure 6.12b) are equivalent to the left and right north–south scan chords but shot at right angles to compare the east–west with the north–south patterns. Since the hydraulic patterns of the trays were being examined, they were only scanned in unflooded conditions. A highly stable tower operation was sustained.

To avoid excessive costs, only about one-third of the total number of trays was scanned. Even though the number of chords and trays selected may appear lean, it was sufficient to provide the desired diagnostic that led to a successful solution to the operating problem.

One place where we did not compromise was the repeatability and consistency checks. Two of the six chords were repeated along part of their lengths to assure repeatability. Ensuring the data are valid is critical as described earlier.

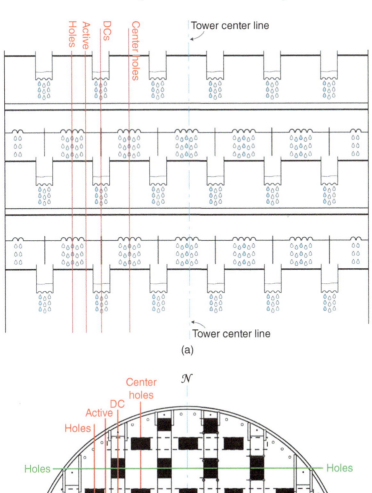

(a)

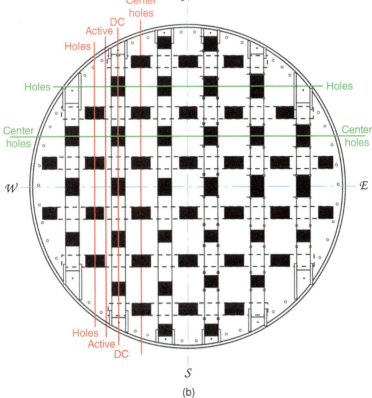

(b)

Figure 6.12   Scans of a tower containing high-capacity sieve trays with multiple truncated downcomers. Successive trays are rotated 90 degrees to each other (a) Elevation (b) Plan. (From Kister, H. Z., Chemical Engineering Progress, p. 45, April 2013. Reprinted Courtesy of Chemical Engineering Progress (CEP). Copyright © American Institute of Chemical Engineers (AIChE).)

6.2 QUANTITATIVE ANALYSIS OF PACKING GAMMA SCANS

Qualitative gamma scanning (Chapter 5) has been successfully applied to troubleshoot tens of thousands of packed towers for over five decades. Scan data have traditionally been interpreted by visual examination of the scans to detect flooding, missing packings, crushed packings, plugging, foaming, and liquid maldistribution. In most troubleshooting investigations qualitative observations suffice.

Qualitative packing gamma scans are more difficult to analyze and interpret than tray gamma scans because the radiation is absorbed both by the packing metal and the liquid held in the packing. Additional absorptions can be due to the plugging material. In many cases, it can be very difficult to decouple these from each other and correctly interpret the scan data. This may lead to subjectivity and incorrect diagnostics.

A recent breakthrough by Pless (371, 376) applies quantitative analysis to decouple the packing metal absorptions from the liquid absorptions and from absorptions due to plugging. Unlike the quantitative analysis of trays, which requires additional chords, judicious selection of scan chords, and sometimes tedious analysis, all of which run up the costs, the quantitative analysis of packed beds is simple, achieved at a relatively low additional cost and effort, and can be incorporated with any scan.

6.2.1 Packed Tower Quantitative Analysis Techniques

Pless (371, 376) did not detail his basis for the calculations. The analysis below is by the author, but he believes that Pless' technique follows at least somewhat similar principles. This analysis is based on Eq. (5.1). As stated in Section 5.1, when the energy of the gamma ray exceeds 200 keV, the absorption coefficient μ, which depends on the gamma-ray source and the medium material, becomes independent of the chemical composition of the medium, and the absorption becomes a function of the product of the density and thickness of the medium (66, 67). For a horizontal chord of fixed length, the intensity of radiation received at the detector is therefore a function of the density of the medium. Equation (5.1) can therefore be expanded to give

$$I = I_0 \exp\left[-\mu\left(\rho_w x_w + \rho_p x_p + f\,\rho_L x_L\right)\right] \tag{6.4}$$

where I is the radiation intensity in counts per second, as seen by the detector; I_0 is the radiation intensity of the source, in counts per second; ρ is the density of the medium; x is the thickness of the medium; μ is the absorption coefficient discussed earlier; and f is the volumetric fraction of the liquid holdup. The subscript w signifies column wall, p packing, and L liquid.

The first term, $\rho_w x_w$, is the wall thickness times the wall metal density (steel is about $500\,\text{lb/ft}^3$). The packing density ρ_p is a property of the packings and can be obtained from the vendor or from Perry's Handbook (245). x_p and x_L are the same length, which is the scan chord length and ρ_L is the liquid density, known from the simulation or design data. The fractional liquid holdup f can be estimated from Figure 6.13 based on Refs. 35 and 458. These estimates tend to be good approximations but may not be accurate (65).

Equation (6.4) has two unknowns, I_0 and μ. To solve these, two transmitted intensity values are needed. One available intensity is that measured at the clear vapor line, where there are no packings and no liquid, so $\rho_p x_p = \rho_L x_L = 0$. The second (usually) available intensity is at normal operation. As the transmitted intensity I is known for each of these conditions, Eq. (6.4) can be applied to each, giving two simultaneous equations that can be solved to give I_0 and μ. If a dry scan is available, a third condition can be solved with $f = 0$.

Often, the gamma scan contractor will perform the above calculations and superimpose a density scale onto the scan data. The baseline of the density scale would be the dry bulk packing

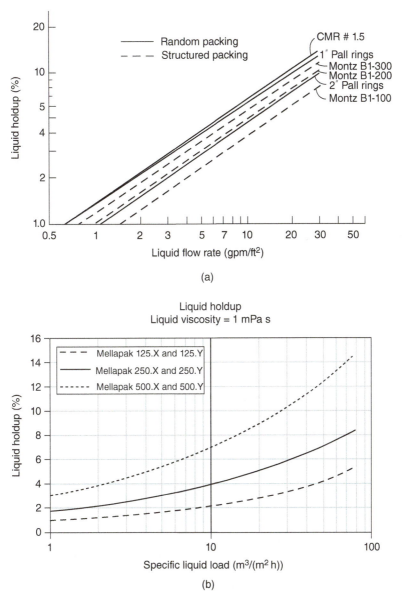

Figure 6.13   Packed tower liquid holdup variation with superficial liquid velocities down the bed, air–water data at ambient conditions, preloading regime (a) for various random and structured packings, data from Billet, R., "Packed Column Analysis and Design," Ruhr University, Bochum, 1989 (From Kister, H. Z., "Distillation Design," Copyright © McGraw-Hill, 1992. Used with permission of McGraw-Hill); (b) for Mellapak[R] structured packings. Part b Sulzer's data for Mellapak[R] structured packings (Reprinted with permission from Spiegel, L., and M. Duss, "Structured Packings," Chapter 4 in A. Gorak and Z. Olujic (ed.) "Distillation Equipment and Processes," Elsevier, 2014).

density (376). Calculated densities above the dry bulk represent liquid or solids retained in the packing. The fraction of liquid retained can be estimated and compared to the expected from Figure 6.13. If the density is much greater than the expected, solids are implicated. If there is a difference between the scan lines, the retention density gives a measure of the severity of maldistribution.

6.2.2　Flooding at the Bottom or Missing Packing Tower at the Top?

Contributed by Lowell Pless and Tracerco

Figure 6.14a shows the gamma scan from a small-diameter random packed tower (only two scan lines were performed due to the small diameter, per Section 5.2.3, Figure 5.6c). The tower diagram on the right side of Figure 6.14a shows where the packed bed was supposed to be as per the reference tower drawing. There was a reduction in radiation counts from the clear vapor line at the expected elevation for the top of the bed. A short distance into the bed, the radiation counts decreased. The question was, did the lower counts (higher densities) signify flooding (high liquid holdup) at the bottom of the bed, or was the radiation count at the bottom of the bed normal and something had happened to the packing in the top of the bed? A visual or qualitative evaluation of the radiation counts could not answer these questions with confidence.

Figure 6.14b shows the gamma scan results from Figure 6.14a with the liquid retention scale added. The density of the dry packing is 160 kg/m³, and the top of the bed has a density essentially equal to this. However, the tower was operating and there was liquid traveling down through the packing. Where is the additional density from the retained liquid? The overall density at the top of the packing should be the combination of retained liquid and packing density. Since the scan shows the overall density to be nearly equal to that of the dry packing, either there was no liquid

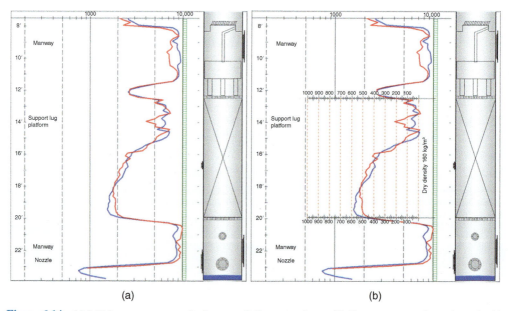

(a)　　　　　　　　　　　　　　　(b)

Figure 6.14　(a) Initial gamma scan results from small-diameter column. (b) Gamma scan results, enhanced with liquid retention scale, showing missing packings at top of bed. (From Pless, L., PTQ Q2, 2018. Reprinted Courtesy of PTQ.)

(obviously not the case) or packing was missing. The quantitative analysis provided conclusive evidence that portions of the packing were missing. The high density in the bottom of the bed was likely due to crushed packing retaining an excess of liquid. Eventually, an inspection of the tower confirmed these results.

6.2.3   Dense Grid or Flooding/Coking in a Wash Bed?

Contributed by Lowell Pless and Tracerco

Figure 6.15a shows the gamma scan of the wash bed of a crude vacuum tower (the fourth scan line was not performed due to limited access). Coking of the wash bed is a common experience in this service, and the gamma scan was performed to assess the situation. As shown in the right side of Figure 6.15a, this wash bed consisted of two different types of packings: structured packing at the top section and grid packing in the bottom section. The scan showed the bottom section to be very dense, as the radiation intensity was less than five counts, essentially background (no radiation passing through the tower). Based on this result, the diagnosis was that the grid section was coked and/or flooded with liquid, but plant management was not convinced. Grid packing is dense, so a question was asked whether the radiation source was not strong enough. In other words, some radiation would pass through the packing from using a stronger radiation source, and then perhaps the grid would not appear to be flooded.

Figure 6.15b shows the gamma scan results from Figure 6.15a with the liquid retention scale added. The dry density of the grid packing was 255 kg/m^3. The density of the process material inside the grid packing, based on the scan results, was calculated to be approximately 300 kg/m^3. Grid packing is a very open packing and the typical liquid rate in a vacuum tower wash bed is

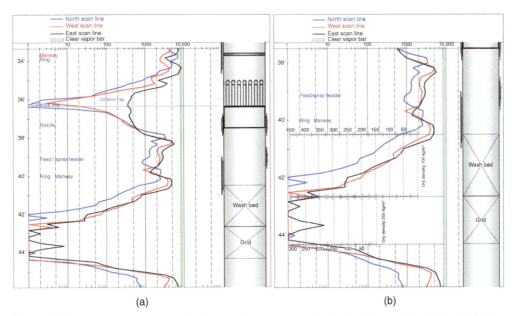

(a)   (b)

Figure 6.15   (a) Initial gamma scan results from crude vacuum tower showing bed placements and high-density (very low radiation intensity) through the wash bed grid. (b) Gamma scan results, enhanced with liquid retention scale, showing wash bed grid either completely coked or flooding with retained liquid. Excess liquid backing up into wash bed packings. (From Pless, L., PTQ Q2, 2018. Reprinted Courtesy of PTQ.)

very low. A typical liquid retention density for grid packing in this service is 80–100 kg/m³ or less. The density calculated from the scan data was far above this, indicating the grid was coked and/or flooded. Therefore, the scan radiation source used on this tower was more than adequate. The quantitative analysis proved that had this wash bed not been coked or flooded, there should have been 30–50 counts passing through it, well above the observed count of less than five.

An additional computation confirmed this diagnosis. The liquid volume fraction was calculated by dividing the liquid retention density of 300 kg/m³ by the process liquid density (800 kg/m³). The resulting liquid volume fraction was 0.38. Figure 6.13 shows typical liquid holdup curves in packings. With wash bed liquid loads of the order of 0.2–2 gpm/ft², the liquid volume fraction is only a few percent. Packings usually reach flood with liquid volume fractions greater than 0.12. Therefore, the grid packing was coked or flooded.

6.2.4 Good or Bad Distribution Quality?

Contributed by Lowell Pless and Tracerco

The overhead product gas of a packed CO_2 scrubber was off-spec for CO_2. The gamma scan (Figure 6.16a) showed no major problems with the scrubber internals other than the possibility of liquid maldistribution. The four scan lines did not match each other very well, indicating that the overall density between the four sets of data were different from each other. The radiation intensity counts varied from 1800 to 3600, as read from the horizontal scale of Figure 6.16a. But what was the severity of the liquid maldistribution? The radiation intensity or counts gave no perspective or evaluation of the actual liquid distribution quality.

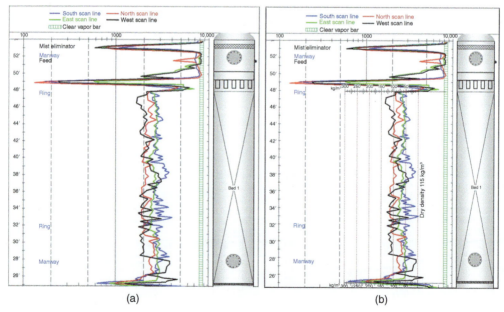

(a)   (b)

Figure 6.16 (a) Initial gamma scan results showing what appears to be maldistribution through nonuniformity of the scanlines through the bed. (b) Gamma scan results, enhanced with liquid retention scale, showing large density differences among the scanlines. One side of the tower (blue scanline) shows almost dry packing as liquid retention density is almost the same as the dry packing density. (From Pless, L., PTQ Q2, 2018. Reprinted Courtesy of PTQ.)

Figure 6.16b shows the same gamma scan as Figure 6.16a, but with the liquid retention density scale added. The dry packing density was $115\,kg/m^3$. The quantitative analysis showed that the liquid retention ranged from 10 to $130\,kg/m^3$. In terms of liquid density across a bed of packing, this was a large density difference. Reinforcing the liquid maldistribution diagnosis was the fact that one scan line (the blue curve in Figure 6.16b) had almost the same overall density as dry packing alone. Thus, this scan line was nearly dry or there was very little liquid on that side of the scrubber. Furthermore, the liquid volume fraction calculated, based on a process liquid density of $800\,kg/m^3$, ranged from 0.01 (very low liquid fraction) to 0.16 (bordering on flooding). Thus, the quantitative analysis conclusively proved that the packed bed experienced very poor liquid distribution.

6.3   NEUTRON BACKSCATTER TECHNIQUES APPLICATION

Neutrons are high-energy particles that can penetrate a thick metal wall while traveling only short distances (typically about 6 in.) into the tower. The fast neutrons are slowed down by collisions with hydrogen nuclei. These collisions transfer energy to the hydrogen atoms, and slow neutrons are reflected back toward the source. This is analogous to the rebound of balls on a pool table. The intensity of the rebounded neutrons is proportional to the concentration of hydrogen atoms in the fluid adjacent to the tower wall and can be measured by a detector. So the backscatter detector reading indicates the concentration of hydrogen atoms near the vessel walls. A more detailed description of this technique is provided elsewhere (66, 503).

Figure 6.17 shows a typical neutron backscatter device along with an illustrative neutron scan of a vessel. The source and detector are mounted on the same handheld sweeper. The neutron source and detector are positioned near the wall of the vessel and are moved up or down along the vessel wall. In the vapor space near the top of the vessel, the reading is low because in the vapor the molecules (and therefore the hydrogen atoms) are far apart and there is very little to reflect the neutrons. The reading increases as the sweeper is lowered to the liquid-rich region in

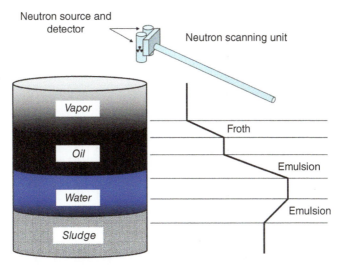

Figure 6.17   A typical neutron backscatter device along with an illustrative neutron scan of a vessel. The source and detector are mounted on the same hand-held sweeper. (From Kister, H. Z., and C. Winfield, Chemical Engineering Progress, p. 28, January 2017. Reprinted Courtesy of Chemical Engineering Progress (CEP). Copyright © American Institute of Chemical Engineers (AIChE).)

the mid-section of the vessel. In this region, the molecules are close together, so the concentration of hydrogen atoms is much higher.

The step change in hydrogen atoms concentration at an oil–water interface makes this technique suitable for detecting a liquid–liquid interface and emulsions. Sludge and solids sitting at the bottom of a vessel generally contain less hydrogen atoms, and this too can be detected (Figure 6.17).

The most extensive applications of this technique in distillation and absorption columns is for liquid level and liquid level interface detection, especially when normal level-measuring devices suffer from plugging. Neutron backscatter is invaluable for detecting overflows from chimney trays, collectors, and distributors, but can only be applied when the liquid is near the tower wall. Neutron backscatter can detect a liquid interface far more effectively than gamma scans, particularly in large-diameter columns and when the densities of the two liquids are similar.

In one case (158), a neutron scan (Figure 6.18) showing unequal liquid heights on the two halves of a chimney tray provided a major lead for diagnosing a tower problem. There was an abundance of liquid level feeding the Light Cycle Oil (LCO) pumparound, but a much lower liquid level feeding the LCO product. The cause was the high weir on the LCO draw compartment, together with a leak from that side.

Neutron backscatter is extensively used for measuring froth densities in downcomers containing hydrocarbons, organics, or water, all of which contain hydrogen atoms. Neutron backscatter is powerful for distinguishing between normal, flooded, near-dry, foaming, or plugged downcomers. A normal downcomer has mostly clear liquid in the bottom and progressively more vapor as one ascends it, so it will give a high neutron backscatter reading near the bottom and progressively lower further up. A flooded downcomer (which is near liquid full) will usually give a high reading along its entire length. Low readings throughout the downcomer indicate an unsealed downcomer. Uniform readings throughout the upper part of the downcomer, too low to be liquid and too high to be vapor, suggest possible foaming. Low readings near the bottom, with high

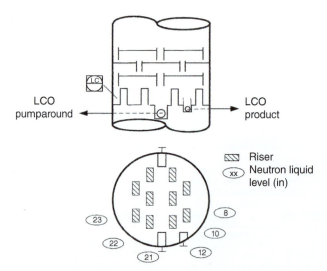

Figure 6.18 Liquid heights (inches) measured by neutron backscatter along the walls of a chimney tray, showing 21–23 in. on the left half, and 8–12 in. on the right. (From Hanson, D. W., J. Brown Burns, and M. Teders, "How Does a Collector Tray Leak Impact Column Operation?," in Kister Distillation Symposium, Topical Conference Proceedings, p. 105, AIChE Spring Meeting, Austin, Texas, April 27–30, 2015. Reprinted with permission. Disclaimer: the information provided in relation to this diagram expresses the opinions of the authors and may not reflect the views of Valero; any use of the information is at the reader's risk.)

readings above, indicate possible plugging with low hydrogen-bearing solids. One case history has been described (66) where downcomer froth density measurements using neutron backscatter led to the detection of downcomer deposits that caused premature flooding of the column.

Neutron backscatter is invaluable for measuring liquid levels in reboilers (217, 230), where gamma-ray transmission is obscured by the tubes. In one of the cited case studies (230), neutron backscatter was applied to measure liquid levels on the condensing side of a vertical thermosiphon reboiler, and in the other (217), on the boiling side of a kettle reboiler.

Neutron backscatter is difficult to apply through wall thicknesses exceeding 1–1/2 in. or insulation exceeding 5 in. Wet or icy insulation can lead to misleading measurements due to the significant concentration of hydrogen atoms in the insulation. Neutron backscatter cannot be applied where hydrogen atoms are absent (e.g., carbon tetrachloride).

6.3.1 Detecting Flood and Seal Loss in Downcomers

Figure 6.19 illustrates the application of neutron backscatter to troubleshoot downcomers. The scan was shot through the downcomers on the left and the vapor spaces right above these downcomers. The black curve shows the highest rates scan, at which the tower experienced flooding.

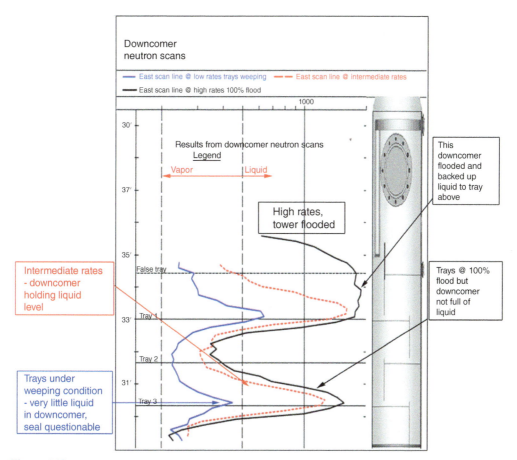

Figure 6.19 Application of neutron backscatter scans to troubleshoot downcomer operation at different loadings. The black pen is at high rates when the tower experienced flooding, the red pens is intermediate rates, indicating normal downcomer operation, and the blue scan is low rates, indicating near seal loss (Reproduced Courtesy of Tracerco.)

It clearly shows that the bottom downcomer operated normally with no flooding, but the downcomer right above it, just below the manhole area, was flooded. The red curve shows that when the rates were lowered, the flood in the top downcomer cleared, and the downcomer operation returned to normal, looking similar to the lower downcomer. This would suggest that at red pen rates, the downcomer was not near a limit.

The blue scan is at reduced rates. At these rates, both the upper and lower downcomers (especially the lower downcomer) were on the verge of losing their liquid seals.

6.3.2 Distinguishing Dry from Full Downcomers

Kister, H. Z., and C. Winfield, Ref. 227. Reprinted Courtesy of Chemical Engineering Progress (CEP). Copyright © American Institute of Chemical Engineers (AIChE)

A US refiner wanted to extend its run length, but after a few years of operation, was experiencing severe throughput limitations. A standard active area gamma scan identified tray deck fouling, resulting in isolated downcomer backup flood on trays.

To devise a strategy to minimize the throughput limitation, the refiner needed to identify the regions where the plugging was most severe. Although downcomer scans can provide this type of information, in this case there was a practical difficulty: this tower had abnormally large stiffening rings, which would have produced too much scatter that would severely interfere with side downcomer gamma scans.

The only reliable way to scan the side downcomers was with neutron backscatter. However, this tower's platforms did not provide adequate access for the scanning equipment. An innovative technique was devised: the neutron scanning unit (Figure 6.17) was taken off the pole and suspended on a standard scanning cable. A crane lowered a technician in a man-basket along the sides of the tower, and he guided the neutron backscatter device on the outer wall as it was being lowered. To the best of our knowledge, this is the first time such a technique has been employed.

The innovative neutron backscatter downcomer scans clearly identified the plugged and dry downcomers. The neutron scans in Figure 6.20 revealed that five downcomers (numbered by the tray they bring liquid to) had lost their liquid seals:

- Zone B–Trays 13, 15, 17: south-side downcomers (blue scan)
- Zone D–Trays 7 and 9: north-side downcomers (red scan)

The neutron backscatter scans also show that liquid seal loss in downcomers can occur in some zones of a tower (in this case zones B and D), while other zones (A, C, and E) operate normally. The dry downcomers suggest severe fouling on the side panels right above them. Vapor, unable to ascend through the plugged active area, broke the seal on the downcomer and ascended through it.

The neutron backscatter technique provided a clear picture of the side downcomers in a situation where a traditional gamma scan would not have been possible. The refiner was able to use the data obtained from the neutron backscatter to identify the severely plugged areas and perform online cleaning of the tower. As a result of this work, the refiner avoided an immediate shutdown and could continue operation at acceptable throughput rates.

6.3.3 Detecting Maldistribution in a Kettle Reboiler

A depropanizer kettle reboiler bottlenecked the capacity of a gas plant (217). When feed rates were raised, tower flooding occurred, initiating at the kettle reboiler. A pressure balance showed that

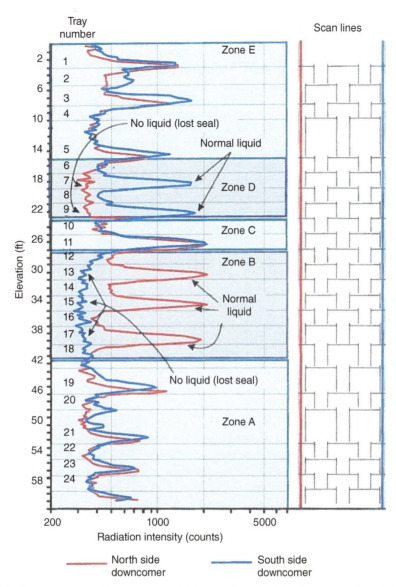

Figure 6.20   Neutron backscatter downcomer scans identify the plugged and dry downcomers. In zone B, trays 13, 15, and 17, north side downcomers (blue pen) lost their seal. In zone D, trays 7 and 9, south-side downcomers (red pen) lost their seal. (From Kister, H. Z., and C. Winfield, Chemical Engineering Progress, p. 28, January 2017. Reprinted Courtesy of Quantum Technical Services, LLC.)

to make the pressure drop in the kettle circuit high enough to raise the bottom sump level above the reboiler return nozzle (and initiate flood) would take significant entrainment from the kettle. A multitude of tests, including reboiler neutron scans, were conducted to evaluate the likelihood of such entrainment.

The tower and reboiler were operated at rates just below those at which the tower was observed to flood. Six neutron backscatter scans were performed across the length of the ket-tle (Figure 6.21). In each scan, neutrons were shot every 2 in. circumferentially from the kettle

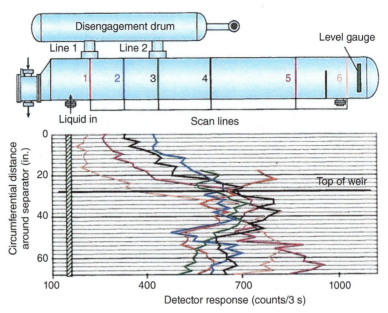

Figure 6.21 Neutron backscatter of the kettle reboiler shows heavy entrainment in the inlet region vapor space, gradually declining from line 1 toward the liquid draw nozzle. (From Kister, H. Z., and M. A. Chaves, Chemical Engineering, p. 26, February 2010. Reprinted Courtesy of Chemical Engineering.)

top centerline to the kettle bottom centerline. Figure 6.21 plots the detector count (x-axis) against distance along the circumferential distance from top to bottom.

Figure 6.21, scan line 1 (red pen), near the kettle inlet, shows a large amount of liquid in the vapor space above the kettle overflow weir. Ten inches (circumferentially) above the weir had higher liquid concentration than below the weir. Scan lines 2 and 3 also show high liquid concentration 10 in. (circumferentially) above the weir, much the same as the liquid concentration below the weir. Along scan line 2 (blue pen) in the reboiler inlet region, there is significant holdup of liquid in the vapor space all the way to the top of the kettle. After vapor line 2, the concentration of liquid in the vapor spaces gradually declines along the kettle length.

This pattern indicates a large amount of entrainment in the inlet region, between the kettle liquid inlet and vapor line 2, with gradually declining entrainment as one moves from vapor outlet line 2 toward the liquid draw nozzle. This is well in line with findings from other tests (217) that used entirely different techniques.

Below the weir, most scan lines showed a fairly uniform liquid content. The liquid content was higher for scan lines 5 and 6, especially lower in the exchanger, indicating less frothiness. The liquid content below the weir was the lowest in the inlet region.

The entrainment separator drum above the kettle was also neutron-scanned, with three different scan lines circumferentially from the top to the bottom centerline. The liquid level in the drum was low, less than about 2 in. It appears that the drum did not hold much of a liquid level.

6.4 CAT SCANS

While a grid scan (Figure 5.6) is capable of diagnosing maldistribution, and often provides clues for its nature, a computer-aided tomography (CAT) scan (46, 64, 240, 374, 507, 508, 511) provides the distribution profile of the liquid (in some cases, solids) at a given elevation. The CAT

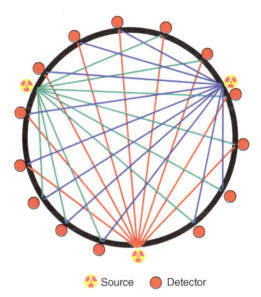

Source Detector

Figure 6.22 Source and detector placement in CAT scans. For clarity, only three source placements are shown; typically there will be nine different placements giving a 9 × 9 matrix. (Reprinted Courtesy of Tracerco.)

scan can also readily detect a center-to-periphery liquid maldistribution that a grid scan may not identify.

A CAT scan sets a gamma scan source and several detectors (typically about nine; for simplicity and clarity, Figure 6.22 only shows three) around the bed at evenly spread marked radial positions, all at the same elevation. Once done, the source is moved to the position of the nearest detector, the detector to the position of the source, and the scan is repeated. This continues until the source is placed in all the radial positions around the bed. The profiles obtained in all positions are then integrated to give the two-dimensional absorption density profile at the elevation. This density profile identifies liquid-rich (high-density) regions and drier (low-density) regions. If desired, the CAT scan can be repeated at additional elevations along the bed.

Figure 6.23 (240) is a CAT scan through a packed bed. The liquid density increases as one moves from the center of the bed to the peripheral regions. There is a circle right at the center of the bed that approaches total dryness. This circle occupied about 15% of the tower cross-section area. The normal grid scan (chords shown in Figure 6.23) showed maldistribution, but it took a CAT scan to define the nature of maldistribution and diagnose the tower problem.

CAT scans are expensive. Scaffolding is often required. Also, they may have difficulty getting a good measure of the liquid density near the periphery, showing a higher density in the wall region even in dry scans (65, 382). The accuracy of CAT scans can be affected by local variations in the packed bed, with dry packed bed density varying over a range of 1–2 lb/ft^3 across the tower area due to different orientations of the packing in the scan paths (371, 507). When high accuracy is required, a dry scan to account for packing orientation and external influences has been recommended (507). With a dry scan, the CAT scan measurements of liquid distribution closely matched those measured in that packed bed using a bucket and a stopwatch (507).

To improve visualization of the liquid distribution, it is common for the scanners to subtract the dry packing density (which is known, at least as an approximation) from the readings of the scans (511). The scan in Figure 6.23 uses this technique. Again, a dry scan can improve the accuracy.

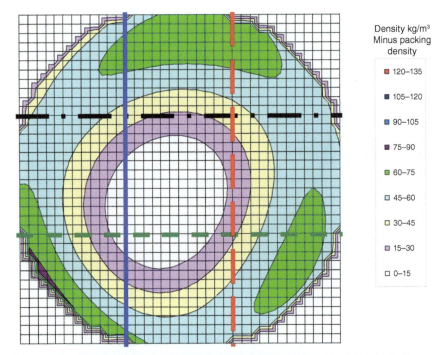

Density kg/m³
Minus packing
density

- ■ 120–135
- ■ 105–120
- ■ 90–105
- ■ 75–90
- ■ 60–75
- □ 45–60
- □ 30–45
- ▨ 15–30
- □ 0–15

Figure 6.23   A CAT scan through the upper packed bed under upset conditions. The liquid density increases as one moves from the center of the bed to the peripheral regions. There is a circle right in the center of the bed that approaches total dryness. (From Kister, H. Z., W. J. Stupin, J. E. Oude Lenferink, and S. W. Stupin, Transactions of the Institution of Chemical Engineers, 85, Part A, January 2007. Reprinted Courtesy of IChemE.)

Before performing the CAT scan, it is a good idea to perform a normal grid scan. The grid scan often identifies the region where the problem is intensified, which guides the optimum selection of the CAT scan elevation(s).

6.4.1   Identifying Unexpected Flashing in Reflux Distributor

A chemical tower experienced two operation modes: "normal" and "upset" (240). In the upset mode, the measured temperature near the top of the upper bed became 70°F higher than normal, and a "reversal" occurred: this temperature became hotter than temperatures further down the bed. The heating up of that temperature was accompanied by a large rise in heavies in the overhead product. A shift to the "upset" mode was triggered mostly by raising the set point on the tower temperature controller, but also by excessively raising feed rates.

Grid scan and CAT scan of the upper bed were shot both at the upset and the normal conditions. The one in Figure 6.23 was shot during the upset condition. It was shot 4 ft below the top of the bed, almost at the same elevation as that of the temperature indicator. The grid scan identified maldistribution, but it took the CAT scan to define its nature, leading to the diagnosis of the tower problem.

The CAT scans revealed the following:

1. Both in normal and upset conditions, a liquid-deficient central region and liquid-rich peripheral regions are observed (Figure 6.23). The liquid content steeply increases as one moves radially from the tower center toward the tower wall.

2. In the upset condition, there is a circle right in the center of the bed that approaches total dryness. This circle occupies about 15% of the tower cross-section area.

3. At upset conditions, the median density was 43 kg/m^3. The equivalent density for the normal condition scan was much higher, 65 kg/m^3.

4. The lower densities throughout the bed in the upset condition would normally mean less liquid flow down the bed. However, since the reflux flow rate was the same in the normal and upset conditions, the liquid flow rate did not change, as verified by the grid scans. This suggests a greater concentration of liquid near the wall in the upset condition. The CAT scans have difficulty measuring liquid near the wall, and the integration of densities was performed roughly over 80% of the tower cross-section area. A large increase in liquid flow near the wall would elude the integration.

5. The grid scan of the upset condition showed no chordal maldistribution in the upper 6 ft of this bed. Below this, two chords showed a liquid bias.

The pattern on the CAT scans, which was well in line with the results from the temperature survey (Section 7.1.7), indicated very severe maldistribution, with liquid descending around the periphery and vapor rising in the center. In the upset condition, this pattern appeared to intensify, with the center approaching dryness and the liquid building in the wall region. This maldistribution was not observed in the grid scans.

These observations, together with the temperature survey, led to the conclusion that the supposedly all-liquid reflux was flashing upon tower entry and boiling in the liquid distributor. It was later found that the reflux flash was caused by steam tracing of the reflux line and promoted by the presence of a low boiling component in the reflux (240). The high velocity of the flashing feed, which entered at the center of the distributor, blew the liquid from the distributor center to the peripheral regions, leaving the center of the tower relatively dry. We theorize that the upset condition was initiated due to nearly complete drying up of the tower center, as observed and confirmed by the CAT scans.

6.4.2 Diagnosing Unexpected Parting Box Overflow

Poor separation was experienced in a chemical tower (508). A grid scan showed fairly good distribution throughout the packed bed. The grid scan also indicated that the parting box was overflowing, but this did not appear to affect the liquid distribution in the bed. The client had concerns about excessive wall flow in the bed, that would not show up in the grid scan.

A CAT scan was performed near the top of the bed, with the chords of the grid scan superimposed on the CAT scan (Figure 6.24). The CAT scan did not show any excessive flow near the wall. Instead, it showed regions of higher liquid density under and on either side of the parting box, with dwindling density as one moves along the troughs away from the parting box. This provides strong evidence that the parting box overflow generated the maldistribution.

Figure 6.24 also shows why the grid scan did not show significant maldistribution. Each of the grid chords passes through roughly equal lengths of high-density and low-density regions, which overall would give the same detector reading.

Grid scans cover 17% of the cross-sectional area. CAT scans cover more than 70% and can therefore see what grid scans may miss.

6.4.3 Monitoring Coking in a Refinery Vacuum Tower

A major refiner has applied CAT scans to develop a unique optimization strategy of their vacuum towers (58, 375).

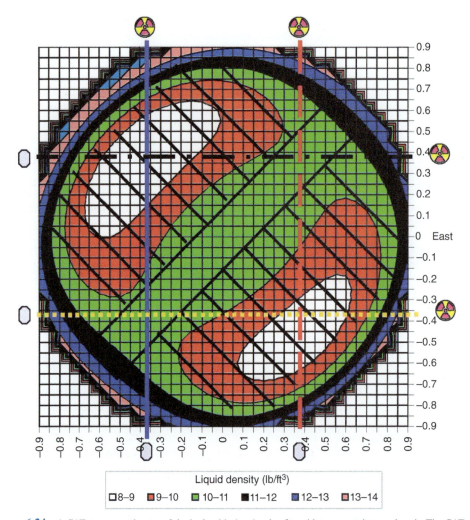

Figure 6.24   A CAT scan near the top of the bed, with the chords of a grid scan superimposed on it. The CAT scan showed higher liquid density under and on either side of the parting box, with diminishing density as one moves along the troughs away from the parting box, suggesting parting box overflow. (Xu, S. X., and L. Pless, "Understand More Fundamentals of Distillation Column Operation from Gamma Scans," in Distillation 2001: Frontiers in a New Millennium, Proceedings of at the Topical Conference on Distillation, the AIChE National Spring Meeting, Houston, TX, April 22–26, 2001. Reprinted Courtesy of Tracerco.)

The function of a refinery vacuum tower (typical arrangement in Figure 6.28, later) is to maximize the recovery of distillates from the crude feed. There is a very large price differential between the distillates and the residue ("resid") tower bottoms. The wash bed is the heart of the tower optimization. It uses distillate reflux ("wash") to keep the lowest distillate side product (heavy vacuum gas oil, or HVGO) on-spec. For low-metal crudes, the wash rate needed for keeping the HVGO on spec is low, and the reflux rate is usually set at the minimum wetting limit of the packing, i.e., the rate needed to prevent the packing from drying up. Drying up leads to aggressive coking and short run lengths, even though the packing used in wash bed is usually highly fouling-resistant. Excessive wash degrades precious distillate to low-value resid with a large economic penalty. Common industry practice keeps the net reflux at the driest spot, which is near the bottom of the bed ("the true overflash"), at about 0.2 gpm/ft^2 or a similar value. When followed diligently, this practice typically yields four to six years of run length at the expenses of some loss of HVGO to the low-valued resid.

The refiner's new optimization strategy (58, 375) minimizes wash below the conventional minimum wetting limit, while applying CAT scans to monitor the rate of coking in the wash bed. Rather than attempting to prevent coking altogether, the refiner's strategy allows coking to proceed at a monitored rate so that at the end of the desired run length period (typically four to six years), the bed will be fully coked. Throughout the run, the wash rate is increased if the coking rate is too fast, or decreased if it is too slow, with the target being achieving the desired run length. The benefit is improved distillate recovery over the run length, which is worth millions of dollars.

Immediately after the return from turnaround, a baseline CAT scan near the bottom of the bed (Figure 6.25a) as well as a grid scan of the bed were shot. To set parameters for aggressive operation, the bed was operated with a minimum wash rate, maximizing the distillate yield. The average and maximum bed densities were determined from CAT scans at the same elevation every three months and logged in Figure 6.26, Cycle 1. Over the 2-year cycle, the bed density increased by 40% due to coking and liquid accumulation (flooding) in the coked regions. A CAT scan near the end of the cycle (Figure 6.25b) shows regions of higher density due to coking. After these two years, an opportunity shutdown due to an unrelated issue allowed replacing the bed by an in-kind. Bed replacement costs are minor compared to the gain in distillate recovery.

The second cycle used higher wash rates to establish a boundary for non-aggressive operation. At the end of two years (Figure 6.26), Cycle 2, the bed density only grew by 20%. The next cycle was tailored to last six years. Figure 6.27 shows the overflash (net wash rate near the bottom

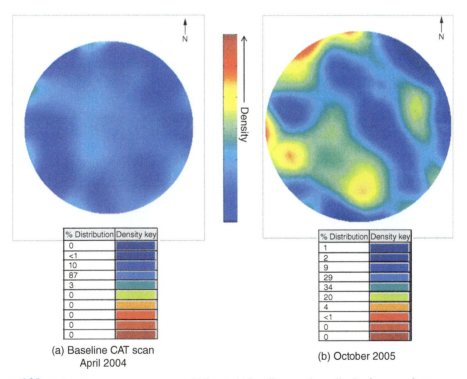

(a) Baseline CAT scan
April 2004

(b) October 2005

Figure 6.25   Refinery vacuum tower wash bed CAT scan. (a) Baseline scan, immediately after return from turnaround; (b) CAT scan at the end of cycle 1, a year and a half later, showing regions of higher densities due to coking. (Reprinted Courtesy of Tracerco.)

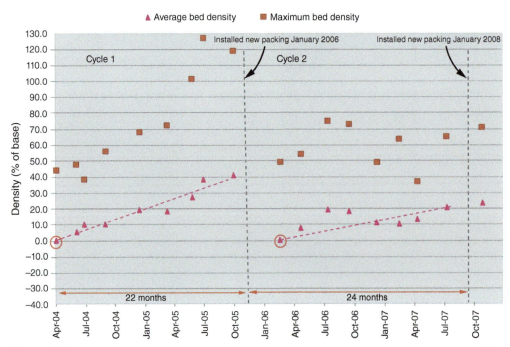

Figure 6.26   Vacuum tower wash bed densities measured by CAT scans. Density rise over the run is due to coking. Cycle 1, wash bed operated aggressively; cycle 2, wash bed operated conservatively. (Courtesy of Tracerco.)

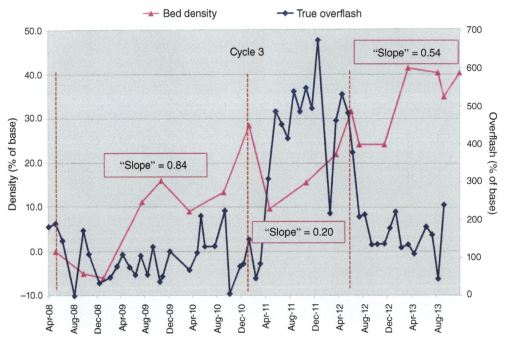

Figure 6.27   Vacuum tower wash bed true overflash (net reflux leaving the bottom of wash bed) optimized according to bed density data measured by CAT scans over 69-months cycle 3. The objective was to minimize true overflash while controlling the coking rate to achieve the desired run length and prevent premature cycle termination. (Courtesy of Tracerco.)

of the bed) and the bed density over this cycle. The first segment (April 2008 to January 2011) used a low wash rate to maximize the HVGO yield. This came at the price of some coking. The wash rate was significantly increased in the second segment to retard the coking rate at the expense of a reduction in HVGO yield. This was successful, and as the desired end of run was approached, the wash rate was reduced again to maximize the HVGO yield.

The principles of this optimization can be extended to many towers that experience packing fouling.

6.5   TRACER TECHNIQUES

These involve the injection of a radioactive tracer into sections of the plant and monitoring its movement with the aid of radiation detectors. Depending on the application, the tracer may be injected either as a pulse or at a constant rate. To minimize contamination, the vast majority of applications use pulses. The tracers involved are predominantly gamma-ray emitters of short half-lives, with some common sources listed in Ref. 67. Tracers can be gas tracers or nonvolatile liquid or solid tracers. The liquid tracers can be water-soluble or hydrocarbon- or organic-liquid-soluble. If operating at high temperatures, it is important that the tracer be stable enough not to decompose at that temperature.

It is useful to test whether the tracer ends up in the desired phase. For instance, if injecting a nonvolatile liquid tracer at the feed, place a detector in the tower overhead vapor to verify a zero reading.

Tracer techniques are often applied for leak detection, flow measurement, packing maldistribution studies, and entrainment studies. For example, in leak detection, a tracer can be injected into the reboiler steam line, and a detector on the process side determines whether any of the tracer has entered the process fluid. Cases where this technique successfully diagnosed reboiler leaks and measured the rate of leakage have been reported (66, 101, 283, 378, 503).

For flow measurement, detectors are placed at two (or more) points on the line, a known distance apart, and a pulse is injected upstream. The distance divided by the travel time between the points gives the velocity, which is then multiplied by the pipe's internal area to give the volumetric flow rate. For accurate determination, the flow needs to be steady, turbulent, the distance large enough to give an accurate result, and the first detector far enough from the injection point to allow good lateral mixing of the tracer with the process fluid. As the radiation pulse crosses each detector, it produces almost a binomial distribution of counts, usually of a narrow width with turbulent flow. Computerized integration performed by the scanners determines the centroid (102, 283). Straight runs of pipes with no control valves or orifice plates are essential for good accuracy (102). Flow measurements as accurate as 1% can be obtained (283, 503). Alternative flow measurement procedures for irregularly shaped conduits are also available (283).

Tracer flow measurement can also be applied to detect a leaking tower relief valve, as described in one case study (102). They can also be applied to diagnose the location of air leakage into a vacuum tower. A marked helium tracer is sprayed near many flanges, one by one, and the vent gas is monitored.

For packing maldistribution diagnosis, a liquid tracer can be injected into the reflux, with detectors placed radially in the bed. The author is familiar with two successful applications of this technique. In one instance, the detectors were mounted in the 6-in. space between the trough distributor and the top of the bed. The other application was in the wash section of a refinery vacuum tower.

Simulated tracer tests were also described to evaluate fouling precursor residence time in a column. In this application, the tracer needs to have similar thermodynamic properties to the precursor (27). Residence times derived from tracer tests can also help fine-tune a dynamic simulation model (27).

Tracer techniques are discussed in detail elsewhere (66, 67, 283, 503). The cases discussed below focus on applications where tracer techniques provide invaluable data on tower operations and are not so well appreciated by many engineers.

6.5.1   Measuring Flow Rates, Internal Leaks, and Spray Entrainment in Refinery Vacuum Tower

Figure 6.28 shows a typical arrangement of the lower section of a refinery vacuum tower. High-velocity, hot flashing feed enters the tower swage via a vapor horn (a flashing feed distributor/deentrainer, Ref. 245). Feed liquid drops to the bottom sump to become the vacuum tower bottoms (VTB) product. Above the feed point is the total-draw slop wax chimney tray that collects the slop wax product from the wash bed above. Reflux ("wash oil") to the wash bed comes from the HVGO total-draw chimney tray above and is distributed to the wash bed via spray nozzles. From the HVGO chimney tray, the HVGO is pumped partly as reflux to the wash bed, partly as product, and the rest is cooled and sprayed to the top of the HVGO bed above, where the cooled HVGO condenses some of the ascending vapor by direct-contact heat transfer to produce the HVGO.

There are a number of potential trouble spots in this system. Both the slop wax and HVGO total draw trays are prone to leakage, which degrades the more valuable upper products into less

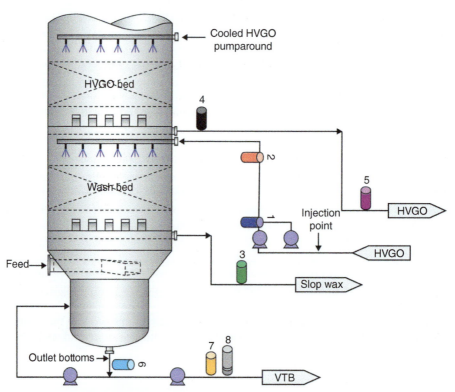

Figure 6.28   Lower sections of the refinery vacuum tower that were investigated using a tracer technique, showing the locations of the detectors. (Reprinted Courtesy of Tracerco.)

valuable lower products at large cost penalties. Drying in the wash bed, caused by excessive vaporization or excessive entrainment from the spray nozzles, is a source of bed coking and short run lengths. Flow meters in this system commonly have reliability issues. A tracer study can shed light on many of these potential trouble spots.

In this investigation (378), a special high-temperature-stable liquid tracer was injected into the wash oil (Figure 6.28), and its path was monitored by eight different detectors. Detectors 1 and 2 were mounted at the two ends of the wash oil line to the wash bed. Dividing the length of the line between 1 and 2 by the time it took for the tracer pulse to travel from 1 to 2 gave the liquid velocity. Multiplying by the pipe cross-section area gave the volumetric flow rate. Similar tests were performed at a later time between points 4 and 5 along the HVGO line, slop wax product, and VTB lines, giving a measurement of each flow rate.

Detector 3 monitored the tracer in the slop wax product. The total amount of the tracer reaching the slop wax product can be obtained by integrating the radiation observed at this detector from its rise until its return to the initial value. Similarly, detector 4 monitored the tracer entrained from the wash sprays, and detectors 6, 7, and 8 monitored the tracer reaching the VTB. As there was no direct path for the tracer to reach the VTB other than a leak at the slop wax chimney tray, the presence of the tracer at the VTB confirmed that this tray was leaking.

A mass balance on the tracer showed that 53% of the injection ended in the slop wax, 41% was entrained from the wash bed sprays into the HVGO, and the other 6% leaked into the VTB. From these figures and the flow measurements, the leaks and entrainment can be quantified. The slop wax was measured at 5300 BPSD, and contained 53% of the tracer, so the leaking amount (6% of the tracer) was $5300 \times 6/53 = 600$ BPSD. The wash oil rate was measured at 31,900 BPSD. 41% of the tracer was entrained, and the other 59% went down, giving 13,100 BPSD entrainment and 18,800 BPSD of net wash flowing down the wash bed. With the slop wax product plus leak from the slop wax chimney tray adding up to $5300 + 600 = 5900$ BPSD, at least $(18,800–5900 =) 12,900$ BPSD, or 69% of the wash oil, were vaporized in the wash bed. The fraction is likely to be higher, as some of the 5900 BPSD is likely to be entrainment of liquid from the feed zone.

One source of inaccuracy is that the tracer-containing entrained wash oil ended in the HVGO, so some of the tracer was recycled into the wash oil after a time lag. Fortunately, the HVGO flow rate of 156,100 BPSD was about five times larger than the wash oil rate, so the recycled tracer was only a small amount.

This case demonstrates the ability of tracer tests to measure tower external flow rates as well as internal leaks and entrainment. The ability to measure internal leaks and entrainment depends very much on the tower and internals configuration and may not be achievable in many towers. Before embarking on a tracer test, it is important to evaluate the information that it would be able to provide and, based on this, decide whether to proceed.

6.5.2 Observing Downward Vapor Flow in Packed Bed

Excessive carryover (0.1–1.5% of the top tray liquid flow rate) was experienced at design rates from an LNG plant packed tower (493). A grid gamma scan showed no obvious problems. Tracer tests were conducted to gain insight. Figure 6.29 shows the tracer locations. Each red dot represents a set of 4 detectors mounted radially at the northeast (NE), northwest (NW), southeast (SE) and southwest (SW). A liquid tracer injection showed equal split between the four quarters. However, it shows that liquid velocity down the south was much faster than down the north quadrant (0.63 ft/s versus 0.29 ft/s) .

A vapor tracer was then injected into the stripping gas, although the feed gas rate was an order of magnitude larger. The stripping gas appeared uniformly distributed at the vapor distributor, but

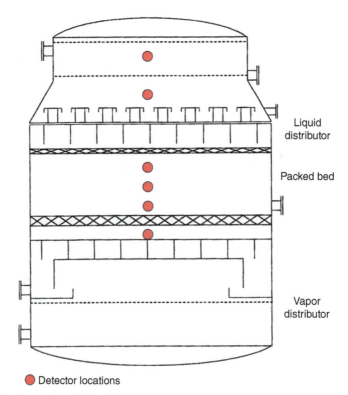

● Detector locations

Figure 6.29   Top bed of LNG plant packed tower that was investigated using a tracer technique, showing the detectors locations. Each red dot represents a set of four detectors mounted radially at the northeast (NE), northwest (NW), southeast (SE) and southwest (SW). (From Vidrine, S., and P. Hewitt, "Radioisotope Technology - Benefits and Limitations in Packed Bed Tower Diagnostics," Paper presented at the AIChE Spring Meeting, New Orleans, Louisiana, April 25–29, 2004. Reprinted Courtesy of Tracerco.)

a severe vapor bias was observed in the bed, with none traveling up the SE. When the vapor reached the top of the bed, it was determined that 35% of it was diverted downward via the SE quadrant. Upon exiting below the bed, it followed the original flow path of the stripping gas, and the cycle repeated with diminishing amounts of recycle gas each recycle pass. This resulted in a vapor mean residence time of more than three minutes.

Figure 6.30a shows the SE detector readings. The detector at the vapor distributor is the first to respond, with the top of the bed the next. A short time later the mid of the bed detector responds, then the bottom of the bed, and lastly the vapor distributor peaks the second time. On the right-hand side of Figure 6.30a, the sequence begins to repeat. Figure 6.30b shows the NW detectors right across. Here, the detector sequence of response is normal: vapor distributor, bottom of bed, mid of bed, and top of bed. But note what happens then: the vapor distributor detector peaks the second time! Then the sequence repeats at a lower intensity.

The tests were repeated at three different process conditions with similar results. Although reverse flow in a packed tower is uncommon, the ability of tracer tests to diagnose such an unexpected pattern is amazing. The cause of this pattern was not known at the time of publication of the reference paper by Vidrine and Hewitt (493).

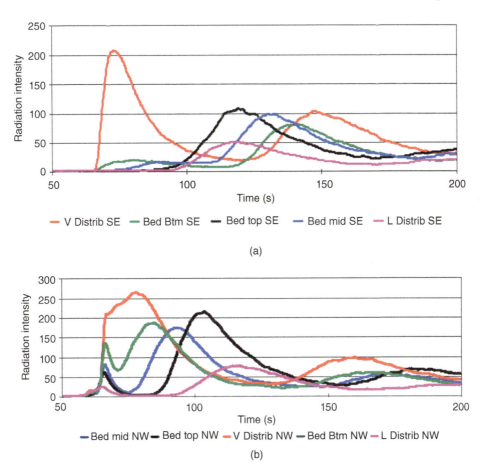

Figure 6.30 Tracer detector readings on the southeast (a) and northwest (b) upon gas tracer injection into the stripping gas. The sequence of response shows downward gas flow on the southeast and re-rise on the northwest. (From Vidrine, S., and P. Hewitt, "Radioisotope Technology - Benefits and Limitations in Packed Bed Tower Diagnostics," Paper presented at the AIChE Spring Meeting, New Orleans, Louisiana, April 25–29, 2004. Reprinted Courtesy of Tracerco.)

6.5.3 Quantitative Determination of Entrainment from a Kettle Reboiler

This is the same kettle reboiler as in Case 6.3.3 (Figure 6.21) and later Case 7.1.8 (217). As stated, significant entrainment from this kettle would lead to a high pressure drop in the kettle circuit, which in turn would raise the liquid level in the tower bottom sump above the reboiler return, leading to premature flooding of the tower. It was therefore critical to evaluate how significant the entrainment was.

The tower and reboiler were operated at rates just below those at which the tower was observed to flood. A nonvolatile tracer was injected into the reboiler feed. Detectors were mounted at the feed line just before reboiler entry, at the two reboiler vapor outlet lines to the disengagement drum (Figure 6.21), at the horizontal vapor outlet line from the disengagement drum to the tower, and at the reboiler liquid outlet line. Figure 6.31a shows the various detector responses to a pulse injection.

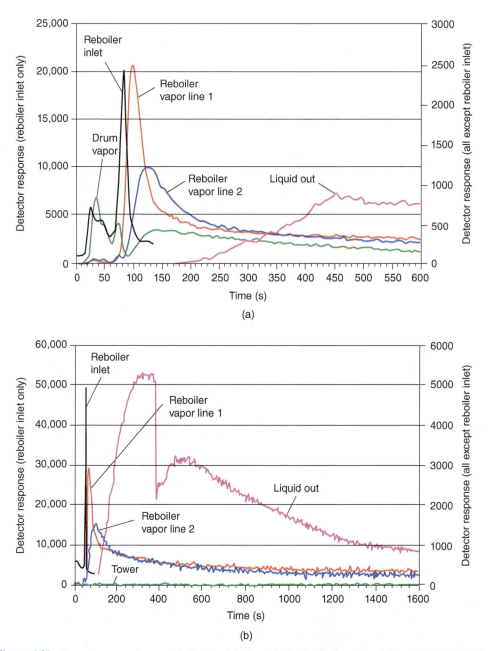

Figure 6.31 Detector response to a nonvolatile tracer injection at the kettle reboiler process inlet. The response conclusively shows heavy entrainment from the reboiler inlet area and provides a means of measuring the entrainment. (a) Drum drain valve to reboiler outlet open; (b) drum drain valve closed for the first 400 seconds, then opened. (From Kister, H. Z., and M. A. Chaves, Chemical Engineering, p. 26, February 2010. Reprinted Courtesy of Chemical Engineering.)

Tracer injected at time zero was detected at the reboiler inlet 80 seconds later, giving a sharp and well-defined peak (black curve in Figure 6.31a). This was followed 10 seconds later by a peak at vapor line 1 (red curve). The tracer did not finish entering the reboiler when it was detected at vapor nozzle on line 1. Tracer began coming out of vapor line 2 (blue curve) about the same time when it peaked in vapor line 1, peaking at about 140 seconds. The peak was much lower in counts but of a wider area, suggesting that the tracer was reasonably evenly split between vapor lines 1 and 2. Trailing edges on all these detectors were long, indicating that the tracer continued to be entrained for some time.

The sharp peaks and the short duration from tracer entry to the peaks in the kettle vapor outlet lines provide conclusive evidence that the entrainment is large and occurs primarily in the reboiler inlet region. Tracer entering the reboiler was immediately sent overhead via vapor line 1, and after a short duration also via vapor line 2. The tracer reached the liquid outlet (pink curve) about 190 seconds after the injection, and peaked at 450 seconds, leveling off after that.

Can the Entrainment Be Measured?   This test repeated the tracer injection, but this time the valve in the 3 in. line that drained liquid from the drum to the kettle liquid outlet line was shut just before the tracer injection. In this test, we also verified that the tracer remained in the liquid by mounting a detector (green curve) in the vapor space below the tower bottom tray. Its zero reading (Figure 6.31b) verifies that no tracer entered this vapor space.

Figure 6.31b shows similar trends to Figure 6.31a in the first 200 seconds. Tracer peaked at the reboiler inlet 60 seconds after injection (black curve), followed immediately by a peak in vapor line 1 (red curve). At about the same time, a tracer was first detected in vapor line 2 (blue curve), peaking 40 seconds later. Tracer was detected at the kettle liquid outlet after 120 seconds, peaking and then leveling off after about 300 seconds.

With the tracer in the reboiler outlet liquid at the high reading of about 5,200 counts, and with the counts in vapor lines 1 and 2 dropping to below 600 counts, the valve on the 3 in. drain line from the drum to the reboiler outlet line was opened at the 400-second mark. Immediately the detector signal dropped, stabilizing at about 3,100 counts. From that drop of counts, the flow rate of liquid from the disengagement drum can be calculated. A component balance on the tracer gives

$$3{,}100\,W = 400\,X + 5{,}200\,(W - X) \tag{6.5}$$

where W is the flow rate of the kettle outlet stream, which is known from the downstream column mass balance (in this case 141 hot gpm), and X is the flow rate of disengagement drum liquid, unknown. The count of 400 was obtained from the drum vapor inlet detectors once the liquid out count leveled at 3,100 counts. This is not an accurate representation for the drum liquid, but being small, the calculation is insensitive to its accuracy. Solving Eq. (6.1) gave $X = 62$ gpm (hot), or 44% of the kettle outlet stream.

This test produced a surprising and unexpected fallout. When the valve in the drum drain line was closed, the liquid level in the depropanizer tower started shooting up. Instead of being removed by the now-closed drum drain line, the liquid was carried over into the tower base. Over the seven minutes at which the valve was closed, the level glass indicated that about 55 ft^3 of liquid accumulated in the tower base, giving an average rate of 59 gpm. This figure was strikingly close to that inferred from the tracer component balance above.

It is important to recognize that the test only measured the entrainment that was knocked out by the drum and returned to the kettle outlet line via the 3- in. drain line that was blocked in and later opened. In addition, there was entrained liquid from the kettle that continued on from the drum to the tower. This entrainment was not measured by this test.

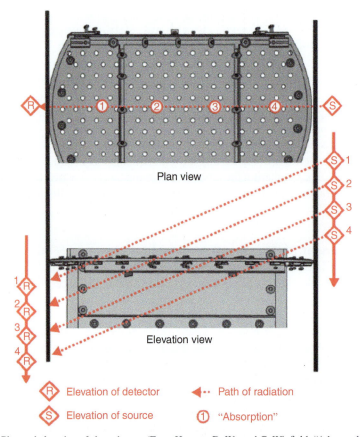

Figure 6.32   Plan and elevation of slanted scan. (From Hanson, D. W., and C. Winfield, "Advanced Scan Techniques Aide in Column Diagnostics," in Kister Distillation Symposium Proceedings, p. 27, AIChE Spring Meeting, Virtual, April 18–22, 2021. Reprinted with permission per disclaimer at the front of Section 6.6.)

6.6   SLANT SCANNING

Disclaimer   The information provided in this section, including Figures 6.32–6.34, expresses the opinions of the authors and may not reflect the views of Valero; any use of the information is at the reader's risk.

Slant Scanning (152) is a patent-pending technique that intentionally skews the angle between the source and the detector to audit the tray integrity. It scans across the different tray parts, providing a sectional view of radiation transmission across the tray.

For good integrity, the tray's transmission profile should be a fairly uniform, with the exception of locations that have slightly higher absorption (e.g., tray down-flanges). A slant scan showing nonuniform transmission indicates damage or fouling and provide insight to their nature. The key to interpreting slant scans is:

- No tray damage–uniform scan pattern.
- Tray fouling–uniform scan pattern; lower transmission readings across affected trays.
- Tray damage (damage, corrosion, missing manway)–nonuniform transmission pattern.

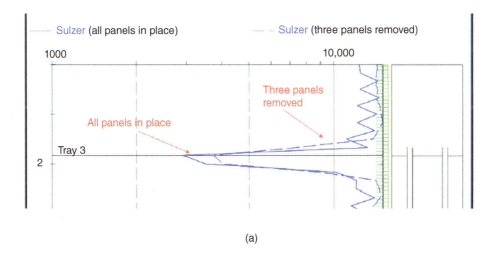

(a)

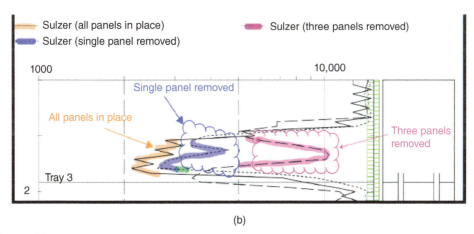

(b)

Figure 6.33   Scans of a two-pass tray in an empty 10 ft ID Sulzer training tower (a) normal scan; (b) slanted scans. (From Hanson, D. W., and C. Winfield, "Advanced Scan Techniques Aide in Column Diagnostics," in Kister Distillation Symposium Proceedings, p. 27, AIChE Spring Meeting, Virtual, April 18–22, 2021. Reprinted with permission per disclaimer at the front of Section 6.6.)

Figure 6.32 shows slant scans through the four numbered locations on the tray deck. Larger-diameter trays require additional locations. The source (S) is placed above the tray deck, and the detector (R) below the tray deck. Both are lowered simultaneously at the same rate of descent. At position 1, the scan will see the tray panel closest to the detector. Subsequent readings at positions 2 and 3 will see the middle panel at two locations. Subsequent reading at position 4 will see the panel closest to the source. In the absence of damage, radiation transmission differences between the positions should be minimal so that the relative transmission comparison will yield information on tray integrity.

Figure 6.33 shows scans shot in Sulzer's training tower, containing two-pass trays with the tower empty. The upper diagram shows conventional (not slant) scans. The scans show very little difference between having all the panels in place and having three panels missing. The bottom diagram is the equivalent slant scans. The elevation plotted is that of the detector. The slant

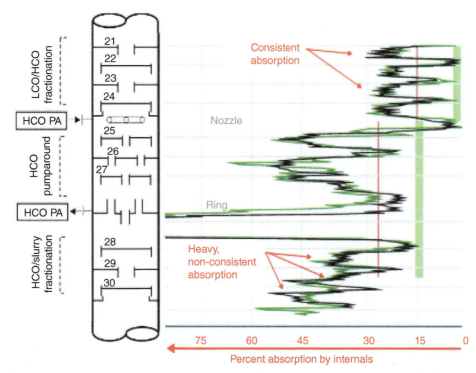

Figure 6.34   Slanted scan of an empty 12.5 ft ID FCC main fractionator tower positively diagnosing fouling on trays 28–30. (From Hanson, D. W., and C. Winfield, "Advanced Scan Techniques Aide in Column Diagnostics," in Kister Distillation Symposium Proceedings, p. 27, AIChE Spring Meeting, Virtual, April 18–22, 2021. Reprinted with permission per disclaimer at the front of Section 6.6.)

scans show major differences between having all panels in place, one panel missing (simulating uninstalled manway) and three panels missing (simulating major damage). At detector elevations, when the scan goes through the missing panel, there is less absorption.

Figure 6.34 shows a slant scan in an empty 12.5 ft ID FCC main fractionator tower. Trays 21–24 (LCO/HCO fractionation zone) showed normal and consistent transmission. Since trays 21–24 are similarly laid out to trays 28–30, their transmissions are directly comparable. Figure 6.34 clearly shows that:

- Transmission on trays 28–30 (HCO/slurry fractionation zone) is about half of that on trays 21–24.

- Transmission on trays 27–30 is not very consistent by comparing tray to tray. Also, transmission across a single tray deck is varying, highlighting fouling and/or damage.

- Trays 25–27 were fouled.

The above findings were fully confirmed by tray inspection at the outage shortly later, with some photographs shown in Ref. 152.

To date, slant scans have been used to determine extent of damaged tray decks and to identify tray fouling locations and severity to enhance productivity during turnarounds. Slant scan technology is undergoing extensive development and has the potential to gather additional invaluable information about troubled columns.

6.7 USEFUL CASE HISTORIES LITERATURE

Several case histories involving radioactive scanning have been reported in the literature. Below is a partial list, far from comprehensive, of valuable literature references. There are many more valuable literature references that are not listed. These sources illustrate, with the aid of scan graphics, the application of these techniques to diagnose the following abnormalities:

6.7.1 Qualitative Gamma Scans of Trays

1. Flooded trays (44, 45, 52, 96, 137, 148, 160, 164, 180, 185, 189, 226, 227, 238, 265, 302, 323, 327, 363, 380–383, 387, 412, 423, 424, 483, 485, 508–510)
2. Rate of flooding a tray (160, 227)
3. Time study to determine exact rate or location of flooding (45, 146, 485, 508, 509)
4. Distinguishing jet from downcomer flood (1, 44, 45, 180, 189, 227, 302, 363, 380, 381, 383, 485, 508)
5. Plugged trays and/or downcomers (45, 52, 96, 101, 180, 189, 227, 363, 372, 379, 383, 412, 483, 485, 493)
6. Plugging causing maldistribution, unsealing in two-pass or multipass trays (180, 189, 227, 412)
7. Maldistribution in multipass trays (234, 327)
8. Differences between panels of 2-pass trays (180, 310)
9. Identifying and bypassing a downcomer restriction (227, 363, 383, 493)
10. Monitoring tray plugging (66)
11. Foaming (1, 44, 66, 114, 185, 227, 424, 469, 485, 508)
12. Spray regime operation (508)
13. Variation of tray dispersion with vapor loadings (185, 371, 380, 510)
14. Dry tray panels (424)
15. Unsealed downcomers (227, 246)
16. Missing trays (67, 310, 483)
17. Damaged trays (114, 310, 412, 423, 424, 483, 485)
18. Weeping (483, 485)
19. Entrainment (185, 236, 485, 509)
20. Hydraulic gradient, channeling (214, 226, 508)
21. Evaluating the effectiveness of a modification by changing a few trays (114, 160)
22. No abnormality where one was expected (246, 423, 424)
23. Two or more different types of abnormalities (or floods) in one column (52, 67, 327, 508, 509)
24. Not showing flood when other testing showed flood (18)
25. Shed deck scan (412)
26. Slant scanning (152)

6.7.2   Qualitative Gamma Scans of Packings

1. Liquid maldistribution in packed beds (2, 13, 45, 46, 89, 137, 151, 160, 240, 323, 324, 332, 373, 374, 376, 377, 483, 485, 508, 509)
2. Rate-related maldistribution (46)
3. Out-of-level distributor (483, 508)
4. Distributor or collector overflow (151, 323, 377, 387, 509)
5. Parting box overflow (508)
6. Plugged, damaged sprays (2, 89)
7. Foaming (332)
8. Variation of packed bed liquid holdup with bed loading (64, 65, 382)
9. Variation of scan profile with loadings (64)
10. Packing gas distributor problem (332)
11. Flooded packing (2, 52, 58, 64, 65, 89, 90, 323, 332, 372, 375, 376, 485, 509)
12. Plugged packing (2, 52, 58, 90, 323, 372, 375–377, 480, 485)
13. Monitoring plugging progress in packed beds (58, 90, 323, 372, 480)
14. Packing distributor fouling (89, 332, 509).
15. Collapsed, uplifted packed beds (46, 89, 283, 485)
16. Damage to liquid collector or distributor in packed tower (46, 89, 485)
17. Packing damage, crushing, missing, bed shrinking (89, 137, 376, 377, 509)

6.7.3   Gamma Scans of Points of Transition and Inlet and Outlet Lines

1. Identifying a point of transition (feed, reflux, draw) problem (118, 238, 381, 483, 489)
2. Base liquid level above reboiler return or bottom vapor feed (66, 160, 164, 377, 485, 509)
3. Entrainment of seal pan overflow liquid by reboiler return vapor (163)
4. Excessive liquid level on a chimney tray (118, 231, 238, 330, 485, 489, 507)
5. Chimney tray causing a bottleneck (44, 118, 236, 412, 489)
6. Draw tray damage (310)
7. Plugging of internal seal pan at a draw (379)
8. Plugging or restriction in an outlet line (44, 412, 503)
9. Subcooled feed causing flood (160, 485, 508)
10. Subcooled feed not causing flood (508)
11. Reboiler return line scanning for liquid content (217, 352)
12. Draw line scanning for vapor content (118)

6.7.4   Quantitative Gamma Scans of Trays

1. Hydraulic gradients, channeling (203, 204, 231, 244, 252)
2. Flooded trays (160, 203, 204, 244, 252, 371, 380)
3. Identifying flood location (160, 204, 244, 371, 380)
4. Kistergram application (203, 204, 231, 244, 252)

5. Monitoring modification effectiveness (160)
6. Blowby in truncated downcomers (26, 205, 427)
7. Drying of active panels in multi-downcomer trays (205, 252)
8. Maldistribution in multipass trays (205, 252, 427)

6.7.5   Quantitative Gamma Scans of Packings

1. Maldistribution in packed beds (370, 371, 376)
2. Packing damage, crushing, missing (376)
3. Flooded packing (323, 376)
4. Plugged packing (323, 376)

6.7.6   Neutron Backscatter

1. Level measurement, tower base or drum (101, 352)
2. Level measurement, reboiler (230, 352)
3. Level measurement, chimney tray (158)
4. Level in draw pan (243)
5. Reboiler pipes (352)
6. Kettle reboiler distribution and entrainment (217)
7. Unsealed downcomers (227, 381)
8. Vapor in draw pans (379)
9. Time study to verify flood initiation (146)

6.7.7   CAT Scans

1. Liquid maldistribution in packed beds (46, 64, 240, 374, 507, 508, 511)
2. Parting box overflow (508)
3. Flashing in liquid distributor (240)
4. On-line fix to a liquid maldistribution issue (507, 511)
5. Monitoring bed plugging progress (57, 58, 90, 372)
6. Variation of CAT scan profile with loadings (64, 65, 382)
7. Tendency of liquid to flow toward the wall (64, 65, 382)

6.7.8   Tracer Technique

1. Exchanger (reboiler, condenser, preheater) leak quantification (101, 283, 503)
2. Identifying which of the bottom coolers was leaking (102)
3. Chimney tray (total draw) leak (378)
4. Relief valve leak (102)
5. Packing vapor maldistribution (493)
6. Packing liquid maldistribution (493)

7. Vapor downflow in a packed bed (493)
8. Entrainment from spray nozzles (378)
9. Entrainment from kettle reboiler (217)
10. How much each of parallel columns is entraining (46)
11. Flow rate measurement (102, 378)
12. Residence time of fouling precursors (27)

Chapter 7

Thermal and Energy Troubleshooting

"If something doesn't look right... it probably is not right"

—(Ross Vail, presentation to AIChE)

One of the very first checks at the doctor's office is the patient's temperature, which is central to the doctor's diagnosis. The same applies when diagnosing the ills of distillation columns. The main difference is that in columns there are many more temperatures that can be measured, all of which provide invaluable input to the diagnosis.

Decades ago, troubleshooters were totally dependent on column thermocouples for temperature measurement. Local indicators often read incorrectly, and contact thermocouples required long thermal stabilization periods to yield reliable information. Rapid developments in infrared (IR) thermography changed this, by bringing to our industry pyrometers ("temperature guns") and IR cameras, which now enable troubleshooters to measure as many temperatures as they wish, even shoot "thermal videos." There is only one problem: most engineers do not take advantage of this tool to its full extent. It is the author's hope that this chapter will familiarize the reader with this important resource and demonstrate how valuable it is for diagnosing and solving column problems.

Energy balancing is another invaluable technique. A correctly compiled energy balance does not lie, but can invalidate incorrect theories and lying measurements. The author was once humbled when an intern applied an energy balance to diagnose a problem that the author and his colleagues struggled with for months (216). The author then learned how important it is to compile these energy balances. It is his hope that future tower doctors will appreciate their importance too.

7.1 WALL TEMPERATURE SURVEYS

Holes can be cut in the insulation around the tower, its auxiliaries, or its piping, and wall (skin) temperatures measured by an IR surface pyrometer or a thermal camera. These wall temperatures provide a wide range of invaluable diagnostics, such as testing theories, validating simulations, detecting packed tower maldistribution, and detecting liquid levels in reboilers, condensers, bottom sumps, and chimney trays. They can also provide invaluable information on flood determination, transient behavior, the presence or absence of a second liquid phase, accumulation of non-condensables in condensers, and many more. Such applications are discussed in this section.

Distillation Diagnostics: An Engineer's Guidebook, First Edition. Henry Z. Kister.
© 2025 American Institute of Chemical Engineers, Inc. Published 2025 by John Wiley & Sons, Inc.

7.1.1 Dos and Don'ts for Temperature Surveys

Successful temperature surveys require valid and reliable measurements. It is crucial to verify the data by extensive consistency checks. Invalid or unreliable temperature measurements lead to incorrect diagnostics. In one case, variations of more than 100°F along the circumference of a crude tower wash bed, which were interpreted as maldistribution, were caused by incorrect surface temperature measurements. The key to a successful temperature survey is reliability, repeatability, and consistency. Here are some guidelines for achieving them:

1. Check the expected temperature profile in the tower. If the temperature does not change much (less than 15–20°F or so) over a section of the tower, inaccuracies in wall temperature measurement may be large compared with the temperature changes, and a temperature survey may provide limited or no benefits. For instance, near a pure product end of a column, where boiling point temperatures change very little with composition, a temperature survey is a waste of time and effort.

2. Explore features that may render wall temperatures non-representative of the tower's internal temperatures, for instance, if a section of the tower is glass-lined or cladded with corrosion-protective material on the inside that does not conduct heat well. In cryogenic columns, holes cut in the insulation ice up rapidly and often do not permit reliable measurement.

3. Confirm with your health, safety and environmental regulations (HSE) department that holes in the insulation can be safely cut and whether there are any special requirements. When the insulation is fabricated from asbestos, cutting holes is not permitted or requires complex procedures and special protective clothing that may make the survey unattractive or even impractical. In this case, the temperature survey may need to be abandoned.

4. Water ingress into the insulation must be avoided, as it can lead to severe corrosion with both carbon steel and stainless steel columns (e.g. Ref. 183). Holes cut in the insulation should have tight-fitting caps that tightly close and positively prevent water ingress when not in use.

5. Keep the holes in the insulation capped when not taking readings and definitely when the test is complete. This is especially important if the temperatures are close to or below ambient. If left open, ambient condensation from humid air can lead to rapid corrosion.

6. Steady tower operation at the desired operating conditions is imperative for a successful temperature survey. Process temperature swings during the survey can easily be misinterpreted as tower issues, leading to incorrect diagnostics. The one exception here is when temperature surveys are used to study the temperature variation at a given point as a function of time ("time studies," Section 7.1.4). This technique can be applied to study the source and nature of an instability, as has been demonstrated in case studies (145, 237).

7. Calibrate your temperature measurement device. Measure surface temperatures next to every thermocouple in the tower. Typically, the inside temperature is a little higher than the wall temperature. For example, when the inside temperature is 200°F, the wall temperature is about 180–190°F. A plot of wall temperatures at the thermocouples versus the thermocouple-measured temperatures should be a smooth curve, which is used for calibration. This calibration curve will be applied to convert any measured wall temperature into an inside fluid temperature. In the above example, when a wall temperature measures 185°F, it will be converted to an inside temperature of 200°F.

Caution is required where thermocouples stick some distance into a packed bed and may therefore read temperatures that are different from those at the wall. Check the thermocouple specs and do not include them in the calibration if there is a reason to suspect that the wall and inside temperatures are different.

There is often a discussion on what emissivity to set the pyrometer at. Usually, a high value, between 0.9 and 1.0, is set. In the author's experience, this value does not make much difference when calibration is properly done, as calibration will make up for any errors in emissivity setting.

8. Audit the repeatability of your measurements. Measure wall temperatures proceeding from point to point in one direction and then repeat, without looking at the previous results. Assuming the tower is stable, differences between measurements at any given spot should not exceed about 5°F. If they do, there are some unresolved measurement issues. Keep doing it until your measurements are fully repeatable.

9. Another useful consistency check is going up the tower in a team of three (or at least two). Involve an operator if possible. Each team member takes a reading without telling the others. Compare notes only after all members are done. The differences should be less than 2–3°F.

10. Check that temperatures consistently decline as you climb the tower. The temperatures should smoothly change with elevation. Investigate any "bumps" or "inversions" by a thorough study of the region where they are observed. If real, an inversion can signify a chemical reaction, the presence of a second liquid phase, or packing maldistribution.

11. People often express concern about whether wind, rain, and sunlight affect wall temperature measurements. In the author's experience, when insulation holes are not larger than about 3 in. in diameter, the impact of sun, wind, and light rain (which is not blown into the hole) has been minimal. Temperature surveys should be avoided during strong rain or other extreme weather conditions. Not only can the measurements be affected, but towers often become unsteady or water can get into the insulation (see steps 4 and 5).

12. Several other consistency checks can be implemented. Additional consistency checks enhance the confidence in the validity of the data.

13. In packed towers, temperature surveys are most valuable for checking liquid and vapor distribution. To achieve that, holes are cut radially at a desired elevation and the wall temperatures in these holes are measured by a pyrometer (Figure 7.1). Typically, six to eight holes are cut radially at the desired elevation. If distribution is good, the wall temperature variations will be less than 10–20°F. Higher variations signify maldistribution.

Caution is required when calibrating packing temperature surveys. As stated in step 7, wall temperature may differ significantly from thermocouple temperature deep inside the bed. Also, maldistribution can be local. The best calibration points are on collectors or distributors (Figure 7.1), where the temperatures tend to be more uniform, or in vapor spaces.

14. In packed towers, circumferential temperature surveys should be conducted along at least two, preferably three, elevations in the investigated bed. This procedure, introduced by Duarte (89), provides excellent consistency checks of the measurements. For each radial point, data collected at one elevation need to be consistent with those at the other elevations. This procedure also yields the trends in the propagation of maldistribution, which give invaluable insights into the cause of maldistribution. Application of this procedure is demonstrated in the temperature surveys in Sections 7.1.2 and 7.1.3.

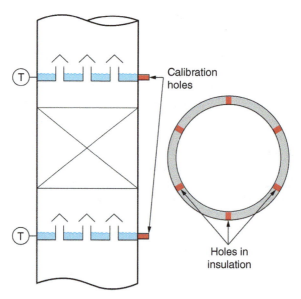

Figure 7.1   Circumferential temperature survey for packed towers.

15. To perform an adequate packed tower temperature survey (per steps 13 and 14 above), scaffolding is usually required. Scaffolding is expensive and requires special safety precautions, yet in the author's experience, necessary for obtaining reliable packing distribution data. It has been the author's experience that shortcut procedures such as measuring fewer points or using a crane to access inaccessible spots lead to misleading data. It is not recommended to proceed with a packed tower temperature survey without adequate scaffolding and access.

   A preliminary test that may minimize the scaffolding requirements is to perform a radial temperature survey above each bed, or at least above some of the beds, or at elevations partially accessible from platforms. This survey can show which beds should be singled out for more detailed testing. At times, this survey will suffice to establish maldistribution, and eliminate the need for further tests and scaffoldings. One such survey above each of the three fractionation beds was reported for a refinery fluid catalytic cracker (FCC) main fractionator (134), with high to low temperature differences ranging from 48°F to 85°F. These high numbers established the existence of major maldistribution issues. In another FCC main fractionator (129), a temperature survey of one side of the tower was conducted near the top of a fractionation bed, and that of the opposite side near the bottom of the bed. Both showed high to low temperature differences of 100°F or more, which clearly established the existence of major distribution issues, eliminating the need for more detailed measurements.

16. With IR pyrometers, it is important to keep in mind that surface temperature is not uniform. Point-to-point temperature variations of 20–30°F, or even more, in the same insulation hole are common due to variations in surface texture and emissivity. We have had an excellent experience with taking a multitude of measurements in each insulation hole and picking the highest consistent temperature measurement as the true wall temperature. For instance, if all measurements fall between 150°F and 200°F, and there are at least a few near 200°F, we will pick 200°F. If we have one reading of 220°F,

but this occurred only once, and the next lower reading was 200°F, we will ignore this odd reading.

17. With IR pyrometer surveys, it is important that each point reading captures as small an area as possible rather than spreading over a wide area containing higher and lower temperatures. For this reason, the measurements should be performed with a pyrometer that has a focusing mechanism such as laser guidance or camera focusing.

    Even though the pyrometer laser spot projects on one point in the hole, the pyrometer averages the wall temperatures over a circle centered at that point. The pyrometer specs list the diameter of the circle. Circles 1.5 in. in diameter provide unreliable tower wall temperature measurements. Standard pyrometers draw circles that are ¾″ in diameter and are suitable. High-accuracy pyrometers draw circles as small as ¼″ in diameter and are recommended. The circle diameter varies with the distance between the pyrometer and the tower wall. Generally, the pyrometer is held 6–12 in. from the tower wall (follow the pyrometer specs for the best distance) to achieve the smallest circle. Some pyrometers actually show a light ring around the focal point to define the circle drawn on the tower wall.

    Do not hold the pyrometer closer to the wall than needed. The author has seen loss of measurement reliability when the pyrometer was heated up by the wall.

18. With IR pyrometer surveys, shiny surfaces have high reflectivity. This reduces surface measurement values, generating a high and erratic difference (which can exceed 100°F) between inside and surface temperature measurements. When high differences occur, spray-paint the shiny insulation holes with dull (non-shiny) black paint before the survey. If the wall is fabricated of stainless steel, make sure the paint is low in chlorides as chlorides cause stress corrosion to stainless steel.

    Some of the author's colleagues report success with sticking dull black low-chloride tape. The author has no experience with the black tape technique.

19. Beware of changes in tower wall metallurgy. If the metallurgy changes, each section of the tower needs to be calibrated separately, since emissivity changes with metallurgy.

20. The limited experience available with temperature surveys on plastic-walled towers has not been good. Wall temperatures in such towers tend to be much lower than inside fluid temperatures. For instance, when the tower temperature is 200°F, the wall temperature is typically 100–150°F due to the good insulation qualities of plastic. The author expects a similar argument to extend to glass columns.

7.1.2 Diagnosing Internals Damage

This is the same tower discussed in Section 5.5.14 (89). As part of the troubleshooting endeavor, surface temperatures were measured along the wash bed (Figure 7.2a) circumference at three elevations. The three elevations selected were just under the wash reflux spray distributor, in the structured packing, and in the grid. The temperature profile (Figure 7.2b) showed severe liquid maldistribution, with a strong liquid bias (lower temperatures) on the northeast and a strong vapor bias (higher temperatures) on the south, west, and east. At the grid and structured packing elevations, the temperature differences exceeded 100°F. This cold spot on the northeast persisted at all three elevations, confirming the measurements' validity. There was a temperature reversal on the northeast, with the structured packing temperature 20–30°F colder than the temperature at the distributor elevation above. On the southwest, northwest, and southeast, a structured packing temperature approaching the grid temperature indicated the drying up of the wall region below the structured packing measurement point.

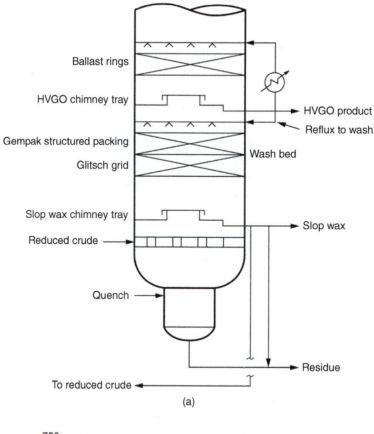

(a)

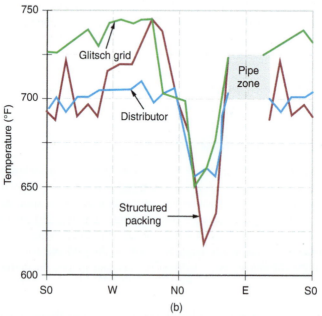

(b)

Figure 7.2   (a) Refinery vacuum tower that experienced wash bed packing damage. (b) Temperature survey in the damaged wash bed, showing a clear cold spot on the northeast and a hot zone on the south, west, and east. (From Duarte Pinto, R., M. Perez, and H. Z. Kister, Hydrocarbon Processing, April 2003. Reprinted Courtesy of Hydrocarbon Processing.)

The described pattern can be indicative of either damage or severe maldistribution, or both. In this case, gamma scans conclusively showed damage (Section 5.5.14). Also, the column history described in Section 5.5.14 pointed to damage.

7.1.3   Diagnosing Packing Maldistribution

The upper part of an atmospheric crude tower (Figure 7.3) separated a full-range naphtha (FRN) top product from a heavy naphtha (HN) side product drawn from the HN chimney tray below (89). A small side stripper D-124 stripped lights from the HN and returned them to the fractionator. The lower of the two packed beds in Figure 7.3 fractionated the FRN from the HN. The upper top pumparound (TPA) bed drew a hot liquid stream from the chimney tray underneath, cooled it, and returned the cooled stream to the top of the bed to perform direct contact condensation, providing reflux to the fractionation bed below. This reflux was supplemented by additional reflux condensed in the overhead condenser.

In 1993, the tower was modified to lower the FRN end point, i.e., to move the heavy components in the FRN into the HN product. This lowered the crude tower top temperature by about 30°F and raised the HN product flow rate. Hydraulically, these adjustments loaded up the beds and raised the TPA heat duty. To handle the higher loadings, the trays in the lower fractionation bed were replaced by 6 ft of 250Y structured packings fabricated from 0.1-mm-thick 410 stainless steel, and the trays in the upper TPA bed were replaced by 4 ft of 125Y structured packing fabricated from 0.1-mm-thick monel.

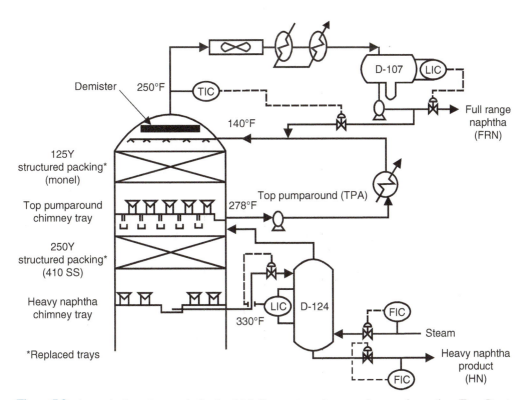

Figure 7.3   Atmospheric crude tower in Section 7.1.3. Temperatures shown are for normal operation. (From Duarte Pinto, R., M. Perez, and H. Z. Kister, Hydrocarbon Processing, April 2003. Reprinted Courtesy of Hydrocarbon Processing.

Following restart, the TPA monel packing experienced severe corrosion due to an NH_3 injection problem. The heat removed at the TPA, and the TPA draw temperature declined. In November 1994, all the TPA packing was replaced by in-kind. Amine injection into the reflux was started to prevent the recurrence of the corrosion.

Beginning in June 1995, the TPA draw temperature again gradually declined from 278°F to 229°F. The heat removed in the top bed was down by 23%. To keep the FRN on-spec (end point at 295°F), the reflux needed to be raised from 12,000 to 15,000 barrels per stream day (BPSD). The following possible causes were considered:

- Packing damage due to corrosion (like in 1994)
- Liquid spray distributor plugging or damage
- TPA chimney tray damage or overflow.

To diagnose the cause of the heat transfer loss, a sprays pressure drop test, a temperature survey, and gamma scans were performed.

Sprays Pressure Drop Test   This test, described in detail in Section 4.15, should always be the very first test when dealing with a spray distributor.

The measured pressure drop of the TPA sprays at the operating flow rate was 17 psi, close to the 20 psi calculated from the spray vendor catalog. This ruled out either plugging or distributor damage as the sole root cause. For the sprays pressure drop to be close to the design, as was measured, either the sprays were functioning properly or both plugging and damage occurred simultaneously. So the second possible cause above has now become "simultaneous plugging and damage of the spray distributor."

Temperature Survey   Surface temperatures measured around the tower circumference near the top, middle, and bottom of the TPA bed per step 14 in Section 7.1.1 are shown in Figure 7.4. Mid-bed temperatures were intermediate between the top and bottom at any vertical line (Figure 7.4a), verifying excellent consistency. Figure 7.4b is a more illustrative representation of the temperature data. For clarity, only measurements for the top and bottom elevations are shown in Figure 7.4b.

Figure 7.4 shows severe maldistribution with a strong liquid bias on the north and northwest and a strong vapor bias on the southeast. On the north and northwest, liquid appears to descend with little heat pickup from vapor. On the southeast, vapor appears to ascend without any cooling.

This test ruled out heat transfer loss due to TPA chimney tray damage or overflow. A chimney tray issue would have entrained liquid into the TPA bed or maldistributed vapor to it, but the bed would have still cooled vapor on the southeast and warmed the liquid on the north and northwest.

Of the remaining causes, this test argued against corrosion. Corrosion can explain the cold region on the north and northwest; the corroded packing may have opened a gap through which the liquid fell with little contact with the vapor. But on the southeast, the rising vapor would have cooled down at least some by contacting the descending spray liquid. The observation that the vapor on this side rose without any cooling argues against the corrosion theory.

This leaves the second cause as the front-runner, with plugged sprays on the southeast and header or spray damage on the northwest. As the evidence was not conclusive enough to rule out corrosion, it was decided to gamma-scan the bed.

Gamma Scans   The temperature survey guided setting the gamma scan chords. With the vapor and liquid biases peaking on the southeast and northwest, a grid scan with southeast and northwest chords (Figure 7.4b) is most informative. Further, to test the broken spray theory, the space between the spray distributor and the top of the bed (about 2 ft) was scanned at 1″ intervals (instead of 2″) per guideline 15 in Section 5.3.

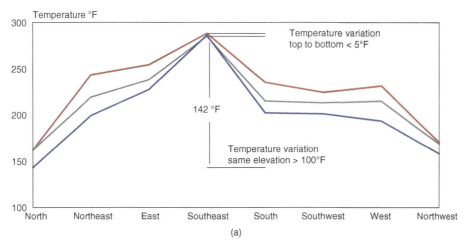

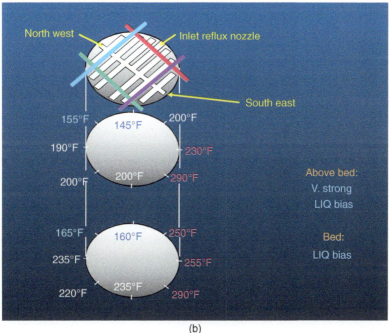

Figure 7.4   Temperature survey in the TPA bed (a) showing the bed top, middle, and bottom measurements. The consistency between the three elevations validates the observations. (b) An alternative, more illustrative representation. The middle measurements omitted for clarity. Also shown are the gamma scan chords. (Part (a): From Duarte Pinto, R., M. Perez, and H. Z. Kister, Hydrocarbon Processing, April 2003. Reprinted Courtesy of Hydrocarbon Processing.)

The scan (Figure 7.5) showed the packing in place with no apparent large-scale damage. In the vapor space between the spray distributor and the top of the bed, the southeast and southwest chords showed very little liquid. The northeast chord showed a strong liquid bias throughout the packed bed and, most importantly, in the vapor space between the spray distributor and the packing. Along this chord, as well as along the northwest chord, some of the spray appeared to be entrained above the spray distributor. The northwest chord showed a very strong liquid bias just above the bed.

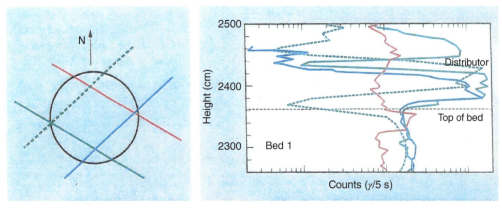

Figure 7.5   TPA bed gamma scans, showing liquid bias on the northeast. In the space between the distributor and the top of the bed, there is a strong liquid bias on the northwest and northeast and a strong vapor bias on the southwest and southeast. (From Duarte Pinto, R., M. Perez, and H. Z. Kister, Hydrocarbon Processing, April 2003. Reprinted Courtesy of Hydrocarbon Processing.)

These observations verified the temperature survey findings. The bed integrity visible in the scans, together with the liquid and vapor biases in the vertical space between the spray distributor and packing, ruled out packing corrosion. This left only one plausible root cause: spray distributor damage on the north–northwest, with spray plugging on the southeast.

Inspection   Based on this diagnosis, no replacement packing was ordered for the TPA. When the tower was opened up, the nozzles on the southeast were found heavily plugged, while the spray header was found broken on the north–northeast, as predicted by the temperature survey and gamma scans.

Takeaways   A thorough temperature survey, supported by judicious gamma scans and plant tests, led to a correct diagnosis. Prior to the temperature survey, the recurrence of the corrosion problem was the leading theory, the solution being 2–3 million US$ (1995 values) worth of packings replacement. The temperature survey and gamma scans disproved the corrosion theory, saving the refinery the cost of new packings.

For maximum effectiveness, the temperature survey should be applied at three different elevations: near the top of the bed, near the bottom of the bed, and near the middle of the bed, per step 14 in Section 7.1.1. This procedure tracks the variation of the maldistribution pattern along the bed height and provides a superb consistency check for the measured data.

7.1.4   Time Study for Diagnosing the Nature of Instability

An atmospheric crude tower (Figure 7.6) was revamped for clean fuel production and to process lighter crudes (237). In anticipation of additional naphtha (top product) production, the heat removal capacity at the TPA and mid-PA (MPA) was increased, and trays 21 and 22 in the MPA were replaced by high-capacity trays. Details of the changes and the reasoning behind them are in Ref. 237.

Following the revamp, the TPA could not reach its target heat duty. Attempts to raise the TPA duty led to temperature and pressure drop swings over four to five minute intervals. The stove oil draw temperature to the side stripper swung 15–25°F peak to valley, and the MPA draw

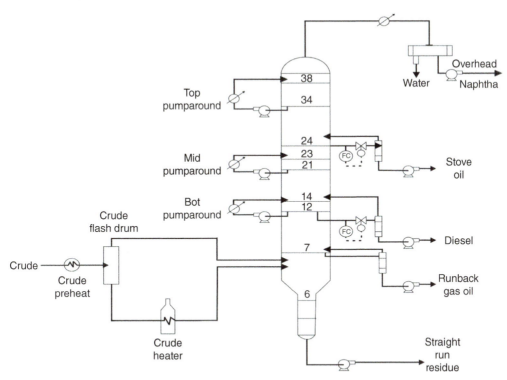

Figure 7.6 The refinery crude tower experiencing instability. (From Kister, H. Z., D. W. Hanson, and T. Morrison, Oil & Gas Journal, February 18, 2002. Reprinted Courtesy of The Oil & Gas Journal.)

temperature 10–20°F, with smaller swings in other temperatures above the MPA. The pressure drop between trays 14 and 24 swung 0.1–0.15 psi, out of a total of 1–1.2 psi. These swings were never observed previously and were stopped by limiting the TPA heat duty at the expense of constrained naphtha recovery.

Simulated stove oil temperature was about 15–20°F hotter than measured. This was explained by mixing the liquid from the trays above with some cold MPA return, which could enter the stove oil draw sump. To gain insight, wall temperatures were measured by an IR surface pyrometer (Figure 7.7) and compared with the control room thermocouples (all temperatures prefixed by "CR," meaning control room, in Figure 7.7). Where both were available, the agreement between most was excellent.

The wall temperature profile (Figure 7.7) looked normal, with temperatures continually rising down the tower, with one exception: the stove oil draw temperature was 20–30°F colder than tray 24 above and 10–20°F colder than tray 26, three trays above. Tray 23 liquid and vapor wall temperatures were measured, and appeared normal and consistent with those on trays 22 and 24. The only conceivable cause for the colder stove oil draw temperature was the mixing of hot liquid from tray 24 (425°F) with cold MPA return liquid (325°F). This was in line with the simulation findings.

A heat balance or simple mixing calculation determined the fraction of cold MPA return mixed with the tray liquid. To give a draw temperature of 400°F, 75% of the draw liquid needs to come from tray 24 above (at 425°F), 25% MPA return (325°F).

A wall temperature time study (Figure 7.8) provided insight into the nature of the swings. Upon increasing the TPA heat duty, tray 23 vapor temperature and tray 24 downcomer liquid temperature cycled with the stove oil draw temperature, peaking when the draw temperature valleyed

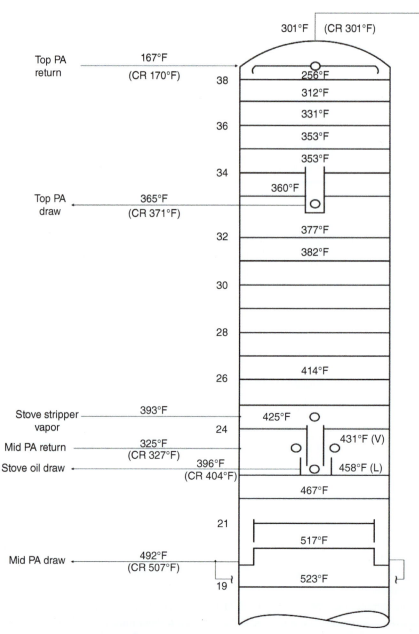

Figure 7.7 Steady-state temperature survey of the refinery crude tower experiencing instability. (From Kister, H. Z., D. W. Hanson, and T. Morrison, Oil & Gas Journal, February 18, 2002. Reprinted Courtesy of The Oil & Gas Journal.)

and vice versa. The difference between the tray 24 downcomer temperature and the draw temperature changed from 10°F to 30°F during a cycle.

Application of a heat balance or simple mixing calculation to the hot tray 24 downcomer liquid and the cold TPA return temperature led to an important finding: during a cycle, stove oil draw varied from 30% MPA and 70% downcomer liquid when the stove oil was 30°F colder than the downcomer liquid, to 10% MPA and 90% downcomer liquid (when the difference between

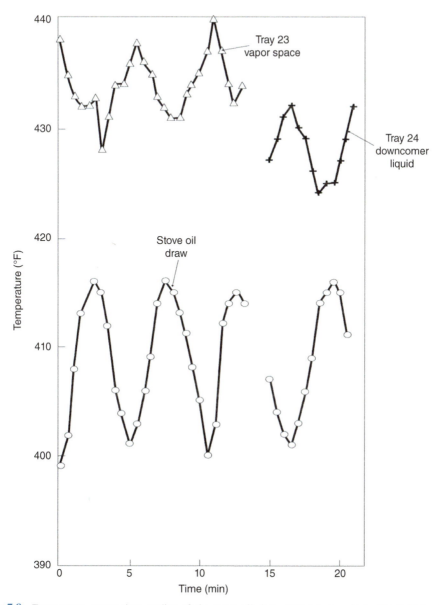

Figure 7.8   Temperature survey time studies of the stove oil draw temperature swings. (From Kister, H. Z., D. W. Hanson, and T. Morrison, Oil & Gas Journal, February 18, 2002. Reprinted Courtesy of The Oil & Gas Journal.)

the temperatures was 10°F). The temperature cycles therefore reflected changes in the fraction of MPA return liquid in the stove oil draw.

This explained the temperature swings on tray 23 and in the rest of the tower. When more MPA return was drawn at the stove oil draw, the temperatures above warmed up. When less MPA was drawn, the temperatures above cooled down.

A key observation was that at lower TPA duties, the stove oil draw was steady at 400°F, indicating 25% MPA return and 75% tray 24 downcomer liquid in the draw. When TPA duty increased beyond a certain value, the stove oil draw temperature would slowly rise. An increase in draw

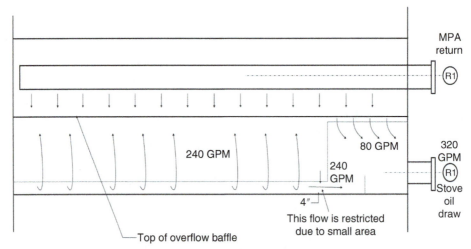

Figure 7.9 Root cause of the instability. (From Kister, H. Z., D. W. Hanson, and T. Morrison, Oil & Gas Journal, February 18, 2002. Reprinted Courtesy of The Oil & Gas Journal.)

temperature would indicate imminent swings. Swings would initiate when the draw temperature reached about 423–425°F (as measured by the control room thermocouple). As the downcomer liquid from tray 24 was 425°F, this suggested that at the onset of the swings, the stove oil draw mostly contained downcomer 24 liquid and little MPA return.

Figure 7.9 shows the stove oil draw sump, with a detailed description elsewhere (237). The center downcomer bringing liquid from tray 24 to the sump extended the full tower diameter for the first 22 in. of elevation, then the downcomer was stepped down, so its first 30 in length near the tower wall on the draw nozzle side was closed off with a solid side wall. The 18 in. long stepped-down downcomer terminated 4 in. above the sump floor.

Liquid exiting the downcomer flowed under the downcomer walls. The path toward the nozzle was restricted both by the relatively small opening in the draw direction and by a 5 in. weir (Figure 7.9). Due to the restriction, some downcomer liquid turned upward, ascended, and overflowed the sump walls (the upward arrows in Figure 7.9). It was replaced by the MPA return liquid that entered the sump in the zone between the downcomer end wall and the tower shell (the downward arrows curving to the right in Figure 7.9). Flows shown in Figure 7.9 are based on 75% of the draw being tray 24 downcomer liquid and 25% MPA return liquid.

The entry of the MPA return liquid into the sump kept the draw liquid subcooled at 400°F. This persisted over a wide range of TPA duties. Beyond a certain critical TPA duty, a small rise in the TPA duty caused the stove oil draw to rise 25°F and lose subcooling.

Subcooling is lost very gradually with sensible heat changes. A sudden loss of subcooling indicates the entry of latent heat. The only conceivable source of latent heat in the sump is vapor breakthrough from the downcomer. Such breakthrough may be caused by excessive liquid velocities (302), which in turn are caused by the higher TPA duty. This is why the first step in the cycle was always a rise in the draw temperature. Once the subcooling was gone, this breaking vapor would aerate the sump liquid. The aerated liquid between the sump wall and downcomer walls would have less density and would preferentially rise rather than flow to the nozzle. Less hot liquid would be available for the draw. It would be replaced at the draw by the cold MPA return liquid. The subcooling at the draw will return. The high-density unaerated subcooled liquid would back some liquid up the downcomer, reducing its velocity and interrupting the vapor

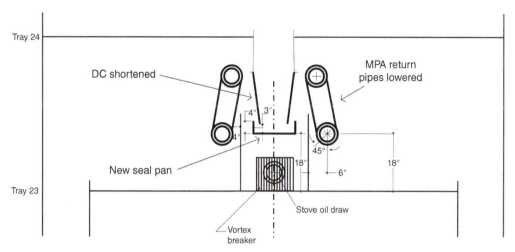

Figure 7.10   Crude tower stove oil draw modifications (boldface indicates modification). (From Kister, H. Z., D. W. Hanson, and T. Morrison, Oil & Gas Journal, February 18, 2002. Reprinted Courtesy of The Oil & Gas Journal.)

breakthrough. After a while, more liquid would accumulate in the downcomer, the velocity would increase, and the vapor breakthrough would return.

Based on this diagnosis, the downcomer from tray 23 was shortened, terminating in an elevated seal pan and the MPA return pipes were lowered so they discharge outside the stove oil draw sump (Figure 7.10). This low-cost modification eliminated the swings and permitted the desired TPA duties to be achieved.

Takeaway   Temperature survey time studies can give invaluable insights into tower instabilities, leading to correct diagnoses and inexpensive solutions to challenging problems.

7.1.5   Diagnosing an Unexpected Second Liquid Phase

Detailed temperature profiles are key to diagnosing the presence or absence of a second liquid phase in a tower. A second liquid phase is recognized by a temperature drop of the order of 10–20°C. Gans et al. stated (117), "If a distillation column temperature profile is completely off design, water is probably leaking in." The author believes they were referring to columns processing organics that are water-insoluble or partially water-soluble, as in the following cases.

Case 1 An azeotropic distillation column used a cyclohexane entrainer to remove water from an organic (351). The organic exited in the bottoms, and water and cyclohexane were the tower overhead. After condensation, cyclohexane was decanted from water and returned to the tower. After performing well for 15 years, the tower was replaced by a larger tower. Since then, instability, reduced capacity, organic and cyclohexane losses, and inability to control the temperature were experienced.

A simulation using a VLLE model produced the temperature profiles in Figure 7.11a. The diagram shows that as the water content of the reflux increased by as little as from 3.5% to 6%, the temperature profile in the middle of the tower flattened. This flattening was due to the presence of two liquid phases in the tower, reaching well into the middle and bottom sections. The upper curve matched past operation, when the tower operated well. The bottom two curves matched the temperature profiles of the revamped tower. The change in the temperature profile identified the cause of the problem to be excessive water in the cyclohexane entrainer. This was confirmed

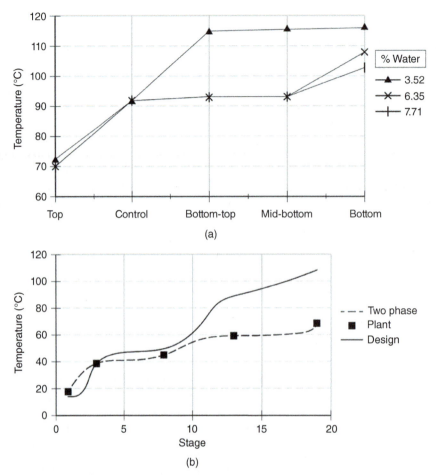

Figure 7.11   Temperature profiles showing two liquid phases in the tower. (a) Case 1, temperature profile detects excess water (heavy phase) in the reflux. (b) Case 2, temperature profile indicating excess acid (light phase) in the reflux. (From Opong, S., and D. R. Short, "Troubleshooting Columns Using Steady State Models," in "Distillation: Horizons for the New Millennium," Topical Conference Preprints, p. 129, AIChE Spring National Meeting, Houston, TX, March 14–18, 1999. Copyright DuPont 1999. Used under permission.

by sampling the cyclohexane. An undersized line from the condenser to the decanter caused a buildup in the condenser all the way to the midpoint vent, from where it drained directly to the cyclohexane side of the decanter.

Case 2 A packed tower recovered acids as a top product from a mixture of organics (351). The tower had an internal condenser. A collector under the condenser separated the light acid-rich product from the heavy organic-rich phase, refluxing only the heavy phase back to the tower. The tower operated normally for a few days after commissioning, but then the acid recovery became very poor, the temperature profile flattened, and the pressure drop rose.

A simulation using a VLLE model produced the solid line temperature profile in Figure 7.11b. The solid line matched the measured temperature profile in the initial few days, when the tower operated well. The data points were the temperature profile when the tower became abnormal, and the dotted line was generated by assuming that some of the light acid-rich phase was refluxed into the tower. The agreement between the two confirmed a malfunction of the collector that led to the refluxing of some light acid-rich phase into the tower.

Case 3 A refinery reformer debutanizer separated C_3 and C_4 hydrocarbons as the top product from gasoline (C_5–C_9) hydrocarbons bottom product. The tower had 30 trays, 15 above the feed and 15 below. In the top section, temperature indicator readings in the control room were present just above the feed and at the tower overhead.

The tower performed normally at moderate rates. However, when the rates were pushed up, the temperatures both above the feed and at the tower top jumped by about 10–15°C. This jump was accompanied by a large rise of gasoline (C_5+) components, as much as 15%, in the top product, making the C_4 product off-spec. There were no signs of flooding.

While just looking at the two TIs showed nothing unusual, a wall temperature survey (Figure 7.12) revealed an unexpected behavior. For on-spec operation, temperatures gradually rose from the top to the feed tray, as they should, but the off-spec operation showed a temperature inversion. Here, the temperature gradually rose from 70°C at tray 30 to about 80°C at tray 22 below and then reversed the direction, going down to about 70°C on tray 18. Below tray 18, the temperature rose again going down the column.

This behavior has been observed in chemical separations, where it often signifies the formation of a second liquid phase as described in Cases 1 and 2 and Figure 7.11. In hydrocarbon towers like debutanizers, it is far less common. Small amounts of water often enter in the feed, but for as long as the water content of feed is small, it will vaporize and end up in the overhead system, where it will be condensed and removed via a water removal boot. In this case, the water content in the feed was indeed small.

Usually, the boot is an integral part of the reflux drum (192, 286, 288), which allows free movement of water into the boot and hydrocarbons back to the reflux drum. In this tower, the boot was a separate vessel, connected to the drum via a 1–1/2″ valve (Figure 7.13). This line was far too restrictive, restricting the water descent and causing poor water removal from the drum. Water was refluxed back into the tower, accumulating in the tower and causing the formation of the second liquid phase. The presence of nonvolatile water-soluble components like salts or caustic could have induced the water pocket to move down to just above the feed.

Other Cases Reference 175 presents a case of ethanol–water distillation, with intermediate alcohols (fusel oil) drawn as a side product. Initially, the fusel oil could not be drawn. Experimenting with various steam rates in both the extractive distillation and the rectifying towers

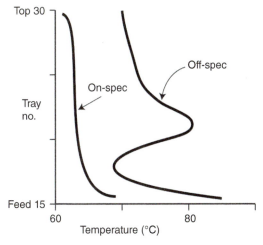

Figure 7.12 Temperature survey of the top 15 trays above the feed showing a temperature inversion during the upset conditions.

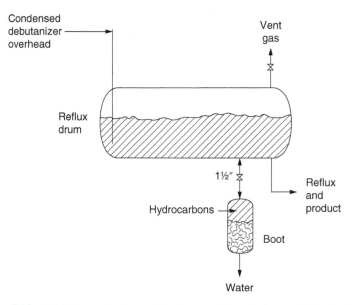

Figure 7.13 Debutanizer reflux drum with water removal boot in the case 3 in Section 7.1.5.

and comparing the measured with simulated temperature profiles provided the key to diagnosing the problem and solving it. In this case, a temperature inversion was expected in the extractive distillation tower, and reducing the boilup to establish it was a milestone toward the solution.

Takeaway Temperature surveys can reveal unexpected behavior and diagnose problems.

7.1.6 Diagnosing Poor Mixing in a Packing Distributor

The top bed of a refinery FCC main fractionator (Figure 7.14) fractionated gasoline from light catalytic cycle oil (LCCO) (134). Liquid leaving the bed at 420°F was collected on a chimney tray, some of it drawn as the LCCO product and the rest descended to an orifice pan distributor via a center downcomer. Also entering this distributor was a cooled intermediate PA (IPA) return stream at 135°F. The orifice pan distributor distributed the liquid to the IPA packed bed below. Liquid from the bottom of the IPA bed was collected by a second chimney tray. Some of this liquid was cooled to 135°F, becoming the IPA stream returning to the orifice pan distributor, and the rest of it was used as reflux to the bed below (not shown).

The gasoline–LCCO bed experienced poor fractionation, with a HETP of more than 11 ft where 3 ft was expected.

As part of the troubleshooting, a radial temperature survey above the IPA bed was conducted, which showed high to low temperature variations of 85°F. While most temperatures were around the 450°F mark, one temperature was as high as 512°F. It was realized that very little cooling was done in at least one location.

A close examination of the feeds to the orifice distributor (Figure 7.14) showed that the hot center downcomer reflux from the chimney tray was fed into the middle of the orifice pan distributor, while the cold IPA return stream was fed by two off-center pipes. The hot reflux did not adequately mix with the cold IPA return. A liquid with a nonuniform composition and temperature was therefore distributed across the IPA bed. Vapor leaving the central zone was

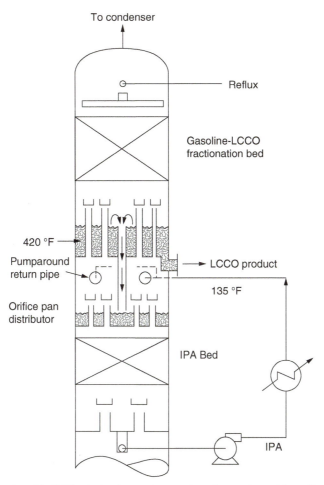

Figure 7.14 Top section of the FCC main fractionator that experienced separation problems. (Based on Golden, S. W., D. W. Hanson, J. Hansen, and M. Brown, Oil & Gas Journal, March 15, 2004.)

hotter and contained a lot more heavies than the vapor rising in the regions to which the cold PA liquid was distributed. This resulted in poor fractionation in the bed above. Additional problems with this orifice pan distributor were described in the original Ref. 134.

7.1.7 Diagnosing Flashing in a Packing Distributor

This is the same tower that was described in Case 6.4.1. The temperature survey was conducted before the computer-aided tomography (CAT) scan (240).

A chemical tower experienced two operation modes: "normal" and "upset." In the upset mode, the measured temperature near the top of the upper bed became 40°C higher than normal, and a "reversal" occurred: this temperature became hotter than temperatures further down the bed. The heating up of that temperature was accompanied by a large rise in heavies in the overhead product. A shift to the "upset" mode was triggered mostly by raising the set point on the tower temperature controller, but also by excessively raising feed rates.

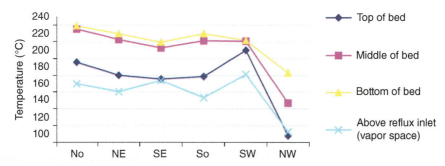

Figure 7.15   Temperature survey of the top bed of the chemical tower that was experiencing upsets. (From Kister, H. Z., W. J. Stupin, J. E. Oude Lenferink, and S. W. Stupin, Transactions of the Institution of Chemical Engineers, 85, Part A, p. 136, January 2007. Reprinted Courtesy of IChemE.)

Per step 14 in Section 7.1.1, a circumferential temperature survey was conducted at three elevations in the 4.8 m tall upper bed. Holes were cut in the tower insulation 0.5, 2.4, and 4.2 m below the top of the bed (referred to as the top, middle, and bottom rings, respectively). Each elevation had six holes around the circumference (at 60° angles). In the "upset" mode, the following observations were made (Figure 7.15):

1. The consistency between the measurements at the various elevations confirmed the validity of the survey data.

2. There was a cold spot on the northwest that persisted throughout the bed. Temperatures on the northwest near the top of the bed were much closer to the reflux temperature of 87°C than to the tower overhead temperature of 195°C.

3. The temperature of the vapor overhead was 195°C. Most of the wall temperatures at the top of the bed were about 180°C or less, significantly lower than the overhead vapor temperature.

4. There was a thermocouple located 1.2 m below the top of the bed, sticking 300 mm into the bed. It measured 221°C, significantly higher than the wall temperatures of the top and even the middle rings.

To investigate why the tower overhead temperature exceeded the temperatures at the top of the bed, a temperature survey ring was added in the vapor space above the reflux inlet, at 0.100 m below the top tangent line. The reflux was the only source of cold temperatures in the system, so above its inlet point, the temperatures should be the same as the tower overhead temperature.

The "above reflux inlet" curve in Figure 7.15 shows that the cold spot on the northwest persisted in the vapor space above the reflux inlet. Other temperatures in this ring ranged from 140°C to 180°C, well below the overhead temperature of 195°C. Values obtained in this vapor ring were very consistent with the values obtained in the ring at the top of the bed.

Interpretation of the Temperature Survey   Temperature differences exceeding 10°C at the same radial elevation usually give rise to concerns about maldistribution (differences not exceeding 5°C are good). In the upper bed, the differences exceeded 100°C at the same elevation.

In the upset mode, wall temperatures near the top of the upper bed were lower than 300 mm in from the wall (as measured by the thermocouple). This suggests preferential liquid flow at the wall and vapor in the center. There was a cold spot along the northwest that persisted along the entire upper bed.

Besides the maldistributed profile, the temperature survey provided a major clue in the investigation. It showed cold, maldistributed temperatures near the top tangent line, mostly 15–35°C colder, but as much as 80°C colder, than the overhead vapor temperature. This is very unusual and suggests that the cold reflux is likely to have been sprayed upward – with vengeance. This spraying led to the diagnosis of flashing of the reflux. Prior to the temperature survey, there was no basis to suspect flashing of the reflux that was pumped from the reflux drum.

The CAT scan in Section 6.4.1 later confirmed the suspected flow pattern. A subsequent investigation based on the findings of the temperature survey and CAT scans (240) showed that reflux flashing upon tower entry, and boiling in the reflux distributor, generated the bad liquid distribution observed and the upset condition. The flashing was promoted by a 90°C gap between the boiling points of the two major components in the reflux. The flashing was induced by vigorous steam tracing of the reflux line.

7.1.8 Identifying Various Boiling Regions in a Kettle Reboiler

This is the same kettle that was described in Cases 6.3.3 and 6.5.3 (217). The temperature survey preceded the neutron backscatter and tracer tests. The oil-heated depropanizer kettle reboiler bottlenecked the capacity of a gas plant. Tower flooding, initiating at the kettle reboiler, occurred when feed rates were raised. A pressure balance showed that to make the pressure drop in the kettle circuit high enough to raise the bottom sump level above the reboiler return nozzle (and initiate flood) would take significant entrainment from the kettle. A multitude of tests, including reboiler temperature surveys, were conducted to evaluate the likelihood of such entrainment.

Surface temperatures were measured by a laser-guided IR pyrometer, calibrated with measurements of the depropanizer bottoms and the kettle liquid draw temperatures. Figure 7.16 shows the pyrometer measurements, at locations roughly drawn to scale, at rates just below those at which the tower flooded. Temperatures were measured at the top, bottom, east, and west sides of the kettle. For clarity, only the east side temperatures are shown; those measured on the west followed the same trend. The temperatures of the streams entering and leaving the kettle, as measured by thermocouples, are shown encircled in Figure 7.16. Non-encircled temperatures were pyrometer-measured. It is important to recognize that the surface temperatures were measured at the wall, at a varying distance from the tube bundle.

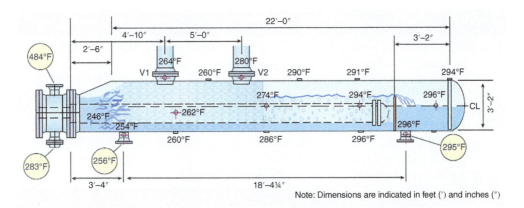

Figure 7.16 Temperature survey of a kettle reboiler shows that the inlet region was thermosiphoning, while the outlet region operated like a kettle. (From Kister, H. Z., and M. A. Chaves, Chemical Engineering, p. 26, February 2010. Reprinted Courtesy of Chemical Engineering.)

Figure 7.16 distinguishes two regions along the kettle length. In the inlet region, which stretched from the channelhead to just to the left of vapor outlet V2, the temperatures were only slightly higher (by no more than 4–6°F) than the kettle inlet temperatures. In the second region, which stretched from V2 to the liquid draw, the kettle temperatures began rising right after V2, finally reaching the liquid draw temperature.

Temperatures measured in the disengagement drum above (drum shown in Figure 6.21) were slightly (about 4°F) higher than the kettle liquid inlet temperature. The temperature in V1 was significantly colder than in V2.

In the kettle, the process inlet region was the vigorous boiling zone. Here, the fraction of the liquid vaporized was low, so less lights were removed from the liquid. This lights-rich liquid persisted all the way to the vapor nozzles. The lower temperatures observed in this region reflected the low vaporized fraction. Vapor line V1 was coldest, because it carried a vapor–liquid mixture from the region of intense boiling. The action in this region had similarities to the action in horizontal thermosiphon reboilers in that the quantity of vapor generated resulted in a low-density froth which promoted liquid flow to this region as opposed to the denser liquid in the less aerated zones.

There was a net flow of liquid from the reboiler inlet region toward the overflow baffle since this was the only manner (other than entrainment) in which the unboiled liquid could permanently exit the kettle reboiler. As one moves closer to the overflow baffle, the liquid contained progressively less lights. This resulted in less vigorous boiling and a denser froth in this region. This was actually observed in the neutron backscatter (Figure 6.21). This was the one region that actually operated like a kettle. In this region, there was not much entrainment, as verified by the neutron backscatter, and there was a lot of disengagement space, both of which promoted true kettle behavior. As the flow of the reboiler feed liquid through this region was relatively low and the lights gradually disengaged along the path from vapor line V2 to the overflow baffle, the temperatures in this region gradually increased as observed in the temperature surveys (Figure 7.16).

In this case, the temperature survey provided insight into the entrainment mechanism.

7.1.9   Identifying Plugging Zone in a Ladder Pipe Distributor

A coker scrubber tower separated heavy residue tower bottoms from distillates that exited in the tower overhead vapor (49). The tower had a lower PA bed containing shed decks (baffle trays) and an upper refluxed wash bed containing grid packings. About 1.5 years into the run, significant residue carryover from the tower was observed, along with a steep rise of the wash bed pressure drop. Coking and fouling in this service are common and occur gradually over a three-year cycle, resulting in the entrainment of residue and high grid pressure drop. The sudden rise in pressure drop and entrainment midway through the cycle was unusual.

The PA was distributed to the shed decks by a simple H-header ladder-type slotted-pipe distributor (Figure 7.17) with laterals positioned over each top layer of shed decks. It was designed to distribute the PA liquid proportionally to the open area of the slots in the laterals, providing even distribution of the PA liquid to the shed decks below. Normally, shed decks are robust to handle a reasonable degree of liquid maldistribution, but in this tower, due to the very low PA liquid flow rates (very low weir loadings), liquid maldistribution would dry up sections of the shed decks, allowing uncontacted residue to rise through them.

Temperature surveys were at the heart of a thorough troubleshooting investigation, which is described in detail elsewhere (49). Figure 7.17 shows the two temperature survey rings taken above the distributor. The unbracketed values were measured at an elevation immediately above

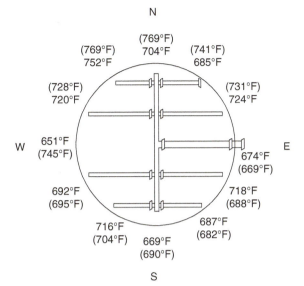

Figure 7.17   Temperature survey showing a hot spot in the northwest quadrant. (Data from Brown Burns, J., D. Hanson, C. Riley, and C. Wicklow, "Targeted Injection Improves Shed Deck Performance," The Kister Distillation Symposium, Topical Conference Proceedings, p. 119, the AIChE National Spring Meeting, San Antonio, TX, April 29–May 2, 2013.)

the distributor, while the bracketed numbers were measured about 6 ft 9 in. above the first ring. The space between the two rings was vapor space with no packings, shed decks, or other internals.

The temperature survey identified a hot spot in the northwest quadrant, suggesting possible plugging of the laterals in this region. Less liquid descended in the northwest region of the distributor, giving poor separation and allowing heavies to ascend up the tower. Reference 49 reports making good use of additional temperature surveys in coming up with an innovative fix to the problem while the tower was operating and optimizing subsequent operation.

7.1.10   Identifying Uneven Quench Distribution in a Bottom Sump

In some hot services, where material degradation occurs at high temperatures, subcooled liquid is injected into the bottom sump liquid to keep the temperature down (72). In other cases, it is injected into the vapor space to desuperheat the vapor. In either case, that injection needs to be adequately distributed. This can be checked by a temperature survey as in Figure 7.18.

The temperature survey in Figure 7.18 shows a case of uneven quench distribution in the bottom of an FCC main fractionator (72). Although the bottom temperature (which is usually monitored by a thermocouple) showed 680°F, due to the poor mixing some of the sump was at 720°F and would experience a much more rapid degradation (in this case, coking) rate that could lead to severe heat exchanger fouling.

7.1.11   Identifying Flood

Temperature surveys can be very effective for diagnosing floods. Ideally, both the normal and flooded surface temperature profiles should be obtained. This is discussed under "loss of separation" in Section 2.6 and is illustrated in Figure 2.14.

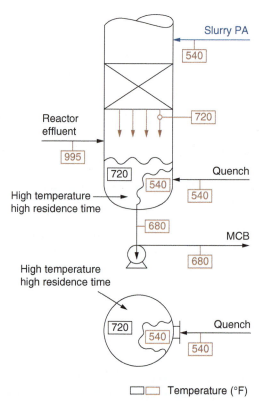

☐☐ Temperature (°F)

Figure 7.18   Temperature survey showing poor mixing of quench liquid in the bottom sump of an FCC main fractionator. (From Clark, D., and S. Golden, PTQ, p. 65, Summer 2003. Reprinted with permission.)

7.2   THERMAL CAMERA (THERMOGRAPHY) APPLICATIONS

While temperature surveys using pyrometers give accurate information on insulated towers and reboilers, they can be laborious and tedious. When towers, condensers, or reflux drums or their piping are not insulated or when it is easy to remove insulation strips, thermal cameras can provide troubleshooters with invaluable information. Thermal cameras are effective for observing temperature changes with time and can provide temperature profiles over large areas, rather than capturing single spots at specific moments of time. With insulated towers, thermal cameras can identify sections of damaged insulation where energy is lost.

Thermography detects and measures variations in heat emitted by objects. Thermal cameras detect radiation in the IR range of the electromagnetic spectrum. The amount of radiation emitted by an object increases with temperature. The IR energy detected is transformed into electrical signals, which in turn are converted to visual signals that form thermal images. These images display a range of colors that correspond to temperatures. The colors identify the temperature distribution in the image. Thermal cameras can be operated in photo mode or video mode, just like cell phone cameras.

7.2.1   Dos and Don'ts for Thermal Camera Imaging

Just like pyrometer surveys, successful thermal imaging requires valid and reliable measurements. It is crucial to verify the data by extensive consistency checks. Invalid or unreliable temperature measurements lead to incorrect diagnostics. In one case, a thermal camera survey through a number of cuts in the insulation in a baffled column bottom sump gave such an inconsistent pattern that it had to be abandoned and successfully replaced by pyrometer readings. The key to successful temperature thermal imaging is reliability, repeatability, and consistency. Here are some guidelines for achieving them:

1. Review the guidelines in Section 7.1.1. Many of these extend to thermal imaging, especially guidelines 1, 2, 6, 7, 8, 10, 12, 16, 19, and 20. If cutting a section of the insulation, also guidelines 3, 4, and 5.

2. Emissivity should be carefully set. Emissivity is the ratio of the amount of radiation emitted from the surface of an object to that emitted from a perfect black body at the same temperature as the observed object (7, 22, 105, 495).

   Emissivity varies with the surface of the object and also with temperature and wavelength (7, 22, 105, 306, 495). Metals, especially polished metals, pose problems due to their low emissivity, with values below 0.2 (495). Emissivity may vary considerably as a result of surface structure. While a polished metal may have a low emissivity, the same but roughened metal can have an emissivity as high as 0.8 (495). Radiance of a surface also depends on the viewing angle; an object that is observed normal to its surface will emit more radiation than when observed at an angle. Emissivity also depends on wavelength; emissivity of metals usually decreases with longer wavelengths (495).

   Several practical methods for setting emissivity are available, with emissivity tables being the most common (22, 105). These are found in thermal camera manuals and in the literature. These values had been recorded under specific measurement conditions and, if they vary from the measurement at hand, may not be accurate for that measurement.

   The influence of errors in emissivity can be minimized by calibration against a surface of the same material at a known temperature. Measure the temperature at a location near a thermocouple. Adjust the emissivity to give that temperature. For shiny surfaces of low emissivity, painting or taping is practiced (495), in a similar manner to step 18 in Section 7.1.1. However, when using thermal cameras, the area surveyed may be large, and painting or taping may not be practical.

3. For a good measurement, it is important to minimize the background noise, i.e., the radiation that does not come from the object of interest but contributes to or detracts from the radiation picked up by the camera. Some guidelines are as follows:

   i) Choose the best position for the camera. Shorten the measurement path length, and/or position it so that it is not directly in front of an object with a higher temperature than the object of interest (22). The camera manual has other guidelines for minimizing the background noise.

   ii) Warm or hot objects in the surroundings, including the person operating the camera, may lead to reflections that will be picked up by the camera. Watch out for them and avoid them.

   iii) Radiation omitted or absorbed by the air is another factor to be considered. For the short paths traveled in distillation applications, the only significant variable is relative humidity. The water in the air can affect the results. Therefore, it is important to provide the camera with good values of relative humidity at the measuring site (495). Avoid foggy conditions.

4. Other input parameters that need to be entered into the camera include the atmospheric temperature, the distance to the object, and the reflected apparent temperature. These factors usually have a small influence as long as there are no intense radiation sources in

the surroundings (105). The camera manual usually provides a procedure for setting the reflected apparent temperature value.

5. Unlike pyrometer readings that are taken in relatively small holes in the insulation, which offer protection to the scanned surface from wind, rain, and sunlight, thermal scans are conducted over areas much more exposed to elements and are more affected by them. Prefer to conduct thermal scans in dry conditions, where the sun is not strong and the wind speed is low. If this cannot be done, check that each side of the tower reads the same as the other. The author has seen cases where there were differences between the two sides. Thermal scans should be avoided altogether during strong rain or other extreme weather conditions.

6. It is beneficial to attend a training and certification class to learn how to correctly use and how to get the most out of thermal scans (22, 495). The author has seen many poor scans that looked a lot better after being redone by a trained person.

7. Liquid temperature should be uniform horizontally, either with trays or with packings. At a given vertical height, radial differences may be indicative of maldistribution or other malfunctions.

7.2.2   Diagnosing Packing Maldistribution

Two acid gas scrubbers were operating in parallel, removing all the water and CO_2 from excess ammonia (22, 23). CO_2 was fully consumed by the excess ammonia, yielding concentrated ammonia as the top product. The reaction between CO_2 and ammonia is exothermic, so the top two random-packed beds were PA beds cooled by an external cooler to remove the heat of the reaction.

One of the two scrubbers was experiencing top temperature instability, with swings of $\pm 10°C$, posing a contamination risk to the overhead product. In the past, the tower was very stable. The parallel tower remained very stable.

Several troubleshooting checks were performed, but the key lead came from a thermal scan of the towers (Figure 7.19, TPA bed). Tower 1 was the one experiencing the problem, while tower 2 was the stable one. The thermal image of tower 1 showed a $27°C$ temperature gradient from left (hot) to right. For tower 2, the difference was about $2°C$. Tower 1 profile conclusively showed liquid and/or vapor maldistribution, while tower 2 showed good distribution.

Turnaround inspection found that the liquid distributor and bed limiter were incorrectly installed in tower 1, causing channeling which promoted fouling in the bed. The fouling was so severe that individual pieces of packings conglomerated and stuck in the bed limiter above the packing. The packings were then replaced by the same type, the distributor cleaned, and the bed limiter replaced. All were inspected for proper installation. Following restart, very stable operation returned to tower 1.

7.2.3   Diagnosing Vapor or Liquid Maldistribution in Insulated Towers

Once the location of the cold or hot region is identified, a strip of insulation can often be removed and a thermal scan taken. This scan is invaluable for confirming or denying a theory and can save the elaborate scaffolding often needed for a detailed pyrometer survey (e.g., as in Section 7.1.3). Below are two examples where this technique led to the correct diagnosis and fix of operating problems.

After a long production run, separation deteriorated in a chemical packed tower. With the hot, reactive vapor feed, plugging at the bottom of the bed was the prime suspect. To investigate, a strip of insulation a few feet wide was removed from the bottom of the bed, about half the

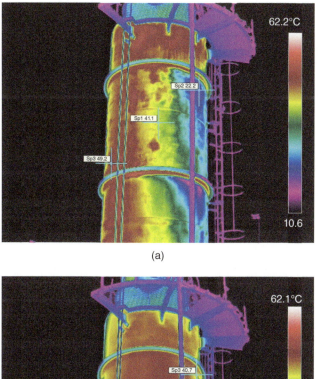

(a)

(b)

Figure 7.19   Thermal scans of two acid gas scrubbers operating in parallel. (a) Tower 1 shows a highly maldistributed profile, with a 27°C temperature gradient from left (hot) to right. (b) Tower 2 shows uniform temperature from side to side, indicating good distribution. (From Barnard, H., Chemical Engineering Progress, p. 21, January 2017. Reprinted Courtesy of Chemical Engineering Progress [CEP]. Copyright © American Institute of Chemical Engineers [AIChE].)

column perimeter, and a thermal scan was shot along the wall. The scan showed that the wall right across from the vapor inlet (Figure 7.20a) was hot, at about the same temperature as the feed. Other sections of the wall were much cooler. This verified the plugging theory, with a vapor jet ascending near the hot wall.

When the tower was opened, deposits were found at the base of the bed near the center of the tower (Figure 7.20a). The plugged packings were replaced by in-kind. Upon restart, the tower worked well. A new thermal scan showed even distribution.

A similar technique can be applied to troubleshoot liquid maldistribution. In one refinery vacuum tower, the spray header bringing cooled liquid to the top (light vacuum gas oil, LVGO)

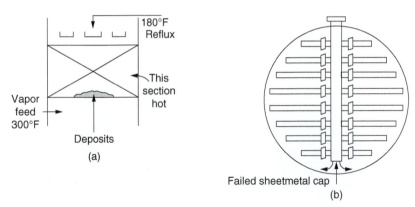

Figure 7.20 Two situations where removing a strip of insulation and shooting a thermal scan helped diagnose an operating problem. (a) Plugging near the bottom of a packed bed, generating vapor maldistribution. (b) Failure of a spray header sheet metal cap causing liquid maldistribution.

bed appeared to have failed, as could have been inferred from the sudden reduction in the sprays pressure drop (Section 4.15). This led to severe maldistribution, poor heat transfer, hot overhead temperature, and slop in the overhead. In an endeavor to implement a fix without a unit shutdown, it was important to identify the failure location. Figure 7.20b shows a multitude of laterals and flanges, each one with the potential for failure.

A temperature survey using contact thermocouples at the top platform showed much colder temperatures at the end of the main header. To learn more, the plant stripped insulation at this end of the tower and shot a thermal scan. The scan showed a cold region propagating throughout the entire bed (a few feet long). This and gamma scans identified the failure of a thin-gauge metal cap at the end of the header (Figure 7.20b) as the most likely cause. Based on this diagnosis, the refinery was able to devise a fix that kept the unit operational and performing well until the next scheduled turnaround.

7.2.4 H$_2$S Amine Absorber Temperature Profile, Foam, and Flooding

The tower shown in Figure 7.21 used a monoethanolamine (MEA) solution to absorb and remove H$_2$S from a gas stream on one-pass trays.

Figure 7.21a presents a thermal scan during normal operation. Tray liquid (yellow) and vapor space above (orange) can be clearly distinguished. This scan shows higher temperatures in the liquid than in the gas. A common design practice in ethanolamine absorbers is to keep the lean amine temperature $>10°F$ hotter than the gas temperature to prevent condensation of foam-promoting heavy hydrocarbons into the amine (192, 286, 476). A close look into Figure 7.21a shows the lean amine entering at approximately 90°F and the gas at 80°F, in accordance with this practice.

Figure 7.22a shows normal operation, but with the reverse pattern, on single-pass trays in a methyldiethanol amine (MDEA) absorber (476). Again, the tray liquid (orange) and gas (yellow) can clearly be distinguished, but here the tray liquid was cooler than the gas. In this tower, the lean amine entered at 83°F, much colder than the gas inlet that entered at $>120°F$, giving this reverse pattern. Lower liquid than gas temperatures were observed in other gas absorption towers (e.g., Figure 3 in Ref. 475). In both Figures 7.21a and 7.22a, the thermal scans show the trays to be in place with froth heights about halfway up the tray spacing.

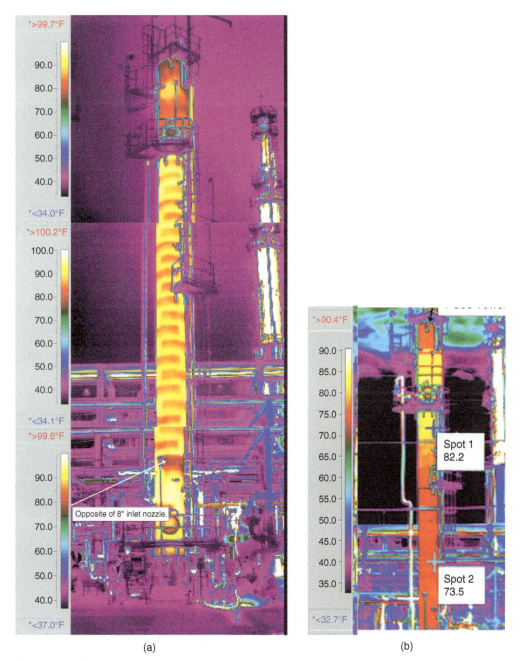

(a)   (b)

Figure 7.21   Thermal scans of an amine absorber: (a) normal operation; tray liquid (yellow) and gas space above (orange) can be clearly distinguished. Note that the scan shows higher temperatures of the liquid than of the gas. (b) The same tower a year later when it was foam-flooded. Note that the liquid and gas spaces cannot be distinguished.

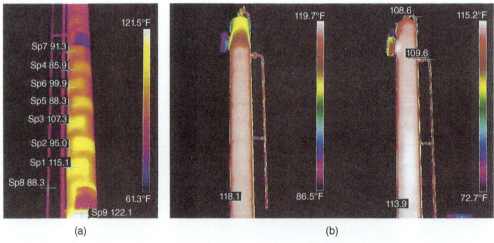

Figure 7.22   Thermal scans of an amine absorber: (a) normal operation; tray liquid (orange) and gas space above (yellow) can be clearly distinguished. Note that this scan shows higher temperatures of the gas than of the liquid, the converse of Figure 7.21a. (b) The same tower during a foaming upset. (From Teletzke, E., and B. Madhyani, Gas 2018, p. 35. Reprinted with permission.)

In both Figures 7.21a and 7.22a, the hottest temperatures appear near the bottom. In Figure 7.21a, the hottest temperature is in the bottom sump; in Figure 7.22a, it is just above. The reaction of amine with H_2S is exothermic, and most occurs near the bottom.

Figure 7.21b shows a thermal scan of the tower as in Figure 7.21a taken about a year later, when it was experiencing foam flooding and had a high pressure drop. Amine absorbers are prone to fouling, and the solids (such as corrosion products) catalyze foam (Section 2.10). The tray elevations can no longer be seen throughout the tower. Instead, two distinct temperature zones can be distinguished, with the upper one about 8–9°F hotter. The reaction between the amine solution and H_2S appears to mostly take place on the upper trays.

The tower in Figure 7.22a experienced foaming upsets daily between noon and the evening. The rest of the time, the tower was stable. Figure 7.22b shows a thermal scan of the tower at a foaming upset. As the foaming upset progressed, the liquid was seen to move up the tower, eventually leaving the tower as entrainment in the overhead gas. Under the foam flood conditions, the tower temperature was quite uniform, just like in the foam-flooded region in Figure 7.21b. In the tower in Figure 7.22, the foaming was caused by condensation of heavy hydrocarbons and was eliminated by improving temperature monitoring of the inlet gas that enabled operators to keep the lean amine hotter than the inlet gas.

In a recent article (477), Teletzke and Roberts extended the application of thermal scans to detect tray channeling, excessive weeping, and tray damage. Weeping and channeling at low rates was recognized by an irregular path of liquid flow down the tower instead of the well-formed liquid paths shown on Figures 7.21a and 7.22a. The scans showed heavy liquid flow on one side of the tower, somewhat similar to the pattern in Figure 7.19a. In another tower where the bottom trays were dislodged or damaged, the thermal scans showed an irregular low temperature pattern across the damaged zone.

7.2.5   Temperature Bulges in CO_2 Amine Absorbers

Ideally, an amine absorber maximum temperature, indicating the location of the exothermic rection, is at the bottom third of the tower (430). In the top two-thirds of the tower, the gas is polished

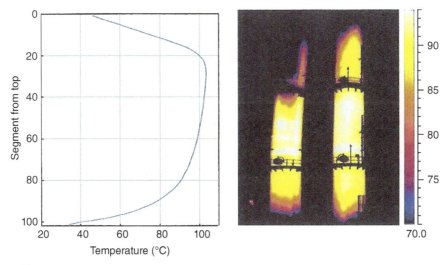

Figure 7.23   Thermal scan of an amine absorber compared with a temperature profile from a rate-based simulation, temperature bulge throughout most of the column, solvent to gas mass ratio 0.87. (From Sheilan, M. H., and R. H. Weiland, "Troubleshooting AGRUs in an LNG Train," LNG Industry, August 2015. Reprinted with permission.)

to the desired specifications (475, 477). Thermal imaging is ideal for checking whether this is achieved and can diagnose damage, foaming, or insufficient amine circulation.

A liquefied natural gas (LNG) plant (430) had two process trains using MDEA for absorbing CO_2, which was 2.4% of the feed gas, to an overhead gas spec of 50 ppmv. There was no H_2S present. The absorbers contained X-type structured packings, and the regenerators were trayed. After years in operation, the regenerator in one train started experiencing foaming, making the operation unreliable and leading to periodic amine discharges. The amine content of the MDEA solution was >50% weight, which is likely to promote corrosion, in turn producing solids that catalyze foam, and its heavy hydrocarbon content was 2–4%, also likely to promote foams.

Figure 7.23 compares the thermal image of the absorber with the temperature profile from a rate-based simulation. This simulation has an excellent track record in amine systems. Both show a temperature bulge throughout most of the tower, indicating significant CO_2 presence in the upper parts. The design solvent to gas mass ratio was 1.1, and the bulge was supposed to be in the bottom, but since the operating ratio was only 0.87, not enough CO_2 was absorbed in the lower part, so the bulge spread out to the upper part of the tower. While the bottom sections had stainless steel cladding, the upper parts were non-cladded carbon steel, prone to corrosion at high CO_2 concentrations and hot temperatures. The authors (430) noted that the temperatures measured by thermal imaging were slightly lower than those from the simulation, possibly due to the temperature gradient through the thick vessel walls.

Figure 7.24 shows the bulge descending to the lower part of the tower after the solvent to gas mass ratio was raised to 1.01. Both temperatures and CO_2 concentrations in the non-cladded upper part of the tower were largely reduced, thus alleviating the corrosion and improving the operation.

Figure 7.25 shows amine absorber temperature profiles as a function of liquid to gas (L/G) ratios from a rate-based simulation (432, similar plots were also produced in Ref. 477). Note the sharp change of the bulge over a relatively small L/G range from 0.91 to 1.12 compared with the much smaller changes at higher and lower L/G ratios. Combining a rate-based simulation of the bulges with thermal scans can provide valuable diagnostics for amine absorbers.

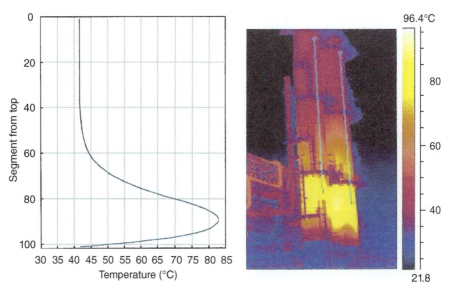

Figure 7.24 Thermal scan of an amine absorber compared with a temperature profile from a rate-based simulation, temperature bulge descending to the lower part of the tower, solvent to gas mass ratio 1.01. (From Sheilan, M. H., and R. H. Weiland, "Troubleshooting AGRUs in an LNG Train," LNG Industry, August 2015. Reprinted with permission.)

7.2.6 Excessive Liquid Levels

Thermal images can show excess liquid levels at the tower base in services where there is a large difference between the liquid and gas temperatures. The same can apply to the liquid level on a chimney tray, as long as the vapor and liquid temperatures differ. When the tower is insulated, a vertical strip of insulation can be removed, and a thermal scan shot.

Figure 7.26 shows the bottom sump of an amine absorber that was experiencing premature flooding (475). The small cylinder to the left of the tower is the guided radar level transmitter that controls the base level. Figure 7.26 shows that the actual liquid level was above the span of the transmitter and rose above the vapor inlet (the purple pocket on the right). The level transmitter showed 55% at the time. Liquid levels rising above the gas inlet are known to cause massive entrainment, which initiates a premature flood in the tower (Section 2.13). Lowering the level setting to 45% on the transmitter was enough to stop the massive entrainment and eliminate the flood (475). A tower in parallel had an almost identical experience.

7.2.7 Diagnosing a Flood due to Bottom Baffle Malfunction

A deethanizer tower in a newly licensed unit experienced premature flooding initiating at its bottom. One lead in the troubleshooting investigation was the observation that the reboiler inlet and outlet process temperatures measured by the control room thermocouples were 30–40°F hotter than the design. Nothing else in that region appeared suspicious; the bottoms temperature was the same as the design, and the bottom level behaved normally.

The bottom of the tower contained a preferential baffle (Figure 7.27), dividing it into a reboiler draw compartment and a product draw compartment. The purpose was to divert all the lower-boiling bottom tray liquid into the reboiler. The seal pan for the bottom downcomer had a tall weir above the bottom draw compartment and a short weir over the reboiler draw compartment, with all liquid overflowing the short weir. The baffle arrangement supplemented the

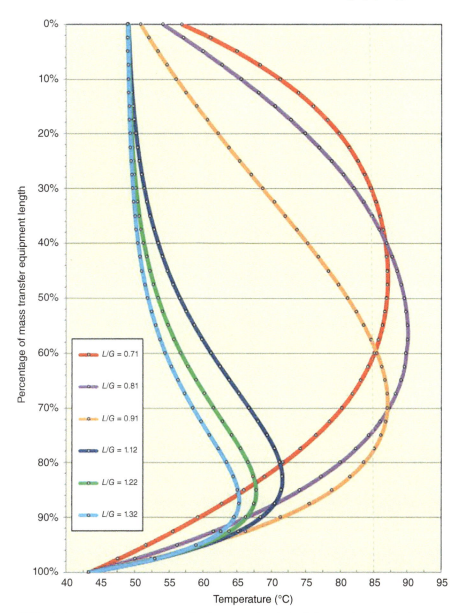

Figure 7.25   Amine absorber temperature profiles from a rate-based simulation as a function of solvent to gas mass ratios. Note the sharp change of the bulge over a relatively small *L/G* range from 0.91 to 1.12. (From Shiveler, G., and H. Wandke, "Steps for Troubleshooting Amine Sweetening Plants", in Kister Distillation Symposium, p. 96, AIChE Spring Meeting, Austin, Texas, April 26–30, 2015. Reprinted with permission.)

reboiler draw with some reboiler return liquid to provide sufficient circulation in the thermosiphon reboiler. The rest of the reboiler return liquid was intended to overflow into the bottom draw compartment. The design flows, with the design temperatures next to them, are shown in non-bold print in Figure 7.27.

To investigate the high temperatures in the reboiler circuit, insulation was removed and thermal camera shots were taken between the top of the cover baffle and the seal pan. Two observations were striking: one, the liquid descending from the seal pan was 25°F hotter than the design bottom

Figure 7.26   Thermal scans of the bottom sump of an amine absorber that experienced flooding due to entrainment from a high base level. (From Teletzke, E, and C. Bickham, Chemical Engineering Progress, p. 47, July 2014. Reprinted Courtesy of Chemical Engineering Progress [CEP]. Copyright © American Institute of Chemical Engineers [AIChE].)

tray temperature, and two, the temperatures between the seal pan and top of the cover baffle were 12–15°F colder than the reboiler return temperature. Combining this with the observation that the liquid out of the bottom draw compartment was slightly colder than the liquid out of the reboiler draw compartment conclusively demonstrated that the baffle arrangement was not working like it was intended and that there was unexpected mixing taking place.

The observation that the liquid coming down from the seal pan was 25°F hotter than the design bottom tray temperature, with the rest of the tower temperature profile close to the design, implies liquid entrainment from the reboiler return region into the bottom tray. To explore, gamma scans were shot through the bottoms draw compartment and above it all the way to the fifth tray above the bottom. The scans showed liquid as dense as the liquid in the bottoms draw compartment filling the space above the cover baffle all the way to the seal pan, with flooding with less dense froth propagating to the bottom tray and beyond. This verified that the flood initiated in the baffle arrangement. Further analysis determined that the flooding was due to flashing occurring where the hot reboiler return liquid met the cold, lights-rich liquid from the bottom tray. The vapor generated upon the flash entrained the liquid attempting to descend in the narrow space between the two vertical baffles, this entrainment flooding the bottom trays.

Again, the takeaway was the value of thermal scanning in providing invaluable troubleshooting information.

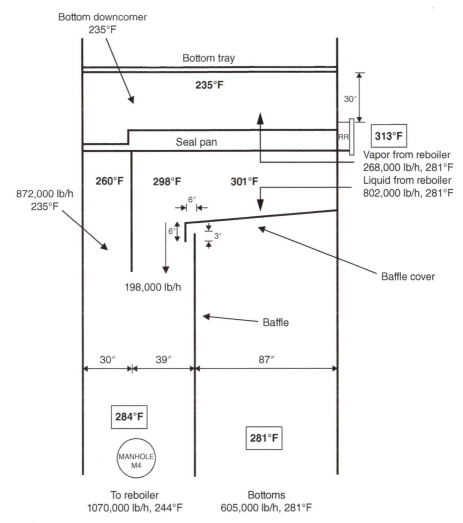

Figure 7.27 Design flows and temperatures at a tower base (non-bold print), compared with actual temperature measurements (bold print). Bold print numbers in rectangles are control room temperature measurements. Bold print numbers with no rectangles are thermal camera measurements.

7.2.8 Diagnosing a Damaged Draw Pan

An electric power failure resulted in an emergency shutdown of an atmospheric crude tower (310). The unit started up two days later, and it was observed that the light diesel oil production was cut by half. The light diesel draw temperature was 18°C less than normal, and the heavy diesel draw, which was the next draw down, was 35°C colder than normal.

Thermograph readings were taken close to the light diesel drawoff nozzle (Figure 7.28). With the draw valve open, temperature right at the outlet was 209°C, while a few inches away in the pipe, the temperature declined by 28°C to 181°C. The lower temperature indicates a gas pocket in the pipe. When the valve was closed (Figure 7.28), the temperature at the top of the pipe became uniform at 200°C, indicating liquid filling the pipe above the valve. This suggested that the draw pan was holding liquid and that the reason for the low light diesel flow was tray damage rather

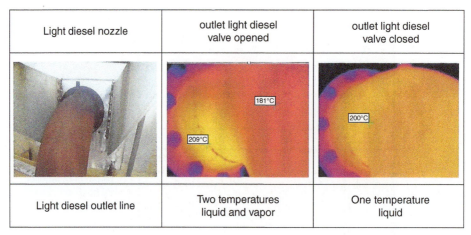

Figure 7.28   Thermal scans of the light diesel draw pipe to check whether the diesel draw pan was leaking. (From Marchiori, A., A. L. Wild, A. Y. Saito, A. L. de Souza, C. Mittmann, C. C. Anton, F. S. Duarte, S. L. A. Pereira, and S. Waintraub, "More Tower Damages Caused by Water-Induced Pressure Surge: Unprecedented Sequence of Events," in "Distillation 2013: The Kister Distillation Symposium, Topical Conference Proceedings," p. 421, the AIChE National Spring Meeting, San Antonio, TX, April 29–May 2, 2013. Reprinted with permission.)

than pan damage. Gamma scans verified extensive tray damage in the region, including the draw tray. The unit was shut down and the damage repaired.

The damage resulted from a pressure surge during restart, caused by water entry via a liquid relief valve discharging into the heavy diesel circuit underneath. The discharge piping contained dead pockets that could have accumulated water and were not drained during the short shutdown.

7.2.9   Liquid Levels in Condensers and Reflux Drums

Thermal images are highly effective in showing the liquid levels in condensers and drums. Figure 7.29 (22) shows a thermal image of a partially flooded tower overhead condenser. The tower experienced unstable overhead pressure, and inability to condense was one theory. The thermal scan proved that condensation was taking place, sending troubleshooters to seek another root cause. The issue was found to be a manual block valve that had been left open since startup. Closing it reinstated good operation.

Figure 7.30 shows the liquid level in a reflux drum using a hot vapor bypass (HVB) control system. The system operated well, and the thermal scan was only used to verify. Notice the sharp change from colder liquid to hot vapor, with a temperature difference of 10°F between them. The temperature gradient took place over a few inches. The sharp change is also seen in Figure 7.29. Process liquids are excellent thermal insulators, hence the very short length over which the temperature gradient takes place.

7.2.10   Troubleshooting Condensers and Reboilers

Thermal images are effective for troubleshooting condensers, reboilers, and other heat exchangers around the tower.

Case 1 Plugged tubes in air condenser Figure 7.31 shows the application of thermal scans to identify plugged tubes in an air condenser. The condenser, switched from another service, was

Figure 7.29   Thermal scan of a partially flooded overhead condenser, showing a liquid level at the bottom and a very sharp temperature transition going from liquid to vapor. (From Barnard, H., Chemical Engineering Progress, p. 21, January 2017. Reprinted Courtesy of Chemical Engineering Progress [CEP]. Copyright © American Institute of Chemical Engineers [AIChE].)

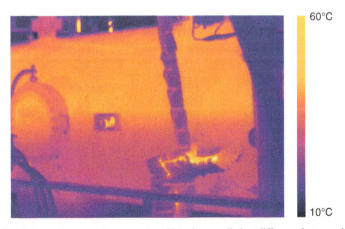

Figure 7.30   Reflux drum in a hot vapor bypass system. Note the very distinct difference between the liquid and vapor temperatures and the sharp transition from liquid to vapor.

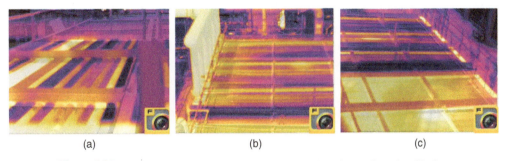

Figure 7.31   Plugged tubes in air condenser. Yellow shows hot tubes, and purple cold tubes.

performing below the expected heat transfer. Figure 7.31a shows why. The purple tubes are cold, and the yellow are hot. For comparison, Figure 7.31c shows an active bay operating well (yellow) next to an inactive bay (purple), which is cold. As all the tubes in the condenser were active, the cold (purple) tubes indicated plugging. This thermal scan changed the solution from buying an expensive new exchanger to inexpensive cleaning of the tubes.

Case 2 Identifying cause of poor heat transfer in air condenser A large chemical tower and its air condensers were started up in 2014 (251). By September 2020, the air condenser performance fell well short of the design. The air condenser was unable to achieve the design heat duty at the design air temperature of 35°C.

Figure 7.32a compares the design conditions to test data at the same tower overhead flow rate and a colder air temperature. The measured condensate temperature was well above the design. The condenser was a total condenser. The boiling range was about 10°C. At 66°C, the condensate was near its bubble point and not 8°C subcooled like in the design. About 10% of the condensation took place in the chilled water condenser, which consumed energy.

The air condenser had five bays, two parallel bundles of finned tubes per bay, arranged in two passes, four rows in the upper pass, two rows in the lower pass. Each bundle had two vapor inlet nozzles, two condensate outlet nozzles, and one vent nozzle.

The test in Figure 7.32a was accompanied by a detailed thermal camera survey (Figure 7.32b). All temperatures in Figure 7.32b were measured by a thermal camera, except for the shaded values that were measured by control room thermocouples. The excellent consistency between the readings as well as between the thermal camera readings and the thermocouples attests to the accuracy of the thermal camera survey.

The vent lines leaving bays 1–5 (marked in black just above the bays, two vent lines per bay) were at 74–77°C, only 2–3°C cooler than the incoming vapor, so the vented vapor had not undergone much cooling. Temperatures measured in the horizontal header that collected the individual vent streams (marked in black in the common vent header) were about 14–17°C colder. The lower temperatures persisted throughout the vent header, all the way to the chilled water condenser. This compares well with the thermocouple measurement of 61°C. The temperature drop was most likely due to the entrainment of condensate from the air condenser into the vent lines. There is little separation in the head of an air condenser, and the vents were likely to drag with them liquid drops, which flashed at the expansion of the vent lines into the vent header.

Condensation was complete at about 10°C below the inlet temperature, i.e., at 66°C. At the measured condensate header temperatures of 65–68°C the condensate was close to its bubble point. All but one of the condensate outlet temperatures from bays 3 and 4 (marked in blue in Figure 7.32b) were in the 67–71°C range, indicating the presence of some uncondensed vapor in many condensate outlets.

Results of heat transfer calculations were expressed as oversurface percent, defined as the ratio of the heat exchange area available to the heat exchange area needed to achieve the condensation less 1. An oversurface of zero percent means the heat exchange area in the condenser equals the heat exchange area required to achieve the condensation and the specified outlet temperature, so the exchanger performs as expected. To get a zero percent oversurface in the air condenser, the condensate would have come out subcooled to 40°C. The high oversurface percent indicated that the exchanger was oversized for the measured condensation and subcooling duties. For the test data, and assuming the design air temperature difference, the oversurface was 99%, showing that the condenser had about twice the area needed to achieve the measured heat duty.

Different fouling factors were tried, but even with a fouling factor five times the design, the condensate outlet temperature at zero oversurface would still have been highly subcooled at

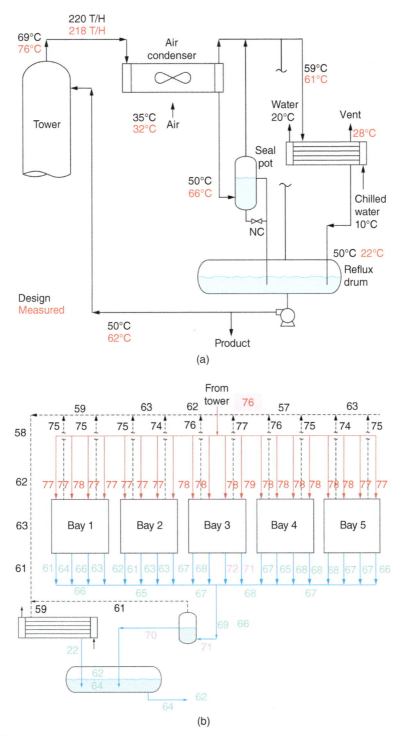

Figure 7.32 Air condenser thermal survey. (a) Tower overhead scheme. (b) Thermal camera survey temperatures, °C. (c) Looking at bay 3 tubes from underneath.

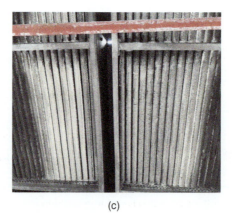

(c)

Figure 7.32   (*Continued*)

46°C. So the reduced heat transfer coefficient due to fouling would not go anywhere near the length required to explain the high oversurface.

The next troubleshooting endeavor explored operating data from the initial running period of the column, shortly after the 2014 startup. At that time, the tubes had been in service for only a short period and therefore were reasonably clean. At the same condensation rate and an air temperature of 35°C, the condensate was subcooled to 55°C, giving an oversurface of 23%, indicating a much better heat transfer.

This raised the question of whether the exchanger ΔT has diminished over time due to airflow restriction. Photographs were taken of the bottom rows of condenser tubes, looking up (Figure 7.32c). The photographs showed extensive plugging of the fins, to the point that they were hardly visible. The passages between the tubes were narrowed. The bottom row of tubes was covered with a layer that would not only form insulation around the tubes, giving very little heat transfer, but would also restrict the air passage, reducing the airflow, which in turn will raise the air outlet temperature and lower the condenser ΔT value. Lieberman and Lieberman (293) stated (in the fourth edition of their book), "A finned tube bundle acts as an air filter. Air is drawn up through the lower two rows of fins where much of the dust, dirt, bugs and moths are filtered out, and this restricts air flow."

Measurements on the air side verified the diagnosis. The measured temperature difference between the air in and air out was 30°C, much higher than the temperature difference of less than 20°C based on no fouling restriction to airflow.

It is likely that the heavy fouling of the bottom row of tubes was due to dust, dirt, and corrosive materials. During plant commissioning and early operation, there were episodes of a nearby grinding workshop releasing dust, some of which is likely to have been inspirated into the condenser air intake, settling in the bottom rows. A lot of powder is believed to have covered the surface of the fins in the first two years. Later in the run, the plant changed the grinder, but by that time, the plugging had already occurred. In addition, in the early years of the plant, chlorinated organics were present, which would have generated HCl upon hydrolysis in wet air. The plant had improved the maintenance and emissions management to the point that this was no longer an issue, but by that time, the plugging had already occurred.

Case 3 Channelhead leaks in horizontal thermosiphon reboiler Figure 7.33 (400) is a thermal image of the channelhead of a horizontal thermosiphon reboiler, showing a sharp change in temperature at the condensate (tube side) liquid level. At this elevation, the temperature difference is

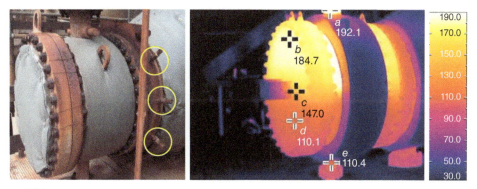

Figure 7.33   Right: thermal scans of the channelhead of a horizontal thermosiphon reboiler, showing a sharp change in temperature at the condensate (tube side) liquid level. Left: circles show the resulting channelhead leak spots. (From Risko, J. R., Chemical Engineering Progress, p. 43, July 2021. Reprinted Courtesy of TLV.)

likely to subject the channelhead flange to severe thermal stresses, causing leakage at the gasket. As the level moved up and down with the change of reboiler duty, the elevation at which severe thermal stresses occurred is likely to shift, and additional leakage points were likely to develop. The circles on the left photograph show the places where the channelhead gasket leaked.

7.2.11   Thermal Video Diagnoses Cause of Pressure Instability

Disclaimer   The information provided in this section, including Figures 7.34 parts a through c, expresses the opinions of the authors and may not reflect the views of Valero; any use of the information is at the reader's risk.

Case History   A hydrocracker unit (HCU) revamp (145) increased the naphtha (overhead product) yield and main fractionator reflux requirement. To accommodate this, the overhead air condenser E-28 capacity was increased by 50% (Figure 7.34a) and the reflux/product pump debottlenecked. No changes were made to the overhead accumulator or its internals nor to the water trim cooler E-30. Liquid from the water cooler entered the drum via a slotted pipe, which was not changed.

Following restart, the tower overhead system experienced severe instability, with large pressure and accumulator level fluctuations accompanied by cyclic banging noise emitted from the trim cooler. The hot vapor bypass (HVB) control system was struggling to respond quickly enough to the drum pressure changes. Troubleshooting was immediately started, focusing on the accumulator, condenser, and cooler.

The temperature changes in the accumulator as the HVB was opening and closing signaled that vapor was mixing with subcooled liquid and collapsing onto it. Physically touching the cooler outlet pipe during a cycle showed that temperatures would increase and then decrease in concert with the banging.

The problem was diagnosed by an innovative application of thermal imaging: shooting thermal videos of the temperature in the liquid line from the cooler E-30 (Figure 7.34b). Six static shots from a typical video are shown in Figure 7.34c over a 72-second time period. The temperature rose from 70.3°F to 83.3°F before dropping back to 67.8°F over about 1 minute and then started rising again.

Subcooling tremendously lowers the vapor pressure of the liquid. The vapor pressure of the subcooled liquid from the cooler was well below the drum pressure, so instead of coming out of the slots in the pipe, the subcooled liquid sucked vapor into the pipe via the slots. The vapor

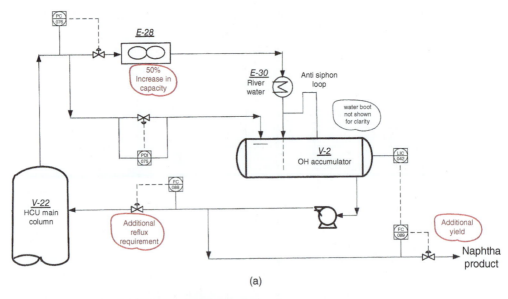

(a)

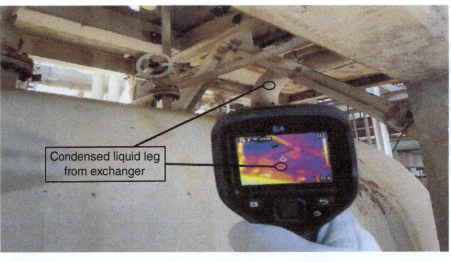

(b)

Figure 7.34   Troubleshooting by thermal videos. (a) Overhead system of a hydrocracker main fractionator that was debottlenecked to increase product yield and tower reflux. (b) Positioning of the thermal camera used to shoot the condensate line temperature video. (c) Static shots from a typical video of the condensate line temperature, showing large temperature cycling over one minute. (From Hanson, D., "Details Matter When Diagnosing Hot Vapor Bypasses," in Kister Distillation Symposium, Topical Conference Proceedings, AIChE Spring Meeting, San Antonio, Texas, March 26–30, 2017. Reprinted with permission per disclaimer at the front of Section 7.2.11.)

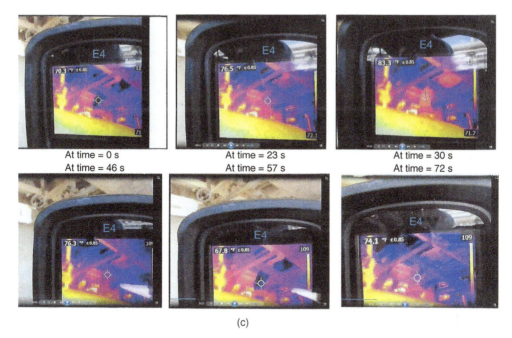

| At time = 0 s | At time = 23 s | At time = 30 s |
| At time = 46 s | At time = 57 s | At time = 72 s |

(c)

Figure 7.34   (*Continued*)

collapsed onto the subcooled liquid, causing the hammering. Once the liquid was heated up by the condensing vapor, its vapor pressure would rise, it would no longer suck vapor from the drum, and liquid would flow into the drum via the slots, as intended. The hot liquid in the pipe returned to the drum and was replaced by the subcooled liquid from the cooler again, and the cycle would repeat.

The problem did not occur before the revamp because the subcooling was small, only 5–10°F. Upgrading the air condenser largely increased the subcooling, causing low vapor pressure and initiating the cycles. The problem was resolved by closing the HVB and adding an inert gas blanket to the accumulator, which would not condense onto the subcooled liquid.

7.3   ENERGY BALANCE TROUBLESHOOTING

Energy balances are invaluable for validating simulations, forming a basis for vapor and liquid loading calculations, and identifying leaks, including internal leaks, entrainment, and insufficient reboil or reflux.

7.3.1   Energy Balance Application for Correctly Validating Simulations and Correctly Diagnosing Tower Problems

Simulations are the primary tool to tell troubleshooters whether the tower performs like it should and where to look for problems. The energy balance is a key tool for validating these simulations. Step 8 in Section 3.1.3 emphasizes the importance of compiling a valid energy balance with a closure of ±5% (4, 192).

Flooding   A reflux rate significantly higher than the rate that would match the measured reboil rate often indicates massive entrainment of liquid from the top of the tower, i.e., flooding. One debutanizer tower returning to service after turnaround (280) had a measured reflux rate the same as normal, but the reboil rate was very low. This signified that reflux liquid was not descending down the tower but had been entrained back to the reflux drum and recirculated.

Poor Separation   When the measured reflux rate is higher than the actual, so is the condenser duty based on this measurement. An energy balance based on the reboiler duty and tower feed will give a lower condenser duty and reflux rate than those inferred from the measured reflux rate. This may identify a reflux measurement issue where insufficient reflux is the root cause of a separation problem. Case 2.6 in Ref. 201 describes a tower that experienced poor separation even though the reflux rate and temperature profile appeared normal. The cause was an undersized orifice plate in the reflux line giving a misleading high reflux reading. The tower had a fired heater reboiler, making it difficult to compile an energy balance. The energy balance would have indicated a reflux flow too high to balance the low reboiler duty and would have identified the metering problem. When the undersized orifice plate was replaced by the correct one, the plant saw a large increase in fuel gas consumption at the same reflux reading (which was correct this time).

Another Poor Separation   An olefin gasoline fractionator produced off-spec gasoline after returning from turnaround (216). During the turnaround, the impulse line to the reflux meter orifice plate was repaired, but no other changes were performed to the tower or auxiliaries.

An energy balance performed by an intern following the turnaround revealed a reflux rate much lower than measured. A check of the orifice by the instrument foreman and technicians did not reveal any problems with the meter.

The mystery was explained after the intern found a flange with the old orifice plate in the scrap heap. Apparently, to repair the impulse line, the installers cut out the orifice flange, replacing it by in-kind. The old flange disappeared in the scrap heap. The installers fabricated a new orifice plate, with the orifice size provided by the instruments department. The orifice size in the instruments records was outdated, so the replacement orifice diameter was smaller, giving a misleading high reflux reading.

One More Poor Separation   In a chemical tower, steam consumption was high, giving the appearance of low packing efficiency. The plant was considering replacing the packings and distributor. While preparing a simulation to provide vapor and liquid loads to the packing vendor, it was noticed that the reboil was disproportionally higher than the reflux. Upon close investigation of the discrepancy, it was discovered that the reboiler lost its liquid seal, so some steam was blowing straight into the condensate header without condensing. This increased steam consumption, but not due to poor packing efficiency.

Good or poor Packing Performance?   In a chemical packed tower, the measured reflux rate was 25% higher than the reflux rate that would have matched the measured reboiler duty. The reason for the discrepancy was not identified, so both cases were simulated. The simulation based on the measured boilup gave reasonable HETPs in the tower, indicating the packing and distributors were working quite well. The simulation based on the measured reflux rates required HETPs that were twice higher, indicating poor performance of the packings or distributor. Until today, we do not know whether the packings performed well or not.

Flashing in the Reflux Distributor   This is the chemical tower discussed in Sections 6.4.1 and 7.1.7. Temperature surveys and CAT scans established that flashing was taking place in the reflux distributor, but the reason for the flash remained obscure. A number of theories were put forward, but they turned out to be either inconsistent with the plant data or thermodynamically flawed. In an endeavor to get to the bottom of the flashing, instruments were closely checked and calibrated, and a simulation was compiled to match plant data. A rather small discrepancy in the energy balance, about 7%, was noticed, with the boilup being lower than the matching reflux. This signaled the possibility of an additional heat source and led to the evaluation of the heat tracing on the reflux line, which accounted for this discrepancy. This heat tracing was under heavy insulation, and until the energy balance was compiled, no one in the team considered it.

Proprietary Trays Not Meeting Their Guaranteed Efficiency   Reference 234 describes a case where a tower achieved excellent separation but was unstable when operating at maximum rates. The problem was thought to be a control problem.

A simulation was prepared, based on the vendor's guaranteed tray efficiency. A comparison of the simulation with the plant data (Figure 7.35a) indicated that the simulated reflux was 18% higher than the measured, while the simulated boilup was 19% lower than the measured. The simulation values are always energy-balanced, so they provide an excellent benchmark for the closure of the measured energy balance. As the reflux and boilup deviated in different directions from the energy-balanced solution, Figure 7.35a indicates about 30–40% nonclosure of the measured energy balance, with reboiler heat duty well in excess of the heat duty that would match the measured reflux rate. Such a large discrepancy pointed out a measurement issue.

The instrument engineers were invited to join the task force, and a detailed investigation of the key flow meters (reflux and reboil) was initiated. The investigation identified a valving problem on the reflux meter that was undetected in previous instrument checks due to the inaccessibility of the meter. In addition, meter density correction factors needed updating. Once both problems were rectified, it was found that the reflux in the tower was 37% higher than previously thought and 16% higher than previously simulated. The simulation was revised to match the correct reflux

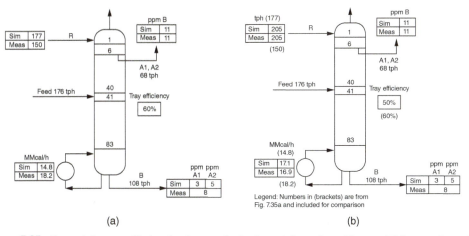

(a)   (b)

Figure 7.35   Energy balance troubleshooting that completely changed diagnosis on this tower (all flows are in tons/h). (a) Initial simulation compared with plant measurement, showing a mismatch between reflux and boilup. (b) Revised energy balance, after measurements revised, showing much higher reflux and poor tray efficiency. (Reprinted with permission from Kister, H. Z., S. Bello Neves, R. C. Siles, and R. da Costa Lima, Hydrocarbon Processing, p. 103, August 1997.)

measurement (Figure 7.35b), which also gave a good match to the revised boilup measurement. The only way that the simulation could have matched these higher reflux rates is by using a low tray efficiency, well below the guaranteed. The higher reflux and boilup rates (needed to achieve the good separation) hydraulically loaded up the trays to the verge of flooding. The instability had little to do with controls and a lot to do with operation right at the flood point. So getting the energy balance to close changed the diagnosis of the problem from that of a control problem to one of low tray efficiency, below the vendor's guarantee.

Unnecessary Shutdown   Case 25.2 in Ref. 201 describes a large C_3 splitter tower that experienced poor separation, with the propylene product being only marginally on-spec and high propylene content in the bottoms. The reflux orifice plate was pulled out (which required tower shutdown) and found good. Damage or incorrect assembly was suspected. The tower was shut down. It took three weeks to purge and vent this huge tower and to remove its 700 manways. Neither damage nor installation problems were found, and the trays were clean. The reflux orifice plate was again checked and again found good.

Eventually, heat and mass balances were compiled. An imbalance was found, with the reflux higher than the flow rate needed to match the reboil. An accurate kilowatt/hour meter was installed in the power supply to the reflux pump motor, and accurate pump curves were obtained from the pump manufacturer. Data collected showed the reflux flow meter to read 24–33% higher than the flow obtained from the kilowatt/hour meter. A subsequent investigation found that the reflux pump discharge piping was a spiral-wound (seamed) pipe, even though a seamless pipe was specified. Spiral flow in the proximity of an orifice meter can yield readings that are out by up to 50%.

As replacing the pipe would have been expensive, the plant decided instead to calibrate the reflux flow meter to match the flow rate calculated from the pump curves. The operators raised the actual reflux to the design value. Tower operation was tremendously improved, and the propylene product came on-spec even at above the design rates and at the design reflux ratios. The energy balance now closed within ±5%.

Incorrect Internal Flowrates   This system is similar to the one shown in Figure 7.36. The measured liquid flow rates in the upper PA loop did not match the simulation (41). The cause was that a small portion of the lower (heavy vacuum gas oil, HVGO) product was cooled, used as a seal liquid to the vacuum pump (not shown), and then recycled to the tower via the lower (HVGO) return sprays. This stream was ignored in the simulation. Since the ignored seal oil return was much colder than the rest of the PA, there was a significant error in the heat balance.

7.3.2   Energy Balance Troubleshooting to Detect Internal Leaks

Energy balances are the primary tool for identifying internal leaks and overflows in direct contact cooling services, such as cooling circuits removing the heat of reaction or absorption, PAs in most refinery main fractionators, and chemical, olefins and sulfur plant quench towers (201).

To demonstrate this technique, an application to the top section of a refinery fuels vacuum tower is shown in Figure 7.36. This section contained a PA, which was a direct contact total condenser (negligible vapor product) consisting of a packed bed and a total draw chimney tray underneath. Some of the liquid collected on the chimney tray was drawn as the LVGO product, whereas the rest was cooled and returned to the top of the packed bed where it cooled and condensed the ascending vapor.

The tower had difficulty achieving the desired vacuum. During troubleshooting, an energy balance was compiled on the upper section. Since it is a total condenser (the chimney tray is a

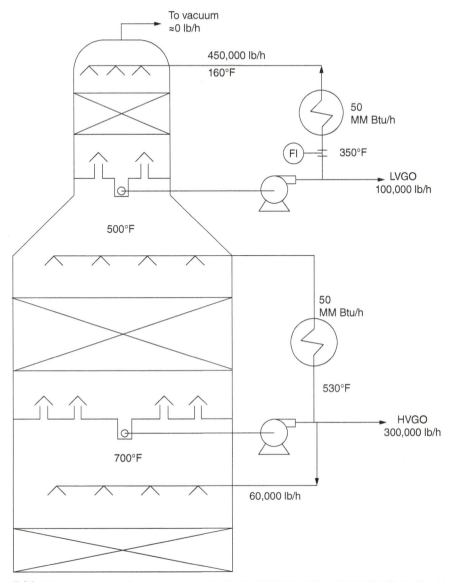

Figure 7.36 Top section of the refinery vacuum tower in Section 7.3.2. (From Kister, H. Z., "Distillation Troubleshooting," John Wiley and Sons, NJ, 2006. Reprinted by permission.)

total draw tray), the flow rate of vapor into this section equals the flow rate of the LVGO product out. The LVGO section heat duty is therefore:

$$Q_{\text{LVGO}} = \left(H_{\text{V,500 F}} - h_{\text{L,350 F}}\right) M_{\text{LVGO}} = 200 \text{ Btu/lb} \times 100{,}000 \text{ lb/h} = 20 \text{ MM Btu/h,}$$

where Q_{LVGO} is the heat duty in Btu/h; H_{V} and h_{L} are vapor and liquid enthalpies in Btu/lb, with the subscripts indicating the temperature basis for each enthalpy; and M_{LVGO} is the LVGO product flow rate in lb/h. Altogether, 20 MM Btu/h was removed from the vapor in this section. The LVGO section heat duty can also be calculated from the heat removed in the coolers:

$$Q_{\text{LVGO}} = M_{\text{PA}} \, C_{\text{P}} \left(350°\text{F} - 160°\text{F}\right) = 450{,}000 \times 0.6 \times 190 \approx 50 \text{ MM Btu/h,}$$

where M_{PA} is the PA flow rate (as measured on the flow meter, Figure 7.36) in lb/h and C_p is the specific heat in Btu/h lb°F.

The heat removed by the coolers was much larger than that required to condense the LVGO product. Instruments were checked. Flow rates were compared with values calculated from control valve openings, pump curves, and spray nozzle pressure drops. The checks verified that none of the measured numbers was in gross error.

The first heat duty calculation above, from the difference between the entering vapor and leaving liquid enthalpies, assumes that all the vapor entering the LVGO section exits as the LVGO product. This assumption is valid only if the condensed liquid does not escape in some alternate route. There are two plausible alternate routes: entrainment from the top of the tower or leakage/overflow from the chimney tray. Entrainment from the top of a refinery vacuum tower is readily detectable as liquid product ("slop") in the ejector steam condensate. In this tower, little slop was produced. This leaves leakage/overflow from the chimney tray as the only plausible explanation.

The energy balance permits the calculation of leakage/overflow. At 200 Btu/lb enthalpy difference between the entering vapor and condensate (above) and a heat duty of 50 MM Btu/h calculated from the heat removed by the coolers, the total condensate make in the LVGO section is

$$M_{CONDENSATE} = 50 \times 10^6 / 200 = 250{,}000 \text{ lb/h.}$$

So a total of 250,000 lb/h LVGO was condensed. Of this, 100,000 lb/h became the LVGO product. The balance leaked or overflowed down the chimney tray. In this tower, the leaking or overflowing LVGO ended up in the lower (HVGO) section, where it reduced the bubble point and HVGO draw temperature, causing a heat transfer bottleneck in the HVGO coolers. Seal welding of the chimney tray at the next turnaround solved the problem.

Overflow/leakage over time can be logged. The above energy balance is repeated in regular intervals, typically once per day, and the leakage logged against time (Figure 7.37). An upward trend may indicate progressing corrosion or operators running a higher level. An upward trend can be reversed by lowering the level control set point, as was done on week 15. The log shows how effective the action is.

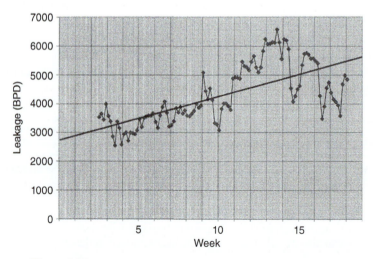

Figure 7.37   Leakage from an LVGO collector tray logged against time.

Other Cases References 128 and 130 describe very similar cases, where a chimney tray leaking cold LVGO generated heat removal problems in the HVGO circuit, which in turn led to difficulty in achieving the desired vacuum. Reference 443 described lower LVGO temperature due to overflow from a uniquely designed collector limiting preheat in the crude unit.

Case 9.2 in Ref. 201 describes a similar case of leakage from the LVGO chimney tray in another refinery vacuum tower. In this case, the leak was so large that the level on the chimney tray was falling even when the LVGO product route was blocked in. To avoid losing prime on the pump, the refinery needed to pump a liquid bleed from the section below (the HVGO in Figure 7.36) into the LVGO section that would establish an LVGO draw and level control on the chimney tray. This bleed solved not only the pump problem, but also the vacuum problem. The HVGO bleed served as absorption oil that absorbed lighter components from the overhead vapor and unloaded the ejectors. In that case, product purity was not a concern.

Case 9.4 in Ref. 201 describes a case of a refinery FCC main fractionator where a major leak in the bottom chimney tray was diagnosed by a heat balance. The leaking liquid was revaporized in the hot slurry section underneath, inducing excessive vapor recycle in the tower that flooded the upper trays. This diagnosis led to lowering the level set point on the chimney tray, which effectively reduced the leak and the vapor recycle and eliminated the flooding.

7.3.3 Energy Balance Troubleshooting to Eliminate Overflows or Leaks in the Upper Parts of Chimneys

When a chimney tray overflows or liquid leaks from the upper part of the chimneys, an energy balance can lead to completely eliminating the leak, as demonstrated by the case history below (158).

Figure 7.38 shows the bottom section of an FCC main fractionator. Superheated feed from the reactor is first cooled in a packed bed by a circulating slurry PA stream by direct contact heat transfer to remove the superheat. The vapor leaving this bed ascends to the light cycle oil (LCO) PA bed, where part of it is condensed. The condensed liquid is collected by a chimney tray. Some of the chimney tray liquid is cooled and sent to the top of the LCO PA bed, some of it is sent to the top of the slurry PA bed as reflux, and the balance is drawn as the LCO product.

The FCC main fractionator in this case worked well until halfway in the five-year run. At that time, there was a step change where the bottom slurry product became lighter, with no major change in the quality of the side product drawn above it. This was accompanied by severe operational upsets in which the bottom level was lost abruptly and there were also steep temperature changes near the vapor feed and smaller changes in other temperatures in the region.

An extensive troubleshooting investigation combining operating charts analysis, product sampling, gamma scans, and wall temperature measurements yielded invaluable information and is described in detail elsewhere (158). A major lead was the input from operators that one action that helped restore the loss of the bottoms level was to reduce the LCO pumpdown (Figure 7.38). This was counterintuitive, but could make sense if there was a leak from the LCO draw tray. An energy balance was compiled around the LCO draw tray. It showed that an additional 3000 BPD liquid was condensed in the LCO PA and not accounted for in the product or pumpdown meters.

To independently quantify the leak, transient heat balance data were collected in a special test. The data showed that the slurry section heat duty was 86.3 MM Btu/h when the LCO collector tray level was above 50% and sharply rose to 89.9 MM Btu/h when the level was reduced below 50%. The difference in duties corresponded to a leak of 2500 BPD, which was close to the figure obtained from the LCO PA heat balance at above the 50% level. This verified that lowering the level below 50% would eliminate the leak. The most likely cause is a leak consistent with a weld seam failure that resulted in a concentrated liquid stream pouring out into the lower

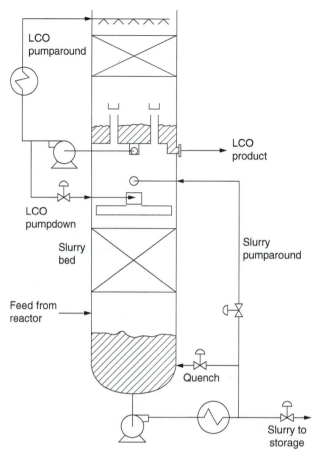

Figure 7.38   Bottom section of the FCC main fractionator that experienced separation problems and upsets. (Based on Hanson, D. W., J. Brown Burns, and M. Teders, Hydrocarbon Processing, 96, p. 37, September 2017.)

section, violently vaporizing there and causing the operational upsets. A detailed explanation of the mechanism is provided in the original Ref. 158.

Other Cases   Case 9.3 in Ref. 201 describes liquid overflow from the HVGO chimney tray in a refinery vacuum tower. The level transmitter showed the level to be well below the top of the chimneys, but the energy balance identified 9000 BPD out of the 22,000 BPD condensed in the PA unaccounted for in the product or reflux meters. Experimenting with reducing the level on the chimney tray from 40% to 30% completely eliminated the overflow, allowing more than 90% of the condensed liquid to be recovered. The observation that reducing the level from 40% to 30% completely eliminated the unaccounted for 9000 BPD indicates that the liquid descended by overflowing rather than leakage at the bottom of the chimney tray. If it were a leak, it would have been reduced when the level was lowered, but not totally eliminated.

Case 3 in Ref. 127, which is also described in Refs. 128 and 130, is similar to Case 9.3 in Ref. 201. The energy balance showed that out of the condensed 282,000 lb/h in the HVGO PA, 63,000 lb/h was leaking or overflowing from the HVGO chimney tray, reducing the HVGO product yield. Repairing the tray eliminated the leak and raised the HVGO yield by about 63,000 lb/h.

In Case 1 in Ref. 127, overflow from the reflux distributor to an LVGO/HVGO fraction-ation section caused poor separation and low LVGO yield. An energy balance on the LVGO PA, identical to this in Section 7.3.2, determined that the reflux rate was 1358 gpm, well above the design 850 gpm, causing the overflow. Cutting the LVGO PA duty reduced the reflux rate to 750 gpm, eliminating the overflow. Within 8 hours, good separation and the design LVGO yield were achieved.

7.3.4   Energy Balance Troubleshooting of a Two-Compartment Chimney Tray

The jet fuel chimney tray in an atmospheric crude tower (Figure 7.39) had a central "hot" com-partment, receiving liquid from the tray above, which was meant to be a total draw of the liquid from the tray above (395). Some of the drawn liquid was stripped externally and became the jet

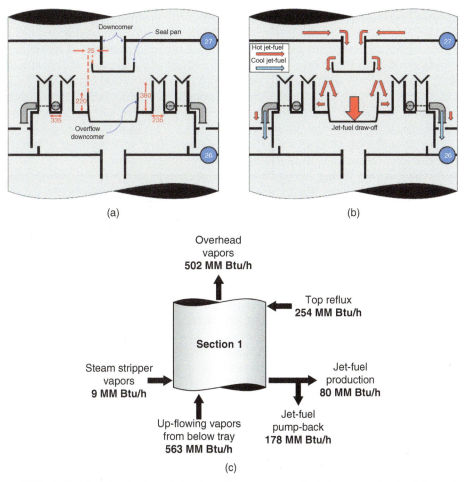

Figure 7.39 Jet fuel draw tray having a hot and a cold compartment. (a) Chimney tray sketch, all dimensions in mm. (b) Flow patterns on the chimney tray. (c) Energy balance. (From Rendón, G. T., L. A. Quintero Rivera, J. A. Recio Espinosa, and R. A. Loubet Guzmán, "Troubleshooting Jet-Fuel Production on Crude Distillation Unit in Pemex Cadereyta Refinery," The Kister Distillation Symposium, Topical Conference Proceedings," p. 119, the AIChE National Spring Meeting, San Antonio, TX, April 29–May 2, 2013. Reprinted with permission.)

fuel product, whereas the rest was cooled and piped to the peripheral "cold" compartments of the chimney tray via the two floor nozzles. From the cold compartments, the liquid descended via side downcomers to the tray below. Each downcomer contained a horizontal impingement plate with two 210 mm × 140 mm openings that permitted liquid descent. Two 220-mm-tall weirs, one on each side of the hot compartment, extending from wall to wall, prevented the hot and cold jet fuel from mixing.

The tower experienced an inability to maintain the liquid level on the chimney tray, which led to reflux, product, and temperature fluctuations, lights in the jet fuel, and loss of jet fuel yield to a lower product.

One theory was that the openings in the impingement plates caused the overflow of liquid from the cold to the hot compartment. This theory was denied by the observations that the level on the hot compartment of the chimney tray tended to be low and often lost and that the jet fuel yield was low.

To troubleshoot the problem, a heat balance was compiled on the upper section of the tower (Figure 7.39c). There were no PAs at or above the jet fuel draw, so all the heat entering and leaving the section was contained in the streams shown. The enthalpies were compiled from the mass balance. The overhead stream flow rate equaled the top product (naphtha) flow rate plus the top reflux rate, and its enthalpy calculated from this flow and the overhead temperature. The steam, jet fuel, and reflux flow rates were all metered, and from the flows and temperatures, their enthalpies were calculated. The upflowing vapor flow rate was calculated from a mass balance and its enthalpy from that flow and the measured temperature. The energy balance showed that the heat entering the section was 792 MM Btu/h, and the heat going out 760 MM Btu/h, an imbalance of 32 MM Btu/h. Like in the cases in Sections 7.3.2 and 7.3.3, this indicates unaccounted for liquid going down past the chimney tray. This was also consistent with the lower temperatures below the chimney tray and the loss of jet fuel yield to the lower product.

The first suspect was leakage or overflow, like in the cases in Sections 7.3.2 and 7.3.3. Leakage tests were performed prior to the run, and showed no leakage. The next turnaround was close, and the leakage test was repeated, again showing no leakage.

The root cause of the problem was identified with the help of the sketches in Figure 7.39a and 7.39b. Liquid from the seal pans above was meant to drop into the central hot compartment. However, the horizontal distance between each seal pan overflow weir and the top of the weir separating the compartments below was only 25 mm, so some of the hot liquid landed in the cold compartments (Figure 7.39b) and went to the trays below unmetered and unaccounted for. The amount descending into the cold compartments would vary, leading to level fluctuations in the hot compartment and in the reflux and product flow rates.

Based on this diagnosis, the weirs separating the hot and cold compartments were installed further apart, ensuring all the hot liquid descended into the hot compartment. Also, the openings at the downcomer impingement plates were widened to prevent future bottlenecks. This completely eliminated the problem.

7.3.5   Leaks from Heat Exchangers: The Role of Mass and Energy Balance in their Troubleshooting

Many towers have preheaters, precoolers, and other heat-integrated reboilers and/or condensers. Each one of these may be prone to leakage. The leak will always proceed from the high-pressure to the low-pressure side. However, in many systems, the pressure on one or both sides may alternate, and so would the direction of the leakage.

Exchanger leaks into or out of the tower cause product or utility contamination and, in many cases, also safety and/or environmental hazards that may be severe. A complete book can be written on the safety and environmental hazards of leaks, and these are dealt with at length in most safety and environmental texts. The few pages here cannot do justice to this topic, so these hazards are outside the scope of this book.

The discussion below focuses on some contamination considerations, although it may also be useful for diagnosing the hazards mentioned above. Also discussed, but not so well appreciated, is that leaks into the tower from heat exchangers can induce premature flood or other instability issues in the tower.

Component Analysis   Analyzing both the process and the heat transfer streams leaving the exchanger is invaluable for identifying exchanger leaks. Focus on analyzing a sample from the lower-pressure stream(s) leaving the exchanger. Alternatively, when the heat transfer fluid is at a higher pressure than the tower, analyze the tower top and bottom products for components in the heating or cooling fluids. In one refinery vacuum tower (443), detecting light components in the naphtha and kerosene range in the slop top product, and having the slop rate doubled, diagnosed the leakage in a vacuum gas oil PA cooler that preheated crude oil. If a steam or cooling water leak is suspected, check for water in the bottoms or top products and for the presence of cooling water chemicals. When the tower pressure is higher than the heat transfer fluid pressure, analyze the heat transfer stream leaving the exchanger for process components. In some cases, these components are gases, which can be detected where the heat transfer stream is vented.

With steam-heated (or other vapor-heated) reboilers controlled by a valve in the steam (or vapor) line, the condensing pressure in the reboiler varies with the heat load and with the reboiler fouling, so the direction of the leak may vary with the process conditions. In contrast, when the control valve is in the condensate line from the reboiler, the condensing pressure is that of the steam or vapor supply, so the direction of a leak is well defined. However, with the condensate control scheme, a variable portion of the tubes is submerged in the condensate. With condensate density being orders of magnitude higher than that of steam, the quantity of condensate leaking will be much higher if the leak is in the tube portion submerged in the condensate (208). Similar considerations may apply to a condenser, depending on whether its control valve is in the inlet or outlet.

In vacuum towers, leakage of air or other non-condensable gases (including steam, which may act as a non-condensable under deep vacuum) can destabilize the tower pressure control, especially when using an internal condenser (210). Theunick's statement (479) "Effectively managing the vapor inventory is the key to controlling the pressure" summarizes this issue, as has been demonstrated in a number of case studies (210, 479, 491). Accordingly, destabilization of a pressure control scheme may be a symptom of leakage of non-condensables or steam. An analysis of the vent gases may sometimes provide a valuable clue.

Field Tests   As the leak will always proceed from the high-pressure to the low-pressure side, a valuable field test, if feasible, is changing the pressure on one of the sides, so it rises above or falls below that of the other side. A change in the performance or in the composition of the streams exiting the exchanger indicates a change in the direction of the leakage. The case study by Lieberman (286, 295) cited in Section 1.2 is a classic illustration of the effectiveness of this technique.

Tests in which the heat duty of the exchanger is varied can help diagnose a leak as demonstrated in Case 1. In a few cases (like in Case 2), it may be practical to temporarily cut the heat transfer stream and check for process components on the heat transfer side. Tracer techniques (Section 6.5) have been very effective for leak detection.

Energy and Mass Balances   A mass and heat balance can be valuable for diagnosing, as illustrated in the cases below.

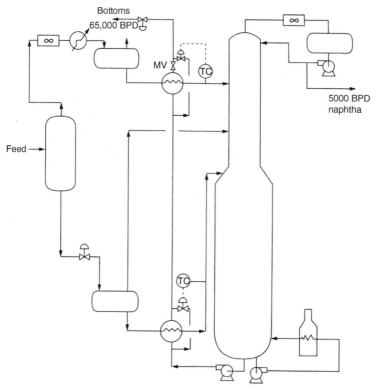

Figure 7.40   Hydrotreater fractionator with three feeds and two feed-bottoms interchangers. (From Kister, H. Z., "Distillation Troubleshooting," John Wiley and Sons, NJ, 2006. Reprinted with permission.)

Case 1 (201) A series of flashes split the feed to a hydrotreater fractionator into three streams (Figure 7.40). The lower feed was the main tower feed. The upper and lower feeds were preheated by the tower bottoms, and their temperatures automatically controlled by manipulating bypasses on the heating sides of the preheaters.

Following a debottleneck in which the trays above the top feed were retrofitted by high-capacity trays, the tower flooded at well below the design loads. The flash point of the bottom product was low, while the end point of the top product was high, indicating poor separation.

A tower simulation based on operating conditions gave a good match between all measured and simulated values, except that the simulated reboiler duty was much less than the measured duty.

The possibility of a feed preheater leak was considered. The fractionator bottoms was analyzed for feed components that usually end up in the overhead, but those were not found. The possibility of an exchanger leak was discounted.

At one time, the feed from the upper preheater was running hot even though the bypass around the preheater was fairly wide open. In an attempt to lower the feed temperature, the manual valve MV downstream (Figure 7.40) was throttled. This should have reduced the upper feed temperature, but instead, this temperature increased. The only plausible explanation is massive leakage from the bottoms side into the feed side of the upper feed preheater. A tower resimulation with a leak of 19,000–20,000 BPD fully matched plant data, including reboiler duty.

At the next opportunity, the system was taken off-line and the leak repaired. Following the fix, the flooding disappeared and separation was good.

Case 2 (286) This case is described in detail in Section 1.2. In this case, a debutanizer reboiler was heated by hot oil from the main fractionator in an FCC unit. When the debutanizer pressure exceeded 130 psig, the debutanizer bottoms leaked into the fractionator. The light naphtha debutanizer bottoms leaking into the hot fractionator vaporized, which doubled the rate of naphtha production from the fractionator and flooded the top of the fractionator. At pressures less than 130 psig, the hot, heavy fractionator oil leaked into the debutanizer and would have been detected upon analysis.

Case 3 (described in detail as Case 20.4 in Ref. 201). A rosin tower operated at 37 mmHg abs. in a tall oil refinery had two separate PA cooling loops to remove heat and generate reflux. At one time, following a reconnection of one side-draw line, the top PA low-flow alarm sounded. Initially, it was thought that its strainer was plugged, but switching to the spare pump (that had its own strainer) did not help. It was then realized that the line reconnection raised tower pressure to 50 mmHg abs., indicating a steam leak into the tower, with steam acting as a non-condensable at these low pressures. The steam leak was from a nipple connected to the draw line. Disconnecting the draw line returned the pressure back to 37 mmHg abs. The top PA came good and remained good after the leak was fixed and the draw line reconnected.

The reason for the low top PA flow was the shift of heat duties between the PA loops. At the higher pressure, the temperatures at the lower PA loop were higher, which condensed more of the vapor. The upper PA loop was thus starved of the vapor needed to condense and sustain the PA flow.

Chapter 8

Point of Transition Troubleshooting: You Do Not Need an Expert, You Need a Sketch

"Sketching is the tool for innovation, and is so vital to the engineering process that it should be taught and used as an essential part of engineering education and professional practice"

—(David McCormick, Ref. 317)

8.1 GUIDELINES FOR POINTS OF TRANSITION SKETCHES

Our tower malfunction survey (198), summarized in Table 1.1, lists points of transition (tower base, packing distributors and redistributors, intermediate draws, chimney trays, and feeds) as a primary cause of column malfunctions. A good troubleshooting investigation must closely review the relevant points of transition for a possible root cause of the problem.

Generating a sketch (or sketches) to scale that will give a clear picture of how this point of transition is supposed to work is the best technique for diagnosing points of transition flaws. Use arrows and colors to show the likely vapor and liquid flow patterns and address the question, "Will it work as expected?"

The author does not recommend just looking at the tower drawings. Drawings contain multitudes of details, of which only a few are relevant to the functionality of the point of transition in hand. Too many trees hide the forest. Worse, the important details are often scattered among several drawings, and these details need to be put together to clearly see the full picture. Without a clear picture in your mind, a good evaluation of a point of transition is likely to remain obscure. Clear sketches omit the irrelevant details and focus on the important details. Sketches can be computer-generated or hand-drawn.

The sketches need to be drawn to scale. There is no need for high-accuracy, but it is imperative to see the proximity of items to each other. Out-of-scale sketches may mislead. The sketches may include elevation and plan sketches, side views, or whatever is needed to get a good picture of how the point of transition is supposed to work and what the obstructions are in the path of the vapor and/or liquid. In complex situations, a three-dimensional sketch, isometric, a paper, or cardboard model is invaluable.

Distillation Diagnostics: An Engineer's Guidebook, First Edition. Henry Z. Kister.
© 2025 American Institute of Chemical Engineers, Inc. Published 2025 by John Wiley & Sons, Inc.

It is important to include all the relevant pieces in the sketches. For example, if a feed pipe is feeding a distributor, it is important to have both on the sketches. Often, the feed pipe as well as the distributor are of good design, but when put together, they generate a problem. Some examples are presented in this section.

There appears to be a great reluctance among process engineers to draw sketches. Many clients throw it back at the author. "You do it." So I draw the sketch. The client sees the sketch and immediately identifies the problem. So what do I get paid for? Drawing sketches. Draw the sketch yourself, and you will figure the problem out and not need an expert. Or if experts are already involved, a sketch will save many of their hours that you are paying for.

Drawing sketches should not be used as a "standalone" technique. Once the sketch is drawn, and leads to a suspected issue, it should be combined with other troubleshooting techniques such as field tests, column history and operation experience, hydraulic analysis, gamma scans, temperature surveys, and others. Many of the cases in this chapter illustrate the combination of the sketches with one or more of these troubleshooting techniques. For instance, in the case in Section 8.2, the diagnosis derived from the sketch was combined with hydraulic analysis, gamma scans, and addressing why previous fixes were unsuccessful.

8.2   FLASHING FEED ENTRY CAUSING A 12-YEAR BOTTLENECK

Kister, H. Z., T. C. Hower, P. R. de M. Freitas, and J. Nery, Ref. 233. Reprinted Courtesy of the American Institute of Chemical Engineers.

In this case study, a flashing feed entering a downcomer bottlenecked an entire olefins plant for 12 years. This bottleneck survived three unsuccessful fix attempts by a major engineering contractor who failed to study plant data, look at gamma scans, and most importantly, draw a sketch.

An Olefins plant demethanizer (Figure 8.1a) separated methane and a small amount of hydrogen as the overhead product from C_2 and heavier hydrocarbons as the bottom product. The tower had a small-diameter upper section containing 23 valve trays, and a larger-diameter bottom section containing 44 valve trays and also an interreboiler. The demethanizer received four feeds. Three of these entered the upper section above trays 7, 11, and 17. The fourth feed entered the tower swage, just below tray 23.

The demethanizer started up in 1978, with the plant producing 46 tons/h ethylene. It operated well at full reflux and an overhead pressure of 30 bars. As the plant's throughput was increased to 50 tons/h ethylene, the pressure drop in the upper section of the demethanizer (tray 23 upward) rose and the column became unstable. The demethanizer became a plant bottleneck.

In an attempt to debottleneck the tower in 1983 and 1986, the following modifications were performed (Figure 8.1a) at the recommendation of a major engineering contractor:

- Downcomers from trays 17–23 (and later those from trays 10–16) were expanded by 14%.
- A new 15-tray rectifier was added in series with the existing demethanizer to reduce ethylene losses.

The modifications gained nothing. The demethanizer remained bottlenecked at 50 ton/h ethylene. The ethylene losses remained the same.

At the capacity limit, there was a rise in differential pressure, massive entrainment (observed as the liquid accumulation in the bottom of the rectifier), and a loss of separation (seen both by an increase in ethylene losses and by warming of the tower top section). These were symptoms of flooding (Section 2.6), initiating somewhere above tray 23 and propagating upward.

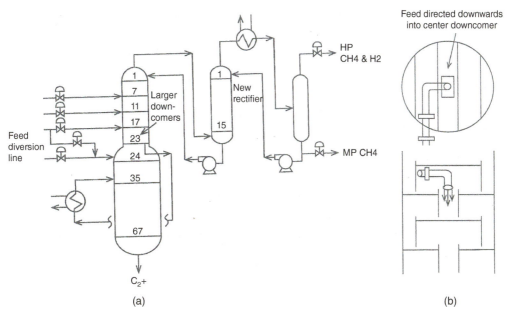

Figure 8.1   Olefins plant demethanizer bottleneck. (a) Process scheme, showing the fixes attempted; (b) flashing feed arrangement for trays 7, 11, and 17 that brought flashing feeds into downcomers, causing the bottleneck at tray 17. (From Kister, H. Z., T. C. Hower Jr., P. R. de M. Freitas, and J. Nery, "Problems and Solutions in Demethanizers with Interreboilers," Proc 8th Ethylene Producers Conf, New Orleans, Louisiana, 1996. Reprinted with permission. Copyright © American Institute of Chemical Engineers (AIChE).)

As the feeds were mostly liquid (by weight), the peak hydraulic loads in the upper section of the demethanizer were between trays 17 and 23. To debottleneck this section at the 1986 turnaround, a line was connected from the tray 17 feed into the tray 23 feed (Figure 8.1a). A portion of the tray 17 feed was diverted into the lower feed point, so that it bypassed the small-diameter section. With the diversion valve 50% open, a plant throughput of 57 ton/h ethylene was achieved.

Although the feed diversion line was effective, it was a control nightmare and an inefficiency. In the next debottleneck, the plant wanted to raise throughput by about 20% with the feed diversion line closed.

A debottlenecking study by a major engineering contractor recommended replacing all the demethanizer trays by random packing. A critical evaluation by the plant revealed that in the large-diameter (bottom) part, the existing trays had plenty of capacity and there was no need to replace them. On the other hand, the top section was a bottleneck, so replacing trays by packing appeared justified.

Just prior to the modifications, the plant became aware of field data (229) of low random packing efficiencies at the top section of a demethanizer, under conditions of good vapor and liquid distribution. The low efficiency was attributed to the high hydrogen concentration (Case 4.8 in Ref. 201). The plant naturally became concerned. A task force was formed to critically evaluate the need for packing in all upper sections of the demethanizer.

A hydraulic analysis showed that at the debottleneck throughput, trays 1–10 and 11–16 would operate at 40% and 55%, respectively, of the closest flood limit. They were capable of handling double the current throughput. Replacing trays 1–16 by packings, therefore, would achieve no capacity gain, would lose separation, and would cost money. Plans to replace trays 1–16 by packings were therefore scrapped (even though the packings and distributors have already been purchased).

At the debottleneck throughput, trays 17–23 approached the downcomer choke (Section 2.3). With the packings already purchased, it was decided to proceed with replacing these trays by packings. A minimal loss of separation was expected both because this section contained only a few stages and because this section had the least hydrogen concentration.

The hydraulic analysis reopened the search for the root cause of the bottleneck. While the downcomer choke on trays 17–23 was the closest flood limit, it was not expected to be reached until the throughput was raised by 20% and the feed diversion line was closed. In practice, the bottleneck was observed without raising throughput and with the feed diversion line 50% open. Calculations showed that at these loads, trays 17–23 should have been operating at 60% of the downcomer choke and nowhere near any other flooding limit.

Another strong argument against the downcomer choke was that in 1983, enlarging the downcomer top area by 14% did not affect the column bottleneck. Had the bottleneck been a downcomer choke, the larger downcomer top area should have led to an improvement.

Tower gamma scans were closely examined. The scans showed flood initiation around tray 17, not tray 23. This was unexpected since the hydraulic loadings on tray 23 were higher than on tray 17. Had the tower encountered a tray or downcomer limitation, it should have initiated in the higher loading region, i.e., tray 23.

Tray 17 was a feed tray. Points of transition, like feed trays, are spots where major tower bottlenecks often initiate. A drawing review (Figure 8.1b) showed flashing feed entering the downcomer. The literature (192, 198, 285) recommends against this feed entry for flashing feeds. For the tray 17 feed, the vapor content may appear small, about 0.6 wt%, but, 0.6 wt% is 10 vol%, a vapor fraction far too large to be entered into a downcomer. This vapor disengaged from the downcomer liquid and impeded the descent of tray liquid into the downcomer. This in turn brought about a premature downcomer choke limitation.

The tower bottleneck, therefore, turned out to be neither the tray nor downcomer capacity. It was a poor introduction of feed. The problem was not unique to tray 17. The feed entries on trays 11 and 7 were just as poor. However, trays 7 and 11 had not encountered a bottleneck yet, possibly due to their large margin away from the downcomer choke.

The feed arrangements on trays 7, 11, and 17 were replaced by well-designed feed arrangements. In addition, trays 17 to 23 were replaced by 2-in. modern random packings. Upon return to service in May 1995, the bottleneck completely disappeared. The feed diversion line never needed to be opened again. The 12-year-old feed entry bottleneck ceased to exist.

Takeaway   This bottleneck could have been eliminated 12 years earlier had a sketch of the relevant point of transition (feed to tray 17) been drawn.

8.2.1   Upward Component of Flashing Feed Entry Causes Damage and Flooding

In one amine regenerator (63), an upward slot in the flashing-feed distributor caused impingement on the bottom of the seal pan that collected liquid from the trays above. A crack developed at the clamp and propagated, leading to shearing of a portion of the seal pan. Seal loss led to flooding in the top trays and reflux surging into the reflux drum. These were solved by a new seal pan and adding stiffener plates at both sides of the clamping lengths.

Upward velocity components of flashing feeds should be avoided. At the least, strong shields should be added to protect internals from their impingement.

8.3   FEED MALDISTRIBUTION TO 4-PASS TRAYS CAUSING POOR SEPARATION

In the following case study, a participant in the author's seminar class drew the point of transition sketch that correctly diagnosed the separation problem in a debutanizer tower in his refinery gas plant. This tower separated C_3 and C_4 hydrocarbons as the top product from C_5, C_6, and other heavier gasoline components as the bottom product at a pressure of about 150 psig.

The tower could not achieve a satisfactory separation. There was a strict purity specification (Reid vapor pressure; essentially, C_4 content) on the bottom product which they could not violate. Due to the separation issue in the tower, keeping the bottoms on-spec came at the expense of over-reboiling and sending heavies to the top product. They tower was experiencing 8–10 mol% C_5's in the overhead product, which was an economic loss, and gave a C_4 product that was difficult to sell.

Figure 8.2 shows the sketch of the feed point of transition, with dimensions in mm. The bottom section of the tower was 10 ft in diameter and contained 4-pass trays. The top section of the tower was 8 ft in diameter and contained 2-pass trays. The region shown is where the feed entered and also where the number of passes was switched from two to four.

The following describes how this design was meant to work with reference to one half-tower. The other half was a mirror image of the first. The side downcomer from the last upper tray

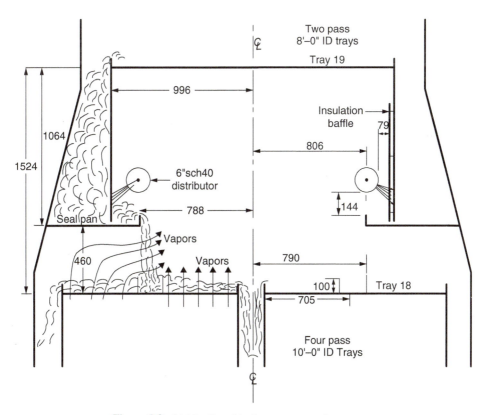

Figure 8.2   Maldistributed feed arrangement to 4-pass trays.

(tray 19) terminated a short clearance above the floor of a seal pan mounted 460 mm above the deck of the first lower tray (tray 18). The seal pan was extended all the way to about midway between the tower wall and centerline. Flashing feed from the 6 in. pipe, 2/3 liquid by weight, was discharged via slots directed at 45° toward the downcomer external wall. The downcomer wall redirected the feed vapor upward and the feed liquid downward. Liquid from the downcomer would flow under the downcomer wall, join the feed liquid, and together overflow the seal pan weir into a trough underneath. The trough would split the liquid equally between the side and center panels of the 4-pass tray 18 below. At least this is the way it was meant to work.

With the sketch in Figure 8.2, it does not take long to see why this arrangement would not work. The seal pan would deflect the vapor from the side panel of tray 18 horizontally, and its horizontal momentum would blow the descending liquid into the center panel, just like wind blows the rain sideways. The additional liquid reaching the center panel would generate a higher liquid head there, causing more vapor to break through the side panel. This will intensify the wind, which will blow even more descending liquid to the center panel. A self-accelerating process will initiate, resulting in most of the liquid blown to the center panel of tray 18 and most of the vapor from tray 17 below rising through the side panel of tray 18.

The question is whether this maldistribution would only affect tray 18. To answer, consider that the larger liquid flow through the center panel of tray 18 will continue to the center panel of tray 17, again channeling the vapor to the side panel of tray 17 and therefore 16. So this maldistribution would propagate through many trays before it evened itself out, if it evened itself out. This channeling would give uneven L/V ratios on the panels of the lower section, and poor efficiency, resulting in the poor separation experienced in the tower.

Another, more subtle issue can also be seen by a close review of Figure 8.2. The bottom of the feed pipe is vertically only 144 mm above the top of the seal pan overflow. This short distance does not give the vapor enough length to change direction and disengage, and it would blow on the liquid in the seal pan. This would shoot more liquid into the center panel, aggravating the maldistribution. Alternatively, some of this vapor may even penetrate under the downcomer wall. Vapor penetrating the downcomer is unlikely to have been the issue here, as it would have caused downcomer flood, which was not observed in this tower. However, vapor penetrating a downcomer by this type of mechanism was observed in another tower (381).

In a new tower, it is best to avoid this design. A good practice is to rotate the upper trays 90° compared to the lower trays (192). In an existing tower, rotating is usually not practical, and the least troublesome solution is replacing tray 18 by a chimney tray, adequately designed to provide good liquid distribution to the 4-pass trays below. The disadvantage is the loss of 1 tray, as a chimney tray does not perform mass transfer. However, in the author's experience, losing 1 tray out of 40 hardly ever makes a difference in performance. When it is important to avoid losing this tray, a clever piping arrangement can be devised and can be feasible with a vertical height of 5 ft like in this tower. However, such piping arrangement can be tricky, especially when the vertical height is 3 ft or less.

Additive Maldistribution   Many towers contain nozzles to add anti-fouling or anti-corrosion inhibitors. These should be reviewed to provide effective distribution of the additive to 4-pass trays. In one 4-pass C_2 splitter tower (266), inadequately distributed methanol injection just above the tower feed was only partially effective for dissolving hydrates.

Takeaway   The author has been using the Figure 8.2 sketch as a class exercise in his seminars, asking participants to identify the problem after describing the tower. The author has not had a class in which the participants (who never saw this sketch before) have not correctly identified the problem. This shows you the power of a sketch.

8.4   FEED PIPES BLOCKING LIQUID ENTRANCE TO DOWNCOMERS

Figures 8.3 and 8.4 show situations in which feed pipes too close to the downcomer entrance obstructed liquid descent, which stacked up liquid on the tray, prematurely flooding it. The Figure 8.3 case was part of a tower revamp with high-capacity trays that failed to achieve high capacity because the downcomer obstruction by the feed pipe bottlenecked tower capacity. The fate of the tower in Figure 8.4 was much better, because the bottleneck was discovered by an engineer that drew this sketch, used it to identify the problem, and changed the feed pipe design before it would run into the same bottleneck as in Figure 8.3.

Reference 104 describes one more similar case in which feed pipes too close to the down-comer entrances became capacity bottlenecks. The tower had five alternate feeds, each with

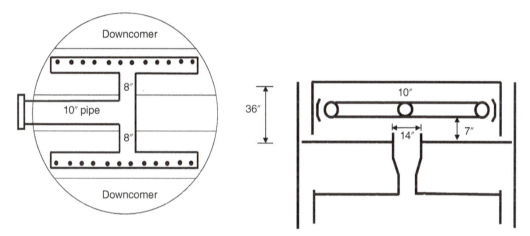

Figure 8.3   Feed pipe blocking entrance to downcomer.

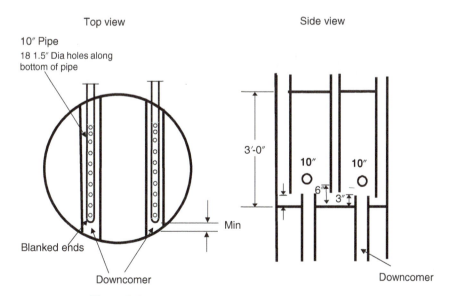

Figure 8.4   Another feed pipe blocking entrance to downcomer.

the same piping arrangement. The problem was diagnosed by gamma scans that noticed liquid stackup not only on the tray where the feed entered during the scans, but also on the four alternate feed trays that had no feed going in at the time.

Reference 406 describes an even worse scenario in which the feed pipe distributor was inside the downcomer, reducing its cross-section area by 60% and bottlenecking the tower.

8.5 DRAW SUMP BLOCKING LIQUID ENTRANCE TO DOWNCOMERS

Case 1 (based on Ref. 18) Following a revamp in which internal vapor rates were raised in a fluid catalytic cracker (FCC) main fractionator (MF), fractionation throughout the column degraded. The gasoline top product had an excessive and erratic concentration of heavies (end point varied erratically between 450°F and 520°F). Flooding was suspected. One tower gamma scan showed the column flooding from tray 13 all the way to tray 1 at the top of the column, while other gamma scans showed no flood. The revamp changed all the column internals except for trays 12–14, which were determined to have sufficient capacity.

A thorough pressure survey measured a pressure drop of 0.23 psi/tray across the top 13 trays. No pressure points were available to narrow it further. However, vapor sensitivity tests, similar to those described in Section 2.7.1 identified the problem area.

In the first test, the vapor load to trays 1–9 was lowered by raising the condensation rate in the light cycle oil (LCO) pumparound (PA). The LCO PA was drawn from tray 12 and returned to tray 10, and increasing condensation there would lower the vapor load on trays 1–9. This did not reduce the pressure drop. In the next test, the vapor rate entering tray 14 was lowered by condensing more in the lower pumparounds, and this effectively reduced column pressure drop and the heavies in the gasoline. The tests clearly indicated flooding starting on trays 10–14, that could be eliminated by reducing the vapor flow to these trays.

Both the flooded gamma scan and the field tests focused on trays 10–14. The relevant point of transition, the LCO draw arrangement, was examined using a sketch (Figure 8.5a). It clearly showed the LCO draw box on tray 12 restricting the active area on tray 13 and choking the liquid flow into the tray 13 center downcomer. This was the likely root cause.

Case 2 One 8.5 ft ID tower experienced flooding initiating one tray below the 4″ draw in Figure 8.5b as seen by the gamma scans. The draw sump extended to an elevation of about 6″ above the outlet weir of the tray below and choked the entrance to the downcomer. The draw was sized for self-venting flow. Here again the draw sump choked the liquid flow into the downcomer.

Case 3 The likely cause of premature flooding in a tower containing 4-pass trays was obstruction of the downcomer entrance by the seal pan and I-beam (Figure 8.5c). The problem was diagnosed by thorough gamma scanning (327). The author is familiar with a retray of a tower handling a foaming system in which lengthening the support I-beams hanging above the center downcomers restricted the downcomer entrance areas and lost tower capacity. Re-shortening the I-beams recovered the lost capacity.

Other Cases Ref. 236 and Case 9.9 in Ref. 201 describe another case in which a downcomer entrance choked by a draw sump caused a premature entrainment which bottlenecked a tower. Reference 112 describes a related case in which trusses over the outlet weirs restricted vapor disengagement from downcomers in a high-pressure system, causing premature flooding.

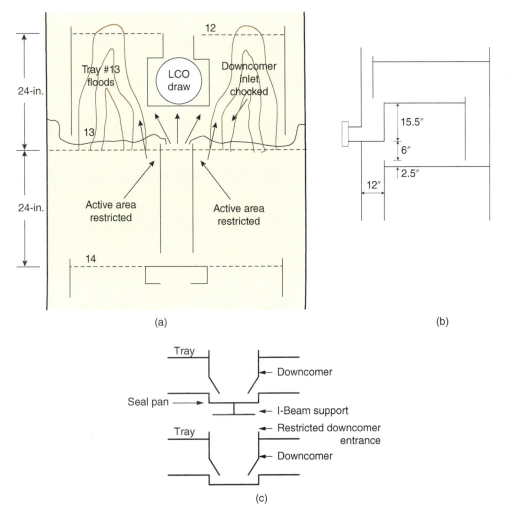

Figure 8.5   Draw boxes choking downcomer entrance. (a) Draw box restricting the active area above tray 13 and choking the center downcomer entrance; (b) draw box choking side downcomer entrance; (c) seal pan and I beam choking downcomer entrance. (Part (a): From Barletta, T., Hydrocarbon Processing, p. 69, November 2003. Reprinted with permission. Part (c): Based on Mosca, G., E. Tacchini, S. Sander, D. Röttger, A. Wolff, A. Rix, and G. Schilling, "Upgrading a Super-Fractionator C4s Splitter with 4-Pass High Performance Trays at Evonik Degussa, Marl, Germany", in Distillation 2009: Topical Conference Proceedings, p. 229, AIChE Spring National Meeting, Tampa, Florida, 26–30 April, 2009.

8.6   UNSEALED DOWNCOMERS OR OVERFLOW PIPES CAN LEAD TO PREMATURE FLOOD

Lack of liquid seal on downcomers or overflow pipes can lead to premature flooding or entrainment propagating up the column.

In the froth and emulsion regimes, the outlet weirs usually hold a frothy liquid pool on the trays that keeps the downcomers from the tray above sealed, preventing vapor entry into downcomers. In the spray regime, where the liquid pool may not form due to the low liquid rate

(Section 2.11), and in many high-capacity trays that have truncated downcomers, downcomer sealing is achieved by restricting the downcomers exit areas.

To seal the bottom downcomer, or the downcomer above a chimney tray, a seal pan is usually used. Alternatively, but far less frequently, this downcomer is extended into the bottom sump or chimney tray liquid and immersed in the liquid.

When reviewing bottom sump and chimney tray sketches or inspecting them, it is important to check for adequate sealing of bottom downcomers. When the downcomer is submerged in the bottom sump or chimney tray liquid, it is important to check for excessive submergence. Unsealed downcomers or excessive submergence can lead to premature flooding as illustrated in the following.

Case 1 (Figure 8.6a) During the startup of a sour water stripper column (385), the feed flow rate was gradually increased toward the 35 m³/h (154 gpm) design rate. At approximately 60% of design feed rate, the pressure drop suddenly rose and the reflux drum flooded. These are symptoms of flood (Section 2.6). The flood was sensitive to vapor loads but not to liquid loads. Reducing the feed rate or the stripping steam were the only ways of bringing the tower back under control.

The column was shut down and inspected. Tray 18 was found bowed downward and had detached from its support clips, as shown by the dashed lines in Figure 8.6a. Other trays in the tower top section were also bowed but remained in place. The damage is believed to have been caused by an excessive step-up in the cold water feed during the startup while the column was full of steam. The cold step-up rapidly condensed steam at the inlet, causing a low-pressure region that sucked the upper trays downward.

At the inspection, it was observed that the seal of the tray 19 downcomer had been lost as the side panel of tray 18 (under the downcomer from tray 19) was also bent downward. It is believed that the steam flow from tray 17 traveled preferentially through the tray 19 downcomer rather than through the fixed valves of tray 19. This explained why the top section of the column quickly flooded (evident from a sudden increase in tray differential pressure) and flooded the reflux drum, as well as the vapor sensitivity of the flood.

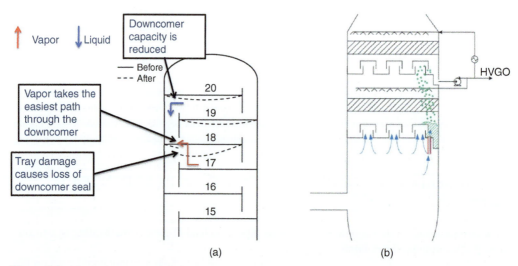

(a)   (b)

Figure 8.6   Floods due to unsealed downcomers or downpipes. (a) Downcomer unsealed due to tray damage; (b) unsealed chimney tray overflow due to shortened overflow pipe. (Part (b): From Vail, R., and G. Cantley, "Pinch Points: Commonly Overlooked Pieces of Equipment which Result in Flooding and Unit Constraints," Topical Conference Proceedings, Kister Distillation Symposium, AIChE Spring Meeting, Houston, Texas, March 12–16, 2023. Reprinted with permission.)

Case 2 (Figure 8.6b) A refinery vacuum tower (489) had a history of heavy gas oil (HVGO) showing entrainment breaking through the packed ("wash") bed below the draw. Entrainment was consistent throughout the operating cycle and not associated with coking. A pressure survey showed a liquid head of 450 mm of liquid on the chimney tray, suggesting that the liquid was backed up and covered the chimney opening.

A turnaround inspection showed that the overflow pipe (shown in red) was short and unsealed. This allowed vapor to ascend and back up the liquid on the chimney tray. The high liquid level on the chimney tray resulted in entrainment through the bed above into the product, as illustrated in Figure 8.6b.

Other cases of seal pan issues observed during inspections are described in Section 10.3.7 and Figure 10.21.

8.7   EXCESSIVE DOWNCOMER SUBMERGENCE CAN LEAD TO PREMATURE FLOOD

Figure 8.7 shows a water removal tray in a gas plant stabilizer (based on Ref. 472). The feed rate to the tower was bottlenecked at 7800 BPD (design 15,000 BPD) by severe carryover from the top of the tower.

The liquid on the chimney tray is degassed, and its density is that of the actual liquid. In contrast, the liquid in the inlet downcomer is aerated with vapor bubbles. With a typical downcomer aeration of 60%, the hydrostatic head of the submerged height (about 31 in.), will lift 31/0.6 = 52 in. of aerated liquid in the downcomer. In addition, the pressure drop of the tray above plus the friction of liquid flowing under the downcomer bar will add another 10 in. or so of aerated liquid, causing a downcomer backup to exceed the 60 in. tray spacing and initiating downcomer backup flood. Shortening the outlet weir height of the chimney tray from 33 to 11 in. solved the problem and the stabilizer reached its design capacity.

Reference 265 describes a similar experience. This was an olefins condensate stripper with an internal head forming a sump for water removal. The top of the overflow weir was about 3 ft below the tray above, and the submergence of the downcomer in the tray liquid was about 5.5–6 ft. There was enough density difference between the degassed sump liquid and the aerated downcomer liquid to back up liquid all the way to the tray above. Removing the tray above, which raised the spacing above the overflow from 36 to 57 in., along with tray changes in the section above the chimney tray, solved the problem.

In a third case, a primary absorber in a refinery coker unit experienced premature flooding right above a large chimney tray that was collecting liquid to be drawn for a cooling loop. Here, the downcomer submergence in the chimney tray liquid was 23 in., and the space between the top of the overflow weir and the tray above was 27.5 in. The aeration of the downcomer liquid in this case may have been slightly less than 60% as the system tends to be slightly foamy. Here too, the degassed liquid on the chimney tray pushed the aerated liquid in the downcomer to the tray above.

A fourth case (125) is described in Section 9.2.6 and in Figure 9.20b. Here, the high submergence of a downcomer in a seal pan drawing liquid to an intercooler loop in an FCC hydrocarbon absorber caused tower premature flooding.

The author has a number of other similar experiences.

Another type of problem is liquid vaporization in downcomers or downpipes submerged in hotter bottom sump liquid. In one case (489), such vaporization choked a line from a chimney tray to a once-through thermosiphon reboiler that passed through the hotter bottom sump. The choke backed up liquid above the chimneys, which in turn flooded the column.

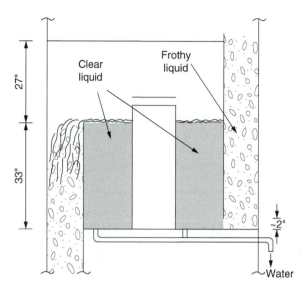

Figure 8.7   Downcomer submergence in chimney tray or sump liquid can bring about premature downcomer backup flooding on the tray above.

8.8   "LEAK-PROOF" CHIMNEY TRAYS IN AN FCC MAIN FRACTIONATOR

Kister, H. Z., B. Blum, and T. Rosenzweig, Ref. 236. Reprinted Courtesy of Hydrocarbon Processing.

As part of a revamp to maximize capacity of an FCC main fractionator, the light cycle oil LCO and heavy cycle oil (HCO) draw/pumparound off-take trays were replaced by total-draw chimney trays. The purpose was to minimize reflux to the tower section below. Excess reflux induces additional liquid and vapor recycle that consumes capacity. The reflux was minimized by careful monitoring and control while avoiding any fluctuating leakage or overflow from the chimney tray above. Each chimney tray was to be seal-welded.

A schematic of either chimney tray is shown in Figure 8.8a. Liquid, from the two-pass tray above, descended via side downcomers, which terminated in seal pans. All liquid from the chimney tray was drawn from a sump (not shown) located right beneath the chimney tray. The downcomers from the chimney tray to the tray section below were converted to overflows by raising outlet weir heights to 612 mm. Normal liquid level on the chimney trays was about 300 mm and the overflow downcomers were inactive. However, should an upset occur and the chimney tray liquid level exceed 612 mm, the liquid would overflow into the downcomers.

At the design stage, the seal pans and the chimney tray were on different drawings, which had been approved for fabrication. A last-minute drawing review put together the sketch in Figure 8.8b, which revealed a major flaw. The vapor issuing from the outside chimneys and blowing toward the tower wall would blow liquid descending from the seal pan directly into the overflow downcomers. Thus, despite the seal welding of the chimney tray, liquid would bypass it.

Figure 8.8c shows how the problem was circumvented. The openings of the outside chimneys that would blow vapor toward the wall were closed. A 25-mm vertical drain lip was installed at the bottom of each seal pan to prevent the issuing liquid from crawling underneath and ending in the overflow downcomers.

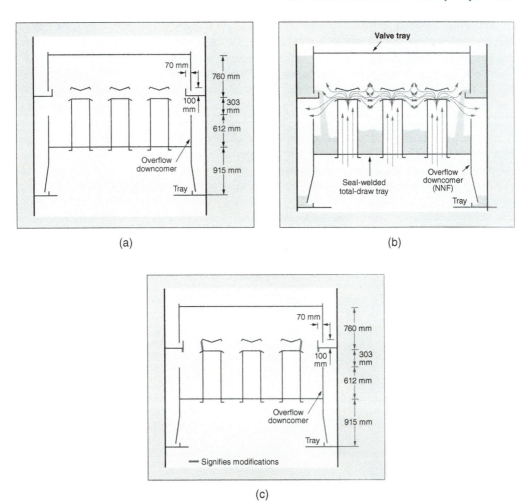

(a)   (b)

(c)

Figure 8.8   "Leak-proof" total draw chimney tray. (a) Initial design; (b) expected flow patterns; (c) modifications to circumvent liquid bypassing the chimney tray. (From Kister, H. Z., B. Blum, and T. Rosenzweig, Hydrocarbon Processing, p. 101, April 2001. Reprinted Courtesy of Hydrocarbon Processing.)

Takeaway   For troubleshooting points of transition (feeds, drawoffs, bottom sumps, chimney trays) you do not need an expert. You need a sketch. Without a sketch, even an experienced troubleshooter can miss a key issue.

8.9   MORE "LEAK-PROOF" CHIMNEY TRAYS

Case 1 In an FCC main fractionator (see Figure 8.9a), the tower bottoms ("decant oil") was a small heavy, carbon-rich stream used as feedstock for battery production. The quality of this stream is often expressed as API gravity, with a typical value of −1. The first distillate stream drawn from the fractionator was LCO, which was also the most valuable product from the fractionator. In between the LCO draw and the bottom, there were two fractionation sections (one trayed, the other packed) as well as two PA sections in which heat was removed by direct contact with externally cooled circulating streams.

Following a revamp in which the trays in the HCO PA section were replaced by packings to increase capacity, the fractionator experienced low LCO yield. The trays below the HCO draw appeared to work well and did not dry up, as seen by the temperature indicator at the bottom tray, which was always relatively cold. The decant oil product flow rate, about 3000 BPD before the revamp, more than doubled to 7000 BPD after the revamp, and its API gravity was around 6 to 7, making it too light and unsuitable as battery production feedstock. It appeared that the lost LCO yield ended in the decant oil.

Figure 8.9b shows the HCO PA draw. It was a chimney tray with submerged side sumps from which the HCO pump took suction. Liquid to the distributor below descended via a center downcomer. The downcomer rose 24 in. above the chimney tray floor, so that it only took the overflow. The chimneys were 29 in. tall. The center downcomer stretched from one wall to the opposite wall, so liquid could not cross it from side to side. The chimneys had considerable gaps between them, allowing liquid equalization on each half-tray.

Attempts to raise the LCO draw rate led to cavitation of the HCO PA pump. Cavitation of the pump means that the level on the chimney tray was lost, but during this time, liquid still appears to be descending down the tower, as suggested by the cold wash temperature and the high decant oil API gravity.

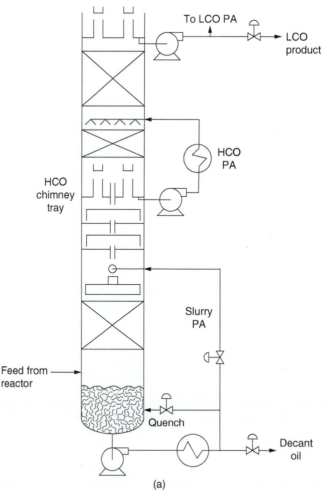

(a)

Figure 8.9 Liquid getting past another "leak-proof" chimney tray. (a) Bottom section of FCC main fractionator; (b) how the liquid bypassed.

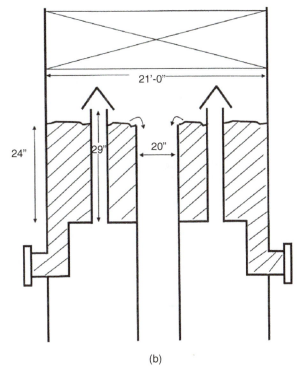

(b)

Figure 8.9   (*Continued*)

A close observation of the sketch in Figure 8.9b shows that while the chimneys have hats, the downcomer does not. This means liquid from the bed above rained into the downcomer. The cross-section area of the center downcomer was 10% of the tower cross-section area. Assuming the liquid rained equally from the packed bed above, 10% of the flow at the bottom of the bed would rain directly into the downcomer without overflowing the weirs. The PA circulation rate was 60,000 BPD, not including condensation, so at least 6000 BPD directly rained into the downcomer, and most would end in the decant oil.

There are a number of ways of solving this problem. An umbrella over the downcomer is a simple solution, but its engineering can be tricky. Alternatively, since the LCO is far more valuable than the decant oil, there is incentive for maximizing it. This will best be achieved by bringing reflux back as flow-controlled pumpback from the discharge of the HCO PA pump, and eliminating the downcomer.

A participant in the author's seminar class struggled with the problem for a few months. At the author's recommendation he drew the point of transition sketch overnight, and the next morning had the problem correctly diagnosed. Magic.

Case 2 A refinery vacuum tower had problems with liquid getting past the light vacuum gas oil (LVGO) collector chimney tray. As a result, much of the LVGO ended up in the lower middle vacuum gas oil (MVGO) cut.

The chimneys were rectangular, about 5 ft long, and had sloping hats (Figure 8.10). The hats did not have any overflow weirs on any side. So liquid dripping on the hats flowed along the slope, with nothing stopping this liquid from wrapping around the hats and dripping into the chimneys. The slopes were very small, about 1 in. per 5 ft.

The high, narrow edge of each hat, as well as the sides, were open. During a turnaround, the engineers noticed water marks showing liquid flowing down the slope of one of the middle

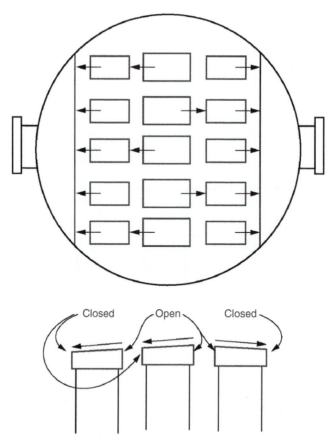

Figure 8.10   Liquid getting past yet another "leak-proof" chimney tray. Liquid collected on one hat flowing into the space between the chimney and hat in front.

chimneys and pouring into the open high, narrow end of the chimney in front. The narrow end openings were then closed, but this gave only a small improvement. The main path of liquid descent was by wrapping itself around the hats and pouring into the chimneys.

Case 3 A (Section 7.3.4) A refinery atmospheric crude tower experienced instability at the jet fuel draw and loss of jet fuel to a lower product. The root cause was some of the liquid overflow from the seal pan above falling into the wrong compartment. This was identified by the sketch in Figure 7.39.

Case 4 Watching out was recommended (429) for liquid dumping through overflow pipes located right beneath seal pans, overflow pipes with no hats under packed beds (similar to Case 1 above), chimneys with flat hats (Case 9.5 in Ref. 201), or hat diameters not larger than the chimneys.

8.10   HYDRAULIC GRADIENTS GENERATING CHIMNEY TRAY OVERFLOWS

Chimneys (risers) blocking liquid flow toward the draw sump on chimney trays incur excessive hydraulic gradients. This may lead to overflows and allow liquid bypass getting past the chimney tray to the section below.

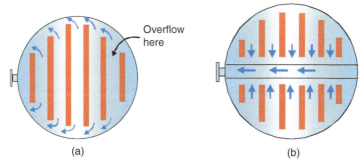

Figure 8.11   Overflow due to hydraulic gradients on chimney trays. (a) High hydraulic gradient causing overflow; (b) good design, no hydraulic gradients and no overflow. (From Kister, H. Z., B. Blum, and T. Rosenzweig, Hydrocarbon Processing, p. 101, April 2001. Reprinted Courtesy of Hydrocarbon Processing.)

Case 1 (236) In a packed refinery vacuum tower, valuable HVGO was drawn from a total draw chimney tray. The tower experienced HVGO loss to the residue leaving from the bottom. The lighter gas oil lowered the viscosity of the residue, making it unfit for asphalt. Penetration (a test in which a standard ball is dropped into a container of product, and the distance it penetrates is a measure of the viscosity) was high (~20 in., compared to ~2 in. for good asphalt).

Figure 8.11a shows the chimney tray. The long edges of the chimneys perpendicular to the liquid flow narrowed the flow area around each chimney. This generated a large hydraulic gradient. On the opposite side from the outlet draw, the liquid rose to the chimney height and overflowed into the chimneys and then onto the tower bottom, lowering the viscosity of the residue.

A good chimney tray design (Figure 8.11b) minimizes hydraulic gradient. In this existing tower, an easier and cheaper solution was to rotate the tray in Figure 8.11a by 90°. This solved the problem, eliminating the HVGO loss to the residue, and restoring good penetration of the residue.

In a related case, the same issue occurred and was resolved by raising the heights of the chimneys on the side opposite of the draw at the high end of the hydraulic gradient.

Other cases The author was involved in numerous incidents where chimneys obstructed liquid flow, causing excessive hydraulic gradients and overflows. An additional case is described in Section 4.6.3.

8.11   LOOK FOR THE POSSIBILITY OF A SYSTEM LIMIT SETTING IN

The system limit flood (Section 2.2) occurs when the upward vapor velocity exceeds the terminal velocity of the large drops. At a point of transition, piping and column internals consume a significant portion of the tower cross-section area, leaving a restricted passage area for the rising vapor. In the restricted space, the vapor velocity may exceed the system limit. For as long as only vapor flows through this restricted passage, there is no problem. However, if liquid attempts to descend into this passage, the vapor velocity will prevent it from doing so. Instead of going down, the liquid will accumulate above, initiating flood.

A system limit flood at a point of transition will give the same symptoms of a regular flood (Section 2.6): pressure drop rise due to the liquid accumulation above, massive entrainment from the top of the tower, a possible reduction in bottom flow rate, and loss of separation. Symptoms unique to this type of flood include the trays or packings in the tower operating well below the calculated flood point, and gamma scans showing the initiation of flooding at that point of transition.

System limit flood at a point of transition can be vapor- or liquid-sensitive. Being a system limit flood would inherently make it vapor sensitive. However, often the vapor velocity at the restricted passage area is well above the system limit. As long as no liquid enters the restricted passage region, there will be no system limit flood. So the flood will occur only when liquid attempts to enter the area, regardless of the vapor rate. In this case, it will be liquid-sensitive with no vapor sensitivity. So there may or may not be a liquid sensitivity. This will be further discussed in relation to the cases described below.

The possibility of a system limit flood should be checked whenever there is a restricted passage space at the point of transition for which both liquid and vapor compete. The check is a system limit calculation based on the restricted passage area. The system limit criterion in Section 2.16, i.e., a C-factor of 0.45 ft/s, is an excellent approximation for hydrocarbon and organic systems at less than 100 psia and when the liquid flow through the restricted opening is less than 20 gpm/ft^2. Outside of this range, it is best to check using the Stupin and Kister correlation (463), which can be found in Perry's Handbook (245).

It is important to note that the above criteria for system limit flood are based on large liquid drops. However, in many points of transition, the liquid downflows not as drops but as sheets of liquid along the wall. Sheets are more difficult to entrain. In this case, both the Stupin and Kister correlation and the C-factor criterion above will give conservative predictions. When looking at the sketch, it is important to evaluate whether the liquid descent is more likely to be drop-like or sheet-like, and apply the criterion accordingly. The Diehl and Koppany correlation (85) is for flooding in vertical heat exchanger tubes, where the liquid tends to descend as a film along the tube walls, and is therefore suitable for sheets of liquid descent pattern. The author has had excellent experience with this correlation, which is also found in Perry's Handbook (245).

Case 1 Packing distributor overflows When a packing distributor, or redistributor, or a chimney tray overflows, liquid enters the vapor passages. The typical vapor rise area is 50%, 25%, and 25% for trough distributors, pan distributors, and chimney trays, respectively. With C-factors of high-capacity packings often approaching 0.4 ft/s (Section 2.16), a pan distributor or chimney tray overflow is likely to induce system limit flooding. Trough distributors have a better chance to remain unflooded when an overflow occurs, especially when they contain well-designed overflow guide tubes that keep the overflow out of the vapor passages. Further, the overflowing liquid often descends as sheets along the trough walls, which is beneficial. So trough distributors may or may not flood upon overflow.

Figure 4.6 shows a turnaround water test of a packing pan redistributor in a caustic absorber that overflowed at below the design liquid rates. In the absence of vapor, the overflowing water had no problem descending into the riser. During operation, the vapor velocity through this riser was far too high to allow the liquid descent, and upon overflow, the tower flooded with caustic carryover from the top.

A flood resulting from distributor or redistributor overflow is always liquid-sensitive. Without an overflow, there is no flood. Once the liquid rate is raised, the liquid head in the distributor rises and overflows upon reaching the top of the risers. Lowering the liquid rate makes the flood go away. With trough distributors and redistributors, there is often also vapor sensitivity. If the vapor load is not very high, the overflowing liquid will descend, but raising the vapor load will initiate flooding.

Case 2 Vapor space at a seal pan overflow A tower flooded at 80% of the calculated tray flood rates, seen by an increase in tower pressure drop. Gamma scans also showed flood, starting at the bottom tray.

Downcomers from the bottom tray terminated in two wide seal pans, from which the liquid flowed to two once-through thermosiphon reboilers (Figure 8.12a). The reboiler return nozzles were located below the seal pans. The wide seal pans blocked much of the tower

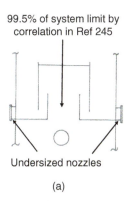

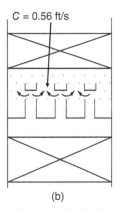

Figure 8.12   System limit troubleshooting at points of transition. (a) Premature flooding in tray tower, due to system limit between the overflowing seal pans; (b) premature flooding initiating at redistributors hats, as C-factor between hats exceeded the system limit.

cross-section area, leaving a restricted passage space between them. The reboiler draw nozzles were undersized, and unable to take down all the liquid descending from the bottom tray, so some of the bottom tray liquid overflowed the seal pans. Ascending vapor and descending liquid competed for the restricted passage space between the seal pans. Applying the system limit correlation (463) showed that the tower flood coincided with the exact rate at which the vapor velocity at the restricted passage space reached the system limit flood velocity.

The bottleneck was eliminated by increasing the size of the draw nozzles and rundown line. The larger pipes took all the liquid to the reboilers, eliminating the overflow and, therefore, also the system limit.

Case 3   Large redistributor or chimney tray hats   Figure 8.12b shows a redistributor between two packed beds. Liquid from the upper bed descends into the redistributor, no collector. The redistributor risers are covered with hats a short distance above to prevent liquid from the bed above from raining into the risers. A common hat design is shown on the lower plate in Figure 4.22. Liquid collected on the hats descends over the short edges, with a short lip protecting it from being blown by the vapor. Short weirs (typically 2 in. tall) avoid liquid downflow along the long edges of the hats to prevent it from being reentrained by vapor exiting the risers.

Vapor from the risers competes with the liquid descending from the bed above in the restricted passage space between the hats. The rectangular hats occupy some of the cross-sectional area, often 20–30% of it. This means the C-factor between the hats will exceed the C-factor in the bed by 20–30%. When the bed is operated at high C-factors, premature flooding may initiate between the hats. A similar situation will develop above chimney trays in packed towers, where the hats consume a considerable fraction of the cross-section area.

There is a moderating factor in this situation. The weirs on the hats are short, so some liquid can find its way down. In addition, unlike the two previous cases, some of the accumulating liquid upon flooding will be directly above the hats. The hat channels are not full, so they can catch the accumulating liquid and run some of it down. Both effects result in additional capacity compared to the usual system limit criteria of the C-factor of 0.45 ft/s or the value calculated from the system limit correlation (463). The author's experience as well as that of many of his colleagues has been that the system limit between the hats above a redistributor or a chimney tray usually occurs at C-factors between 0.56 and 0.7 ft/s based on the most restrictive open area between the hats. Other cases of premature flooding due to oversized hats were reported (112), but the C-factor was not specified.

Case 4 This case is similar to those described in Case 3 above and Figure 8.12b. A packed tower was erected to strip a small amount of carbon dioxide and organics from a wastewater stream, purifying it into a cooling water makeup. The entering wastewater poured above the oversized hats of the liquid distributor. The oversized hats consumed most of the cross-sectional area, leaving little space for vapor ascent simultaneous with liquid descent. The resulting high vapor velocities between the hats blew the incoming water upward, causing flooding. Two different consultants studied the problem, the first offering an incorrect diagnosis and a failed fix, the second, after an extensive simulation study, incorrectly concluding that a larger-diameter tower was needed (a puzzling conclusion, the C-factor was 0.12 ft/s in 1.5″ Pall rings, compare to Section 2.16). The plant gave up and junked the tower, with the wastewater still going to the sewer. Good troubleshooting and a nickel-and-dime job of modifying the hats or feed pipe would have produced good water and avoided turning a good tower into scrap metal.

8.12   VAPOR MALDISTRIBUTION AT THE TOWER BASE AND CHIMNEY TRAY

Case 1 Figure 8.13a shows a monoethanolamine (MEA) contactor vapor inlet arrangement. The tower experienced a premature capacity bottleneck. The sketch clearly shows that more vapor ascended on the half tray near the vapor inlet nozzle. This maldistribution created a capacity bottleneck and did not help separation either. Lieberman (285) reports a similar case in a packed tower, but instead of the center downcomer, an I-beam supporting the bed caused the vapor maldistribution. Section 4.11 item 8 provides a guideline for sufficient vertical height between the bottom of the downcomer or I-beam and the liquid level.

Case 2 (Ref. 393) Massive I-beams were used to support the packed bed above a chimney tray in a 16 ft ID refinery vacuum tower. Figure 8.13b shows that the I-beams would interfere with the vapor distribution to the bed. Vapor from the side chimneys was blocked by the I-beams and preferentially ascended around the periphery. The maldistributed vapor profile was displayed as a carbon deposit on the surface of the bottom packing. The deposit formed an annular ring about 5 ft wide, extending about 1 in. into the bed. In that case, liquid was known to overflow the chimneys for several months due to an incorrect location of level taps. The overflow caused liquid entrainment, and some entrained droplets ultimately carbonized on the base of the bed. Had the vapor profile been uniform, entrainment and deposit laydown would have been more uniform. An uneven liquid overflow could have further aggravated the maldistribution.

Cases 3 and 4 (Ref. 275) A recommended rule of thumb by Lieberman (285) for the distance from the chimney tray hats to the packed bed above is to fully cover the bottom of the packed bed above with vapor expanding at a 30° angle (Figure 8.13c). Lee (275) describes a case where the angle for vapor to fully cover the bottom of the bed was 70°. In reality, the vapor expands at 30°, not 70°, which leaves large portions of the bed inlet with little vapor, causing the observed poor separation.

Rao et al. (390) described another case where the 30° rule was violated. The authors stated that "the chimney tray collector below the top bed was installed with little space between the top of the risers to the packed bed above." From the diagram in their article, it looks like the angle was wider than 70°. Needless to say, the bed above performed inefficiently.

Case 5 (Ref. 144) The feed region ("flash zone") of a refinery atmospheric crude tower (Figure 8.13d) receives high-velocity, partially vaporized hot crude feed, typically about 50% vapor by weight (>90% vapor by volume). Often, the feed enters via a tangential entry ("vapor horn"). The trays above the feed ("wash section") operate at very low liquid flow rates.

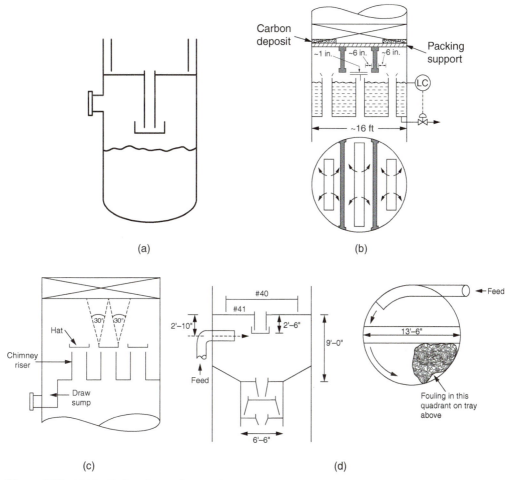

Figure 8.13   Maldistribution of vapor feed. (a) Center downcomer causes vapor maldistribution and capacity bottle-neck at the tower bottom. (b) Long I-beams cause vapor maldistribution at a chimney tray. (Courtesy of D. W. Reay, ret., BP Oil, private communication.) (c) Lieberman's (285) 30° minimum hypothetical angle rule for good vapor distribution above chimney tray riser hats. (d) Center downcomer deflecting high-velocity flashing feed toward the bottom right quadrant of the tray above, causing fouling there. (Based on Guarda, C. F., E. Lieberman, and N. Lieberman, "The Lost Art of Tower Internal Inspection: A Practical Guide for the New Generation," in Kister Distillation Symposium, Topical Conference Proceedings, p. 64, AIChE Spring Meeting, Virtual, April 18–22, 2021.)

Figure 8.13d shows feed obstruction by the center downcomer and seal pan from tray 41 above. Some of the feed would be deflected upward by the seal pan, but most of it would underflow the seal pan and attempt to rise on the opposite side. The downcomer and seal pan would prevent the rising vapor from returning to the inlet side. The feed vapor would then preferentially channel into the shaded quadrant of the tray above. The high-velocity hot feed would dry up this quadrant and deposit some entrained crude there, both of which would cause severe fouling and plugging even of plug-resistant trays, as was observed (144). The solution was to take out tray 41 and its downcomer and install seal pans on the tray 40 side downcomers.

Takeaways   Look for obstructions to vapor passages. Troubleshoot points of transition (feeds, drawoffs, bottom sumps, chimney trays) using to-scale sketches.

8.13   ENTRAINMENT FROM A GALLERY FLASHING FEED DISTRIBUTOR

Gallery distributors are frequently used for flashing feeds into packed towers. A common gallery distributor design is shown in Figure 4.22 and described in Section 4.13.

Following a revamp, one stripper experienced instability initiating at its gallery distributor. The stripper seemed to puke every one to two hours. The problem was initiated upon return to service, so fouling was an unlikely cause.

The tower diameter was larger at the bottom than at the top (Figure 8.14a). The flashing feed entered via an 8″ nozzle into a gallery distributor. The centerline of this nozzle was 22 in. above the floor of the gallery (Figure 8.14b). The feed entered via a rounded V-baffle.

Gamma scans showed buildup of frothy liquid right above the gallery distributor. Calculations using the orifice equation (4.2) (Section 4.6.2) gave a liquid level of 10 in. at normal operation, assuming pure liquid and no aeration.

A sketch (Figure 8.14b) shows the geometry of the entry together with the gallery distributor. The V-baffle width was twice the nozzle diameter (a common design practice), so the bottom of the V-baffle was 8 in. below the centerline of the feed nozzle, or 14 in. above the gallery floor. This is only 4 in. above the liquid level. With turbulence and aeration likely, the gallery liquid would rise above the bottom of the V-baffle, and the incoming vapor would entrain it or cause waves on this liquid, initiating the observed instability.

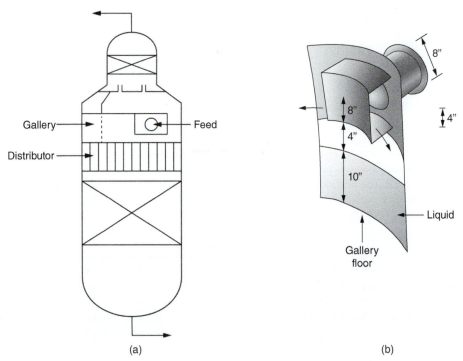

(a)   (b)

Figure 8.14   Tower experiencing instability initiating at the gallery distributor. (a) Overall arrangement; (b) gallery distributor sketch, showing vapor entry only 4 in. above the liquid level assuming no aeration.

8.14   VAPOR IMPINGING ON LIQUID AT THE TOWER BASE

A 78″ ID debutanizer experienced some quick cycling of the overhead liquid product rate. Figure 8.15 is the to-scale sketch that diagnosed the root cause.

Liquid from the bottom tray descended into the tower base via two side downcomers, each terminating in a double seal pan. An overflow baffle divided the bottom sump into a reboiler feed compartment and a bottom draw compartment. All the liquid from the left-side downcomer, as well as the liquid from the reboiler return, would be collected by the reboiler draw compartment. The reboiler return entered the tower via a 12″ feed pipe with an 18″ half-pipe cutout,

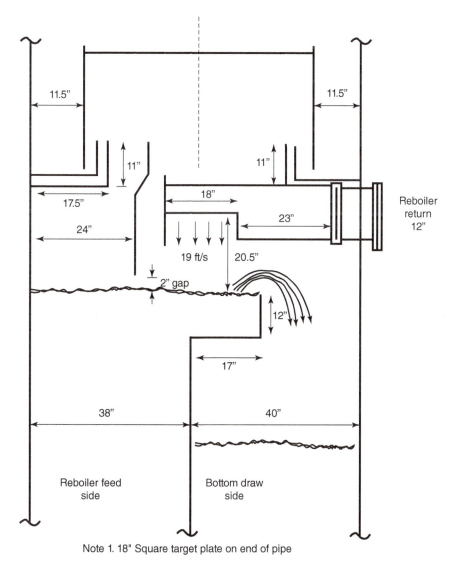

Note 1. 18″ Square target plate on end of pipe

Figure 8.15   Downward deflection of reboiler return causes instability in a debutanizer.

terminating in a vertical 18″ square target plate that defected the return mixture downward into the reboiler draw compartment. Only the liquid from the right-side downcomer would descend directly into the bottom draw sump. With most of the liquid going into the reboiler compartment, the baffle would keep a constant liquid head for the fired-heater reboiler. Liquid not consumed by the reboiler overflowed the baffle into the bottom draw sump. At least this is the way we believe it was meant to work.

Figure 8.15 shows the flaw in this reasoning. The reboiler return two phase mixture traveled at 44 ft/s through the 12″ pipe, then was deflected downward at 19 ft/s. This means that it would travel 19 ft downward in the first second if not stopped. The liquid level at the top of the baffle is only 20.5 in. below the top of the cutout. This strong downward wind would blow the liquid from the reboiler compartment into the bottom draw compartment. This would starve the reboiler of liquid, reducing the amount of vapor produced. The starvation of reboiler liquid is likely to increase the fractional vaporization, possibly leading to mist flow in the hotter tubes, which would further reduce the amount of vapor generated. Reducing vaporization would lower vapor flow through the tower and the overhead product rate. Once the level in the reboiler compartment falls, less liquid would be blown into the bottom draw compartment, the reboiler liquid would return, the boilup would pick up, and both tower vapor rate and overhead product rate would rise. When the reboiler compartment filled up again, the same cycle would repeat.

Takeaway   A to-scale sketch is a powerful analytical tool for diagnosing point of transition issues.

8.15   MORE VAPOR IMPINGING ON LIQUID AT THE TOWER BASE

Case 1 Figure 8.16a shows the bottom sump of a chemical tower. A vapor feed entered near the bottom. A baffle split the tower bottom sump into a reboiler draw compartment and a bottom draw compartment. All the liquid from the bottom 2-pass tray went to the reboiler draw compartment. Reboiler return entered the tower via a pipe that passed through the reboiler draw compartment. It then discharged against a downward-directed channel baffle.

The tower flooded prematurely, as shown by a pressure drop rise and by gamma scans. The gamma scans showed that the flood propagated right through the tower, having a foamy appearance, characteristic of the system. Calculations and gamma scans showed that the trays were not close to flooding. The downcomer velocities were of the order of 0.11 ft/s, which were about two to three times less than the design criteria of the major chemical company.

The cause of the premature flood was the impingement of reboiler return vapor on the liquid surface, generating liquid entrainment that flooded the bottom trays. Gamma scans showed that indeed the flood initiated in the bottom sump, supporting this possibility. Further, when the tower operated without the reboiler, there was no flooding.

Case 2 A 12-ft ID refinery deethanizer stripper had two thermosiphon reboilers in service, one heated by hot naphtha, the other by steam. The tower could not remove enough ethane from the bottom product. When the reboiler hot naphtha flow was raised, the ethane concentration in the bottom went up. The ethane ended up in the downstream tower, the debutanizer, and overpressured it. To prevent overpressuring, the naphtha flow needed to be cut back well below design duty. The steam reboiler had its steam valve fully open to try to make up the duty.

Figure 8.16b shows the naphtha reboiler return entry. The sump contained a baffle that preferentially diverted the tray liquid to the reboilers, and kept a constant liquid head in the reboiler compartment. Liquid overflowing this baffle became the bottom product. The naphtha reboiler return discharged its vapor–liquid mixture against a second baffle that was inclined

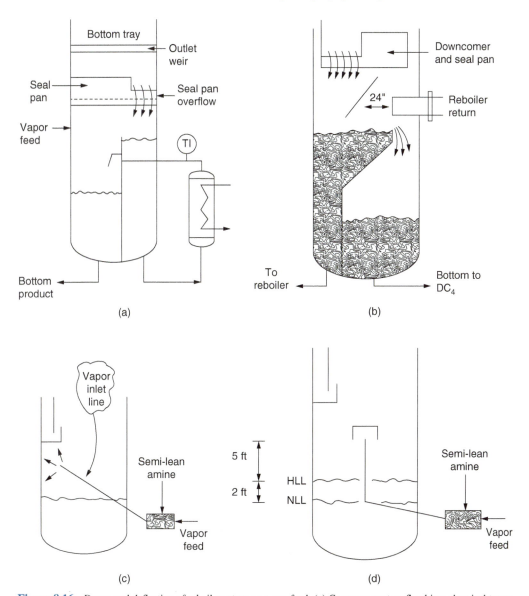

Figure 8.16   Downward deflection of reboiler return or vapor feed. (a) Causes premature flood in a chemical tower. (b) Induces lights into the bottoms of a refinery deethanizer. (c) Causes hydrogen blistering in an amine absorber. (d) Causes premature flood and level instability in an amine absorber.

at 45°. This inclined baffle generated a vertical downward velocity component, which blew some of the reboiler compartment liquid into the bottom draw compartment. This starved the reboiler of liquid, inducing excessive flow into the bottoms product. Per the component balance, as the bottoms flow rate exceeded the amount of heavies entering the column, the balance of the bottoms flow rate would be ethane.

   Case 3 Vapor feed to an amine absorber was saturated with semi-lean amine in a static mixer before entering the tower (Figure 8.16c). At high vapor velocity, the vapor feed line caused impingement of this amine on the tower wall, leading to corrosion.

To correct the problem, the vapor feed line was directed upward (Figure 8.16d), terminating in an inverted bell that deflected the vapor downward. The idea was to prevent impingement on the wall. The new arrangement caused level instability and premature flooding in the tower. Stopping the lean amine injection into the static mixer stopped the flood, but the level instability remained. Here, vapor impingement on the liquid level caused the problem. The problem was finally solved by revamping the vapor feed line with a vane distributor.

8.16   V-BAFFLES PRODUCE UNEXPECTED FLOW PATTERN AT THE TOWER BASE

V-baffles   (Figures 8.14b and 8.17) are used to reduce the inlet velocity of the reboiler return, a vapor feed, or a vapor/liquid feed, and to redirect it along the tower walls. The baffle is closed at the top and at the bottom, open on the sides, and is usually sloped downward at a small angle. The baffle can be rounded (Figure 8.14b) or segmental (Figure 8.17a–f). A good troubleshooting investigation should explore for undesirable flow patterns produced by the baffle using a to-scale sketch. The sketch can identify interferences with other tower internals as illustrated in the following.

Case 1 Following a retrofit with high-capacity trays, a tower experienced excess bottom product, reboil shortage, and high lights content in the bottoms, which bottlenecked the reboiler and tower.

A sketch (Figure 8.17a) identified the problem. An overflow baffle divided the bottom sump into a reboiler draw compartment and a bottom product draw compartment. Liquid from the bottom tray descended directly into the reboiler draw compartment. The reboiler return liquid also entered the reboiler draw compartment. Whatever liquid was not needed by the reboiler would overflow the baffle into the bottoms product draw compartment.

Initially, it was thought that the problem was due to tray weeping into the bottom product draw compartment, but gamma scans showed no significant weep. The scans showed the product draw compartment was full while the reboiler draw compartment was not.

The reboiler return nozzle had a V-baffle in front. At low reboiler return rates, the liquid dropped into the reboiler draw compartment as intended. At high rates, however, the mixture's momentum carried it to the bottom draw compartment. The problem was cured by adding a cover above the product draw compartment that diverted the reboiler return liquid into the reboiler draw compartment (Figure 8.17b).

Case 2 (279, 280). Following a revamp from 1-pass trays to 2-pass high-capacity trays, a solvent recovery tower experienced extremely unstable reflux drum level, inability to control tower pressure and reflux flow, and unsteady operation. The problem was traced to the addition of a V-baffle in the revamp (Figure 8.17c,d), which directed reboiler return liquid into the upper level tap. The seemingly high liquid level in the draw sump would lead to a boilup increase or reflux decrease, and unsettle the tower. The solution was to install a shielding baffle in front of the level tap.

Case 3 Case 6.1 in Ref. 201, contributed by R. F. Olsson, describes an irregularity in the temperature profile of a packed bed as measured by a surface temperature survey. On the north and west, temperatures were steady and in line with expectations. In the south, temperatures were 10–20°F lower than expected. In the east temperatures varied ±50°F in cycles.

Figure 8.17e shows the cause. The tower had two equal reboiler return inlets, each equipped with its own V-baffle. The baffles were separated by an 8 in. gap. Assuming an equal split of the reboiler return fluid by each V-baffle, about half the reboiler return vapor would attempt to rise into the restricted space between the baffles on the east side (Figure 8.17e) of the tower, causing a vapor jet to rise up the tower in the east. This was the side of the tower where the temperatures cycled.

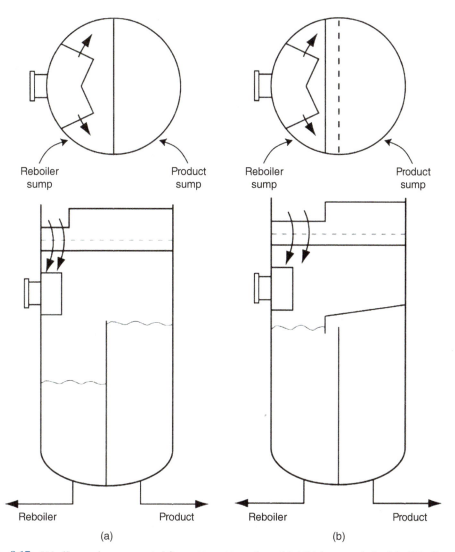

Figure 8.17   V-baffles produce unexpected flow pattern at tower base. (a) At high rates, reboiler inlet V-baffle would deflect liquid into the bottoms product draw compartment, starving the reboiler of liquid. (b) Problem solved by adding a cover over the bottoms product draw compartment. (c, d) Adding a V-baffle in a revamp from one to two pass trays led to reboiler return liquid entering the upper-level tap, which destabilized the tower. (e) Original V-baffle arrangement that led to temperature instability in the packed bed above. (f) Modification that overcame the problem. (g) Absence of a reboiler return impingement plate that caused the tower to flood. (Parts (c, d): From Lenfeld, P., and I. Buttridge, "High Costs of Quick Turnarounds and Erroneous Procedures," in Distillation 2013, Kister Distillation Symposium, p. 455, AIChE Spring Meeting, San Antonio, Texas, April 28–May 2, 2013. Reprinted with permission. Parts (e, f): From Olsson, R. F. in Kister, H. Z. "Distillation Troubleshooting," Wiley, 2006. Reprinted with permission. Part (g): Based on Lieberman, N. P., "Understanding Process Equipment for Operators and Engineers," Elsevier, Amsterdam, 2019.

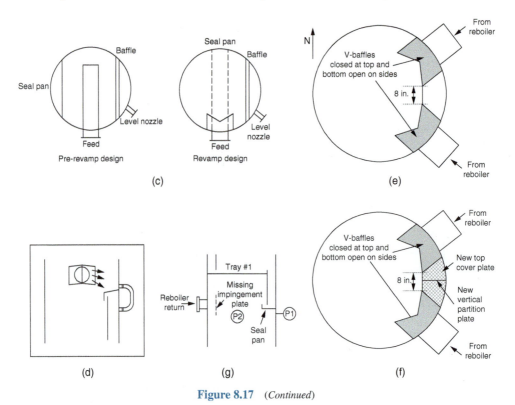

Figure 8.17 (*Continued*)

A uniform temperature profile was achieved after the V-baffles were modified by installing a vertical partition plate between them and a plate over the top of the gap (Figure 8.17f) to keep the two reboiler return streams from jetting up the east side.

Case 4 (290) In this case, the absence of a V-baffle (or an impingement plate, as it was referred to in Ref. 290), which was part of the design but was not installed, led to premature flooding in a naphtha stabilizer tower. The reboiler return nozzle was 180° away from the bottom tray seal pan, and at a slightly lower elevation (Figure 8.17g). The impingement converted the velocity of the vapor–liquid reboiler return mixture into a static pressure at the seal pan that backed liquid up the downcomer, causing the tower to flood. Pressure measured at point P1, right across from the reboiler return, was 165 psig, while at point P2, it was only 160 psig, measured by the same gauge. The localized pressure difference was more than enough to push liquid up the downcomer. Installation of the impingement plate solved the problem.

A similar impingement was reported (292) even though the upper level tap was 2 ft above the reboiler return inlet.

Case 5 A V-baffle in a packed steam stripper experienced repeated damage episodes at the shell attachment. The tower liquid was much colder than the steam. The temperature cycles caused the damage.

8.17 BAFFLING TOWER BASE BAFFLES

With recirculating thermosiphon reboilers, preferential baffles are often used to split the bottom sump into a reboiler draw compartment and a product draw compartment. The bottom tray liquid is preferentially diverted into the reboiler draw compartment. The reboiler return liquid also enters the reboiler draw compartment, but near the baffle, it overflows into the product

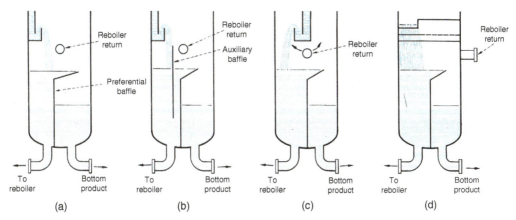

Figure 8.18   Preferential baffle arrangements for bottom sumps. (From Kister, H. Z., Chemical Engineering, July 28, 1980. Reprinted with permission.)

draw compartment. The idea is to recirculate some of the reboiler return liquid to the reboiler, with the balance overflowing into the bottom product compartment. Figure 8.18 shows common arrangements. Many additional variations together with good practice guidelines are described in Ref. 109.

These baffles should be avoided where the bottoms product stream is much smaller (<100 times) than the liquid to the reboiler. This is because for every gallon overflowing into the product compartment, there are >100 gallons going to the reboiler. Under these conditions, the overflow is likely to be intermittent and cause instability in the bottom sump. If this is troublesome, the bottom hatchway in the baffle can be removed, and the product and reboiler compartments become one.

Case 1 Is the baffle configured correctly? Figure 8.19 (133) shows a poorly configured baffle arrangement that did not work. A temperature survey (Figure 8.19) showed the cold tray liquid preferentially going to the product compartment, with most of the reboiler return liquid going into

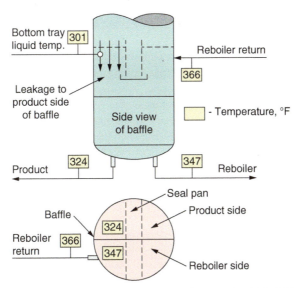

Figure 8.19   Preferential baffle arrangement that worked poorly. (From Golden, S. W., J. Moore, and J. Nigg, Hydrocarbon Processing, p. 75, September 2003. Reprinted with permission.)

the reboiler draw compartment. Presumably, the center downcomer deflected the reboiler return liquid into the reboiler side of the baffle. At the same time, it deflected the reboiler return vapor preferentially to the tray panel on the right-hand side of Figure 8.19, generating a vapor-deficient region with liquid weeping on the left-hand panel in the elevation sketch. Much of this weeping ended in the product side.

Case 2 Does the Baffle Work as Intended? (Ref. 51) The H_2S concentration in the bottoms of a debutanizer was 500 ppmw versus <5 ppmw as per the design. The H_2S concentration was reduced to 10 ppmw by spiking the feed with 2000 BPD of butane (light key, about 5% of the feed) to establish adequate reflux/lift as described in Section 3.2.8. 10 ppmw was still above the acceptable concentration.

Temperature measurements and sketching the bottom sump baffle (Figure 8.20a) are key to diagnosing this problem. The debutanizer bottoms product and the reboiler outlet temperatures were both designed for 620°F, but the measured bottoms temperature was about 20°F colder. This can only happen if some cold liquid from the chimney tray at a temperature of 520°F finds its way to the bottoms draw compartment. Likewise, the bottom tray temperature was designed for 520°F and the reboiler draw for a slightly higher temperature. The measured reboiler inlet temperature was much higher, 570°F, indicating considerable mixing with the hot reboiler return liquid. The high temperature frequently caused seal failure of the reboiler circulating pump. The baffle therefore did not work as intended.

Figure 8.20a shows that the downcomer from the bottom chimney tray terminated only about 2 ft below the top of the reboiler compartment overflow baffle. The 2 ft difference was not enough to prevent mixing of the cold fluid from the chimney tray with the hot reboiler return liquid. Consequently, the bottoms product became cooler and the reboiler feed became hotter than expected. Extending the downcomer by about 8 ft (Figure 8.20b) minimized the hot/cold liquid mixing, brought the two temperatures close to design, and reduced the H_2S concentration to the acceptable <5 ppmw range.

Case 3 Does the Baffle Work as Intended? Section 7.2.7 and Figure 7.27 describe another baffle dividing the bottom sump into reboiler and product draw compartments not working as intended.

8.18   LIQUID MALDISTRIBUTION AT A FEED OR A PRODUCT DRAW

Case 1 The naphtha bottoms product from a naphtha H_2S stripper had 5–15 ppm H_2S, which was above the acceptable concentration. By maximizing preheat, the plant managed to reduce the H_2S in the bottom to about 5 ppm. The stripper was a licensed unit, and the licensor reported experiencing no similar problems in their other units.

The 10 ft ID tower had one packed bed above the preheated feed and two 20 ft tall beds below the feed, all containing random packings. Between the two bottom beds there was an orifice redistributor.

The tower flashing feed entered via a gallery distributor. Liquid from holes in the gallery floor dripped into the middle bed liquid distributor (Figure 8.21) around the periphery. Liquid leaving the top bed was collected on a chimney tray from which it descended via three downpipes into the middle bed distributor. The pipes were generously sized for self-venting flow.

A close study of the sketch of the middle-bed liquid distributor (Figure 8.21) identified the problem. The distributor had closely spaced square chimneys, so liquid movement between them was severely restricted. This resulted in high liquid heads where the pipes entered, and much lower heads elsewhere, generating maldistribution and highly uneven liquid flow down the bed. Gamma scans confirmed gross maldistribution in the middle bed.

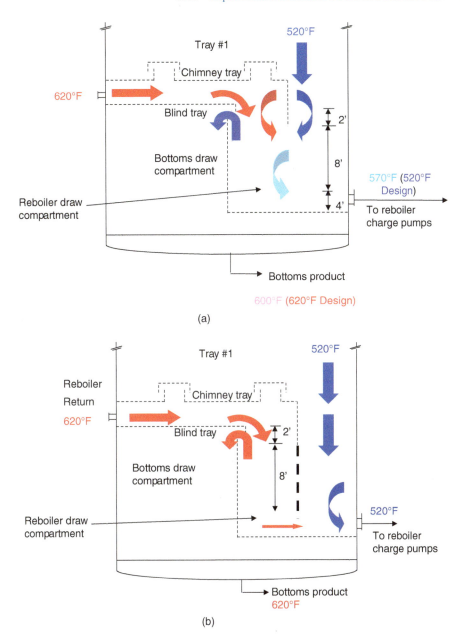

Figure 8.20   Debutanizer bottom sump baffle. (a) Before modification; (b) modified. (From Bu-Naiyan, A., and K. Kamali, "Troubleshooting Debutanizer at Ras Tanura Refinery's Hydrocracking Unit," Asian Refining Technology Conference (ARTC) 6th annual meeting, Singapore, 2003. Reprinted with permission.)

Case 2 Figure 8.22 (based on Ref. 417) shows the chimney and down box arrangement in a collector. The down boxes fed a main parting box that supplied liquid to a trough distributor underneath. Separation in the packed bed was poor. During an outage, an in situ water test measured nonuniform water heights in the distributor troughs, with troughs in some regions, especially toward the liquid draw, having twice the liquid heights of others. A CFD study verified that this

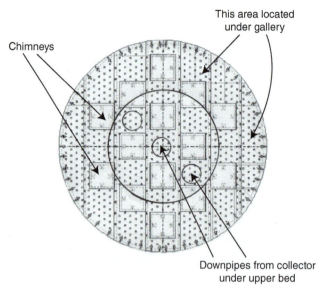

Figure 8.21   Flow obstruction causes liquid maldistribution and poor separation in a packed tower.

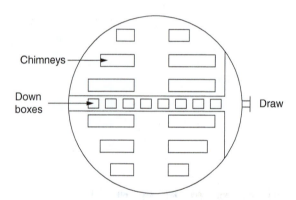

Figure 8.22   Long edges of long chimneys obstruct liquid movement toward adjacent down boxes, causing a skewed liquid flow profile. The generated maldistribution continued to the bed. (Based on Schnaibel, G., A. C. Bellote, A. V. S. Castro, C. N. Fonseca, M. E. P. L. Marsiglia, G. A. S. Torres, K. Ropelato, A. S. Moraes, and F. Feitosa, "The Jet Fuel Problem Strikes Again," in Distillation 2009: Topical Conference Proceedings, p. 89, AIChE Spring National Meeting, Tampa, Florida, April 26–30, 2009.)

was caused by the long chimneys closeness to the down boxes, creating "shadow" regions. The down boxes in these regions had flows far below the average.

Case 3 In a 16 ft ID refinery fractionator, there was a loss of distillate to the lower, less valuable products below and poor separation in the packed bed below. The cause was excessive hydraulic gradients on the collector (Figure 8.23). The center sump contained tall down boxes through which liquid descended to the parting box below. These interfered with the forward flows. There were also supports that were perpendicular to the liquid flow and obstructed forward flow. As a result, the down boxes near the end of the sump on the left of the sketch received more liquid. The draw nozzle and the down box on the right were starved of liquid. The parting box below could not adequately even up the maldistribution.

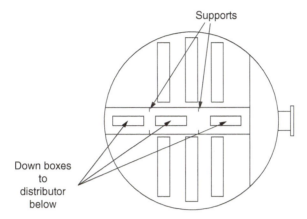

Figure 8.23   Collector that experienced excessive hydraulic gradients, causing maldistribution in the distributor below.

Case 4 A chemical packed tower with an upper and lower beds containing structured packings experienced poor separation. The bottom bed distributor was a trough distributor (Figure 8.24). Liquid to the distributor was supplied by a parting box. Liquid from the top bed was collected by a collector, and descended into the center of the parting box via a rectangular downcomer. The subcooled feed entered at one end of the parting box.

Figure 8.24 shows very poor mixing in the distributor between the subcooled feed and the liquid from the bed above, which led to the separation problem.

Case 5 Figure 8.25 shows the liquid distributor to the slurry bed of an FCC main fractionator described in Refs. 124, 132. There were two parting boxes that received cooled, nonvolatile

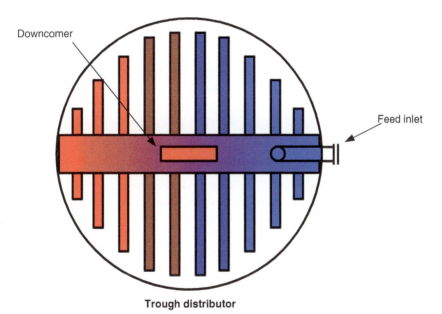

Trough distributor

Figure 8.24   Poor mixing of feed and reflux in a tower, showing reflux liquid from packed bed above descending into middle of parting box, while cold feed entering the end of the box. (From Kister, H. Z., W. J. Stupin, and S. Stupin, TCE, p. 44, December 2006/January 2007. Reprinted with permission.)

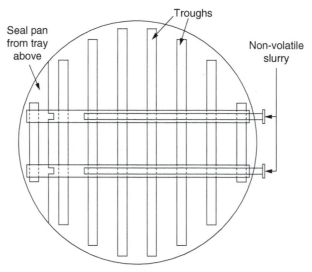

Figure 8.25   Poor mixing of cooled slurry feed and volatile liquid from tray above in the parting boxes to a slurry bed distributor. The region on the left that preferentially received the volatile liquid dried and coked. (Based on Golden, S. W., N. P. Lieberman, and G. R. Martin, Oil & Gas Journal, 92, p. 72, November 21, 1994.)

circulating slurry feed at 238°C via two perforated pipes. Entering the opposite end of the parting box was bubble point volatile liquid at 343°C from the tray above. Figure 8.25 shows that the two liquids had no chance of mixing. The volatile liquid descended on the left-hand side of the diagram, quickly vaporized in the hot bed below, and the dry, hot region coked up as was observed during an inspection at the next turnaround. There was no coke in the rest of the bed, only on the left side that preferentially received the volatile tray liquid. The column internals were modified to properly mix the liquids, after which no further coking occurred (124, 132).

8.19   POOR SOLVENT/REFLUX MIXING GIVES POOR SEPARATION IN EXTRACTIVE DISTILLATION (ED) TOWER

Case 1 One ED tower in chemical service was revamped with new packings and distributors. The solvent distributor gave poor mixing, and as a result, the separation was poor and led to an undesirable component in the overhead.

Liquid to the distributor was provided by two parallel parting boxes. The solvent was fed into the middle of the parting boxes, while reflux from the tray above entered one end of the parting boxes (Figure 8.26a). This created the distribution pattern shown in Figure 8.26b. The region that did not receive the solvent did not extract the component.

Figure 8.26c shows the modified arrangement that solved the problem. The feed pipe and parting boxes were not changed. The reflux piping was modified by adding a central trough with laterals spreading it evenly along the length of the parting boxes.

Case 2 One chemical ED tower used to have trays and performed well. In a debottleneck, the trays below the solvent entry point were replaced by packings. The result was a very poor separation.

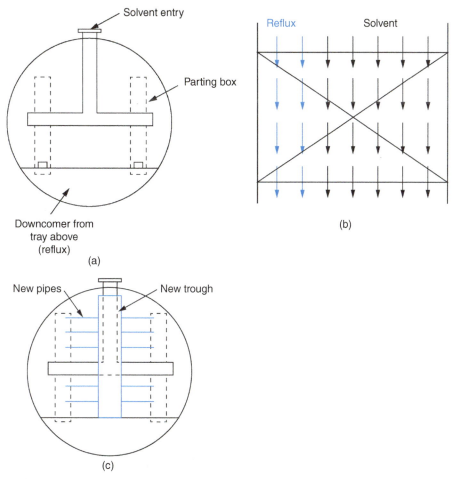

Figure 8.26 Poor mixing of solvent and reflux in ED tower. (a) Reflux entering parting boxes at one end, solvent in the middle; (b) flow pattern generated in packed bed below; (c) problem solved by adding a trough and laterals to spread the reflux along the parting boxes.

The solvent distributor was of a design similar to Figure 4.1b. The vapor openings between the troughs had downward-sloping hats to divert any liquid descending from above into the troughs. The solvent entered from pipes into two parting boxes (instead of the one parting box in Figure 4.1).

Reflux from the trays above was not distributed, but allowed to drop directly from the seal pan (Figure 8.27; for clarity, the parting boxes are not shown). The sketch shows that the liquid preferentially descended into the two troughs right under the seal pan and parallel to it. As a result, the reflux/solvent mixing was very poor, causing poor separation.

Case 3 (257). Solvent feed and reflux from the bed above entered via two parallel distributors without mixing in one benzene–toluene–xylene (BTX) ED tower, so the reflux and solvent did not mix until several feet down the packed bed. Liquid from the bottom of this bed rained directly into the redistributor of the next bed without remixing. That redistributor was a pan type,

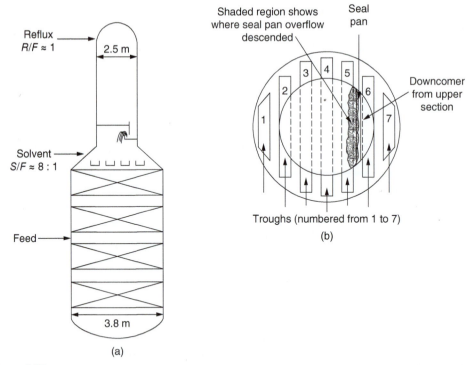

Figure 8.27   Poor mixing of solvent and reflux in ED tower, showing reflux liquid from trays descending into only two troughs. (a) Elevation; (b) view from the top.

unsuitable for distributing two liquid phases that could have separated in the pan. The unit failed to meet design rates and recoveries. Rectifying these deficiencies enabled the tower to achieve its objectives.

8.20   TWO SEEMINGLY WELL-DESIGNED PIECES MAY NOT WORK WELL WHEN COMBINED

Contributed by M. E. Harrison, Eastman Chemical, Kingsport, TN.

A 2-ft diameter scrubber was designed to remove acetic acid from a process off-gas stream. Water, the scrubbing fluid, was fed to the top of a single packed bed. The tower experienced excessive acetic acid emissions that caused unacceptable losses and odors.

Varying the water flow to the scrubber between 20% and 200% of the design did little to improve scrubbing efficiency. There were no indications of flooding. A higher scrubbing efficiency was expected based on packing heights and water rates of similar columns. Consequently, poor liquid distribution was suspected.

A separate review of column specifications and internal drawings did not suggest any problems. Figure 8.28a is a simplified sketch of the feed pipe entry, showing liquid discharged

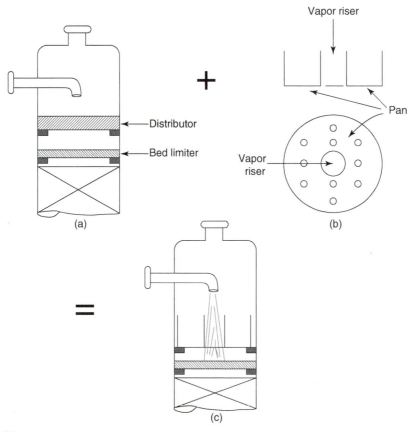

Figure 8.28   Two seemingly well-designed pieces combine to make an arrangement that does not work. (a) Feed pipe, OK; (b) distributor plan, OK; (c) combining (a) and (b), not OK. (From Harrison, M. in H. Z. Kister, "Distillation Troubleshooting," Wiley, 2006. Reprinted with permission.)

onto the center of the distributor. Figure 8.28b is a simplified sketch of the distributor plan, featuring a center, hatless chimney. Each individual sketch shows a sound design. Combining the two (Figure 8.28c) readily shows the design oversight. The scrubbing liquid was discharged right down the chimney and bypassed the distributor. The inspection confirmed this conclusion.

Solution   Calculations showed that the annular vapor space around the distributor was capable of handling the vapor flow. Therefore, the riser was blanked off. Of course, alternative solutions included relocating the feed pipe or the riser or putting a hat on the riser. When started up again, the scrubber removed essentially all the acetic acid with the design water rate.

Takeaway   When putting together the sketches, include all the relevant details, both of the internals and the feed piping, on the same sketch(es).

8.21   ANOTHER TWO SEEMINGLY WELL-DESIGNED PIECES THAT DID NOT WORK WELL WHEN COMBINED

This was an amine absorber that was retrofitted from trays to random packings. Upon restart, the absorption was poor, requiring a higher amine circulation, which in turn bottlenecked the regenerator.

Figure 8.29a is a sketch of the lean-amine entry point when the tower contained trays. A feed pipe dumped liquid behind an inlet weir, which distributed the amine to the top tray. This worked well, confirming that the feed pipe was of good design.

Figure 8.29b shows the orifice–pan packing distributor plan. The circles are the risers. The distribution holes were in the floor of the distributor. This was a standard orifice–pan distributor. Nothing suspicious here.

Figure 8.29c and d shows what happens when the tower was retrofitted with packings. The new distributor was installed, but the feed pipe was not changed. Instead of going into the distributor, the lean amine went straight down the chimney. It would have been a miracle if that tower worked. A sketch or a process inspection would have revealed this problem prior to startup.

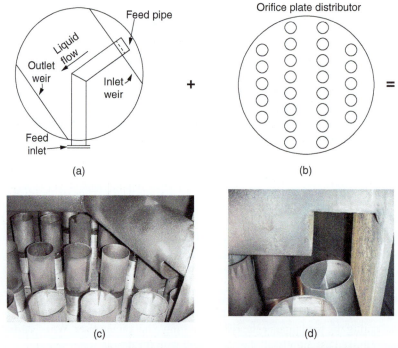

Figure 8.29   Another two seemingly well-designed pieces combine to make an arrangement that does not work. (a) Feed pipe, as was used to bring lean amine to the top tray, OK; (b) orifice–pan distributor plan, used when trays were replaced by packings, OK; (c) combining (a) and (b), not OK; (d) a close-up of (c).

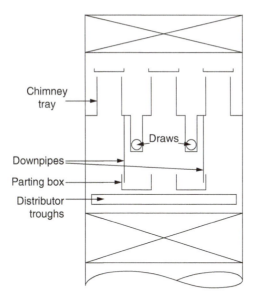

Figure 8.30   Maldistribution of internal reflux below a side draw. (Based on Lee, S. H., PTQ, Q2, p. 59, 2016.)

8.22   LIQUID MALDISTRIBUTION OF INTERNAL REFLUX BELOW A SIDE DRAW (275)

Figure 8.30 shows a side draw configuration from an atmospheric crude tower. Liquid from the upper bed was collected on a chimney tray, which contained two product draw sumps. Liquid not drawn descended via a multitude of downpipes, all flush with the chimney tray floor, to two separate parting boxes that fed liquid to the troughs below (similar to Figure 4.1, except with two parting boxes). The troughs distributed the liquid to a packed bed below. The packed bed experienced poor fractionation efficiency.

Two flaws were identified. One flaw has been described earlier as Case 3 in Section 8.12 (this is the same chimney tray). The second flaw was that the downpipes were flush with the tray floor, so liquid rate to each downpipe was uncontrollable. The descending liquid was therefore nonuniformly split among the downpipes and, therefore, among the distributor parting boxes and troughs. The problem was solved by redesigning the chimney tray with raised overflow pipes.

8.23   WOULD YOU BELIEVE THIS WAS A REAL TROUBLESHOOTING ASSIGNMENT?

A chemical plant experienced a separation problem in one of their packed towers. Their gamma scans showed massive maldistribution below the feed. The author's recommendation of drawing a sketch of the feed inlet met with reluctance from the client, who preferred to contract the author's company for a consulting assignment.

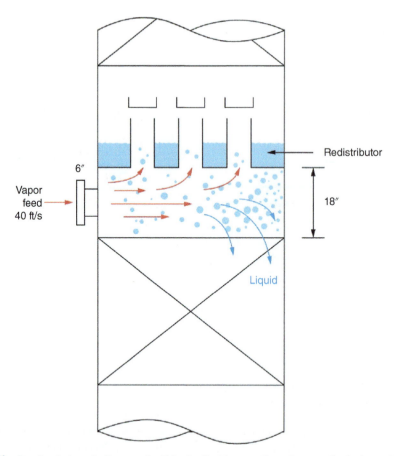

Figure 8.31 One sketch shows it all – vapor feed blowing liquid onto wall causing poor distribution and entrainment.

So the author drew the sketch (Figure 8.31) and emailed it to the client. Upon reviewing the sketch, the client quickly figured out the issue. The vapor feed will blow the liquid from the distributor to the opposite tower wall. The liquid arriving at the bed will never be properly distributed.

So the client paid consulting rates – solely for drawing a sketch.

The takeaway from this story and all other stories in this chapter: when it comes to troubleshooting points of transition (feeds, drawoffs, bottom sumps, chimney trays) you do not need an expert. You need a sketch.

Chapter 9

Making the Most of Field Data to Analyze Events and Test Theories

"Nobody wants failures. But you also don't want a good crisis to go to waste"
—(Professor Henry Petroski, in his book "Success Through Failure", Princeton University Press, 2006)

9.1 EVENT TIMING ANALYSIS

9.1.1 General Application Guidelines for Event Timing Analysis

Items 8 and 9 in Section 1.3 highlight the importance of learning about the column and/or event history. The question, "What are we doing wrong now that we did right before?" is perhaps one of the most powerful troubleshooting tools available.

The most common technique is identifying the time when the problem initiated and simply plotting synchronized operating charts for key variables at around that time on a single plot. With the aid of this plot, carefully review the sequence of events. This type of analysis can reveal not only what the problem is but also what initiated it and how it progressed to give the apparent symptoms. Below are some guidelines:

1. The main application of event timing analysis is to study operating charts and determine the exact sequence of steps taking place for a given event. The starting point is the last stable operation condition. The operating charts are then examined step-by-step at least until the full upset took place. Often it pays to also examine the charts during the recovery from the event.

2. Another type of application of event timing analysis compares current operating charts with charts derived before a change was made to the tower internals or operating conditions, for example, pre-turnaround to post-turnaround.

3. Whenever possible, derive quantitative information from the charts and translate it into changes in flow rates, liquid heights, boilup, and condensation rates. The quantitative information can confirm or deny theories.

4. While analyzing operating charts, closely interact with the operating team. It is very important to obtain operators' input on each step. Operators may be aware of other key events that

---

Distillation Diagnostics: An Engineer's Guidebook, First Edition. Henry Z. Kister.
© 2025 American Institute of Chemical Engineers, Inc. Published 2025 by John Wiley & Sons, Inc.

did not show well on the charts, or of pump trips, noises, or alarms, or of some actions taken that may have led to certain changes. All of these may be central for a correct analysis.

5. Often, there are several repetitions of the event under investigation. When choosing the repetition to analyze in detail, it is advantageous to select the sequence of events that maximizes the amount of available data. Consider an event involving a fast reduction of bottoms level. When the level is fully lost, there will be no data on how material is depleted or accumulated. It is better to select an event at which the level falls but not lost, so there are level readings available throughout the event.

6. Pay attention to select the most informative scales. For instance, when plotting several temperatures, attempt to use the same scale. In as much as possible, keep the same range for flows or temperatures.

7. Avoid cramming too many variables on a single chart. They clutter the picture and make it difficult to keep track of the key variables. The author prefers to use a maximum of four variables, or at the most six, on a single chart. If desired, two charts can be developed. After a preliminary review, some of the variables can be removed and others inserted until the key variables are identified and placed on a single chart.

8. Fill in the blanks with crude estimates. The amount of liquid on trays, in a packed beds, the vapor accumulation upon a pressure spike, and the amount of vapor generated upon a sudden temperature increase can be roughly estimated. Compare these estimates with the operating charts. These estimates are useful to rule out highly unlikely theories. However, keep in mind that these are rough estimates, often giving only a ball park number which may be significantly off. Do not attempt to read any accuracy into such estimates.

9. Split the charts into distinct time intervals. Analyze each interval separately, then the connection with other intervals (see Section 9.1.9 for a superb illustration).

9.1.2   Diagnosing the Unexpected Cause of Off-Spec Product

Contributed by M. Harrison, Eastman Chemical, Kingsport, Tennessee.

The operation team called the troubleshooter to the control room because "the tower did not make on-spec products."

A review of the operation charts (Figure 9.1a) showed that the tower was flooded. This is evidenced both by the dP rise and the reduction in bottom flow. The operating team was aware of the flooding but stated that the tower products were off-spec even before the column flooded. Figure 9.1a confirms this by showing the reflux increasing well before the tower flooded, evidently with the intent to improve product purity.

Rolling back the charts to the start of the reflux rise (Figure 9.1b) indicates that at that time the tower was again flooded. This is evidenced by the high dP, but this time the base level and bottom flow rate were also high. With an ordinary flood, the bottom flow rate and base level tend to fall as liquid accumulates in the tower.

Figure 9.1c shows the initial event. The bottom pump was temporarily lost, causing the base liquid level to go up. The bottom level indicator initially showed a level increase, but then leveled off as it reached the maximum of its range. The liquid level kept on rising, now into the trays, as evidenced by the increase in tower dP.

A while later, the pump came back on and intensely pumped out the base liquid. The liquid accumulation in the tower was interrupted, and the dP first leveled off, then started to decline (Figure 9.1b). Soon later, the bottom level came back to normal. The dP did not fall back to normal; it fell to a value below normal, suggesting some trays collapsed or were damaged during

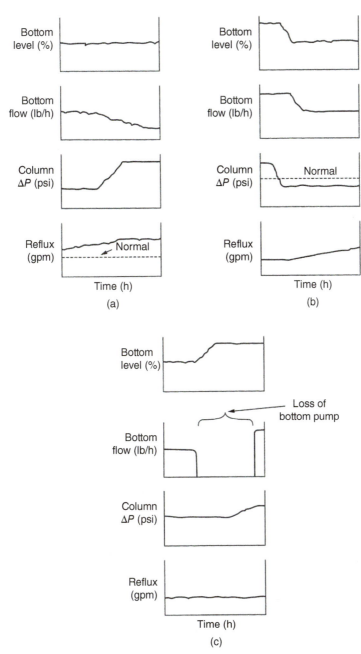

Figure 9.1   Operating charts of a high liquid level damage incident (a) final charts, indicating rise in reflux leading to flood in the tower; (b) intermediate charts, showing rise of reflux following a flood incident; (c) initial charts, showing high liquid level that caused tray damage. (From Harrison, M., in H. Z. Kister, "Distillation Troubleshooting," John Wiley and Sons, NJ, 2006. Reprinted with permission.)

the high liquid level event. At the same time, the reflux started increasing in an endeavor to meet the desired purity with the reduced number of trays. Raising the reflux and boilup brought the tower to flood (Figure 9.1a), still without getting the product back on-spec.

This case provides a classic illustration of the importance of studying the history of an event (or the tower) and the sequence of events.

9.1.3   Diagnosing Reboiler Surging

A high pressure tower (201) equipped with a vertical thermosiphon reboiler removed lights from water-insoluble organic high boilers. Tower bottom at about 180°C contained a small fraction of water. The tower experienced cycling of the heating medium flow and bottoms level.

Figure 9.2a is an event timing chart. The dip in the heating medium flow rate indicates reduced boiling. This was caused by the boiling point of the organics approaching the heating medium temperature. Over a period of time, or due to operation changes, the concentration of low-boiling water in the tower base was depleted, raising the base temperature. The reboiler temperature difference declined, and the boiling was largely ceased. With little boilup, lights-rich liquid from

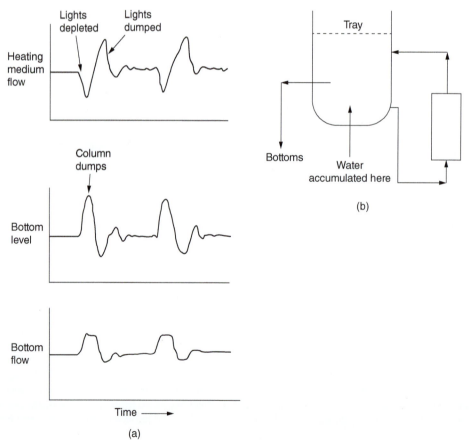

Figure 9.2   Reboiler surge incident (a) Operating charts; (b) raising bottom draw-off to avoid depletion of water eliminated the surging. ((Part (a): From Kister, H. Z., "Distillation Operation," Copyright © McGraw-Hill, 1990. Used with permission of McGraw-Hill. Part (b): From Kister, H. Z., "Distillation Troubleshooting," John Wiley and Sons, NJ, 2006. Reprinted by permission.)

the trays was dumped into the base, which raised the bottom level. The lights restored the reboiler temperature difference, the reboiler took off, and the heating medium flow rate surged. After stabilizing, the lights and water were slowly boiled off and depleted from the base, and the cycle repeated.

To cure, the reboiler drawoff was elevated to about a foot above the bottom (Figure 9.2b), so that water accumulated below the bottom draw nozzle and went back to the reboiler. This ensured the reboiler always received some water and eliminated the surges.

9.1.4   Loss of Condensate Seal in a Demethanizer Reboiler

Kister, H. Z., and T. C. Hower, Ref. 222. Reprinted Courtesy of the American Institute of Chemical Engineers.

A vertical thermosiphon reboiler on a demethanizer was heated by condensing refrigerant vapor (201, 222). Condensate flow from the reboiler was controlled by the control tray temperature (Figure 9.3a).

During low-rate operation, the tower was overreboiled because bottom purity was critical, while the overhead purity was less important. During this operation, the bottom product went off-spec.

An event timing analysis (Figure 9.3b) shows a slight drop in the tower control tray temperature. The controller raised the reboil. This should have raised the control temperature back up, but it did not. The temperature continued dropping. A switch of the controller to manual slowed down, but did not stop the temperature drop. The low temperature induced methane into the bottom stream, getting it off-spec.

The operating charts show that the control system responded as expected. A drop in temperature led to opening the control valve, increasing the refrigerant (heating medium) flow rate. This increase should have brought the temperature back up, but it did the opposite: it caused a loss of heat transfer.

When a control valve is at the outlet of the reboiler (Figure 9.3a), it holds a liquid level in the reboiler shell, which covers a portion of the tubes. This level also liquid-seals the reboiler,

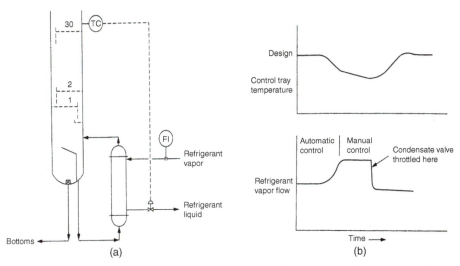

Figure 9.3   Reboiler seal loss incident (a) Demethanizer using reboiler heated by refrigerant vapor; (b) Operating charts. (From Kister, H. Z., and T. C. Hower, Jr., Plant/Operations Progress, 6(3), p. 151, 1987. Reprinted with permission.)

preventing vapor from escaping with the condensate. When the rate of flow through the outlet valve exceeds the rate at which vapor condenses in the reboiler, the liquid level, and therefore the liquid seal, may be lost. When the seal is lost, vapor channels through the reboiler without condensing, and condensation dramatically falls.

To re-establish the condensate seal, the condensate outlet flow was largely reduced by heavily throttling the outlet valve (Figure 9.3b). Once the seal was re-established, column operation returned to normal.

To avoid seal loss, another plant monitored the condensate temperature. When a liquid level is held in the reboiler, the condensate comes out subcooled. A rise in the condensate temperature signals an approaching loss of the reboiler seal, and the operator can take timely action.

9.1.5 Figuring out Hot Vapor Bypass (HVB) Instability

Adapted from Kister, H. Z., and D. W. Hanson, "Control Column Pressure via Hot Vapor Bypass," Chemical Engineering Progress, p. 35, February 2015.

In a hot vapor bypass control scheme for total condensers (no vapor product), the condenser area is partially flooded by condensate. Condensation takes place only in the unflooded area. Tower pressure is controlled by manipulating the unflooded area. Raising condenser liquid level reduces the condensation area, raising tower pressure. Conversely, lowering the condenser liquid level increases condensation area, and lowers tower pressure. The flooded area performs little condensation, but it subcools the condensate. Further discussion is available in Refs. 206 and 221.

Figure 9.4 shows a typical pressure control by hot vapor bypass. In this scheme, the condenser is at ground level, while the drum is elevated, often mounted on the lowest platform. Typically,

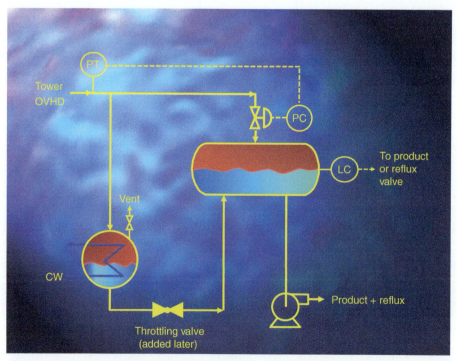

Figure 9.4 Hot vapor bypass control scheme. (from Kister, H. Z., "Practical Distillation Technology", course workbook. Copyright © Henry Z. Kister. Reprinted with permission).

the liquid level in the condenser is 10–20 ft below that in the drum. The vapor pressure difference between the condenser liquid surface, which is at the bubble point temperature, and the drum liquid surface, which is colder due to condensate subcooling, lifts the condensate to the drum.

Even a few degrees of subcooling can lower the vapor pressure tremendously and lift the condensate from 50 to 100 ft, and even more. However, the drum liquid surface is heated by the condensation of hot vapor from the bypass. As process liquids are good thermal insulators, it is the hot surface temperature, much hotter than the subcooled bulk liquid underneath, that sets the vapor pressure, and therefore the pressure, in the drum.

Tower pressure is controlled by varying the liquid lift from the condenser to the drum. Opening the bypass valve heats the drum liquid surface, raising the drum pressure. This pushes liquid from the drum into the condenser, flooding more tubes, reducing condensation, and therefore raising the tower pressure. Conversely, closing the valve cools the drum liquid surface, reducing the drum pressure, and sucking liquid from the condenser to the drum, unflooding tubes, and therefore lowering tower pressure.

In one C_3–C_4 splitter with the hot vapor bypass control of Figure 9.4, the condenser liquid level was 10–12 ft below the drum liquid level. The control usually worked well at a tower overhead flow rate (reflux plus product) of less than 25,000 BPD. At higher flow rates, especially in the summer, there were frequent episodes of the drum suddenly filling up. The drum level controller would increase reflux rate, the condensate temperature rose, and the condenser appeared to lose its seal. The drum level would then sharply drop, the reflux rate plummet, and the reflux pump cavitate. To restore stability, it was necessary to cut back reflux. This restricted the tower and gas plant capacity. A similar instability at times occurred at lower rates.

Per guideline 5 in Section 9.1, we chose to analyze the sequence of events in Figure 9.5, which occurred at lower rates (17,000 BPD of reflux plus product) because at these rates no variables went off-scale, and the condensate seal was not broken, as evidenced from the condenser outlet temperature remaining unchanged.

Referring to Figure 9.5, at about 10:40 p.m. the differential pressure across the hot vapor bypass jumped from the normal 4.5 psi to 7 psi in about one minute, and to 12 psi within five minutes. Over the same five minutes, the tower pressure decreased only slightly from 289.5 psig to 288.5 psig, raising the valve opening from 50% to 80%, while the drum liquid level jumped from 40% to 80%. Over the next 15 minutes, the pressure drop, valve opening, and drum liquid level returned to their pre-event values, while the tower pressure slightly rose to 290.5 psig. Boilup, reflux and product flow rates, and condenser outlet temperature did not change significantly during the event.

To analyze such events, the control valve opening changes as well as the drum level changes need to be translated into flow rate changes (item 3 in Section 9.1). From the control valve characteristic, a plot of flow versus valve opening at various valve pressure drops was prepared and showed that opening the valve from 50% at 4.5 psi pressure drop to 80% at 12 psi pressure drop increased the vapor flow rate through the valve from 5500 lb/h to 31,000 lb/h (221). Over the five-minute event, with approximately a linear increase in valve opening and pressure drop, an additional 1000 lb of vapor entered the drum (above the normal rate).

The drum was of 9 ft diameter, just over 20 ft long (tangent to tangent). To raise the drum liquid level from 40% to 80% on the level transmitter takes 21,000 lb. This is a huge movement of liquid over five minutes, tripling the quantity of liquid entering the drum, from 17,000 BPD to about 50,000 BPD. Determining where this huge amount of liquid can come from is the key to diagnosing the root cause of this event.

From the control valve calculation above, the additional amount of vapor entering the drum would account for about 1000 lb. Condensation of the drum vapor which was replaced by the liquid during the event would account for another 2000 lb. This leaves 18,000 lb unexplained.

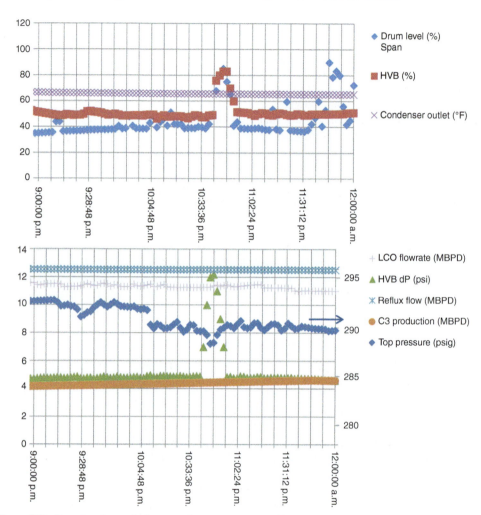

Figure 9.5   Operating charts of a hot vapor bypass upset event. (From Kister, H. Z., and D. W. Hanson, Chemical Engineering Progress, p. 35, February 2015. Reprinted Courtesy of Chemical Engineering Progress (CEP). Copyright © American Institute of Chemical Engineers (AIChE).)

Ambient condensation was considered. Records from the nearby airport show light rain, easing off and intensifying, ambient temperature of about 63°F and wind of about 4 miles per hour with occasional gusts to 20 miles per hour. For a change in heat transfer coefficient of 15 Btu/h per °F per ft^2 upon the onset of a sudden heavy rain, the condensation rate on an uninsulated vessel would rise by about 2000–3000 lb/h. Over five minutes, a heavy rain accounts for additional liquid generation of 200 lb, too little to impact the 18,000 lb liquid movement.

One conceivable source remains: liquid sucked from the condenser into the drum. There were two condensers in parallel, each 46 in. diameter and 20 ft long. The condenser shells held approximately 8000 lb of liquid when full. Based on the condenser areas and heat transfer coefficients, we estimate that during the low rate operation of Figure 9.5 (17,000 BPD reflux plus product) at relatively cold ambient temperatures, only about 25% of the condenser heat transfer area performed condensation. So the condenser shells contained only about 6000 lb of liquid, well short of accounting for the large liquid movement of 18,000 lb. However, emptying all the liquid from the condensers would quadruple the tube area exposed to condensation. The liquid generated by quadrupling the condensation area (from 25% to 100% five minutes later) is calculated at about 16,000 lb. Adding this liquid to the 6000 lb drained from the condenser gives 22,000 lb, in excess of the 18,000 lb increase in the drum liquid. This means that some liquid stayed in the condenser shells, retaining the condenser liquid seal and the subcooled outlet temperature, as observed in Figure 9.5.

The intensified condensation rate reduced the tower pressure, and the pressure controller rapidly opened the hot vapor bypass valve, eventually catching up with the disturbance. The drum liquid surface heated up, the drum vapor pressure rose, the liquid returned to the condenser, column pressure rose back up, and normal operation resumed.

The relatively large liquid inventory in the condenser shells provided a cushion against complete emptying of the condenser liquid during the Figure 9.5 event. This inventory was large due to the colder cooling water (as can be inferred from the low condenser outlet temperature) and the low rate of operation. At warmer ambient temperatures and/or higher plant rates, the liquid inventory in the condensers was smaller, making it much easier to lose the condenser liquid seal. This was the experience described earlier at high reflux plus product rates , especially during the summer.

Initiation It takes a very strong suction to move 18,000 lb of liquid in five minutes. Figure 9.5 shows a fast rise in control valve differential pressure, i.e., of the difference between the tower and the drum pressures. Since the tower pressure changed only slightly (Figure 9.5), the big change was in the drum pressure, eight psi over five minutes. From a vapor pressure–temperature relationship (e.g., the Antoine Equation), an eight psi vapor pressure reduction corresponds to cooling of the liquid surface of the drum by only 2.5°F (221). We searched for the source of a 2.5°F cooling of the drum surface temperature.

A heavy rain would have increased the condensation rate only by 2000–3000 lb/h (see earlier), easily offset by the additional 25,000 lb/h flow rise through the hot vapor bypass. Further, many episodes occurred under dry conditions. This source was therefore ruled out.

Hot vapor impingement ruffling the liquid surface was also ruled out. At the beginning of the event the vapor velocity was only about 6 ft/s, giving a $\rho_v V^2$ of 100 lb/ft s^2, too low to ruffle the liquid surface 5 ft below (where ρ_v. is vapor density, lb/ft^3, and V is vapor velocity at the drum inlet nozzle, ft/s). Also, this vapor flow rate was common during normal operation.

During the first minute, the valve pressure drop rose from 4.5 psi to 7 psi, which corresponds to a drum surface temperature decrease of 1°F. At the drum pressure of 285 psig the corresponding liquid surface temperature was 125°F, much hotter than the 66°F of the bulk subcooled liquid (Figure 9.5). Replacing as little as 1.5% of the hot drum liquid surface by subcooled liquid can explain a temperature drop of 1°F and corresponding pressure fall of 2.5 psi.

This puts the liquid entry in focus. The liquid entered the bottom of the drum via a 6″ nozzle, at an upward velocity of 5.7 ft/s, slightly exceeding the good design practice of 4 ft/s max. With the liquid level only 3.7 ft above, the initial momentum may have raised a jet of the subcooled liquid to the surface, possibly piercing the hot liquid surface. This may even occur during normal operation, with some surface cooling by the rising jet balanced by some of the hot vapor.

A step-up of the subcooled liquid reaching the surface may offset the balance. Just prior to the pressure differential rise, the tower pressure fell by about 0.5 psi and the drum level rose by about 2–3% (Figure 9.5). Both indicate a step-up in condensation rate, possibly due to heavy rain hitting the drum vapor space. The drum pressure fell, sucking liquid from the condenser. A level rise of 2% in the drum is equivalent to about 1000 lb. Sucking 1000 lb from the condensers would raise the area available for condensation by around 50%, quickly reducing the tower pressure. The pressure drop across the hot vapor bypass remained constant at that time, meaning that both the drum and tower pressures fell by the same 0.5 psi. The additional condensate flow, about

1000 lb in two minutes, would increase the liquid inlet velocity to the drum from 5.7 ft/s to 7 ft/s. The higher velocity intensified the incoming jet, raising more subcooled liquid onto the liquid surface. A self-accelerating "fountain" set in. The additional subcooled liquid cooled the surface, the drum pressure fell, sucking more liquid from the condenser, the jet and the fountain intensified, the surface further cooled, sucking more liquid, and so on. At the pressure differential peak, the liquid jet velocity exceeded 15 ft/s, high enough for the jet to break through the liquid surface, even at 80% drum level.

Solution   Field measurements (221) showed that increasing the pressure drop in the lines leaving the condenser can mitigate the self-accelerating process. A manual throttling valve was installed in the line from the condenser to the drum (Figure 9.4). The dP across the valve is measured locally, regularly checked by the operators, and maintained at 2 psi. The valve severely limits the rate of increase in liquid flow to the drum, giving the hot vapor a chance to catch up. This completely eliminated the problem. The modified system can now run the reflux plus product up to even 40,000 BPD during both summer and winter with no instability.

9.1.6   Can Multiple Steady States Occur in a Reboiler System?

Adapted from Kister, H. Z., "Reboiler Circuit Debottleneck with No Hardware Changes," Chemical Engineering, p. 30, August 2017.

A deethanizer stripper reboiler circuit bottlenecked a newly built gas plant. The stripper was a 15'–6" tower equipped with four-pass conventional valve trays with moving round valves. A preferential baffle in the tower base divided it into separate reboiler draw and bottoms product draw compartments. Liquid from the bottom tray was diverted into the reboiler draw compartment (Figure 9.6) using strategically placed notches in the bottom tray seal pan. The reboiler return internal piping had cutouts that directed the vapor–liquid mixture returning from the reboilers toward the bottom draw compartment (Figure 9.6). The purpose is to divert all the bottom tray liquid to the reboiler compartment and all the reboiler return liquid to the bottoms product draw compartment. Any excess bottom tray liquid not boiled by the reboilers overflowed the baffle into the bottoms product draw compartment (207).

Boilup to the deethanizer stripper was supplied by four horizontal thermosiphon reboilers, arranged in two parallel circuits, "A" and "B" (Figure 9.6). The lower reboiler in each circuit (E-4A or E-4B) was heated by hot naphtha liquid, while the upper reboiler (E-3A or E-3B) was heated by high-pressure steam. In each circuit, the design heat duties of the naphtha and steam reboilers were 32 MM Btu/h and 40 MM Btu/h, respectively. Liquid to the reboilers was supplied from the reboiler draw compartment via a 24" pipe that was split into two 18" pipes, each feeding one circuit. From each circuit, the reboiler effluent returned to the tower via a 24" line.

Initial Performance   To stabilize the tower at the low rates during start-up, steam was taken out of reboiler E-3A, creating a steady steam flow through the E-3B. Afterward, the plant had been unable to re-establish flow through the E-3A/E-4A reboilers. For the next six months, the E-3B/E-4B circuit did all the reboiling.

Figure 9.6 compares typical operating data from 6 months after startup to the design. The A circuit had no boiling and no significant liquid circulation, as indicated by the zero steam flow through E-3A and the insignificant temperature difference across the heating side of E-4A. Steam to E-3B was maximized, but this was not enough to make up the lost duty of E-3A.

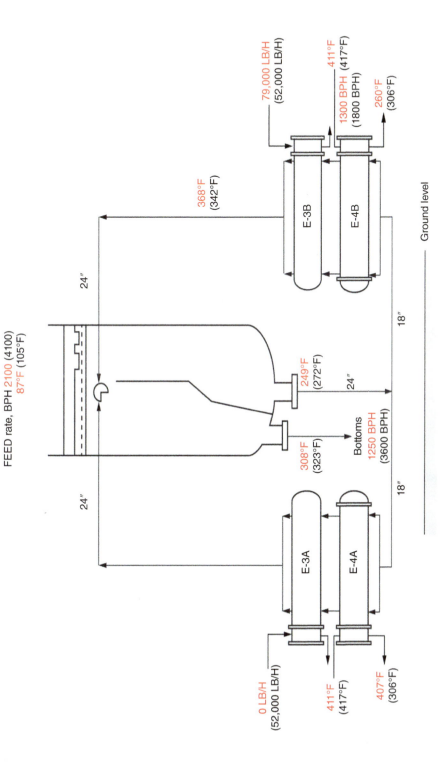

Figure 9.6 Deethanizer stripper reboiler circuits. (From H. Z. Kister, Chemical Engineering, August 2017, p.30. Reprinted Courtesy of Chemical Engineering.)

Troubleshooting   To provide gas lift that would jump-start the thermosiphon through the A circuit, nitrogen was injected for four to six hours through a bleeder valve immediately downstream of E-3A. This action was unsuccessful. This was surprising, as gas lifting in other reboilers (192, 201, 233) had been successful.

An event timing analysis focused on the change from the initial operation, with both the A and B circuits active, to the B circuit only shortly after stripper startup (Figure 9.7). Before 22:30, at a stripper feed rate of about 1700 BBL/H (barrels per hour), both E-3A and E-3B reboilers operated stably with a steam flow rate of about 32,000 lb/h to each. For both E-4A and E-4B reboilers, the naphtha entered at about 400°F and exited at about 320°F, a healthy temperature difference.

At 22:40, the stripper feed rate plunged to about 1200 BBL/H. To improve stability, the steam flow to E-3A was cut to about 8000 lb/h, with the steam flow to E-3B unchanged. The temperatures of naphtha into and out of E-4A became the same, about 360°F, indicating it stopped heating. The A-circuit stopped circulating. Simultaneously, the naphtha outlet temperature from E-4B decreased to about 260°F, indicating that this reboiler was working much harder. The next day, the feed rate to the stripper rose again to 1500 BBL/H, but the A circuit remained inactive. Opening and closing the steam valve to E-3A did not re-establish circulation.

This event timing analysis diagnosed the existence of two hydraulic steady states, with an easy switchover from two-reboiler circulation to one. A likely root cause would be low pressure drop in the reboiler circuits. A review of the licensor's design confirmed a low design pressure drop, about 11 ft of liquid head, which would be less at the low operating rates. With 19 ft elevation difference between the tower tangent line and the bottom of the naphtha reboiler, the liquid level would be in the 24″ reboiler liquid inlet pipe, well below the reboiler draw compartment.

At a low liquid head, resistance to flow is small, so fluids tend to form their preferred flow patterns. Two steady states can form: (1) the two reboilers operating simultaneously, as intended, and (2) all the boiling in one reboiler, the other filling with liquid to the liquid height in the reboiler liquid inlet pipe. Once the one-active-reboiler steady state was formed, the 8000 lb/h still going to E-3A after midnight on the second day in Figure 9.7 would strip the lights out of the near-stagnant reboiler liquid. High-boiling components would remain, making this liquid difficult to boil. This stabilizes the one-active-reboiler steady state, and it will persist.

Gamma and neutron scans can test the above theory. Both the gamma and neutron scans measured 10–12 ft of liquid in the bottom product draw compartment, in full agreement with

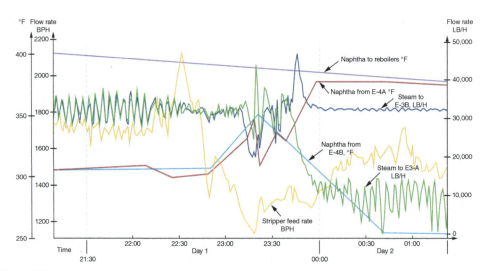

Figure 9.7   Process data for initial operation of stripper reboiler circuits. (From Kister, H. Z., Chemical Engineering, p. 30, August 2017. Reprinted Courtesy of Chemical Engineering.)

the level instrument. Both gamma and neutron scans showed no liquid level in the reboiler draw compartment, supporting the two steady state theory.

Solution A switch from the stagnant "A" reboiler steady state back to the two active reboilers steady state can be achieved by raising the liquid head in the reboiler draw compartment. The higher head will raise the pressure drop through the active reboiler circuit, inducing some of the liquid to follow the easier path (less pressure drop) and circulate through the inactive reboiler. This circulating liquid will bring lights with it, get heated, start boiling, and establish the thermosiphon in the previously inactive circuit.

To build level in the reboiler draw compartment, the level in the bottoms product draw compartment was raised by reducing the bottoms flow rate until the level overflowed the baffle. The bottom product draw compartment was kept full, with the overflowing liquid filling the reboiler draw compartment. The tower differential pressure transmitter was closely watched to prevent the high liquid level from ascending to the trays and causing flooding and possible damage there (Section 2.13).

Once the level in the reboiler draw compartment increased to 13 ft, the steam flow to the E-3B reboiler sharply fell (indicating less liquid to the B side) and the E-4A hot-side temperature sharply decreased, both indicating that the "A" reboilers were starting to circulate liquid. Steam was quickly opened to E-3A to sustain "A" reboiler circulation. Almost immediately, even split and duties were established between "A" and "B" circuits. Bottoms product flow was increased, baffle overflow discontinued, and bottoms level was returned to normal.

The column remained prone to steady state switches, possibly because of some weeping from the bottom tray liquid bypassing the reboiler side of the baffle, and the reboiler continues to stall periodically. The procedure of raising the liquid level in the product draw compartment to switch steady states has been executed several times to regain lost reboiler circulation.

9.1.7 Can a Plugged Packing Distributor Generate Two Steady States?

A refinery kero side stripper (Figure 9.8) had problems meeting the kero flash point spec. The kero flash point indicates the concentration of lights in the kero. Not meeting the flash point spec means that the lights are not adequately stripped from the kero bottom product.

The stripper worked well until an incident in which the unit had to shut down the fractionator top pumparound (TPA) and use more drum reflux to compensate. The drum reflux was cold (100°F) and wet. During the incident, water separation in the reflux drum was poor due to the larger flow through the drum, so water and salts could have entered the fractionator, causing corrosion and fouling. Salts or corrosion products would end in the kero stripper feed from tray 11.

Before the incident, the delta T (kero draw temperature from the fractionator minus stripper bottom temperature) was 40°F, indicating excellent stripping, and there was no problem meeting the kero flash point spec. Increasing the steam flow rate increased the temperature difference and raised the kero flash point as expected. Following the incident, the delta T dropped to 30°F and the flash point decreased. Increasing the steam flow rate no longer increased the delta T nor raised the flash point.

A review of the post-incident bottom level operating charts showed two distinct operating modes (Figure 9.9a). In one mode, the stripper bottom level fluctuated wildly, about 20% peak to valley. With a level transmitter span of 48″, this comes to about 9″. The peak-to-valley period was about five minutes, so the movement came to about 40 gpm, a small fraction of the feed rate of about 730 gpm. The draw line from the tower to the stripper was 14″ diameter, large enough for self-venting flow (Figure 4.9). When the level fluctuated, so did the temperatures of the stripper vapor return and of the stripper bottoms. Before the incident, the vapor return and stripper bottom temperatures were 430°F and 390°F, respectively. Gamma scans with the stripper in a fluctuating mode showed the pan distributor flooded, with liquid accumulation about 10 in. above the risers.

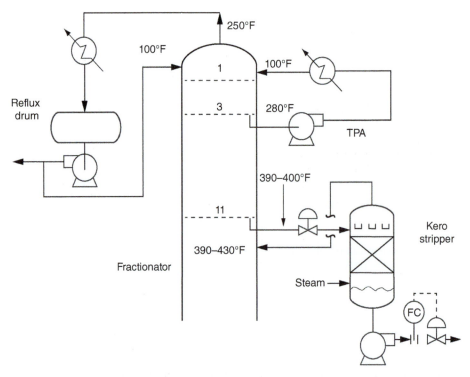

Figure 9.8   Atmospheric crude fractionator and the kero side stripper that experienced flash point problems.

Sometimes the stripper switched to another operating mode (Figure 9.9a). The switch happened very quickly. In the second operating mode, there were no fluctuations but the separation deteriorated. The bottom level became steady (Figure 9.9a). The flash point dived; at times, the stripper could not meet the flash point spec. When this happened, the unit needed to raise the crude fractionator overhead temperature from 250°F to 270°F to reduce the lights content of the stripper feed and thus keep the kero flash point on spec. The stripper vapor return and bottom temperatures became the same.

The stripper was 6.5 ft in diameter and contained random packings. The orifice pan distributor (Figure 9.9b) had 16″ tall rectangular risers. At 730 gpm, we calculated a liquid head of 2.5″ using Eq. 4.2. This means that there should have been no overflow unless more than half the holes were plugged. One thing that looked suspicious on the scan is that the top foot of packing in the stripper was missing. It may well be that these packings were uplifted into the distributor, causing the plugging or aggravating it.

The parting box sat right on top of the risers (Figure 9.9b), there was no clearance between them. The risers were discontinued under the parting box, so that liquid can equalize as it leaves the parting box. The total riser area was about 24% of the tower cross-section area.

The gamma scans showed buildup of liquid level at about 10 in. above the top of the risers, which is near the top of the parting box. This liquid could not descend into the risers, because the risers' vapor velocity exceeded the system limit. We estimated a risers C-factor (Eq. 2.1) of 0.88 ft/s. This is likely to cause liquid buildup above the risers. It is quite conceivable that when enough liquid piled up above the risers it would dump. So the fluctuations are likely to have been caused by intermittent liquid buildup and dump.

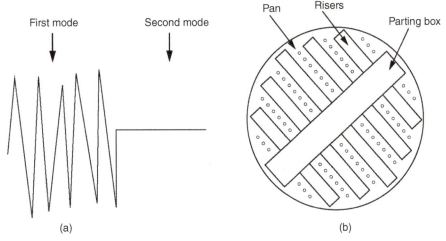

Figure 9.9   Kero stripper two steady states (a) Bottoms level operating chart, showing two modes of operation; (b) Packing distributor plan.

Under certain conditions, the liquid accumulation may have become unequal on both sides of the parting box. When this happened, vapor would tend to channel through the risers that saw the lower liquid head. As the vapor velocity through the risers increased, liquid would dump through the risers on the opposite side of the parting box. The system limit would disappear, and the liquid would descend freely through the risers on one side of the parting box while vapor would ascend through the risers on the opposite side. This would give a very stable flow and eliminate the entrainment, flooding, and fluctuation. However, the channeling would produce very poor separation, as was being experienced.

9.1.8 What Caused Tray Damage in Refinery Atmospheric Crude Fractionator?

During a startup of an atmospheric crude tower (Figure 9.10), five trays below the feed inlet ("flash zone") and six trays above the feed (the trays between the feed and the bottom pumparound (BPA) draw) were severely damaged (Figure 9.11). The intensity of the damage exceeded expectations.

The Pressure Spike Incident   In this refinery, crude feed was pumped by the cold crude pumps via preheat exchangers (to 250–300°F) to a large settling drum (a desalter) where free water and salts were removed. From the desalter, the crude was pumped by booster pumps through additional preheaters and then a fired heater. Effluent from the fired heater entered the flash zone of the crude fractionator, as shown in Figure 9.10.

The key events that led to the pressure spike were:

10:48: The cold crude pumps began cavitating, and at 10:55 they stopped pumping. The desalter feed flow was lost.

10:58: The booster pump stopped pumping and the heater flow was lost.

11:02: The heater outlet temperature spiked at 460°F. At the same time, the cold crude pumps and the booster pump began pumping again.

11:03: The fractionator pressure spiked by 13 psi (Figure 9.12a). The tower bottom level plunged from 39 ft to 31 ft (Figure 9.12b).

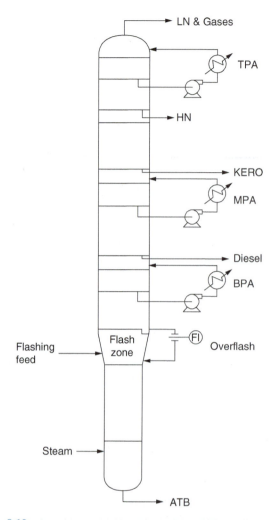

Figure 9.10 Atmospheric crude fractionator that experienced startup damage.

Figure 9.11 One of the severely damaged trays above the feed of a crude fractionator.

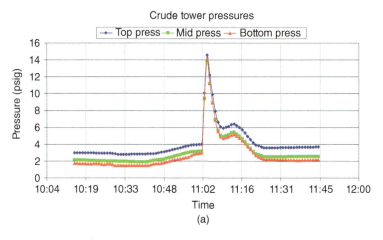

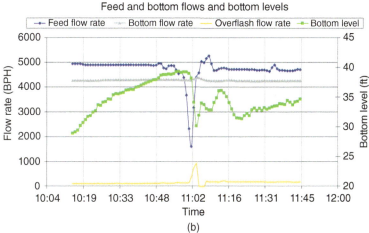

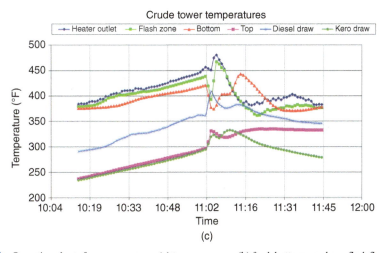

Figure 9.12   Operating charts for pressure surge (a) tower pressure; (b) feed, bottoms, and overflash flows; (c) temperatures.

Event Timing Analysis   Just before the loss of feed flow (10:55), the fractionator established a preliminary temperature profile (Figure 9.12c). Bottom temperature was about 410°F, flash zone about 430°F, heater outlet temperature about 440°F, diesel draw 355°F, kero draw 285°F, tower top 285°F. This suggests the presence of low boiling components on the trays at and above the kero draw, which was 18 trays above the flash zone. When the incident happened, the pumparounds were not yet established.

The reported level at 11:02 was at 39 ft. Normally, this would be about 6 ft below the bottom of the feed entrance nozzle but above the top stripping tray. The location is somewhat uncertain because the startup liquid in the tower bottom had different specific gravity than the normal liquid. At the time, stripping steam (Figure 9.10) was not used, indicating the stripping section contained liquid rather than froth, so the level measurement is unlikely to have been adversely affected by frothing.

At 11:03, the bottom level plunged from 39 to 31 ft over less than two minutes (Figure 9.12b). From the tower geometry, 8 ft of liquid in the bottom of this tower contain about 140 barrels of liquid.

Between 10:58 and 11:03, the feed rate to the crude tower was much lower than that between 10:15 and 10:58. The bottoms flow rate remained unchanged (Figure 9.12b). Roughly, over the five minutes of lower feed rate, the tower received 160 barrels of feed less than it would have received at full feed. This feed reduction fully accounts for the 140 barrels drop in the liquid level (above).

When the booster pump stopped (about 11:00), the bottom and flash zone temperatures were 420°F and 440°F, respectively. By 11:03 (until the pressure spike peaked), these temperatures dropped to 375°F and 390°F, respectively, a 50°F drop (Figure 9.12c). The only source of cooling would have been liquid from the cooler upper trays dropping into the bottom.

Liquid on the upper trays in the crude tower was around 290–350°F (Figure 9.12c). The liquid would dump to the bottom once the feed (vapor source) stopped. Figure 9.12b shows a large spike in overflash flow rate during that time, probably augmented by liquid weeping through the trays. Some of this liquid would vaporize upon contacting the hot liquid in the bottom, resulting in cooling the bottom liquid. The rest would mix with the bottom liquid. Very rough estimates would give 100 barrels at a median temperature of 320°F to cool around 700 barrels of bottom liquid by 45°F. Most of this liquid vaporized.

Assuming a liquid inventory per tray of about 1–2 in., the total liquid inventory on the trays between the feed and the top of the tower would be about 100–200 barrels, which is well in line with the above approximations, suggesting that much of the upper trays liquid dumped between the feed interruption and the pressure spike. This is a not surprising considering a typical time constant per tray is 0.1 minute.

At 11:03, hot feed (heater outlet temperature of 460°F) contacted the flash zone liquid at 390°F. The flash zone quickly heated to 460°F and the pressure spiked. We estimate that the heat in the hot feed liquid was capable of vaporizing about 50 barrels. That is, not accounting for any heating of the bottom liquid. However, the bottom liquid heated by about 30°F, which will account for most of the heat in the hot liquid feed. This suggests that most of the heat in the hot liquid feed was used for heating the bottom liquid, not for vaporizing.

A rough calculation, based on the column volume and gas laws, shows that to give a pressure spike of 13 psi takes a vapor accumulation in the tower equivalent to about 40 barrels. The total vaporization rate for such a pressure spike would be much higher, possibly double (80 barrels), as some of the vapor generated was condensed in the overhead condenser and in heating the upper sections of the tower.

It therefore can be concluded that the heat in the hot liquid feed alone was far too little to effect the vaporization rate required to generate the observed pressure spike.

This leaves water. The conceivable source is dumping from the upper trays during the feed interruption. At 10:00, about an hour before the upset, temperatures as low down as the kero draw, 18 trays above the feed, were only about 210°F, so it is conceivable that pockets of water were present on the upper trays. These pockets would probably have taken some time to descend from the upper trays to the flash zone. The pockets could have got to the flash zone just at the time when the feed resumed. Upon contacting the 460°F feed, the water would vaporize rapidly, generating the pressure surge. Due to the low molecular weight of water, only a small amount of water would be needed to generate the observed pressure surge. The top stripping tray contained special tall weirs for directing the feed liquid to the inlet regions of the top tray. These weirs may have formed ideal pockets for water trapping.

9.1.9 Event timing Analysis of Startup Instability Leads to Improved Startups

Adapted from Bernard, A., "Distillation Tower Startup Following a Plant Revamp with Focus on Reboiler Operation," Topical Conference Proceedings, Kister Distillation Symposium, AIChE Spring Meeting, Houston, Texas, March 12–16, 2023.

Background   The ethylene fractionator is the product tower in an ethylene plant, separating a pure ethylene top product from an ethane bottoms stream that is recycled to the thermal cracking reactors (furnaces). Figure 9.13 is a typical simplified flow scheme. Due to the high ethylene purity and the closeness of the boiling points between ethylene and ethane, ethylene fractionators contain a large number of trays (commonly between 80 and 120), are operated at high reflux to distillate ratios (commonly 2.5–3.5), and are of large diameter (10–20 ft common). Often, the tower overhead is condensed by boiling low pressure propylene refrigeration, while the reboiler is heated by condensing high-pressure propylene refrigeration vapor. With these utilities the tower operates at high pressure, typically 200–300 psig. Here and in Figures 13 and 14, the acronym "C3R" denotes propylene refrigeration.

Unique Features   In this installation, the tower had two parallel symmetrical reboilers. The reboilers process inlet nozzles were about 17 ft below the tower tangent line, and the reboilers process outlet nozzles were 6 ft below the tower reboiler return nozzles.

Normal Startup Procedure   The ethylene fractionator in this plant is started on total reflux. The tower is first freed of nitrogen in a series of steps of pressuring and depressuring with ethylene vapor from the pipeline. Once nitrogen-free, the tower pressure is raised with ethylene vapor to near its normal operating pressure, at which point ethylene can be condensed. The condensate flows to the reflux drum. Once the liquid level in the drum rises high enough, a small reflux flow is started. This liquid descends through the trays and reaches the bottom sump.

In preparation for liquid arrival at the bottom sump, the vapor valve to the reboiler is cracked open to warm up the reboiler. Once the liquid reaches the bottom sump, the reboiler starts boiling. A liquid level in the condensate pot indicates that the reboiler has started working. As more liquid arrives at the bottom sump, boilup is increased. The tower internal circulation rate is driven by the reboiler duty. The reflux drum level control adjusts the reflux in response to changes in reboiler duty. Managing the reflux drum and bottom sump inventory is critical for ensuring smooth startup. The boilup is increased stepwise, with each step small enough to avoid a large drop in tower sump level, and allowing sufficient time for the reflux to compensate. Aggressive raises in C3R vapor to the reboiler drop the sump level, which may affect, and even stall, the reboiler and the startup.

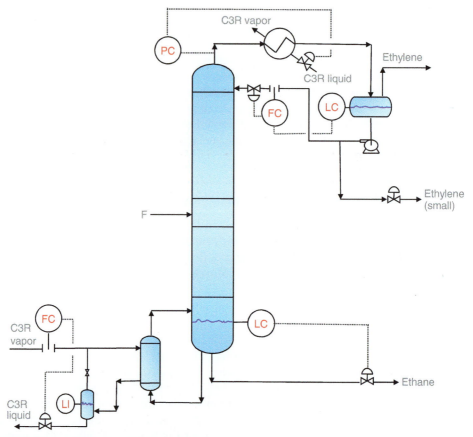

Figure 9.13 Simplified scheme of a typical ethylene fractionator, showing the controls relevant to event timing analysis in Section 9.1.9. (from Kister, H. Z., "Practical Distillation Technology", course workbook. Copyright © Henry Z. Kister. Reprinted with permission).

As vapor condenses and the condensed vapor fills up the trays, downcomers, and sumps, liquid inventory in the reflux drum and bottom sump is depleted, so ethylene makeup needs to be added to the tower to maintain the inventories.

Event Timing Analysis At one startup, instability delayed establishing a stable total reflux operation for several hours. An event timing analysis was performed to understand the root cause and to improve future startups. Figure 9.14 shows operating charts for the key variables, plotted against time in minutes. For good analysis, the charts are subdivided into seven intervals, as advocated in Section 9.1.1.

Interval 1 The C3R valve to the reboiler was cracked open to warm up the reboiler (the valve was cracked open 1.5 hours before time 0). Reflux was going into the tower. With little compensation from boilup, the reflux drum level dropped. The level in the tower sump was below the lower level transmitter tap. Right at the end of the interval, the reboiler kicked in, which is indicated by the sharp rise of level in the condensate pot.

Interval 2 Reducing the reflux flow rate and a small increase in reboiler flow rebuilt the level in the reflux drum. With reflux arriving in the bottom sump, its level rapidly rose.

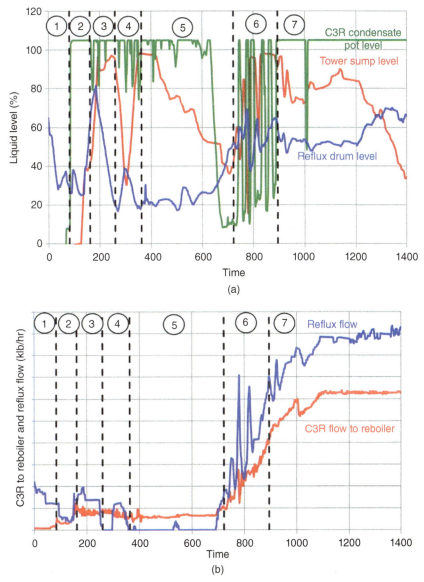

Figure 9.14 Ethylene fractionator transition to total reflux (a) Levels (b) Flows. (From Bernard, A., "Distillation Tower Startup Following a Plant Revamp with Focus on Reboiler Operation," Topical Conference Proceedings, Kister Distillation Symposium, AIChE Spring Meeting, Houston, Texas, March 12–16, 2023. Reprinted Courtesy of Andre Bernard.)

Interval 3 With levels rising in both the reflux drum and bottom sump, reflux as well as C3R flows were increased. However, the boilup was not enough to keep the level up in the reflux drum, and it dropped rapidly, forcing initially a reflux cutback. At the end of the interval, it was necessary to totally shut the reflux to retain the drum level. The level in the condensate pot showed periodic drops, indicating a reduction in heat transfer.

Interval 4 With no reflux, the sump level rapidly fell. Reflux flow was resumed once the level in the reflux drum started to build up. This replenished the bottom sump level. The boilup failed to compensate for the fall in the reflux drum level, and again it was necessary to totally shut down the reflux to avoid loss of drum level. The condensate drum level took a few drops, again indicating low heat transfer.

At this point, the levels between the reflux drum and bottom sump were cycling 180° apart, signifying the liquid inventory was cycling between the reflux drum and bottom sump. This indicates insufficient liquid inventory, and additional ethylene makeup was required to balance both.

Interval 5 With zero reflux flow, the bottom sump level fell due to ethylene vaporization in the reboiler. This vaporization was quite low, as can be seen from the slow drop in bottom sump level and the slow buildup in reflux drum level. When the sump level fell to about 50%, the reboiler heat transfer appeared to completely die, as indicated by the rapid drop in condensate drum level.

The right corrective action, bringing liquid makeup to raise the liquid inventory in the reflux drum, was taken at about 650 minutes. Once this level built up, reflux was resumed, and the bottom sump level started to rise again.

Interval 6 Both reflux and C3R to the reboiler were raised. The reflux drum and bottom sump levels oscillated, cycling 180° apart, indicating the need to keep adding external liquid makeup to build the liquid inventories. A major fraction of this liquid inventory was needed to build up levels on trays and downcomers. During the initial phases of the startup, tray and downcomer levels could not be established due to the low boilup, which kept the trays dumping.

Interval 7 Sufficient external makeup and further increases in reflux and C3R flow to the reboiler enabled stable conditions to be achieved.

Improving Startup Procedure   The event timing analysis paved the way for understanding the problem and for an improved procedure that would shorten startup by about 10 hours. Intervals 3, 4, and 5 can be eliminated by bringing up a large external makeup once the cycling between the reflux drum and bottom sump starts and by ensuring enough C3R flow to the reboiler to overcome its turndown limitation.

Another lesson taught was the criticality of keeping a bottom sump level high enough to permit good thermosiphon action. At low sump level, the circulation of liquid in a thermosiphon reboiler stops (192), and so does heat transfer. In this case, the limitation was particularly severe due to the low process velocities at the reboilers at low sump levels and the 6 ft long height between the reboiler process outlet and the tower reboiler return nozzle. Bernard's article (30) presents a detailed analysis of the heat transfer turndown limitation.

Last but not least, this case illustrates the value of event timing analysis to understand interaction of variables and to improve startup and plant operation.

9.2   FIELD TESTING

9.2.1   General Application Guidelines for Field Testing

Item 11 in Section 1.3 highlights the importance of "checking the patient's responses." Making small, inexpensive changes, one at a time, and monitoring the column's response provides invaluable information for diagnosing the problem. The key is to perform each of these changes judiciously, one step at a time. Performing several steps simultaneously makes it difficult, even impossible, to identify the step that made the difference, and may lead to an incorrect diagnosis.

Previous chapters of this book identified the variety of diagnostic techniques available to troubleshooters. This section demonstrates how these techniques can be combined to reach the correct diagnosis. The cases illustrate the stepwise approach and how to learn the most from field tests. Lessons from plant tests include:

1. Start with the diagnostic technique that is the simplest and can narrow down the number of theories. Generally, exploring for leaks, learning about column operation and history, hydraulic calculations, and comparing data to simulation should be considered prior to any evolved testing.

2. Listen to experienced people. If they say something that does not sound completely right, do not discount it. It may be a major clue. Seek tests that will verify or deny it.

3. Collect people's past field experience with the column in question. This can eliminate unnecessary tests while at the same time provide invaluable input for testing.

4. Be cautious of past tests that have not been properly documented. Unless well-documented, you cannot be sure that conditions were the same or that they were adequately performed.

5. If at one time the column worked well and then stopped, attempt to establish a past operation for a test and see if it still works. Otherwise, seek ways of testing factors that changed.

6. Check every theory against the past experience and initial findings. When this comparison is inconclusive, think about which tests can best evaluate it.

7. Change one variable at a time during the tests. Changing a number of variables simultaneously is likely to lead to inconclusive results.

8. Avoid shortcuts that can reduce the conclusiveness of a test. Shortcuts can lead to misinterpretation and turn the investigation in the wrong direction.

9. Analyze the results of each test. Unexpected results may suggest a change of direction or new tests. Follow these leads. See Sections 9.2.5 and 9.2.8 for such changes.

10. Beware of the limitations of the testing techniques used. In Sections 9.2.5 and 9.2.8, limitations in scanning techniques affected the test interpretation.

11. Tower testing remains incomplete until all tests and field observations are complete.

12. Do not trust expert opinions. One good measurement is worth more than a thousand expert opinions.

9.2.2 Narrowing Down from 12 Theories to the Root Cause

This was the largest debutanizer tower the author has ever seen (216). It was serving three fluid catalytic cracking (FCC) units in a refinery. The tower was built in the 1970s. Years later, as the plant rates were raised beyond the original design, a reboiler bottleneck was encountered. In the 1990s, the reboiler was replaced by a new larger one, and feed rates were raised. At the higher rates, the tower ran into flood initiating near the tower bottom at 80% of the calculated flood point. Observations and checks by the plant bred many alternative theories. The author was invited to join the troubleshooting team.

At the kick-off meeting, the unit process engineer jotted down the 12 theories on the white board, briefly describing each. Some theories suggested it was a vapor flood, others a liquid flood. The author spoke:

"It may be a good idea to perform some vapor and liquid sensitivity tests" (Section 2.7; the column control system was as per Figure 2.4). For the sake of the unfamiliar, during the meeting, the author described these tests, as explained in Section 2.7. But as the author was speaking, he noticed one of the operations team members getting uncomfortable. It was not long before he stopped the author.

"Let me tell you something about OUR column. Right now, I have 40,000 BPSD reflux going into the column. If I add another 1000, make it 41,000, it's going to flood like you ain't seen a column flood before. So I don't need to do no sensitivity tests. You ain't gonna get more than 40,000 BPSD without flooding it."

The author turned to him and asked, "Does the experience that you are describing happen only when the temperature control is in automatic, or does the same thing happen when the temperature control is in manual?"

Had the flooding occurred at 40,000 BPSD when the temperature control was on manual, then it would have pointed to a liquid flood, in which case there would have been no need to perform the liquid sensitivity test. His plant observations would have sufficed. The vapor sensitivity test would still have been needed.

He replied, "I don't need to run no temperature control on manual. You ain't gonna get more than 40,000 BPSD out of it without flooding."

We talked to other operation people. No one had any recollections of having the temperature control on manual. The team decided to proceed with the tests.

With the temperature control on manual, we raised the reflux to 61,000 BPSD. The column was drawing straight lines. There were no signs of flooding whatsoever.

We then reduced the reflux back to 40,000 BPSD. The steam rate was 150,000 lb/h. As soon as we raised it to 155,000 lb/h, the tower flooded like the author had never seen a tower flood before. He was right about that.

The test conclusively established that the flood was a vapor flood, with no sensitivity to liquid rates. We increased the liquid rate quite a bit, but the tower did not respond. We went back to the white board, Out of the 12 theories we scratched seven out. Every theory arguing a mechanism that has liquid sensitivity was invalidated. We went down from 12 to five.

Per Section 2.2, a vapor flood can be a jet flood, a downcomer backup flood, a system limit flood or a downcomer unsealing flood. Gamma scans can give insight as described in Sections 5.5.2 and 5.5.3. We therefore gamma-scanned the tower. During this time the temperature control was run in automatic. As recommended in Section 5.3, item 6, we performed flooded and unflooded active area scans. In the unflooded scans, froth heights in the loaded areas were 16″ out of 24″ tray spacing, and the vapor peaks were near the clear vapor line. The vapor and liquid rates were raised by as little as 3% and the trays flooded. Per reasoning in Section 5.5.2, this denied jet flood. At 97% of jet flood, the dispersion would have been close to 24″ tall, and no clear vapor would have been observed.

In the absence of tray damage, downcomer unsealing flood is highly unlikely, as this tower operates at high liquid loads. As system limit is an ultimate jet flood, it too was denied. So this left us with a vapor sensitive downcomer backup flood. This finding ruled out two more theories that argued a jet-flood type mechanism. We went down to three theories.

One theory that is well consistent with a vapor-sensitive downcomer backup flood is plugging (Section 2.2). To evaluate this theory, it is important to learn about the tower history. There are two types of history: ancient and modern. Both need to be evaluated.

Ancient history: the tower has been in existence since the 1970s. Has tray plugging been experienced? We checked with the operations team, some of them have been in the plant all these years. The answer was negative. Neither fouling nor plugging has ever been experienced in this tower.

Modern history: When did the problem initiate? Immediately after startup or many months into the run? Many months into the run would support a plugging or fouling theory. The answer was immediately after startup, as soon as the limiting rate was reached. The bottleneck did not get better or worse since. The trays were well inspected at the turnaround, they were clean and have not been changed since the 1970s.

Not done yet. Next question: Were there any events during the startup or initial operation that could have plugged or damaged the trays? Look for high base levels, pressure surges, feed contamination, high dP at startup rates. We talked with the operations team, and critically reviewed the startup records. None. It was a very smooth startup.

Our history study taught us that plugging or damage is unlikely. One more theory crossed out. The number of theories was down to two.

The two remaining theories were liquid maldistribution in the four-pass trays and issues due to a poor design of the chimney tray below the bottom tray. This tray collected liquid from the bottom tray and sent it to the reboiler. It was oriented at 45° to the tray above and was poorly designed.

To check on the liquid maldistribution theory, we gamma scanned the active areas of two adjacent side and center passes, as described in detail in Section 5.5.7 (this is the same tower). The gamma scans provided non-conclusive evidence supporting this maldistribution theory.

There was no need to go further. The corrective action would be to replace the trays, including the chimney tray, with well-designed ones. The question of whether the chimney tray was a bottleneck or not is of academic interest only.

This case demonstrates how a step-wise testing procedure can narrow in on the likely root cause.

9.2.3   Making Sense of Plant Data and Operation Experience in a Packed Tower

This case demonstrates the importance of using plant data and experience in developing effective theories (218).

A chemical tower processed different liquid feeds when running different campaigns. At the rate-controlling campaign, feed to the tower (Figure 9.15a) was organic chemicals, plus 2–3% water, 2000 ppm dissolved NaCl, and 100 ppm dissolved $CaCO_3$ salts. The 23″ ID tower had five packed beds, the bottom three filled with a proprietary small random packing, and the top two with 1″ Pall rings.

After six years of smooth operation, the feed and bottom distributors were replaced by high-performance distributors to improve separation. A year and a half later, instability set in, forcing a feed rate cut from 130 lb/min to 100 lb/min. Following a startup, the column would operate well for a day or two with the normal expected temperature profile. Then the temperatures below the feed would gradually decline. When the temperature about 3 ft above the bottom (T1) dropped to 90–95°C, the bottom pressure started increasing, eventually reaching 42 in w.g. Simultaneously, temperature T2 would decline to 85°C. The local pressure gauges indicated that most of the increase in pressure drop was in the stripping section, but on more than one occasion it was observed in the section above the feed.

As a temporary cure, operators cut boilup, which lowered the bottom temperature from normal 148°C to 100°C. During this time, they diverted the off-spec bottoms to the feed tank. An hour later, they raised the bottom temperature back to 148°C and diverted the bottoms back to the product tanks. The tower stayed good for a day or two, after which the procedure needed repeating. Rinsing the tower with water from the top stretched the well-behaved run to two to three days.

After a month, the tower was water-rinsed and then opened. The rings in the bed just below the feed were taken out of the tower. The surface of some of them was covered with calcium carbonate. No deposits were found in the distributors. Any water-soluble deposits would have been washed in the water rinse.

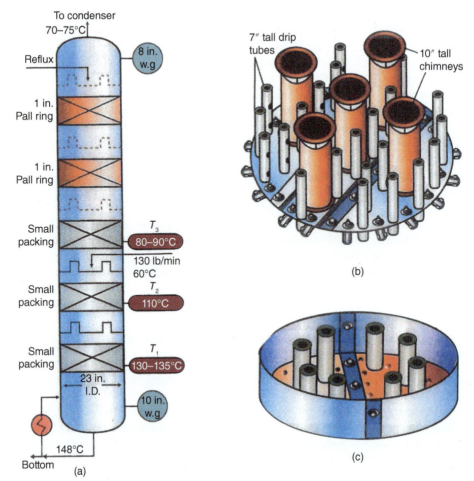

Figure 9.15   Chemical tower that experienced salt plugging (a) the tower (b) the feed distributor (c) the previous feed distributor. (From Kister, H. Z., and S. Chen, Chemical Engineering, p. 129, August 2000. Reprinted Courtesy of Chemical Engineering.)

A pressure survey during one of the runs measured 50 in. w.g. pressure at the bottom, 42 in. w.g. between the two bottom packed beds, and 10 in w.g. above the feed distributor. This indicated flooding below the feed, which was believed to be due to plugging of the small packings. The plant decided to replace the small packings below the feed by a larger packing of the same type. The tower had more separation stages than needed (when clean), so losing some separation due to the larger packings was not an issue.

The result was disappointing. Run times between plugging episodes did not change. This argued against packing plugging, as the larger-size packing should have helped. Hydraulic calculations showed that with clean packing, the bottom bed operated at 65% of flood, and the next bed operated at 54% of flood.

This shifted attention to the distributors (Figure 9.15b). Each distributor had 30 drip tubes, 7/8″ ID and 7″ tall, and five 10″ tall chimneys. Each drip tube had a 0.2″ hole at 1″ above the bottom plate and a second 0.35″ hole 4 in. above the plate. Each of the 30 drip tubes extended 2″ below the plate. The feed liquid entered via an elbowed-down 1.5″ feed pipe that discharged liquid from a height of 2″ onto the center of the hat covering the center chimney of the feed distributor.

Calculations using the orifice equation (Eq. 4.2) showed that the distributor liquid levels should be 4.5–5 in. at 100 lb/min feed, rising to 5.5 in. at 130 lb/min feed. The actual levels may have been a little higher due to frothing, but in the absence of plugging, they should have remained below the top of the drip tubes. The calculations also showed that had the lower 0.2″ holes plugged, the liquid level would exceed the top of the drip tubes at feed rates just below 115 lb/min, and the liquid above the top of the tubes would be entrained into the next bed. Plugging of the 0.2″ tubes was quite plausible. When the tower was opened to change packings, many of the 0.2″ tubes were plugged, even after the water rinse.

The important next step is to check the distributor plugging theory against field observations, as follows:

- As distributor holes plug up, liquid distribution will deteriorate, making separation below the feed poorer. The lights will not get stripped, and temperatures T1 and T2 will decline, as was observed.

- As long as the distributor liquid level stays below the top of the tubes, there will be no entrainment and the pressure drop will not rise. Once the plugging intensifies and the liquid level exceeds the top of the tubes, vapor will entrain the liquid, initiating flooding, and the pressure drop will escalate as was observed.

- The deposits plugging the tubes are mainly NaCl with some $CaCO_3$. The tower distills the water upward from the feed, and the heavy salt will stay behind and travel downward. The salts are insoluble in the organic mixture below the feed and will deposit out.

- During the temporary cure, reducing the bottom temperature from 148°C to 100°C allowed water from the upper part of the tower to descend and dissolve some of the deposits. This was the same as a water rinse, except a lot slower and less effective, as the concentration of water in the feed was only 2–3%. This is well in line with observations.

- Water-rinsing during the turnaround dissolved the NaCl, which was most of the deposits, and destroyed the evidence for plugging. The only witnesses were the water-insoluble $CaCO_3$ deposits and these were found.

- The old distributors (Figure 9.15c) had larger floor holes (0.33 in. diameter), which are far more resistant to plugging than the 0.2″ holes in the new distributor. This explains the stable operation with the old distributors.

With the theory matching all the field observations, the solution was to modify the two old lower distributors below the feed (those that worked before their replacement). The modifications were to redrill the 30 0.33″ holes to eighteen 0.625″ holes to enhance plugging resistance. As the risers in the old distributors were only 3.5″ tall, reducing the liquid head to 1″–1.25″ for the 130 lb/min feed would give comfortable margin from overflow. It was appreciated that some separation would be lost, but the stripping section had more height than needed, so it was acceptable.

With the modifications, the bottom product remained on spec, and the instability totally eliminated, never to return.

9.2.4 Downcomer Unsealing: A Correct Diagnosis Brings a Correct Cure

From Kister, H. Z., "Tower Doctor: Downcomer Unsealing: A Current Diagnosis Brings a Correct Cure," Chemical Engineering, p. 32, September 2022. Reprinted Courtesy of Chemical Engineering.

This case illustrates the importance of analyzing operating experience, performance data, testing theories, and most of all, asking questions, rather than relying on hunches.

Following a distillation conference hosted by a simulation vendor, the delegates were bussed to a venue where the simulation vendor put a delicious BBQ. A gentleman sitting next to the author introduced himself. After exchanging a few greetings, he asked

"Do structured packings work in the wash section of an atmospheric crude tower?"

"They do as long as you have good liquid and vapor distribution and they do not plug" I replied. "Some refiners swear by them. They experience occasional plugging episodes, but are comfortable living with them."

"I was afraid you'll say that. We have to put these in the wash section of the tower."

"Why? What is wrong with trays?"

"Trays just do not work for us. We are struggling with dark gas oil" (this means poor separation in this section).

"Had the trays ever worked for you?"

"At one time they did. But not any more."

That reply got me.

"You went up in rates and then ran into a bottleneck?"

"Not really. We have always run 50,000 BPSD crude"

"This is getting interesting. Why would trays all of a sudden stop working?"

"If you really want to know, I will be happy to tell. At one time trays worked for us. Figure 9.16a shows the tray details. The 12″ nozzle at the bottom is the tower feed. Nozzle 14 is the atmospheric gas oil (AGO) draw. It is a total draw. Reflux is pumped on flow control and returned above it. We ran 1000 BPSD reflux, which is about 2% of the crude feed."

"I would not complain about this. This is a 1–1.5% overflash on the crude, which is excellent. You won't do much better with packing"

(footnote: "overflash" is the term used for the reflux exiting the wash section into the feed zone, as illustrated in Figure 9.10. The temperature there is hotter than the reflux, so on the way down some of the 2% reflux entering vaporizes, so the remaining overflash liquid is 1–1.5% on the crude)

"Yes, but these days are gone and we cannot get back to them"

"Why?"

"About six years ago a vendor reviewed our tower. They were alarmed by the small feed nozzle, so they enlarged it from 12″ to 24″, and added a vane vapor horn distributor (similar to the feed in Figure 9.17) on it. They also added splash baffles on the trays to avoid them drying up, and downpipes on the spent wash to avoid re-entrainment (Figure 9.16b). When we came back on line, we needed to go to 6000 BPSD reflux to get the AGO on color. Actually, once it got on color, we could cut the reflux down to 1500 BPSD. But if we went down below this, it would go dark and we will need to go up to 6000 BPSD to get the AGO back on color."

"We complained to them. They then reconsidered the vapor horn, and in the next turnaround removed it. They also changed the trays. The ones we had before were the venturi moving valves that have smooth orifices and give low pressure drops. They were concerned that these trays will excessively weep, and changed them to moving valves with sharp edged orifices (Figure 9.16c) that are more weep-resistant."

"Did this improve things?"

"Quite the contrary. Now we cannot get the AGO on color, even if we run a reflux as high as 15,000 BPSD. We went to different vendors. They all say we need structured packings."

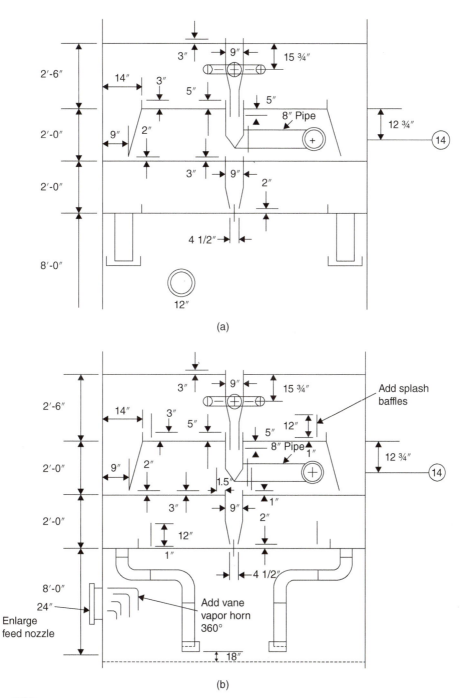

Figure 9.16   Atmospheric crude tower wash section (a) Initial (b) after first set of modifications (c) after second modification. ((From Kister, H. Z., Chemical Engineering, p. 32, September 2022. Reprinted Courtesy of Chemical Engineering.)

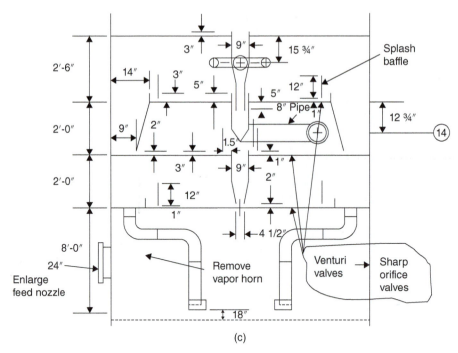

(c)

Figure 9.16   (*Continued*)

"There are many towers that have trays working in the wash zone. And the trays worked for you at one time. What I believe you need is troubleshooting to find why the current trays are not working. I smell a downcomer unsealing problem, but need to take a close look at the dimensions and do a few sums."

"If you can find a way to keep trays you will be a hero. But you'd better come up with a good proof that trays will work, because most people in our refinery, including management, are believing the vendors and have given up on trays."

Once we drew the sketches to scale (Chapter 8), the problem became apparent. The 2″ clearances under the center downcomer are just too large for the low liquid rates experienced in the wash section, which typically operates at very low liquid loads. With this large clearance, vapor will break through into the downcomers. With the splash baffles 1.5 in. away from the weirs, and with the draw pan restricting the vapor exit from the center downcomer, the vapor breaking into this downcomer would not allow the liquid in. The middle tray will flood, and the bottom tray will not receive liquid and dry up. There will be no separation, and the AGO will go dark.

The problem was introduced by adding the splash baffles. The vendor was right about the venturi valves weeping. This weep was actually beneficial, as it allowed the liquid a path downward by weeping through the valves. Once enough weeping occurred, it would seal the center downcomer, preventing vapor rise through it. This happened at about 6000 BPSD reflux. Once the downcomer was sealed, they could go back down to 1500 BPSD.

In the next revamp, the venturi valves were replaced by the sharp-orifice valves that are resistant to weeping. Without weeping there was no way for liquid to go down, so even increasing to 15,000 BPSD reflux would not bring the AGO on color.

"Nice theory, but can you prove it?"

"Absolutely. We can run a test that will get your AGO on color. You we will need to reduce the charge rate to lower the vapor velocity in the downcomer and allow liquid to descend. We will need gamma scans to watch for liquid in the center downcomer."

We proceeded with, the test in which we used gamma scans to continuously watch the center downcomer ("Time Studies," Sections 5.2.4). Initially, the center downcomer contained vapor, no seal. Then when the charge rate was sufficiently reduced, the gamma scans saw liquid in the downcomer and the AGO came on color.

In the next turnaround, we removed the splash baffles and installed inlet weirs . We also made some modifications to the AGO draw. After these changes, the tower operated at a wash rate down to 700–800 BPSD with the AGO always on color.

The takeaway when the patient's health suddenly deteriorates for an obscure reason, always go back to the last time the patient was well, and investigate all changes from then on.

9.2.5 Well-Targeted Field Testing Brings a Correct Diagnosis and an Unexpected Simple Solution

In memory of Neasan O'Shea, a brilliant engineer, a great leader, and a real gentleman and friend.

A vital element of troubleshooting highlighted in this section is proper testing of each theory with valid field tests. One good field test can humble thousands of expert opinions (243).

In this case, an atmospheric crude tower lost capacity following a retray (by others) with high-capacity trays. A systematic investigation utilizing extensive field tests identified two completely independent root causes for the capacity loss. The diagnosis led to a very simple solution that overcame both bottlenecks, required no hardware changes, circumvented the need to shut down the crude unit, cost little to implement, and recovered most of the losses almost immediately. In the longer term, the diagnosis led to low-cost modifications of tower internals that produced capacities well above the design.

Tower Revamp   A 4 m ID atmospheric crude tower (Figure 9.17) separated 52,500 BPSD (7070 t/d) crude into an overhead light naphtha (LN) product, a heavy naphtha (HN) side product drawn from tray 23, a diesel product drawn from tray 11, and an atmospheric residue ("resid") bottom product. The tower contained two pumparounds, which are direct contact cooling circuits: BPA and TPA.

The heavies contents of each side product was controlled by manipulating the duty of its adjacent pumparound. The TPA was drawn from a sump just under the downcomer from Tray 21 to Tray 20, the balance overflowing onto Tray 20 to become reflux to the naphtha–diesel fractionation section below. The BPA was drawn from a total draw chimney tray below Tray 9. The level control on the chimney tray manipulated the wash-back, which was the reflux to the wash section below.

Initially, the tower contained 34 single-pass jet tab trays. In the early 1990s, the tower sections from tray 20 down were revamped to increase fractionation efficiency and to permit intermittently drawing a kerosene side-cut while maintaining the same throughput of 52,500 BPSD (7070 t/d). The changes are color-marked on Figure 9.17. Trays 11–20 were replaced by single-pass high-capacity Nye trays. The trays in the wash and stripping sections below were replaced by structured packings. A new kerosene side draw was installed at the inlet to Tray 16, and a side stripper was added to strip lights from the kerosene. Initially, the kerosene side draw was not operated because the rundown system was incomplete.

After the retray, the tower flooded at throughputs exceeding 38,000 BPSD (about 5200 t/d), well short of the previous 52,500 BPSD, with a rise in pressure drop and excessive heavies in the

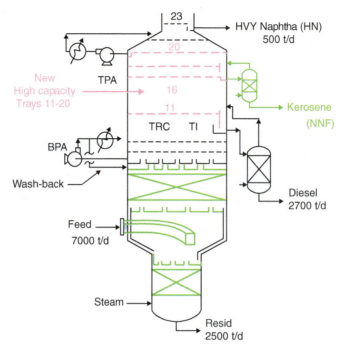

Figure 9.17   Revamp of the lower section of the crude tower in Section 9.2.5, design conditions.

HN product. Gamma scans showed flood initiating near the unused kerosene draw (around Trays 14–16) and building up to tray 23. The TPA trays 21–23 were jet tab trays and were not changed in the revamp.

   In addition, a severe instability, "*Excursion*," set in at about the same rates as the flooding initiated. A relatively small reduction in tower feed temperature (coil outlet temperature, COT) or increase in BPA duty would trigger a massive reduction in tray 10 temperature, usually accompanied by a HN yield loss, with the lost yield showing up in the residue.

Nature of the Excursion   In an atmospheric crude tower, reducing COT or increasing the BPA duty cools the tower, often inducing distillate loss to the residue. The excursion was an intense, sudden version of this phenomenon that took place at feed rates exceeding 36,000 BPSD (4800 t/d), especially at high diesel product rates. There was an instantaneous large drop in Tray 10 temperature, usually (but not always) accompanied by loss of naphtha yield and gain in residue yield, and by about 5°C cooling on trays 11–23.

   Two sensors measured tray 10 temperature, one connected to the controller (TRC), the other to an indicator (TI). Normally, both TRC and TI read about 300°C. During excursions, the TRC dropped sharply by about 11°C, but the TI dropped much more, as much as 30–40°C. Historically, the TRC on Tray 10 manipulated the BPA heat duty to control the diesel cloud point at 0°C. After the revamp, the response of the cloud point to the TRC became erratic, and the diesel cloud point could no longer be controlled from the TRC.

   Early simulation work correlated the excursion with a leak of around 20% of the liquid from tray 11. The tower was briefly shut down and both the diesel draw sump and the BPA chimney tray were water-tested. Neither showed any leaks; both were effectively seal-welded in a previous turnaround. A few popped-out valve caps on tray 11 were reinstated and the joints of tray 11 were seal-welded to minimize leakage. Some minor modifications were made to the diesel draw (below). There was no appreciable improvement in performance upon return to service.

Diesel Draw Geometry and Modification Figure 9.18 shows the diesel draw revamp modifications (in pink). Tray 11 downcomer terminated in a seal pan that overflowed into the diesel draw pan. The draw pan floor was 380 mm beneath the seal pan. The draw pan outlet weir rose 290 mm above the top of the seal pan outlet weir. The diesel was drawn by a 12-in. nozzle almost flush with the draw pan floor. The retray reduced the downcomer width from 585 mm to 370 mm. The downcomer clearance was increased from 50 mm to 90 mm, and the seal pan weir height was raised from 50 mm to 115 mm.

Figure 9.18 (roughly to scale) shows the location of the TI and the TRC sensors on tray 10. The TI was close to the top of the draw pan overflow weir. During the brief inspection shutdown, in an endeavor to minimize splashing of liquid from the draw pan on the TI, the draw pan outlet weir was raised from 720 to 770 mm and in the vicinity of the TI sensor to 820 mm.

Theory 1: A Hydraulic Bottleneck at the Diesel Draw The excursion seemed to take place at a diesel flow rate of about 2000 tons/day (t/d). At 2000 t/d, the liquid velocity at the downcomers entrance was about 0.06 m/s, well below the downcomer entrance flood (downcomer choke) velocity (Section 2.16). The downcomers residence times were well within good design practices (192, 245). A downcomer choke flood was therefore ruled out.

Adequate degassing of liquid requires a residence time exceeding 30–60 seconds (Section 2.15, item 1). The degassing time in the revamped downcomer (Figure 9.18) was only 12 seconds. While there is no need to degas downcomer liquid exiting to the tray below, degassing can be important for liquid draws. Without adequate degassing, the diesel draw nozzle and rundown lines need to be sized for self-venting flow (Section 2.15). Figure 4.9 shows that the 12-in. draw line could handle up to 2000 t/d of non-degassed liquid, which coincided with the maximum observed diesel draw rate. Any additional liquid would back up in the draw pan, overflow its weir, and cool down the TI and the TRC. Being closer to the draw pan, the TI cooled down more.

The original (pre-revamp) downcomer had twice the volume, doubling the residence time to about 24 seconds, quite close to the time needed for adequate degassing. While the diesel draw nozzle and rundown line were undersized for self-venting flow, they were adequate for degassed liquid. It was therefore theorized that the excursion reflected a diesel draw bottleneck resulting from the reduction in downcomer volume at the revamp.

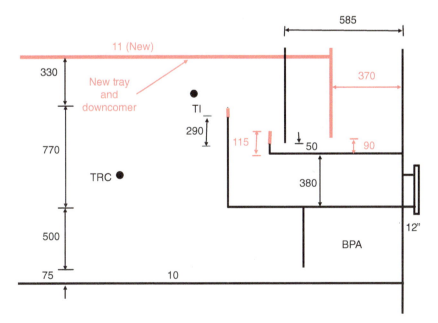

Figure 9.18 Diesel draw revamp modifications.

Test 1: Did the Diesel Draw Pan Overflow? To test the above theory, the diesel draw was neutron-scanned at excursion conditions with the diesel draw rate near 2000 t/d. The scans showed that the diesel draw pan contained about 300 mm of liquid or very dense froth, with lower density froth for 255 mm above. The 210 mm above this were clear, showing neither liquid nor froth. The draw pan did not overflow. Theory 1 was disproved.

Theory 2: Excursion Linked to Flooding Initiating on Trays 14–16 Some people linked the excursion with the flooding that initiated on trays 14–16. A multitude of gamma scans showing no flood on trays 11–13 argued against this theory, so it was considered unlikely. Nonetheless, good troubleshooting should test both the likely and the unlikely theories (item 15, Section 1.4).

Test 2: Would the Excursion Persist with the Flood Eliminated? The tower was set in the "no excursion" mode with the diesel flow near 2000 t/d. Then the BPA duty was raised to about 15%. The TPA duty was reduced by an equal amount, with the feed rate kept constant. This drastically lowered the vapor traffic on trays 11–23 and positively got them out of flood.

The result: tray 10 TRC dropped by 23°C, and tray 10 TI by 45°C. The HN product rate dropped by 300 t/d with an equivalent rise in resid production. An excursion took place. In this test, the flooding was eliminated, yet the excursion occurred. Had the excursion been linked to the flooding, it should not have occurred in this test. Theory 2 was disproved.

Theory 3: Excursion Caused by Tray 11 Leakage The observation that the excursion occurred upon vapor flow rate reduction, bred the theory that leakage or weeping, not flooding, caused the excursion. With all the liquid from tray 11 removed by the diesel total draw, a mass balance on the BPA gives:

$$\text{wash-back} = \text{Tray 11 leak} + \text{BPA condensation} \tag{9.1}$$

Based on equation (9.1), the leakage rate can be measured by monitoring the wash-back flow rate while maintaining the BPA condensation rate (i.e., the BPA heat duty) constant.

Test 3: Tray 11 Leakage Test The tower was set at the excursion mode. The tower feed temperature (COT) was raised from 365°C to 370°C. The BPA duty was kept constant by manipulating the BPA flow rate.

Raising the COT dropped the wash-back by 160 t/d and the residue make by 100 t/d, and raised the HN make by 120 t/d. Tray 10 TRC rose 14°C, tray 10 TI rose 35°C, and temperatures on trays 16 and 19 rose by about 5°C. The tower got out of the excursion mode.

Per equation (9.1), the large observed change in wash-back flow rate at no change in BPA duty was proof that the excursion was due to changes in leakage from tray 11.

Theory 4: Excursion Caused by Tray 11 Weep With tray 11 leakage verified, liquid weep from tray 11 raining on the TI and TRC, causing them to drop, became the prime theory. Raising the COT stopped the weep, the TI and TRC dried and heated up. The weep at the TI was larger than at the TRC because the vapor rate at the TI was obstructed by the nearby draw pan.

This theory gave simple explanations for all the observations. The difficulty of explaining the suddenness of the change was its only weak link. Also, it needed testing, like any other theory.

Test 4: Tray 11 Weep Theory Test Lockett and Banik (300), as well as Colwell and O'Bara (74), found the tray weep rate to primarily depend on the vapor hole velocity. This is consistent with the Test 3 finding that small changes in COT, that change the vapor hole velocity and strongly affect weeping, can get tray 11 in and out of excursion relatively fast. The same authors (74, 300) also found the weep rate to be a much weaker function of liquid flow rate. A "devil's advocate" test of the weep theory would monitor the response of the excursion to changes in liquid load at constant vapor rate. No response, or a minor response of excursion to mild changes in the liquid load would validate the tray 11 weep theory.

For this test, the tower was set in the excursion mode. Both the BPA duty and the COT were kept constant to maintain constant vapor rate to tray 11. The TPA duty was reduced by turning off a fan on the TPA air cooler. This change reduced the tray 11 liquid flow rate at constant vapor rate, and was expected to have no effect on weeping, and therefore, the excursion.

The result was amazing. The TRC on tray 10 rose by 12°C and the TI on tray 10 by 35°C. The wash-back rate dropped 260 t/d, the residue and diesel make dropped 100 t/d and 50 t/d, respectively, and the HN make rose 140 t/d. The column coming out of excursion upon liquid flow rate reduction disproved the tray 11 weep theory.

The observation that higher HN draw rates got the column out of excursion pointed to a liquid-sensitive, not a vapor-sensitive, bottleneck. This test sent us back to reconsider the initial diesel draw bottleneck (Theory 1). Further testing was needed to verify the strong liquid rate sensitivity.

Test 5: Kerosene Mode Tests By this time the kerosene draw was commissioned and provided an invaluable clue. The operators noticed that there were no excursions while the tower operated in the "kerosene mode"—i.e., when kerosene was drawn. Drawing kerosene reduced the diesel make, so the diesel draw bottleneck would not be encountered.

To test the revived diesel draw bottleneck theory, we attempted to force an excursion while in the kerosene mode by lowering the COT. In the no-kerosene mode, when the COT was below 363°C the tower seldom got out of excursion.

Test 5A: Can Lower COT Bring an Excursion in the Kerosene Mode? The tower was set in the non-excursion mode with 350 t/d of kerosene drawn. The COT was slowly lowered from 371°C to 357°C. The BPA duty and tower reflux were kept constant. To maintain the heat balance, the TPA duty was allowed to decrease.

No excursion took place. Both tray 10 TRC and TI cooled down by 15°C. The wash-back did not change. The diesel make dropped by 175 t/d, and the residue make rose by 250 t/d. This test determined that there was very little weep from tray 11, and confirmed that unloading the diesel draw would keep the tower out of excursion.

Test 5B On/Off with the Kerosene Draw At different times during Test 5A, we shut (or nearly shut) the kerosene draw-off valve . This could only be maintained for 0.5–1 minute because the surge volume of the kero stripper was small. The results are listed in Table 9.1. Excursion occurred only when the diesel plus kerosene flows exceeded 1960 t/d. It occurred immediately upon valve closure.

This test demonstrated the bottleneck to occur at a sharp diesel draw rate, between 1955 and 1985 t/d. This test conclusively verified that the diesel draw rate was the only motive for going in and out of excursion. With a typical time constant of 0.1 minute per tray, and only a few trays between the kerosene and diesel draws, the excursion occurred rapidly, as observed. The kerosene tests proved that the excursion reflected a hydraulic limit at the diesel draw at about 1970 t/d.

Table 9.1 Tests of shutting the kero drawoff (243).

COT (°C)	Diesel draw (t/d)	Diesel draw when kero draw stopped (t/d)	Excursion?
371	1650	2000	Yes, twice
368	1620	1985	Yes, little
363	1600	1955	No sign
357	1490	1840	No sign

Excursion Cause   The extensive field tests returned us to Theory 1 of a hydraulic limit at the diesel draw at about 2000 t/d. At this draw rate, the flow through the diesel draw nozzle and rundown line switched from self-venting flow to choked flow.

Before the revamp, the downcomer from Tray 11 provided 24 seconds of degassing time, sufficient to degas the diesel draw. Both the draw nozzle and rundown line were sufficiently large to handle the *degassed* diesel flow rate. The revamp halved the downcomer volume, halving the residence time. Twelve seconds of residence time was insufficient to degas the liquid, and it remained *aerated*. Per the Figure 4.9 correlation, the diesel draw nozzle and rundown line encountered a self-venting flow limit at 2000 t/d. Beyond this diesel draw rate, the vapor bubbles choked the line, backing up liquid in the diesel draw pan.

Rain pouring from the diesel draw pan caused the large drop in tray 10 temperature upon excursion. The TI was closer to the draw pan than the TRC (Figure 9.18), so it got wetter and experienced a much larger temperature drop. At diesel draw rates below the self-venting flow limit (about 2000 t/d), there was no rain, and both the TI and TRC read the same.

The neutron scans in Test 1 disproved overflow from the diesel draw pan but did not rule out entrainment. Neutron scans provide excellent measurement of liquid levels and froth densities, but only near the tower wall (Section 6.3). To address the flooding in the tower, several gamma scans were previously shot, and were now revisited focusing on the region below tray 11. One scan chord detected a large quantity of liquid in the vapor space right above the draw pan wall in the excursion mode, consistent with extensive entrainment from the diesel draw pan.

A plausible explanation for the extensive entrainment is as follows. At non-excursion conditions, flow in the diesel rundown line was self-venting. The aerated liquid comfortably descended, leaving an 80 mm vapor gap between the top of the liquid level in the diesel draw pan and the floor of the seal pan above (Figure 9.19a). Vapor returning from the rundown line spread horizontally through this vapor space. The low vapor velocities generated little entrainment.

Upon reaching the self-venting flow limit in the diesel draw nozzle and rundown line, liquid/froth backed up in the diesel draw pan and filled the vapor space between the liquid in the draw

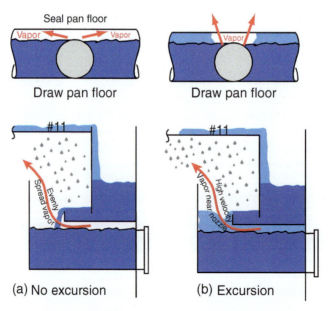

Figure 9.19   How the excursion caused entrainment.

pan and the floor of the seal pan above (Figure 9.19b). Vapor returning from the rundown line formed a high-velocity jet that broke through the liquid right near the nozzle. This jet entrained the liquid, which caused excursions.

Solution   The same field tests that diagnosed the root cause of excursion taught that operation in the kerosene mode prevents excursions. The kerosene mode became the normal mode of operation. The economic penalty was minor, as the kerosene product could be blended with the diesel when not desired as a product. An unexpected windfall was that the kerosene mode operation also alleviated the flood, as described elsewhere (244). With the excursions eliminated and the flooding alleviated, the tower throughput soared from 38,000 BPSD to 49,100 BPSD, just a little short of the design capacity.

9.2.6   Field Testing Brings a Correct Diagnosis where Engineering Analysis Failed

A primary absorber in a refinery fluid catalytic cracker (FCC) gas plant operating at about 200 psig flooded when the lean oil flow rate was 25% below design (125). This 25% needed to be bypassed around the absorber, at the penalty of deficient absorption. Attempts to increase the lean oil rate resulted in a pressure drop increase in the tower and lean oil carryover to the downstream secondary absorber.

An engineering contractor study using computer simulation and vendor hydraulic calculations, but without field tests, identified vapor flowing up the downcomer and restricting its capacity to be the root cause. The fix was a retray with fewer valves on the trays and shorter downcomer clearances. Upon restart, the tower flooded when the lean oil rate was 50% below design, worse than before. A postmortem by the contractor concluded that the tower may experience very severe foaming. However, as this service is usually considered mild or non-foaming, the plant doubted the validity of this conclusion and hired a consulting company to conduct a proper troubleshooting investigation. The approach strongly advocated by this consulting company (125, 126, 129, 133, 153) is strategic field testing. This approach is strongly endorsed by the author.

The absorber had two intercooler loops of similar design. Figure 9.20a depicts this design. Each intercooler loop condenses some of the tower vapor, so the vapor and liquid loadings below an intercooler loop are usually higher than above. This means the highest hydraulic loadings in the absorber are below the lower intercooler loop. So a regular flood is likely to initiate below the lower loop. Additionally, points of transition (inlets, outlets) are common trouble spots where flooding may initiate (Chapter 8).

Test 1. Is a high base level the root cause of flooding? A high base liquid level is a common cause of premature flooding (Section 2.13). To check, the base level was reduced and 50% of the lean oil was bypassed, so the tower was out of flood. The lean oil flow rate was then increased, keeping the level low. The tower dP rose at the same lean oil rate, indicating flooding. This ruled out excessive base levels as the root cause.

Test 2. Identifying the flood location With the lean oil 50% bypassed (no flood), a pressure survey was conducted to provide a baseline. The lean oil rate was then increased, flooding the tower, and the pressure survey was repeated. The survey determined that the flooding started at the lower intercooler draw (Figure 9.20b). The trays above the draw had liquid accumulation, causing the high pressure drop.

Test 3 Does the intercooler draw rate affect the flood? The draw to the lower intercooler was closed. This routed all the liquid from the tray above via the seal pan overflow. With the draw closed, the tower started flooding earlier, at lean oil rates even lower than 50%. This test confirmed that the flooding was caused by a problem in the intercooler draw.

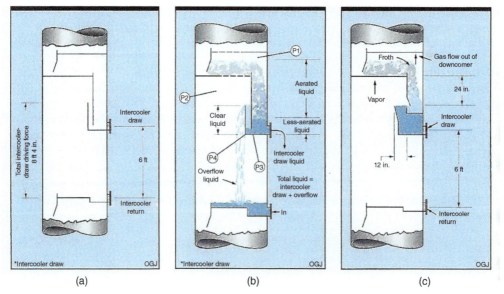

(a)   (b)   (c)

Figure 9.20   Liquid side draw to intercooler that initiated premature flood (a) arrangement (b) same as (a), but showing flow patterns (c) modification that fixed the problem. (From Golden, S. W., Oil & Gas Journal, p. 62, April 7, 1997 and April 14, 1997. Reprinted Courtesy of The Oil & Gas Journal.)

Analysis   Ideally, all the liquid entering the downcomer should go into the intercooler. This will happen if the liquid head (6 ft, see Figure 9.20a) is high enough to exceed the pressure drop in the intercooler and its piping, and the draw line is sized for self-venting flow. Liquid exiting the downcomer contains entrained vapor bubbles, and these need to vent back through the downcomer, hence the need for self-venting flow in the draw line. If the intercooler pressure drop exceeds the liquid head, or if the draw line is not sized for self-venting flow, liquid will back up in the downcomer and seal pan and overflow the seal pan weir.

Theory   The theory postulates that some liquid was overflowing the seal pan weir. The liquid in the downcomer is aerated, while the liquid in the seal pan is mostly degassed, and therefore has a higher density, as depicted in Figure 9.20b. So 2 ft of liquid in the seal pan will raise the downcomer froth to about 3–4 ft. This is almost to the next tray. Adding the pressure drop of the bottom tray causes the downcomer backup to spill over to that tray, initiating column flood.

This theory explains why the retray made things worse. The original valve trays had more valves, so they tended to weep, bypassing some liquid around the downcomer, seal pan, and intercooler loop. Eliminating the weep by using fewer valves increased the intercooler loop pressure drop and the liquid level in the seal pan. In addition, the fewer the valves, the higher the tray pressure drop. All these would bring the flood earlier.

Solution   Figure 9.20c shows the solution. The downcomer was separated from the deep seal pan, terminating in its own seal pan, so that there will be only a few inches (instead of 2 ft) of clear liquid causing backup in the downcomer. The valve tray above the intercooler draw was replaced by a chimney tray to reduce the aeration in the downcomer as well as the pressure drop that backed up liquid in the downcomer. This solution permitted the absorber to operate well at 100% of the lean oil flow rate.

9.2.7 Operating Team Observations and Field Testing Brings a Correct Diagnosis where Engineering Analysis Failed

An atmospheric crude tower (282) had two pumparounds (PAs), a TPA and a BPA, similar to those shown in Figure 9.17. Both PAs take liquid from a tray, cool it, and return the cooled liquid on an upper tray, two or three trays higher. The trays between them are used as direct-contact condensers. The return piping of the TPA and BPA is sketched in Figure 9.21a (282).

After the startup of a grassroots tower, column vibrations and hammering were experienced, most audible in the PA sections. The hammer could only be silenced by running a colder tower feed temperature, which led to losing diesel product to the tower bottoms residue.

An engineering study by the internals supplier looked at PA pump cavitation, PA heat exchanger vibrations, cavitation of the PA check valves, and flashing at the PA return nozzles and distributors. All these were investigated but were ruled out both by observations and calculations. The supplier did not visit the site or talked to the operating team.

An independent troubleshooting endeavor was initiated by a consultant (282), this time by interviewing plant operators, and putting together their observations:

- There was a possibility of operating the tower with top reflux only, no PAs. No hammering occurred during this operation. This narrowed the problem to the PA circuits.

- Hammering started only when the PA pumps were started (cold PA return liquid).

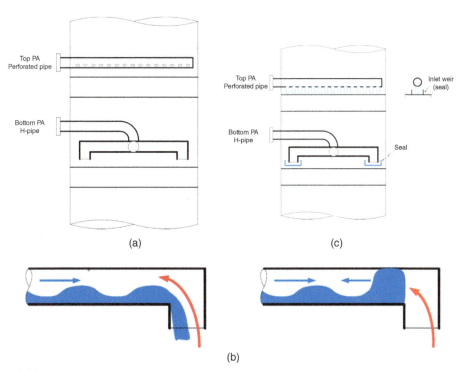

(a)   (c)

(b)

Figure 9.21 Liquid entry piping arrangements of top and bottom pumparound returns in an atmospheric crude tower (a) initial, experienced hammering (b) hammering mechanism (c) Modified, eliminated hammering. (From Lenfeld, P., and D. Lin, "Tricky Liquid Distributors," in Kister Distillation Symposium, Topical Conference Proceedings, p. 428, AIChE Spring Meeting, San Antonio, Texas, March 26–30, 2017. Reprinted by permission.)

- Hammering could be prevented when the PA return temperatures were kept high. (*Footnote: high PA return temperatures reduced the PA heat removal. As the PAs preheated the tower feed, this reduced the tower feed temperature, leading to diesel loss to the residue.*)

- When the PA return temperature dropped below a certain limit, the hammering started again.

- The last three observations verified that high subcooling was the root cause.

These observations suggest a mechanism similar to that in Section 7.2.11 (Figure 9.21b). Due to the low vapor pressure of the subcooled liquid, hot condensable vapor from the tower entered the distributor pipes. It then collapsed onto the liquid surface, generating a local vacuum inside the PA return pipes, causing more vapor to be sucked in, collapsing onto the subcooled liquid. Eventually the condensate was pushed out by the PA liquid, and the cycle repeated. The collapse mechanism initiated hammering in the pipes, as illustrated in Figure 9.21b.

The problem was eliminated after the PA return liquid pipe distributors were equipped with hydraulic liquid seals to prevent the backflow (Figure 9.21c).

Case study 5.5 in Ref. (201) describes a similar hammering issue in a pipe distributor bringing a highly subcooled feed into a chemical tower. In this case, the solution was turning the pipe distributor (similar to the top PA distributor in Figure 9.21a) upward so the sparger pipe always remained liquid-full. A shielding baffle then deflected the issuing feed liquid downward.

9.2.8 Lessons Learnt from Packed Tower History and Testing

Still X and **Still a** had been specialty product finishing columns in the same service. In the early 1990s, a leading packing supplier repacked both stills with the exact same packing heights in each column: two lower packed beds of A meters each and an upper bed of B meters. The stills were about 5 ft in diameter. The packing was similar to BX wire-gauze-structured packing that is commonly applied in fine chemical manufacturing, which requires deep vacuum distillations to prevent compound deterioration. At an expected efficiency of five theoretical stages per meter, each column was expected to have about 50 theoretical stages.

In the early 2000s, a test mixture was run in both **Still X** and **Still a**. The results showed that **Still X** had about 11 effective stages in the lower beds, while **Still a** had >40 stages in the lower beds. Separation efficiency of the lower beds of **Still X** was 25% of what it should be. The test was repeated in the spring of 2015 with identical results. With this low efficiency, **Still X** was unsuitable for fine product purification.

Still X was gamma-scanned in 2001. Typically, packed beds are grid-scanned (Figure 5.6a), but due to the surrounding structures, only two scan lines (a + scan, Figure 5.6c) were shot. The scans found all distributors and collectors at the correct elevations, and all three beds having the proper bed heights. The top bed had a density gradient in the packing and some maldistribution. There was some maldistribution in the top half meter of the middle bed. The liquid distribution in the rest of the middle bed and throughout the bottom bed was good. The packing density was constant throughout the middle and bottom beds. The scanners concluded that there appeared to be no serious mechanical damage, fouling, or major distribution problems that would explain the absence of such a large number of theoretical stages. Neither the gamma scans nor operation saw any signs of flooding.

Still X was a taller column than **Still a**, mostly due to the longer downcomers between the packed beds (about 20 ft long) that brought reflux liquid to the feed parting box below. Some felt that this caused the difference in the stills performance. It was argued that the problem was the high velocity in the pipes blowing out the parting box liquid.

During an outage, **Still X** was inspected through the manholes. It was water-washed with the manholes open and water running through the packed beds. Nothing out of the ordinary was noted.

In the fall of 2015, a world-class expert contractor in mass transfer tested and evaluated **Still X** and came up with a stage count very similar to the previous two tests. The contractor pointed to bed height, drip point density, and other possible issues. The contractor recommended replacement of the internals and three packed beds, using the latest low pressure drop wire gauze packings and contractor-designed internals.

Other options considered included adding a fourth bed, changing the packings but not the internals, and changing the internals but not the packings.

During the fall 2015 test, discussions were held with a long-serving operator. His recollections were that initially both columns had the exact same separation efficiency. "This is an important point, as some believe that the design the of the **Still X** column internals may have an impact on the current performance." In the late 1990s, the separation efficiency of **Still X** degraded, and it no longer performed as an equivalent to **Still a**. Significant efforts to recover the column performance at that time were unsuccessful. "About the time that the performance was lost – the column was unknowingly distilling organic chlorides. Two other columns were also engaged with these distillations and were significantly damaged." However, manhole inspections have not revealed any damage to **Still X**.

To develop a good diagnosis, it was decided to perform a detailed in situ water test. Water was introduced through the reflux line and meter. All the manholes were open and observations made at each redistributor. The flow rate was increased from the minimum liquid loading to the maximum distributors capacity.

At maximum flow, there was no splashing at the parting boxes, all looked well there, denying the argument of issues with the long pipes. However, at the bottom manhole, the water poured right out of one section of the packing (Figure 9.22a). There was definite channeling in the lower section of packing.

Maldistribution below the middle bed was not as obvious as it was below the bottom bed. To get a rough quantitative evaluation, a test using the bucket and stop watch method was conducted (Figure 9.22b) and clearly highlighted channeling in the middle bed as well. Round 1 had the water fed through the reflux distributor, and round 2 through the feed distributor, both showing much the same results. There was heavy flow toward the center (position 1), with section 4 the driest. The in situ water tests proved that the problem was liquid maldistribution.

In situ tests also explained why the gamma scans failed to detect the maldistribution. With a + scan (Figure 5.6c), both scan chords passed through the middle (position 1) where the flow was high. Upon seeing the same high central flow, both chords looked similar, leading to the erroneous conclusion that distribution was good. In situ tests provide a much broader picture of the liquid distribution than regular gamma scans.

Following this test, all of the packing was replaced in all three beds. The older internals (including all distributors and collectors) were reused since the similar column next to it (**Still a**), had adequate separation, and also because these internals have not changed from the time that the tower worked well. **Still X** efficiency was recovered to the expected and continues to perform well today.

Takeaways In situ water tests are invaluable when conducted properly. Incorporating simple quantitative rough measurements like the bucket and stop watch method is very informative. Keeping the test basic helps reach conclusive results.

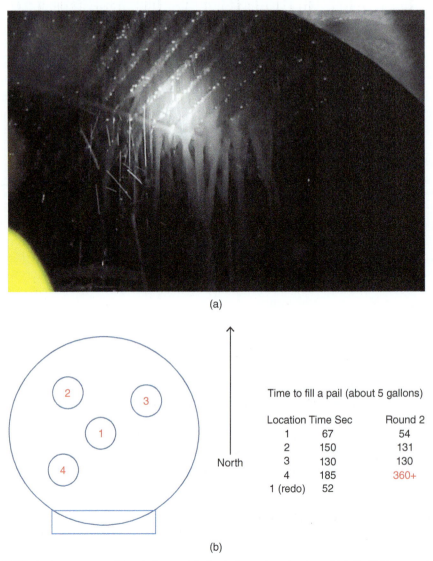

(a)

(b)

Time to fill a pail (about 5 gallons)

Location	Time Sec	Round 2
1	67	54
2	150	131
3	130	130
4	185	360+
1 (redo)	52	

North

Figure 9.22   In-situ water test of chemical tower (a) visual observation, showing a high liquid flow near the center (b) results of bucket and stop watch measurements.

In-situ water tests view liquid distribution over the entire tower cross-section area, whereas gamma-scan chords cover a smaller area. In situ tests may therefore be able to detect some types of maldistribution that gamma scans may miss.

Always do the test yourself. The water test was a repeat of a previous test many years earlier that found the still to be "good." There was no real record from the initial test – other than that it was "good," and there was no record of how well the test was performed.

Always reach out to the operators; often, you can save a lot of expenses and headaches by just listening and gathering information. The long-serving operator's recollection that the column worked and then it started to operate poorly was a milestone in the path to a correct diagnosis.

The world-class mass transfer experts were pointing to bed height, drip density, and other things – they were wrong. One good measurement is worth more than a thousand expert opinions.

Chapter 10

Troubleshooting by Inspection

"You do not get what you expect – you get what you inspect"

—(origin unknown)

Our malfunctions survey (Table 1.1) included installation mishaps among the top five causes of tower malfunctions. Installation malfunctions lead to poor separation, lost capacity, instability, higher energy consumption, and higher carbon footprint, all with negative economic and HSE impact. In some cases, a tower may cease to work, forcing a premature outage. Proper inspection following construction and during turnarounds is the best tool to identify installation mishaps, design oversights, fouling, and damage and correct them before they hurt.

Tower inspection is critical for trouble-free operation. In new and revamped columns, inspection is the last opportunity to detect any design, fabrication, or installation errors prior to startup. In existing towers, it is the last opportunity to detect and repair damage and undetected flaws and clean up any plugged or fouled internals. Remaining flaws are likely to bring about poor performance and even mechanical damage or HSE issues.

Troubleshooting guru Norman Lieberman (290) describes a tower he designed but did not inspect that flooded prematurely due to an uninstalled impingement plate on the reboiler return. The author has a similar story about a revamp that he designed but did not inspect that fell short of meeting expectations due to a minor detail that would have been easily detected had he inspected it. Both of these experiences teach that good hard work can go down the drain due to a lack of inspection. Do not fall into this trap. Remember: you do not get what you expect – you get what you inspect.

Prior to startup, the cost and effort of rectifying flaws is often negligible, becoming enormous after the startup. Additionally, new flaws introduced during the outage are extremely difficult to identify. Unlike the fabrication and design, which are usually well documented by drawings, specifications, and correspondence, few (if any) records are kept of the installation or the differences between the "as-built" tower internals and their drawings. If the tower performs poorly, there is often a scant basis for suspecting an assembly error. Nonetheless, a decision to shut a tower down for reinspection often hinges on this basis. A premature shutdown is extremely costly and may turn out most embarrassing if it fails to cure the problem.

Inspection is one of the few times that process and plant engineers get to see the inside of equipment. It is an invaluable training opportunity. In half a day inside a tower, one can learn more about distillation towers than in a full university semester. So it is an opportunity not to be missed, especially for the inexperienced. To get the most out of such an inspection, it is crucial to know what to look for to prepare ahead of time. The author has seen cases when engineers entered the tower without adequate preparation and learned very little.

This chapter reviews the common issues inspectors should look for and many of the tools that can help them achieve the most out of these inspections. It also describes good and bad practices for troubleshooting tower internals installation and assembly by inspection.

10.1   SAFETY PRECAUTIONS FOR WORK INSIDE THE COLUMN

Working in the confines of a column is potentially hazardous. Appropriate safety precautions are required to mitigate the hazards. Many of these precautions are required by law, The Occupational Safety and Health Administration (OSHA) standards, company standards, and other regulatory bodies. It will take a book just on this subject to describe all these. Such a description is outside the scope of this book. The aim of this section is to provide a brief generic overview of some of the key potential hazards in working inside a tower during installation, removal, inspection, and cleaning work. The list is far from comprehensive. To address each of these hazards, it is necessary to follow statutory regulations, OSHA standards, company standards and practices, safety supervisors' guides, and other requirements. The hazards list is not intended to apply to any specific application, nor to replace any statutory or company standards.

Common safety hazards include:

- Confined space.
- Inadequate isolation of lines connected to the tower, including vent and blowdown lines and lines carrying heat transfer media to and from the column heat exchangers.
- Inadequate blinding and unblinding practices.
- Inadequate ventilation inside the column.
- Inadequate testing of the atmosphere inside the column.
- Inadequate safety training.
- Not being acquainted with all the potential hazards and how to avert them.
- Inadequate protective clothing and protective equipment.
- Inadequate work supervision.
- Inadequate emergency plans and first aid procedures.
- Violation of safety regulations.
- Toxicity: chemicals, reaction products, deposits, decomposition products, nuclear materials.
- Pyrophoric deposits, scale, polymer, tars, which may catch fire.
- Structured packings catching fire when the tower is opened to the atmosphere.
- Hot work and the generated fumes.
- Hazardous materials trapped in dead pockets in the column and piping.
- Inadequate lockout of electrical equipment connected to the column.
- Inadequate removal or shielding of any radiation sources.
- Falling and dropping objects inside the tower.
- Release of materials (e.g., solvents) deliberately brought in by workers to perform a task.
- Climbing inside and outside the tower with tools and protective gear.

- Moving and crawling through tight spaces.
- Heavy lifts.
- Falling while inside the tower.
- Poor foothold, supporting on corroded internals.
- Loose, damaged, corroded tower internals.
- Sharp edges on trays, packings, distributors, and other internals.
- Fogging up of safety glasses or goggles.
- Darkness and inadequate lighting.
- Noise.
- Dust and irritants.
- Claustrophobia.
- Heat stress and dehydration.
- Inadequate or untested communication equipment/system for people working inside.
- Inability to easily get out through the tight spaces during an emergency.
- Lack of a well-thought-out path for moving through the column.
- Inadequate access to the items that need inspection.
- Hazards in the vicinity of the tower, such as falling objects and ignition sources.
- Hazards while climbing on ladders and scaffolds.
- Inappropriate jokes concerning safety.
- Inadequate sanitary facilities for workers on tall columns.

The author strongly recommends declining or aborting a tower inspection if you are not satisfied that all the relevant hazards have been properly addressed or if you have any physical limitations, concerns, or fears that may limit your ability to perform the inspection. Remember also that inspection is a dirty job, which may unsettle your social or family life for a few days. There is a very important rule of thumb presented by Bouck (42):

Know your limits. Your and others' safety depends on you.

10.2   TROUBLESHOOTING STARTS WITH PREVENTIVE PRACTICES DURING INSTALLATION

10.2.1   Preinstallation Dos and Don'ts for Tray Columns

Inadequate preparation of trays prior to installation may prolong the turnaround, adversely affect column performance, and even lead to safety issues. The guidelines below (43, 192, 307, 319) can help minimize these problems:

1. Installation of trays while the tower lies horizontally is generally not a good practice. Saving by shop-installation is often offset, or surmounted, by reinstallation and repair costs. One can get away with it in small (<10 ft ID), short towers. In towers with larger diameters, it is difficult to reach the top of the trays to inspect them. Taller towers (>60–80 ft) often acquire a "banana" shape while horizontal, which throws off dimensions, requiring reinstallation after the tower is raised. Finally, internals can get damaged during transportation and uplifting.

2. Adequately detailed installation drawings need to be available prior to assembly.

3. The tray supplier should be required to clearly identify all parts and pack them separately for shipment.

4. Tray parts should not be removed from the crates prior to installation. Early removal can lead to rusting, dusting, or loss of tray components. The crates should be stored and kept in a dry, covered area.

   Carbon steel trays and internals need protection from a damp atmosphere. Inside storage is recommended even when these internals have surface protection. Stainless steel does not require such protection, so the conventional wooden crate with tarred paper or plastic lining is often considered adequate (25).

5. Valve tray panels should never be shipped or placed "cap to cap" or "legs to legs" in order to prevent interlocking of valve floats. Panels with interlocking valve floats are extremely difficult to separate without damaging the valves (307).

6. Masking tape must not be used as flange covers. In one case (222), an erratic reboiler action resulted from a piece of masking tape left in a reboiler flange. Plastic flange covers are better because they need to be removed before bolting.

7. It is a good practice to order about 10% spares on bolts, nuts, and clamps in case some become lost or damaged. A higher percentage of spares is often advocated for some fouling or corrosive services (192). In such services, spare trays are sometimes justified in order to minimize downtime (319).

8. Construction supervisors should be familiarized with the column internals, their functions, and any unique requirements of the service. For instance, the construction crew should be made well aware of any collectors or seal areas from which leakage must be minimized. They should also be alerted to the common installation traps that deserve specific attention, as discussed later in this chapter.

9. A mock-up tray installation outside the tower is a valuable tool for familiarizing the installation crew with tray parts and training them in the installation procedure (319).

   This is imperative with specialty and proprietary trays. Supplier drawings often omit key details, in some cases due to confidentiality concerns, leaving installers with no idea how certain parts fit in. The mock assembly unveils such details in time to contact the suppliers and obtain their guidance. In one case (43), such a mock installation identified a fabrication error that would have prevented the high-capacity trays from attaining their high capacity. Early detection enabled a remake within the time constraint. The author had similar experiences.

   A mock-up installation on the ground is also useful in other complex tray designs. In one case (43), a mock-up installation identified an issue with a modification of a complex chimney tray that decanted water. Had this modification been installed as shown on the drawings, the tray would have been dysfunctional. In another case, a mock-up installation of specialty trays at the vendor shop allowed a thorough inspection and was invaluable in reducing the tight installation time in a revamp.

10. Prior to any work inside the column, it is essential to implement measures to prevent small parts such as bolts, nuts, and clamps from finding their way into downstream equipment, such as heat exchangers, pumps, and control valves. In one case (192), a piece of rope ladder reached the bottom pump inlet and lodged there, causing the pump to frequently lose suction. Temporary plugs in the tower base and other drawoffs are effective in preventing such incidents. It is imperative to positively ensure that these plugs are removed prior to startup.

In addition, temporary strainers can be installed in outlet lines, especially those feeding pumps. Plugs are more effective than strainers alone, because some debris can get through strainers or damage strainer elements and then pass through them. In one case (434), blockage by debris broke strainers, and pieces of strainer casings damaged the pump. Strainer casings should be mechanically strong enough to withstand pump suction when fully blocked.

11. Keep track of all replacement parts' orders, especially small parts. Keep a record of who ordered them and when, who received them and when, and where they went after receipt. Disappearance of items after receipt is not uncommon, as described by two case studies (42).

12. In one case (360), where many specialty trays were to be installed in a very short time interval, a training column at the vendor shop was used to train the crew.

10.2.2 Preinstallation Dos and Don'ts for Packed Towers

Inadequate preparation of packing prior to assembly may prolong the turnaround, adversely affect column performance, even lead to safety issues. The guidelines below can help minimize these problems. Steps 2, and 4–7 primarily apply to random packings; steps 1, 3, and 8–10 apply to both random and structured packings.

1. Installation of packings while the tower lies horizontally is bad practice. Distributors cannot be properly leveled, and the packings can be easily compressed and damaged in transportation and when the tower is lifted. Wall wipers are one of the weakest links and get damaged most easily. Random packings cannot be randomly installed.

2. New packings often have an oil film coating. The oil film may be lubricants used in the packing press or used to inhibit packing corrosion during shipping or storage. For sea transportation of carbon steel packings, an oil coating is often necessary (e.g., 143). This oil film may inhibit wetting of the packing surface, especially in aqueous systems. Certain lubricants may induce foaming in high-pH aqueous systems. The oil may also be a fire hazard during hot work or hot commissioning/startup operations.

   It is necessary to be familiar with the nature of the oil and, if needed, to adequately remove it. It is best to seek the supplier's advice on the preferred removal procedure. Alternatively, the supplier can be requested to use a water-soluble lubricant in the press, which can be washed during commissioning, or to degrease the packing with solvent after pressing. Early removal of the oil may cause corrosion and should be avoided.

3. Packings should be stored and kept in a dry, covered area well protected from sunlight. Packings left standing in the rain may corrode or oxidize rapidly. Oil-coated packings may collect dust. In the sun, metal packings heat up and the oil coating becomes fluid and flows down until trapped, forming hot oil pools, which are a fire hazard. Plastic packings may be attacked by ultraviolet rays from the sun.

4. Drums used for packing storage should be clean and free of chemicals that may attack the packing or of materials that stick to the packing surface and later inhibit liquid film formation or cause undesirable effects (e.g., foaming). Oversized containers should be avoided as they are hazardous to workers lifting them.

5. New or reused ceramic packing should be screened to remove broken pieces. In some cases (267), up to 40% of the ceramic packings were damaged during transportation. Experiences of damage to ceramic packings during service are abundant (31, 68, 192, 222, 421).

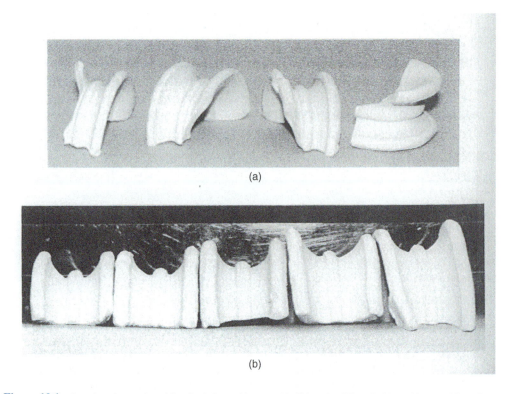

(a)

(b)

Figure 10.1 Samples of ceramic saddles fresh from shipment. (a) Chipped saddles. (b) Nonuniform saddles. (From Kunesh, J. G., Chemical Engineering, p. 101, December 7, 1987. Reprinted with permission.)

Figure 10.1 shows a few samples of ceramic saddles as received from shipment. The breakage (Figure 10.1a) is mainly of chipping at the corners (267). The large pieces can still be used, but the chips must be screened out as they are likely to lower the column capacity and increase its pressure drop. Screening must be carefully performed and closely watched; otherwise, it may cause more particles to break than it removes. In some cases (e.g., 267), it is necessary to pick chips out by hand. Figure 10.1b shows the size nonuniformity of saddles of a single nominal size as received from one shipment.

6. Plastic packings should be checked for brittleness. Plastic can become brittle in some environments and temperatures. Grab a few pieces and squeeze them to see whether they shatter. In case of problems, the supplier should be contacted.

7. About 10% of the spare packing volume (in case of ceramic packings, about 20%) should be ordered. The packing volume supplied is usually based on the supply containers' volume. Typical packings supplies are in 1- or 2-ft^3 boxes or 25-ft^3 shipping containers. When emptied into the column, the total packed height may be less than the specified height (267). The difference may be due to unfilled space near the walls of the box, underfilling of boxes, interlocking of packing particles, compression of packing when the column is filled, and using dry packing techniques. The author is familiar with cases where installed packed beds were shorter than intended.

Instead of ordering spare packings, the supplier can be requested to provide enough packings to fill the bed to the required height. The supplier will then allow for the spare volume. With ceramic packings, additional allowance should be made for any breakage occurring past the point where the supplier's responsibility ends.

8. In reactive chemicals services, the resistance of the packing materials to a chemical attack should be audited by lab testing under simulated process conditions. The ability of a ceramic or plastic packing to weather a chemical attack depends on its texture and composition, and this varies with the manufacturing process. Porous surfaces may be more prone to attack than smooth surfaces. In one case (421), samples of ceramic packings from different suppliers were lab-tested over several days under simulated hot potassium carbonate (hot pot) absorber–regenerator conditions, revealing wide variations in the rate of loss of silica from the packing. The silica loss from one sample was so rapid that the packing was branded unsuitable for the service. Packing deterioration in service can induce fouling, plugging, and poor performance.

    Before selecting a supplier, manufacturers should be requested to submit samples for testing, and any unsatisfactory test results should be discussed. A desirable (or acceptable) performance specification should then be developed and included in the purchase order. Samples of the final shipment should be retested.

9. When replacing trays by packings in a tower section, it is ideal to remove all internal support rings and downcomer bolting bars to approximately ⅜″ to ½″ from the shell (68, 174, 307, 340, 462). Complete removal (i.e., grinding flush) is expensive and time-consuming, can damage the tower wall, and is rarely justified.

    Horizontal support rings left in the tower interfere with liquid distribution and reduce the available open area, with possible adverse effects on packing efficiency and capacity. Straight downcomer supports or other vertical bars are generally far less detrimental to distribution and open area, but make the installation of structured packings difficult. These are usually removed only if they are expected to adversely interfere with the new internals, their installation, or liquid flow through them (461). Some designers (340) prefer to always remove them. Inward projections of manholes can also interfere with the installation of structured packings, in which case they need to be removed (84).

    Strigle (461) states that with random packings, tray support rings need only be removed when they occupy more than 10% of the column cross-sectional area. Strigle's recommendation for the maximum acceptable support ring width depends on column diameter, as shown in Table 10.1.

    In one 13.5-ft-diameter column (413), random packings performed well even though the tray support rings were not removed. In another case (432), the replacement of poorly performing trays by random packings in an amine absorber was only partially successful in improving performance, presumably because the support rings and downcomer bolting bars were not removed.

Table 10.1   Maximum acceptable tray ring width left in a tower retrofitted from trays to random packings (461).

Column diameter (ft)	Acceptable ring width (in.)	Excessive ring width (in.)
4	1¼″	1⅝″
6	1⅞″	2⅜″
8	2½″	3¼″
10	3⅛″	4″
12	3¾″	4¾″

With structured packings, it has been recommended to always remove the support rings prior to packing installation (174). Tests in a 3-ft-diameter column with a 14-ft-tall bed of structured packings showed that leaving in the 2″ tray support rings resulted in a 10% efficiency loss and a 40% rise in pressure drop. In larger columns, the support rings usually occupy a smaller fraction of the column area, and the pressure drop rise is likely to be smaller. The efficiency loss may escalate in taller beds due to distorted distribution profiles and in larger-diameter towers due to the higher ratios of tower diameter to packing diameter, which diminish the effect of lateral mixing (Section 4.3).

In order to minimize irregularities and dead liquid pockets at the tower wall, it has been recommended to internally blind unused nozzles (340) in the packed zone.

10. Closely review possible interference of manholes with distributors, redistributors, and parting boxes. The location of manholes may require distributors or parting boxes to be offset from the feeding nozzle, leading to challenges and errors during installation (84).

11. Steps 2 to 4, 6 to 8, 10, and 11 in Section 10.2.1 also apply to packed columns and their internals.

10.2.3 Removal of Existing Trays and Packings

Inadequate removal of existing trays or packings can endanger workers, damage equipment, and prolong the shutdown. The following guidelines can help minimize the above problems:

1. The likely conditions of the internals, such as degree of fouling, corrosion, toxicity, and potential trouble spots, should be evaluated. This information can be obtained from past inspection reports, by questioning operating personnel, from similar columns, from pressure drop measurements, and from gamma scans. Based on this evaluation, the proper tools, procedures, work and personnel schedules, and safety equipment can be planned.

2. Before removing trays or packings, obtain as much information as safe practices would allow about the condition of the trays, packings, distributors, nature of deposits, corrosion, and damage.

3. The ability to remove tray panels and other internals through the manhole should be carefully reviewed. If not possible, the old trays may need to be burned out. In one case experienced by the author, this burning out placed a relatively small tower on the turnaround critical path. In another case (398), the time required for tray removal in two large towers increased from five to nine days due to the need to oxy-cut tray parts.

4. When the old trays are to be burned out, additional special safety measures and equipment (e.g., fire extinguishers inside the column, explosivity monitoring, good forced-air circulation for removing fumes, water flooding, tarps) are likely to be required. Steps also need to be taken to prevent insulation damage due to overheating. If the tower wall is fitted with wall cladding for corrosion prevention, flammable material may be trapped behind the cladding, and burning trays out is best avoided.

5. When pyrophoric deposits are expected inside the column, all trays should be flooded with water before dismantling. The water level should be progressively lowered to just below the tray that is being cleaned and dismantled. Any freshly exposed tray or wall section should be immediately cleaned.

The air above the water level should be continuously monitored for explosivity and toxicity. When a hydrocarbon or water-insoluble organic is present in a small concentration in the water, the high repulsion of the water phase imparts a very high activity coefficient to it. Its volatility becomes several thousand times that of the pure component, which sends it into the air, where it can exceed the lower explosive or the allowable exposure limit.

6. When removing damaged or bumped trays, special safety precautions are required. Often, pieces are left hanging, and even those in place would not support a worker (307).

7. In corrosive services, workers should bring various sizes of sockets (307) as several bolt heads will be corroded.

8. Some experiences have been reported (307) on blowing highly plugged or coked trays out by explosives. This technique is infrequently used and not always successful (307). There have been cases where the explosions damaged support beams and tower walls.

9. Hand-cleaning and high-pressure water jetting are the common methods of cleaning trays. Water jetting is reported to work well for sieve and valve trays and for cleaning the shell and support rings after the trays are removed (307), but not for bubble-cap trays (307).

10. Removal of random packings from the column is usually performed by opening a manhole or handhole located right above the packing supports and letting the packing pieces roll into a collection drum. Special care is required with ceramic or carbon packings to avoid breakage.

11. Suction equipment is often connected to the manhole (or handhole) via a wide flexible tube to speed up the removal of random packings. This is especially important in large towers. If the packings have sharp edges, the tube should be fabricated from metal, not plastic, and care should be taken not to damage the tube. This operation can damage packing particles, especially with thin-gauge packings, but usually, the damage is slight (407) when done properly. The author had one experience with a Pall-ring type packing where the damage was quite significant, with many pieces caught in each other and unrolled; maybe the operation was not done properly. Care should be taken to minimize packing damage, but having replacement packing pieces on standby is worth considering. This operation is practiced with plastic and metal packings and should be avoided with ceramic or carbon packings where it is likely to cause breakage.

12. With columns constructed in flanged sections, removal of random packings is usually done by dismantling the top distributor (or redistributor) and retaining device, then lifting the packed section (including the support plate), and tipping it over. Extreme caution is required with ceramic or carbon packings as packing pieces may break if dropped from heights exceeding about 2 ft.

13. Care is required with hand-cleaning random packings outside the tower. Thin-walled packing can easily be damaged or deformed by water jets as reported in one case study (290). Thicker-wall packings can also be damaged or deformed, especially when hit with high-pressure water jets, as the author experienced. Placing the packings in baskets before hydroblasting them appears to help at least to some extent. What worked for the author in one case is loading baskets of packings on a flat-top truck and taking them to the car wash (no detergent, takes a few passes through the wash with each load, and makes a special attraction to spectators) or to wash them in a cement mixer.

10.2.4   Tray Installation

The objective of correct tray installation is to follow safe practices while minimizing installation errors and installation time. Practices prior to installation that help achieve the above installation objectives are presented in Section 10.2.1. This section discusses practices that help achieve the above objectives during the installation period:

1. Close contact with the work crew is essential. Lack of communication may result in a bad solution to an installation problem that later adversely affects column performance.

2. Any work that can be performed on the ground is easier and safer than inside the column. It is therefore preferred to preassemble each piece on the ground as it comes out of the crate. Clamps, bolts, nuts, washers, and seal plates should therefore be preassembled to each part (307), with care taken to ensure correct preassembly.

3. To minimize the time spent by the installation crew inside the column and to speed up installation, the preassembly line should be kept ahead of the installation (307). As soon as one tray arrives at the manhole or enters the column, the next tray should be hoisted to the platform, and another tray preassembled on the ground, ready to hoist. Parts of only one tray should be passed on at a time, so that no pieces remain when the tray is installed.

4. Major support beams should be installed first, with the top of the beam flush with the tray support rings. Shims are often added to minimize out-of-levelness.

5. Tray installation usually begins at the bottom and proceeds upward. Tray pieces should be passed to the work area in the order they go in. The preferred order (307) is down-comers first, under-downcomers next, then side-tray floors, center-tray floors, and finally manways.

6. To minimize the impact of support ring out-of-levelness on tray levelness, bolts and clamps that fix tray panels to support beams should be fastened before those that fix the panels to the support ring. To ensure the manway fits in, it should be temporarily set in place before tray panels are fastened.

7. Downcomer panels should be installed on the downcomer side of the downcomer support bars (e.g., the wall side of single-pass trays). With this technique, the weight of the liquid tightens the joints, and the bowing out of the downcomer panels is minimized.

8. To set the required clearance under the downcomer, the length of downcomer panels should be carefully adjusted. Wooden blocks cut to the required dimensions to act as spacers are useful for achieving this (307).

9. The process or operations engineer should periodically spot-check that trays and components are correctly installed and immediately inform the construction supervisors of any features that can adversely affect column performance.

10. Rough handling of valve tray sections can damage valve legs and must be avoided. It can result in valve floats sticking shut, sticking open, or falling out when the column is placed in service.

11. When shutdown time is critical in large-diameter and tall columns, and if manhole location permits, two or even three tray installation crews have been used simultaneously inside the column. This practice suffers from the potential hazards of falling objects and/or poor ventilation and should be avoided whenever possible. If it needs to be used, precautions must be taken to mitigate the risks. Bulkheads are often used in the tower, and these need to be inserted in a way that will permit good ventilation throughout.

12. Ideally, any seal welding of a tray or part of a tray should be performed after completing tray installation (307). This, however, requires large enough tray spacing (at least 24 in) and good ventilation to disperse fumes.

10.2.5 Dry versus Wet Random Packing Installation

Random packings can be either wet-packed (Figure 10.2a) or dry-packed (Figure 10.2b,c). With wet packing, the column is filled with water following the installation of the bottom support plate. Packing pieces are gently poured from a short distance above the water level and are "floated" to the bottom. The water cushions the fall and promotes random settling. With dry packing, the packing pieces are dumped from a certain height above the top layer of packing.

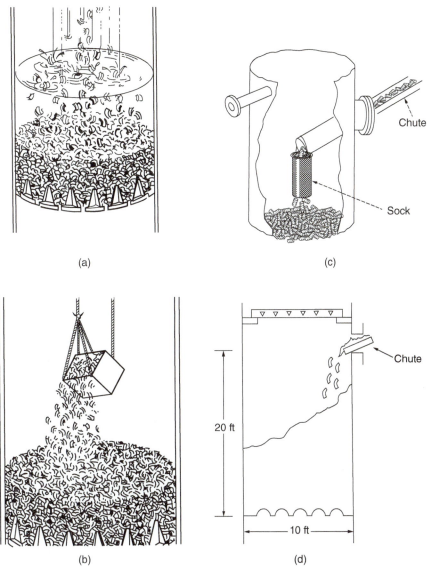

(a)

(c)

(b)

(d)

Figure 10.2 Random packing installation techniques. (a) Recommended wet packing technique; (b) recommended dry packing technique; (c) the chute and sock method of column packing; (d) bad packing technique, promoting hill formation. (Parts (a) and (b): Reprinted Courtesy of Koch-Glitsch LP. Part (c): From Chen, G. K., Chemical Engineering, p. 40, March 5, 1984. Reprinted with permission. Part (d): From Kister, H. Z., "Distillation Operation," Copyright © McGraw-Hill, 1990. Used with permission of McGraw-Hill.)

Wet packing minimizes breakage, compression, and mechanical damage to packings and supports. It also maximizes random settling and reduces packing density. This reduced packing density provides a small capacity gain and pressure drop decrease compared with dry packing and minimizes the required number of packing pieces per unit volume. Billet's experiments (34) with 1 ½" Pall rings in a 20-in-diameter column showed that changing the packing technique from dry to wet increased column capacity by about 5%, lowered column pressure drop by up to 10%, reduced the number of packing particles by about 5%, and had a negligible effect on efficiency. The author also observed a 5% capacity improvement after switching the packing technique of a 3-ft column from dry to wet. Ludwig (305) reported cases in which the pressure drop was lowered by 50 to 60% and even more by switching the packing technique from dry to wet. The author believes that Ludwig's cases are for ceramic packings in which breakage played an important role in increasing the pressure drop of the dry-packed beds.

Dry packing avoids the introduction of water into a dry process, high hydrostatic heads, minimizes rusting of metal packings, and is quicker and less expensive. It is generally preferred in most modern metal and plastic packing applications, especially in the following applications:

1. Plastic packings, as plastic typically floats on water.
2. Large-diameter (>10-ft) columns, as the cost and speed advantage of dry packing are the controlling considerations.
3. Where downtime is critical, as dry packing is faster and also eliminates the need for leak testing prior to packing installation.
4. When the random packings are only in the upper sections of a column. Filling the entire column with water in order to wet-pack the upper sections is rarely justified.
5. When either water presence or some corrosion to the packing cannot be tolerated.

Listed below are applications where wet packing is advantageous. Wet packing is strongly preferred for breakable packings (step 1). For the other steps in the list below (2–4), wet packing offers a smaller advantage:

1. Packings constructed of ceramic, carbon, or other breakage-prone materials.
2. Small-diameter columns (<2 to 3 ft), where lowering boxes of packing into the column is difficult while dry dumping from high elevations may damage packings and supports.
3. Where column capacity is near a limit.
4. Where dry packing offers little advantage. Wet-packed beds perform a little better.

10.2.6   Dos and Don'ts for Random Packing Installation

Poor packing installation may induce maldistribution, flooding, loss of efficiency or capacity or both, instability, and excessive pressure drop (31, 305, 339, 518). Bad packing installation practices may also damage packings and supports.

When correctly done, either wet or dry packing can provide good performance. Both techniques have traps. Failure to avoid these pitfalls, rather than a nonoptimal choice between wet packing and dry packing, is usually responsible for column malfunctions. Several guidelines for avoiding these pitfalls are listed below.

1. The supplier's advice on packing procedure should be sought, reviewed, and in the absence of major issues, followed. Any deviations from the supplier's recommended procedure should be discussed with the supplier.

2. Always remove all the previous packings before adding new ones. Failure to do this can lead to density gradients in the bed as observed in one case (2).

3. Packings should be spread as evenly and as randomly as possible, and formation of "hills" of packing (i.e., buildup of layers sloping away from the dumping region) strictly avoided. Biales (31) reports an experience where forming a "hill" (Figure 10.2d) caused severe maldistribution and lowered efficiency and capacity. Pilot-scale experiments by Billet (34) verify that hill formation during packing installation leads to a lower packing efficiency than when the packing is uniformly spread.

4. To reduce the liquid tendency to accumulate at the wall, it was recommended (305) to lay each horizontal layer of packing starting from the edges and working toward the center.

5. Packing pieces should not be pressed into place. Pressing them may increase bed density, compress them, and cause maldistribution and high pressure drop.

6. In dry-packed towers of larger diameter (>4-ft), it is usually necessary for workers to stand on top of the bed to spread out and level the packing pieces. Standing directly on the packing should be avoided as it may compress and crush the packing. Sheets of plywood, or other rigid material, with an area of 4 ft^2 or higher, should be used to stand on (68). With plastic packing, it is best to avoid standing on the packing altogether. Care should be taken not to introduce dirt into the tower during installation.

7. It is important to positively ensure that plywood and other foreign objects are removed and not buried in the bed. These have caused malfunctions in the past (68).

8. Figure 10.2b shows a frequently recommended technique of dumping the packings from buckets lowered into the column. The buckets should be emptied at several locations, starting from the edges. A worker is often lowered into the column with the bucket and pours the packings out at a height of 6 to 10 ft above the top of the bed. An alternative to this technique (337, 478) is to construct a simple frame that will hold a shipping carton of the packing, with a trip cord attached to the frame. The carton is lowered into the column, and the cord is tripped to dump the packings onto the bed.

9. Figure 10.2c shows the "chute and sock" method (68, 305), which is probably the most popular technique practiced. From an inclined hopper (the "chute"), packing pieces roll into a sheet metal cylinder (the "sock") and from there are dumped at several locations, starting at the edges. A flexible large plastic hose has more movement flexibility and is usually used as the sock instead of the cylinder. This method breaks the packing fall height and evenly spreads the packing. A worker usually stands on top of the bed per step 6, moves the plastic hose to direct the packing pieces to different locations, and uses a device similar to a garden rake to spread the packing. Packings should be raked at frequent heights and definitely no more than every 10 ft (140). This technique is fast and is often preferred in large columns (68). When the tower diameter exceeds 10–15 ft, a number of workers can spread the packings inside the tower to speed up installation.

10. In dry packing, pieces of packing should not be allowed to fall a large distance onto the bed. The fall height for ceramic or carbon packings should not exceed 2 ft (68, 305, 337, 339, 478). Higher fall heights can be tolerated with metal and plastic packing, but should not exceed 10 ft (192, 339, 478). One designer (68), however, believes that with metal packings, a fall height of up to 20 ft may be permissible.

    Caution is required when dumping the first 2 to 3 ft of packings, as excessive fall height can damage the support plate or crush the packings. The support plate is usually of the gas injection multibeam type (140, 192, 407), clamped to a support ring.

The levelness of the ring is not critical, and normal construction tolerances are usually acceptable (407).

In near-freezing conditions, the fall height of plastic packings should not exceed 4 ft. The impact resistance of plastic often diminishes at low temperatures, and it is prone to breakage. In one case (192), polypropylene rings were dropped from a height of 23 ft onto a steel support at about 15°F, and 20% of them were damaged.

11. Dropping packings directly from a column manhole promotes excessive fall heights and hill formation and should be avoided. Instead, the techniques in step 8 or 9 should be used. One designer (68), however, believes that in small-diameter (< 6-ft) columns, dropping packings from the manhole may be permissible as long as excessive packing fall heights are avoided and the packing spread is visually inspected to avoid unevenness.

12. In wet packing, the column shell and supports must be designed to withstand the full hydrostatic head.

13. In wet packing, at least 4 ft of water should be kept above the surface of the packing at all times (337, 339). Ideally, the water level should be kept up to the loading manhole. This should be combined with the bucket dump method mentioned in step 8.

14. When wet-packing reused packings, consider the relevance of step 5 in Section 10.2.3.

15. The distance between the top of the packing and the distributor should be as specified in the drawings. Packing suppliers usually supply 5–10% or more spares in case the bed is too short. The author has seen poor packing efficiency resulting from a foot overlong bed which grew all the way to the distributor as the installation crew kept pouring in the spare packings. The top of the packing should be one ring size below the bed limiter (140).

16. Precautions are required to avoid interchanging when using different packing sizes or types in different beds or column sections.

17. During the loading of metal or ceramic packings into a column, the ground area surrounding the tower should be out of bounds for personnel and should be wired off. Sharp edges of the metal packing particles falling at high speed, or the heavy ceramic packing particles, can cause serious injuries to workers underneath. The author is familiar with one accident where a serious injury to a passer-by was caused by a falling piece of sharp-edged metal packing.

18. Bed limiters should be installed so that they do not interfere with liquid distribution.

19. Thermowells are best inserted before dumping the packings.

20. Debris often remain lodged inside the packings or distributors and can later bottleneck the tower. Throughout the construction, it is important that the debris are removed. In particular, any wooden sheets or planks used must be all accounted for and removed.

10.2.7   Dos and Don'ts for Structured Packing Installation

Incorrect installation has caused numerous failures in columns containing structured packings. Inclined layering, incorrect layer orientation, heavy-footed stepping on structured packings, failure to achieve a snug fit of packings to the tower wall, and excessive compression of packings in an attempt to achieve a snug fit are a few of the many causes of poor performance. Some of the lessons learned are described below:

1. Welding of internals such as packing supports, redistributors, and any other hot work inside the tower needs to be completed before packing is installed (174). Hot work inside the tower while the packings are in place is a serious fire hazard. For both safety and

liability concerns, a number of companies and contractors are now requiring the removal of structured packings before any hot work is started (108). Others (97, 108) strongly recommend taking the packings out before hot work. If hot work above the packings is absolutely necessary, extensive safety measures such as water flooding need to be implemented to protect the packings (97, 108).

2. The edges of structured packings are razor sharp, so any handling of these packings needs to be done with the appropriate protective clothing and cut-resistant protective gloves.

3. Distributors and redistributors used in packed towers can be complex. Installing complex distributors and redistributors on the ground ahead of the turnaround is worth considering. Any issues can then be brought to the supplier's attention.

4. Redistributors that look similar may have subtle differences between them, like different hole sizes. In these situations, special care is required to prevent interchanging. The author has seen poor efficiency caused by interchanges.

5. The support grid should be leveled to the supplier's specifications. If it is not level, the out-of-levelness carries all the way to the top of the bed (407). It was recommended (174) to fasten the sections of the packing support grid only loosely until all parts have been assembled. This allows adjusting the overlap of the rim of the support grid and the support ring before final clamping and tightening (174).

6. Labels, plastic sheets, and plastic bags should be removed before loading the packings into the tower. Care should be taken not to introduce dirt into the tower during installation.

7. The packing is usually supplied in elements, or "bricks," typically 8 to 12 in tall. In smaller-diameter columns, an element can be as wide as the wall-to-wall distance, but to minimize the possibility of damage during handling, it is better to make up this distance in the central portion of the column using a number of elements (Figure 10.3). Ropes should be used to lower the elements into the column (Figure 10.4a) as metal hawsers and chains can damage the packings (174).

8. The workers should not stand directly on the structured packing as this may compress and damage the packings. Sheets of plywood, or other rigid material, with an area of 4 ft^2 or higher, should be used to stand on (68). Any wooden planks or boards used should not be prone to splinter. The installation crew should be instructed not to drop or drag heavy

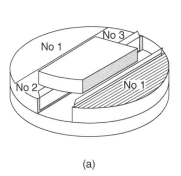
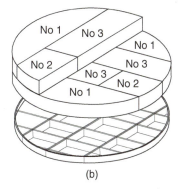

(a)   (b)

Figure 10.3 Installation of structured packing elements. (a) Fitting the elements in one layer; (b) 90° rotation of one layer with respect to the layer below. (From Horner, G. V., The Chemical Engineer, (Supplement), p. 8, London, September 1987. Reprinted with permission.)

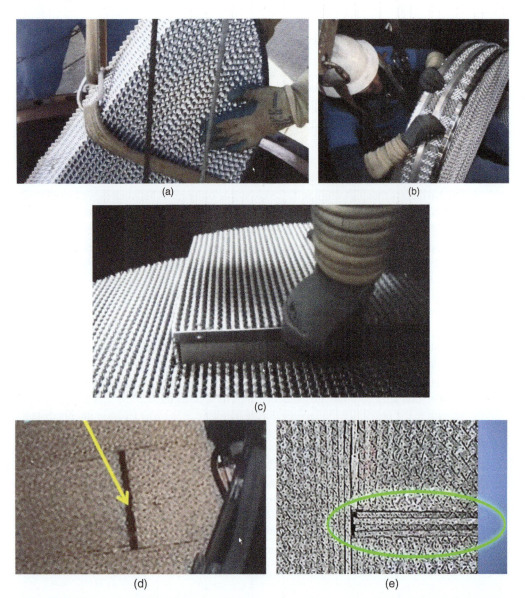

Figure 10.4   Some steps in good practice of structured packings installation. (a) Lowering elements of packings using ropes. (b) Bending wall wipers outward. (c) Using a slide plate to fit centerpieces. (d) A gap between elements due to tower out-of-roundness. (e) Using filler sheets perpendicular to packing layers to fill the gap. (Copyright © FRI. Reprinted Courtesy of FRI.)

objects or elements across the packings. The author is familiar with cases where such mishandling caused indentations, depressions, and other damage to the packing layers.

9. Usually, each layer of packing requires a wall wiper. Some wall wipers are incorporated into the packing elements, whereas others are supplied loose. Any wall wipers should be bent outward prior to element placement (174); see Figure 10.4b. These wall wipers catch liquid flowing down the wall and return it into the packing. It is important to check that the packing reaches the tower walls; in one case (450), gaps as large as ½" were

found in an inspection. The author is familiar with other cases of very poor efficiencies in wire-mesh packings due to ½" gaps between the packings and the wall.

10. Figure 10.3 shows a typical procedure for the installation of the packing elements. The elements are fitted in beginning from the sides. A slide plate (or "shoe horn," Figure 10.4c) is used to fit centerpieces. In Figure 10.3a, element 1 on the left is first inserted. Next, elements 2 and 3 on the left are inserted and pressed against element 1. Element 1 on the right is then inserted, and finally, elements 2 and 3 on the right are fitted in.

11. It is important to avoid vertical misalignment of the installed elements. The author is familiar with cases of misalignment defects, as much as 4 in. minimum to maximum in a single layer, as well as wavy surfaces of the packing layers. Where these were not caught at the inspections, tower performance was poor.

12. The elements of the next layer of packings are rotated 90° so they have flow channels running at 90° to those in the layer beneath (Figure 10.3b). Failure to rotate the layers may result in severe channeling and extremely poor efficiency, as was experienced in one case that the author is familiar with.

13. The orientation of the bottom layer is critical (43, 140, 407) and is set to orient the distributor at a desired angle (usually 90 degrees) to the top packing layer. Further, the bottom packing layer usually needs to be perpendicular to the support bars to prevent sagging into the spaces between the bars. It is therefore mandatory to ensure that the bottom layer is oriented per the supplier's drawings and procedures. With grid packings, each element is usually rotated 45° (instead of 90°) to the one below. In one case (409), grid reinstallation was needed because this was not followed.

14. Column out-of-roundness or other irregularities may make it necessary to reduce or increase the width of some center elements. It may also form gaps between elements (Figure 10.4d). Reducing the width of elements can be accomplished by removing a few strips of the packing material. Increasing the width, or filling gaps, is accomplished by inserting filler sheets of packings. The filler sheets can be inserted perpendicular to the packing layers (Figure 10.4e) or parallel to them. They can also be inserted between the packings and the wall. It is best to seek and follow the supplier's recommendation as incorrect installation may later show up as maldistribution and loss of packing efficiency. In one case (450), gaps as wide as ½" were found in the middle of the bed upon turnaround inspection.

15. The supplier supervisor, or trained process/operation personnel, should be required to inspect each layer of the structured packings after it is installed. Good-quality 360° photographs of each layer of the packings should be taken and clearly labeled to indicate its exact location. This information will be invaluable for a troubleshooting investigation should the packing perform poorly. The inspector should also monitor the packing height as the layers are installed; it is better to identify deviations sooner rather than later.

16. Tight fit of structured packings may result in a "growth" of the overall packed height; in one case (405), a 3-in. growth occurred. The additional height is often sawed off. The supplier's advice should always be sought as sawing may damage the packing and cause maldistribution and/or premature flooding.

17. It is important to keep the exact count of the number of packing layers going into each bed. The author is familiar with cases in which post-startup gamma scans showed layers

intended for one bed installed in another: in one bed leaving a large gap under the distributor, in the other bed leaving no disengagement space under the distributor.

18. Inserting thermocouples into the bed requires special caution. The author is familiar with a case in which packing elements were simply piled up above a thermocouple, forming inclined layers in the bed. Fortunately, this was caught upon inspection, which saved the tower from inferior performance, but required removing the packings and reinstallation. Drilling or bayonetting the packing by sharp-pointed rods to insert the thermocouple is often needed. To minimize packing damage, two or three rods of progressively increasing diameter, the last being slightly larger than the thermowell, are sometimes used. If drilled, there is a question of where the drilling bits will end up. Again, it is best to seek and follow the supplier's recommendation as incorrect installation may later show up as packing damage, maldistribution, and loss of packing efficiency.

19. Steps 16–18 and 20 for the installation of random packings (Section 10.2.6) apply also for the installation of structured packings.

Because of the multitude of installation traps that can impair the performance of structured packings, it is best to have the supplier provide an installation team. If impractical or too costly, the supplier should be requested to provide detailed installation instructions, which must be strictly adhered to, as well as a supervisor (at cost) to supervise the installation. The supervisor's scope of work should include thorough hands-on training for the installation crews and watching that the packings are correctly installed.

10.2.8   Some Considerations for Towers Out of Service for a Time

At times, a tower is to remain idle until there is economics to activate it again. Below are some considerations:

1. Ensure all health, safety and environmental regulations are followed.

2. Keep manholes closed to protect the tower from water and animal entry. In one case (284), the carcass of a dead rat lodged in the kettle reboiler inlet nozzle backed up liquid into the tower upon restart, initiating premature flood.

3. Keeping the manholes closed is especially important in packed columns, where moisture, rain, or dust can lead to rust and particles that plug distributor holes. In one case (264), moisture entering through open manholes of a large tower during an extended outage caused significant rusting on the tower inner walls. Upon startup, the descending liquid washed away the rust particles, which plugged the distributors.

4. Before boxing the tower in, make sure it is dry. If the column was water-washed or chemically washed, make sure it has been adequately dried immediately after the conclusion of the wash. It is amazing how fast carbon steel, and even stainless steel, rust in the presence of oxygen and water.

10.3   TOWER INSPECTION: WHAT TO LOOK FOR

10.3.1   Strategy

In general, any departures from the design and fabrication drawings are a potential source of trouble. Added to these are any details that are not clearly marked on the tower drawings.

Finally, anything that does not make sense should be questioned. "How is this supposed to work?" is an excellent key question.

This section emphasizes some common traps that deserve particular attention, either because of their high frequency of occurrence in practice or because of the severe consequences of overlooking them.

The discussions in this section center on inspections carried out by process and operations personnel. The discussions exclude column inspection for corrosion, fatigue, damage, and equipment mechanical integrity. These inspections should be carried out separately by inspectors specifically trained in these disciplines. However, the operations or process inspectors should be on the look for such issues and call on the right inspection personnel to evaluate them in detail.

The earlier a fault is discovered, the less costly and time-consuming it is to correct, and the less likely that time pressure will prevent the correction altogether. For new installations, it was recommended that the inspection be carried out simultaneously with the assembly of new internals, as demonstrated by field experiences (329). Existing towers should be inspected as early as practical after an entry permit is issued. When safe and feasible, an inspection before cleaning and disassembly (322) can provide valuable data on fouling, corrosion, and internal damage. A second, briefer inspection may be needed after cleaning and repairs.

It was suggested (322, 409) that trained tech service personnel are generally most suitable for performing inspections. The author endorses this recommendation. This helps balance the turnaround workload (operations personnel are usually the busiest during turnarounds). Tech service engineers generally have close familiarity with equipment fluid flow and mass transfer and a good understanding of the consequences of deviations from design dimensions. It has been stated (293), "The most important job of the process engineer working in a refinery or chemical plant is to inspect tower internals during a turnaround."

The inspections are best performed by new graduates and less experienced engineers (or operators). The experts and experienced engineers can follow later with checks of their own. Column inspection is an invaluable training exercise and provides the inexperienced with insights essential for improving their design and operation skills. It has been stated (293) that "Crawling through the trays is the only real way to understand how they work and what malfunctions can be expected." The author concurs.

Prior to the inspection, it is critical to provide the inexperienced inspectors with adequate training in "what to look for." Usually, this is a classroom training 4–8 hours long by an experienced engineer knowledgeable of the towers to be inspected. The author is familiar with cases in which this step was skipped, leaving the inexperienced engineers to "swim or sink," and when they missed some details, they were yelled at. This scenario is unfair, demotivating, and counterproductive and must positively be prevented.

To maximize effectiveness, and to get the most out of the inspection as a training exercise, especially for the inexperienced, it is important to do homework prior to the inspection. What has been the past experience in the tower (plugging, corrosion, damage, where)? What is being changed (trays, distributors, feed pipes)? Why?

Follow-up on the recommended action items is critical. Incorrect implementation is a common occurrence, and under time pressure, the tower is bolted up shortly after. A poorly implemented cure can be more lethal than the original illness and worsen tower performance. Do not let this happen. Keep close to the action, and always reinspect the implementation as soon as it is done. This is also recommended by troubleshooting maestro Lieberman (288).

10.3.2 Should the Tower be Entered at the Turnaround?

The answer to this question should be primarily based on the potential hazards of tower decontamination and, secondly, on history, monitoring key variables, and looking for problems well before the turnaround. Health, safety and environmental considerations prevail. Process-wise, if the tower is in clean service, with no history of, no evidence of, and little potential for corrosion, fouling, or damage; no signs of performance deterioration during service; and no statutory, safety, or environmental requirement to enter, then there is a good case for not entering the tower.

In many cases where the tower ran well, the tower is opened with or without access to the spaces at the manholes. It is important to make the most of the opportunity and look or enter through the open manholes to the extent permitted by safe practices. Attention should be focused on packings, packing distributors and their piping, feed piping, cleanliness, corrosion, and damage. In one case (43), a small void at the top of a structured packed bed gave reason to remove additional layers of packings, which showed a large, corroded void further down in the bed. The entire bed needed replacing with corrosion-resistant packing.

Monitoring the key variables also provides a guide to focus on problematic sections of the tower. For instance, if flooding was experienced near the top of the tower toward the end of the run, the inspection should closely look at fouling or damage near the top of the tower. Differential pressure monitoring (Section 2.8) and gamma scans (Chapter 5) can help identify the sections to be singled out for special attention.

There are three common shortcuts that the author strongly recommends closely reviewing and in many cases avoiding altogether:

1. "Not going into the column at the turnaround." Shutdown planners often argue that the need to remove the chemicals, gas-free the tower, clean it for personnel entry, and later recommission it for service is costly and will prolong the turnaround. This argument is often valid in clean, noncorrosive, trouble-free services as described above. In many services, however, this is false economics. The author has seen far too many cases where upon startup a column experienced poor operation or performance due to plugging, corrosion, damage, and other operating problems that either existed before the turnaround or crept in while shutting down, remained undetected as the tower was not opened, and intensified or just persisted upon restart. An additional outage became necessary weeks after the turnaround. The cost of lost production was orders of magnitude larger than that of entering the tower. In other cases, columns suffered substandard performance for years due to unrectified faults.

2. Partial cleaning of trays or other tower internals. To minimize turnaround work, shutdown planners often limit tower inspection and cleaning work. This may be OK in clean, noncorrosive services with low potential for damage and with no history of operation problems. This can also be justified if the potential trouble areas are inspected, while sections with trouble-free history are skipped. But like in step 1, no-inspection can breed poor performance. In Figure 10.5 (280), for 20 years, the panels of one of the two passes in the tower were inspected and cleaned, but on the other pass, the manways were not removed and the panels were not cleaned. Following one restart, the tower performance was so poor that the unit needed to be shut down again to clean the panels that were not inspected.

3. Turnaround wash without inspection may be counterproductive and needs to be carefully evaluated. Many good experiences with on-line or shutdown washes without tower entry (e.g., Cases 12.3 and 12.4 in Ref. 201, many others abstracted in Section 12.7 in Ref. 201, as well as the cases described in Refs. 353, 384) have been reported, especially where the

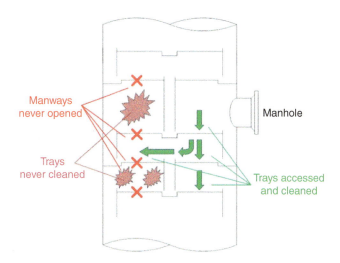

Manways
never opened

Trays
never cleaned

Manhole

Trays accessed
and cleaned

Figure 10.5   Partial cleaning of trays is trouble. (From Lenfeld, P., and I. Buttridge, Chemical Engineering, p. 45, December 2013. Reprinted with permission.)

nature of the fouling was well understood and there was a clear path for removing the foulant-rich spent wash solution.

In contrast, the author is familiar with turnaround chemical washes with no entry resulting in premature flood upon restart. Deposits moved by the washes blocked a couple of downcomers in one case and sumps in another. In the experience cited in step 2 (280), water washing is believed to have moved deposits to the unaccessed tray panels. Lieberman (290) reports water washing a 25-ft-tall bed of 3-in. Pall rings in an 8'-6" scrubber to wash off salts. Some of the salts settled lower in the bed, causing channeling and poor separation upon restart. Sloley (443) reports a case in which replacing a coked vacuum tower wash bed grid took "too long to wait," so instead it was cleaned in place from the top with a high-pressure water lance. Upon restart, the grid pressure drop doubled; the water pushed down the solids, increasing the blockage of the area near the bottom of the bed. In all these cases, the plant needed to be shut down a short time after restart in order to clean the tower.

10.3.3   Inspector's Checklist

Table 10.2 lists the common assembly mishaps reported in the literature as surveyed by our malfunction survey (198). Sections 10.3.4 through 10.3.11 will elaborate on these. Any inspector needs to focus on the items on this list.

During a revamp, it is essential to critically examine the modification areas and to identify the liquid and gas passageways. Several cases have been reported (8, 43, 44) where following a modification, the passage of liquid to reach the section below was blocked.

The prime column inspection tool is a checklist. Table 10.3 provides a generic master checklist. For each tower, a similar list should be prepared, including the relevant items, deleting the irrelevant items, and adding items not included on this checklist. For each satisfactory item, the inspector enters a tick, and for each unsatisfactory item, a number, with a comment of the same

Table 10.2 Assembly Mishaps (198).

		Cases
1.	Packing liquid distributors	13
2.	Packing assembly	13
3.	Untightened nuts, bolts, clamps	9
4.	Tray panel assembly	8
5.	Feed/draws obstruction, misorientation	7
6.	Leaking collectors, low liquid load trays	7
7.	Downcomer clearances	5
8.	Debris left in column	5
9.	Unbolted tray manways	4
10.	Materials of construction	4

number spelled out on a separate sheet. Inapplicable items are left blank. Prior to inspection, design values and relevant tolerances are entered in the right-hand column of the table.

Any alternative checklist is satisfactory as long as it is sufficiently detailed and the user is comfortable with it.

Miller (322) and recently Bouck (42) proposed "inspector survival kits." The author has supplemented this list. Additional items, dictated by safety requirements, need to be on the list, but as these vary with the specific process, they were excluded from the list below. It is mandatory to verify with the safety department whether there are any specific requirements for taking any of the items from the list below into the tower.

The updated list includes:

- Equipment drawings and simplified sketches. The drawings should be used for preparing the sketches. They are too cumbersome to take into the tower. The simplified sketches show what the internals should look like and what the inspector is looking at.

- A checklist.

- A high-intensity flashlight and spare, or at least spare batteries. Losing the source of light while in the tower is a common issue. A well-fastened miner's lamp should be considered (safety permitting) and can help free the inspector's hands. A high-intensity LED light has been advocated by some to give high intensity and long life.

- A color marker pen that writes on metal. Yellow shows up best. The marks should be waterproof. In stainless steel columns, the ink needs to be low in chlorides. Low-chloride pens may require special ordering ahead of the turnaround. Check that the pen does not leak; the paint may be difficult to wash off the skin.

- A bound pocket notebook and two pencils and ballpoint pens. The pens should be waterproof.

- A short (6-in.) steel ruler. The ruler can also be used as a deposit scraper. Bring Ziplock bags to collect the deposits.

- A 6- to 10-ft tape measure.

- An autofocus pocket camera. Photographing damaged, corroded, and plugged regions, as well as flaws and critical equipment items, provides management, designers, suppliers, and experts with factual information essential for immediate decision-making. It also supplies an invaluable record for future troubleshooting. It is important to mark each photograph

Table 10.3 Column inspection checklist.

Column	Inspector										Date
Tray no.	Top								Bottom		Design
	1	2	3	4	5	6	7	8	9	10	Value ± Tolerance
Points of transition											
Orientation of internals											
Passage obstructions											
Possible impingement											
Watermarks											
Instrument location											
Chimney trays											
Riser-hat clearance											
Riser diameter											
Riser number											
Hats correctly mounted											
Flow obstruction to outlet(s)											
Unintended liquid descent											
Hats strong, supported?											
Vapor rise obstruction											
Vapor impingement											
Buckling, riser bow											
Instrument taps											
Trays											
Downcomer clearance											
Weir height											
Downcomer width											
Downcomer brackets/bow											
Hole diameter											
No. of holes/valves											
Valve direction											
Light/heavy valve arrangement											
Panel alignment/overlap											
Valves secure?											
Cracks and crevices											
Holes/valves under downcomers											
Tray levelness											
Tray spacing											
Tray/downcomer materials											
Trays/downcomers secure											
General											
Deposits found? Sampled?											
Corrosion found?											

(Continued)

Table 10.3   (Continued)

Column										Inspector	Date
Tray no.	Top								Bottom		Design
	1	2	3	4	5	6	7	8	9	10	Value ± Tolerance
Bolting firm?											
Double nutting?											
Fastening materials											
Gaskets											
Absence of debris											
Damage: bow, warp, cracked											
Unexpected features											
Sumps											
Inlet weir/seal pan width											
Inlet weir/seal pan depth											
Weep hole area											
Weep hole diameter											
Gaskets: tightness											
Others											

with the location, tray number, and the direction (e.g., east–west) at which it was shot. The author has seen many photographs becoming useless as nobody knew what part of the tray or tower they came from. It is worth considering a camera with a video-taking capability that the inspector can directly talk to. Keep a spare camera handy in your office or locker in case the camera gets damaged.

The camera needs to be approved by the safety department.

- A small backpack or waist pack to carry tools.
- Templates for measuring repetitive measurements.
- A 9/16″ wrench (will fit most tray hardware) and pliers to check tightness.
- Knee pads and/or shin guards.
- A magnetic rod (a good one). Magnets have come off the rods in some cheap ones.
- An unbreakable pocket mirror with a stick to attach to for looking around corners.
- A self-leveling laser light that levels itself with a small internal pendulum (288, 293). This device is small and can be purchased for about US $150 from hardware stores or on the web. Suitable where levelness is not critical. Where levelness is critical, see Section 10.3.4.

A few additional hints from an expert who inspected many a column (288):

- Always verify that all items on your punch list have been corrected to your satisfaction. Review and address any deviations.
- Do not wear loose-fitting coveralls or anything that is likely to catch on tray parts.
- Use the toilet before entering the tower.

Deposits Collection and Analysis   Deposits collection and analysis provide invaluable input to tower troubleshooting. In many instances, deposits of a new foulant appeared the same as those of a past foulant, but were entirely different and required a different mitigation strategy.

Closely observe and photograph (and label the photos) the deposit pattern and watermarks on the trays. In one case (448), the observation that the caked, off-color polymer concentrated at the corners of the downcomers and at the corners of the inlet and outlet weirs led to the realization that the polymer was building up in the peripheral relatively stagnant zones of the tray, probably due to excessive residence times. Directing liquid to these regions brought the product within color specs and an acceptable run length. In another case (137), finding accumulated brown viscous hydrocarbons on the three lower trays in a section of a cryogenic demethanizer explained the premature tray flooding, by either foaming or crystallization.

In a classic case (389), deposits analysis gave the breakthrough that led to the mitigation of a severe polymer fouling that plugged structured packings, caused tower burping, and led to off-spec products and short run lengths. The breakthrough was that the polymer was a cyanide type, which most likely formed due to the decomposition of a compound that was accumulating in the tower (Section 2.14). Increasing the top product flow rate mitigated the accumulation and therefore the polymerization and solved the problem.

10.3.4   Packing Distributor Checks

As shown in Table 10.1, mishaps in packing liquid distributors are foremost among the assembly mishaps. Below are some guidelines for inspection:

Distributor Levelness   Distributor levelness is central to good liquid distribution in packed towers (Section 4.8). Levelness checks are a critical part of distributor and redistributor inspection. The out-of-levelness tolerance used should be that specified by the supplier, with a typical value for high-efficiency separations 1/8″ high to low (43, 140). Lower out-of-levelness values may be difficult to achieve in the upper sections of tall (> about 100 ft) towers due to tower sway. In as much as practicable, distributor leveling, especially in such sections, should be performed under calm conditions.

An adjusting leveling mechanism is built into the troughs and parting boxes for easy leveling. Following the checks and adjustments, it is important to check that all the leveling screws are tight and double-nutted (140).

Three different techniques have been used for checking levelness:

ZIPLEVEL® (Figure 10.6) This technique rose from a complete unknown in the field of packing distributor leveling to perhaps the most popular. Unlike laser methods (below), it does not require line-of-sight, does not amplify error with distance, and does not require calibration. It is simple, accurate, and of low cost, all of which contribute to its success.

ZIPLEVEL® (519) measures the weight of a proprietary liquid sealed within its cord relative to a reference cell in the hub of its reel. It is a high-precision pressurized hydrostatic altimeter that reads elevations by measuring the pressure developed by gravity acting on the net height of liquid between the measurement module and the base unit. This makes it immune to both barometric pressure and altitude changes. Its proprietary liquid is sealed within, does not move in its cord, and is pressurized with a special gas to prevent bubble formation. It is stated not to be damaged by stepping on or kinking the cord though readings will be affected briefly.

The procedures for using ZIPLEVEL in towers are simple and similar to that of a water level or other instruments. Its standard precision of 0.050″ (519) is usually sufficient for distributor levelness measurements, especially considering the tower sway mentioned above. *ZIPLEVEL* is reported to operate from −22°F to 158°F for up to a year of daily use on a single 9-V battery. It is reported to be set up in seconds and offers true one-person operation with no line-of-sight, no error with distance, no factory calibration, and no math.

If the interior of the tower is more than a few degrees different than the outside, ZIPLEVEL requires acclimation to avoid error. The length of the cord to be used needs to be unpacked inside

Figure 10.6   *ZIPLEVEL*® PRO-2000 altimeter. (Courtesy of Technidea Corporation.)

the tower, so the unit acclimates to the tower temperature for at least 10–15 minutes before use and up to 20–30 minutes with subfreezing ambient temperatures. Double-check your starting benchmark zero for repeatability at the end of the measurements.

The popular ZIPLEVEL PRO-2000 consists of a base unit that stores its handheld measurement module, 100 ft of interconnecting cord, and accessories. Accessories include a rubber protective boot for the measurement module, a unipod for measuring without bending, and a pair of stakes to secure the base unit on a hillside. The fully extended unipod doubles as a 4-ft vertical calibration standard for the system. Other models and accessories are also available, allowing sharing and plotting profiles, 3D maps, and data on photos.

Laser techniques create a flat, level reference plane. The distributor level is measured with respect to this plane. A rotating laser is mounted to a tower attachment that projects a 360° light circle on the column wall. Readings are taken in reference to the light circle. Laser techniques can measure 1/16″ out-of-levelness in a large (<30 ft) distributor. Calibration is needed. Accuracy can be checked by marking the walls of a conference room and then rotating the laser and monitoring the reproducibility. Although most experiences with the laser techniques have been good, the author is aware of some that were less satisfactory.

The water manometer technique is the time-honored way of checking distributor levelness. It uses a flexible plastic tube with two vertical steel rulers – one stationary, held by one person, and the other, held by the other person, is moved around with the tube. The tube is filled with water, and the hydrostatic level establishes a level reference plane. To get a good reading, the water should be colored (typically with food coloring). Some of the water often spills while moving around, so it is important to use rubber stoppers and bring some additional water to make up for losses. One needs to watch out for bubbles in the manometer water and eliminate them. Oil and dirt may introduce errors. Accuracy and reproducibility checks are mandatory, including a side-by-side check, as well as reversing the ends of the tubes along the full length. Photographing the readings can overcome meniscus reading problems.

Orientation of Distributor and Collector Pipes   Misorientation of distributor and collector pipes is a common flaw. In Figure 10.7a, an absorber did not absorb because its distributor and

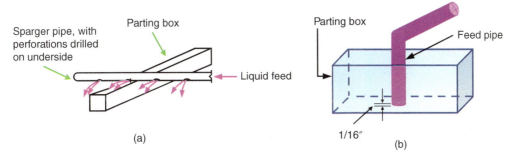

Figure 10.7   Flawed orientation of distributor pipes. (a) Feed pipe installed at 90° to the parting box trough, when it should have been installed parallel to it. (b) Feed pipe too long, leaving a small clearance from the tray floor. (Part (a): From Kister, H. Z., W. J. Stupin, and S. Stupin, TCE, p. 44, December 2006/January 2007. Reprinted with permission. Part (b): Based on Olsson, R. F., Chemical Engineering Progress, October 1999.)

its parting box were rotated 90° from its intended orientation, so the feed pipe sparger became oriented 90° to the parting box. In Figure 10.7b, the feed pipe was too long, leaving a small clearance from the tray floor.

Sections 8.18–8.23 describe additional cases of flawed orientation of distributor and collector pipes. The cases described there were due to design oversights, but a good inspection would have identified the flaws and possibly led to modifications to circumvent poor performance. Only one of all the cited cases (Figure 10.7b) was picked up in inspection before startup; the others were all picked up in inspections after the tower performed poorly.

For effective inspection, it is important to address not only the question "Does the piping orientation match the drawings?" but also the question "Would it work?" Section 10.3.8 addresses this in detail and also presents many other watchouts for points of transition that apply to packing distributors.

Special Checks for Spray Distributors   Special considerations that apply to spray distributors are listed below. Some of these are described in detail in Section 4.15:

1. The checks for spray distributors in existing columns should begin before the turnaround. The first check is by installing a calibrated pressure gauge upstream of the spray header and downstream of any control valve, filter, or other obstruction that gives a high pressure drop. A detailed description of this check is in Section 4.15.

2. Consideration should be given to water-testing spray distributors as described in Section 4.15. Figure 4.27 shows a water test, and Figure 4.29 shows how water testing quickly identifies a plugged nozzle. Figures 4.29 and 10.8a show plugged nozzles. Many cases of plugged nozzles impairing tower performance have been reported (e.g., 89, 209, 211, 409, 505).

3. Plugged nozzles indicate that upstream piping needs cleaning.

4. Spray distributors tend to vibrate, which can loosen the flange and support bolting (43). Bolting should be closely inspected. Double nutting is recommended (43).

5. It was recommended (43) that spray nozzles are removed and inspected (43) because plugging on the upstream side cannot be identified from below. Caution is required during the removal, as pockets of hydrocarbons, H_2S, or other toxic materials trapped behind plugged nozzles may be released when the nozzles are loosened (287). Where liquid distribution is not critical (e.g., pumparound sections in refinery fractionators) and the plugging potential is not high, removing the nozzles and inspecting them is sufficient and there is no need for a water test.

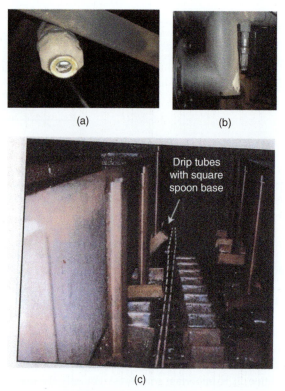

Figure 10.8 Flaws picked by distributor inspections. (a) Plugged spray nozzle. (b) Spray nozzle spraying the pipe instead of the bed. (c) Bent drip tubes with square spoon base. (From Sanchez, J. M., Valverde, A., Di Marco, C., and E. Carosio, Chemical Engineering, p. 44, July 2011. Reprinted with permission.)

6. Check that the nozzles spray the bed and not pipes or supports (see Figure 10.8b).

7. Check that the flanges in the spray header and laterals contain the correct gaskets. One case was reported (443) in which left-out gaskets caused poor performance.

8. Because spray nozzles need to be special-ordered, it is a good practice to have spare nozzles and, in the case of a fouling application, a complete new set, on hand (43).

Other Distributor Checks During inspection, it is also important to check the following:

1. Look for watermarks. Watermarks in one notched trough distributor (Figure 4.26) showed the liquid exiting with a horizontal momentum from right to left. This was neither intended nor expected and created liquid maldistribution.

2. Look for plugging, corrosion, damage, sag, or sloping in the distributor and parting box. When suspected and it is difficult to see due to congestion of internals, consider dismantling parts to properly inspect and clean. Look for signs of bowing in long troughs or chimneys. They may need straightening and bracing to prevent recurrence.

3. Look for any interferences of supports, internal pipes, or other internals. Case 2 in Section 8.12 and another case reported in Ref. 450 are examples of interference of supports. Refs. 257, 261 describe a case where $4'' \times 8''$ distributor supports blocked 8% of the packing cross-section area.

4. Look for broken or bent drip tubes (e.g., Figures 4.30d and 10.8c) or other channels guiding liquid to the packings or distributor troughs.

5. Check that parting boxes properly discharge into the distributor troughs. The openings in the parting boxes should align with those in the distributor. Sometimes the parting box liquid is directed to the troughs by shields and baffles. Check that these are all there and none is damaged or broken.

6. Check that holes in the distributor and parting box are of the correct sizes and punched in the same direction. The side through which the punch entered should feel smoother, and both sides should be free of burrs. Small hole diameters can be checked with taper gauges. These checks are best done at the supplier shop during the water tests, but an additional check during inspection will not hurt. Hole size checks of distributors that have been in service are important in corrosive or scaling services, where the holes may expand or scale during operation.

7. Look for features that obstruct liquid flow or cause excessive hydraulic gradients that lead to maldistribution or overflows. Case 1 in Section 8.18 and Figure 8.21 describe closely spaced large chimneys restricting liquid movement between them, causing maldistribution and poor separation in the bed below.

8. Deck distributors that are supported and sealed to the tray support ring can leak at the joints. It is essential that the deck distributor panels are sealed to the support ring, as well as to one another, with gasketing of the correct materials.

9. Ensure that all gaskets are in place and in good condition. Disintegrated gaskets not only generate leaks that promote maldistribution, but also plug distributor holes. In one case (264), disintegrated gasket pieces falling off from downpipes (Figure 10.12b) were found in six distributors of a large fractionator, where they plugged orifice holes in the main channels and led to poor performance of the entire tower.

10. Check that redistributor hats are correctly installed and are not upside down. Upside-down hats are a common error that can bottleneck the tower, similar to Case 11.12 in Ref. 201. For hats equipped with weirs, the liquid collected between the weirs should pour out over a closed portion of the chimney (as in Figures 10.34 and 4.22) and not descend into the vapor space between the chimney and the hat. Long hats need to be adequately supported; if not, additional support brackets may need to be added. In one case (496), a loose distributor hat fell off blocking a circulation loop, forcing a shutdown.

11. Check that redistributor hats are not oversized. Oversized hats may restrict the vapor ascent area, leading to flood or vapor maldistribution to the bed above. In Case 3 in Section 8.11 (Figure 8.12b), oversized hats led to premature flood. The author has seen similar cases. In case studies 2–4 in Section 8.12 (Figure 8.13b,c), oversized hats and their interaction with support beams led to vapor maldistribution.

12. Critically review drain and weep holes, checking that they are of the correct sizes and located at the correct locations. Seal off any unneeded drain holes. In spray distributors, high-velocity jets issuing from drain holes can damage the packing.

13. Check that all deflector baffles are installed, are firm, and would not bow.

14. Check that all the nuts and bolts are tight and that any gaskets are properly installed and are of the correct sizes and materials. In one case, improperly installed hold-down clamps and missing bolts led to a draw pan being dislodged (259).

15. Check that all the appropriate packing hold-downs are there and do not interfere with liquid distribution.

16. Do not rule out the "impossible." A case was reported in which a distributor was switched from feed to reflux due to a drawing error and another in which a distributor was improperly sized due to a design error (112). Everything looked good except for the separation.

10.3.5 Packing Assembly Checks – Existing Columns

Assembly of packings was discussed at length in Sections 10.2.5 through 10.2.7.

When inspecting a bed of existing packings, the following should also be checked:

1. Where fouling, solids, corrosion, or damage may be suspected, monitor bed condition well ahead of the turnaround. This can be done by pressure drop monitoring (Section 2.8), gamma scans (Chapter 5), or other techniques. Based on this, order any needed replacement. Last-minute ordering of replacement packings may be too costly or even impractical, as proved in some cases (e.g., 443).

2. Are the packings there at the expected height? Cases of disappearance of packings due to corrosion, or shrinking due to compression, are common. Sections 5.5.12 to 5.5.14 describe such cases.

3. Are the packings clean? Corroded? Damaged? Compressed? Sagged (structured beds)? Figure 10.9a, b shows some examples.

4. If damaged, how? Pushed up? Pushed down? Corroded? Eroded? Compressed?

5. Look for flattened packing channels on the top structured packing layer of each bed. This is especially important in services where the distributors above are regularly dismantled for cleaning. In one case (264), maldistribution due to such flattened layers reduced bed efficiency by 25–30%. Similarly, check for crushed or deformed random packings at the top of each bed.

6. For ceramic and carbon, are the packings broken or chipped?

7. For plastic packings, are the packing brittle? Squeeze a few pieces and check whether they shatter.

8. Are the supports plugged or damaged? Figure 10.9c, d show examples.

9. Is the gap between the distributor and the top of the bed per supplier specifications?

10. Are there any watermarks on top of the packings, suggesting dry areas or preferential flow patterns?

11. Is there any mixing of random packings of different sizes?

12. Are there any random packing particles migrated via the bottom support or the top bed limiter?

13. Are there gaps through which random packing particles can escape (e.g., gaps between the support ring and the packing support, bed limiter not a tight fit)?

14. Does the open area of the supports or bed limiters appear restrictive? In metal and plastics, the open area is usually 80–100%+ of the tower cross-section area.

15. For manholes inside the beds, are manhole inserts in place?

16. Are there large gaps between layers of structured packings or between the packings and the wall?

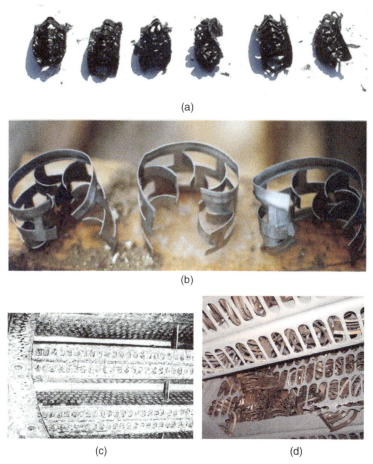

Figure 10.9 Looking for problems in the packings. (a) Fouled and slightly compressed. (b) Deformed. (c) Plugged support plate. (d) Damaged support plate.

10.3.6 Untightened Nuts, Bolts, Clamps, and Downcomer Panel Assembly

Trays and downcomers should be firmly bolted to their supports. Loose bolting may result in excessive deflection, leakage, deficient mechanical integrity, and flow restriction.

Downcomer Panels, Panels Under Downcomers, and Their Fastening Tray active panels containing valves or holes must not be installed under the downcomers from the tray above. This includes false downcomers installed at feeds or refluxes. If present, they are likely to cause premature flooding or excessive inlet weeping, or both, as experienced by the author and others (112). In some cases, column out-of-roundness moves holes or valves from the inlet active panels to the area under the downcomers, and these should be blanked. For each downcomer, carefully look down from above, making sure you cannot see holes or valves.

Loose or improper bolting of the panels beneath downcomers (e.g., Figure 10.10) or holes due to corrosion in those areas are likely to cause excessive leakage, which must positively be avoided. Liquid leaking from this area is likely to completely bypass two trays, making such leaks highly detrimental to efficiency. This area also has the highest leak potential because of the high

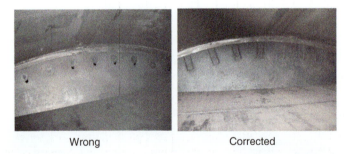

Wrong   Corrected

Figure 10.10   Downcomer gaps that cause leakage will reduce tray efficiency.

liquid head. An inlet weir or a recessed seal pan (if present) further raises this leakage potential, and all its bolting should be firm and carefully checked. The above could not be emphasized more for leak-tight services.

Downcomer panels must be firmly fastened to avoid leakage. They should be installed on the wall side of the bolting bars and not on the tray side to prevent leakage when the downcomer plate is pushed by the hydrostatic head of the liquid. Bolting should be audited for tight fastening. Z-bars, commonly used in revamps that modify downcomers, should fit tightly against the new downcomer panels and the existing weld-ins. Downcomer plates should be checked for firm support, so they do not bow under the downcomer liquid hydrostatic pressure. They should firmly fit into their bottom support brackets. Center downcomers should have spacers to maintain the downcomer width. Figure 10.11a shows a downcomer that bulged under the liquid hydrostatic

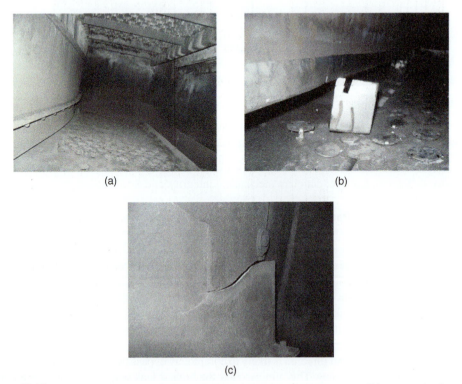

(a)   (b)

(c)

Figure 10.11   Downcomer inspection checks. (a) Bulging downcomer. (b) Downcomer off its support bracket, often leads to downcomer bow. (c) Cracked downcomer.

head. Bulging is especially a problem when the bulge nears the front row of holes or valves, creating a path for vapor to enter the downcomer. Figure 10.11b shows a downcomer that came off its bottom support bracket. A downcomer bracket should be installed about every 3–4 ft (140, 288, 293). When an inlet weir or recessed seal pan is present, an improperly tightened downcomer may bow toward the weir, restricting flow at the downcomer outlet. The downcomer panels should be inspected for cracking (Figure 10.11c), and any cracks repaired.

Tray Panels Fastening   Tray panels and all internal parts should be adequately secured and tightened. Otherwise, they may lift during even minor pressure surges or loosen during operation. All clamps should fully grip the tray support ring or the bolting bar. In one case (42), a piece of an outlet weir was missing probably due to poor tightening of the bolting hardware or loosening of the bolting hardware by vibrations. Clamps are normally spaced about 6–7 in. around the trays' periphery, 5 in. when close to the downcomer, and downcomer clamps are spaced about 4 in. (25).

Double nutting should be used where bolting has shaken itself loose during the previous operation or where there is a potential for vibrations. It is also a good practice to double-nut tower attachments for major support beams and internal pipes. Some experts (140) require double nutting on all bolts to beams and support rings. In one case (254), an entire grid bed fell to the bottom of the column due to poor tightening; in another (254), poor tightening led to a loud banging noise from an operating column. Proper overlaps should be used when fastening tray panels. Washers should be of the correct size in friction fit assemblies; using the wrong size washers is a common error (140).

When workers rush to open the manways at the turnaround, they often hammer the manway clips to the open position, which may bend the clips (42; Figure 10.13f). Clips with a permanent bend need to be replaced. There are special considerations for Nutter-style internal sliding manway clips, and these are discussed elsewhere (42). When dismantling manways, it is a good practice to keep the bottom manway bolted. Workers have almost fallen through an open bottom manway not realizing there were no more trays below.

Special care should be taken to ensure proper tightening of the nuts and bolts at the drawoff trays (particularly for total drawoffs). At least one case (79) was reported where poor column performance was due to failure to tighten the nuts and bolts on a drawoff box. In another case (256), improperly installed hold-down clamps and missing bolts resulted in a draw pan being upset during startup.

In leak-tight services, it is important to ensure that washers and gaskets are installed as specified. In one revamp (192), leaving out gaskets on a total drawoff chimney tray rendered the column inoperable. Gaskets must be carefully checked to ensure they are suitable for the service, are properly cut, and are securely held by the joints. In existing towers, prior to inspection check whether the gaskets are fabricated from asbestos, and if so, seek guidance from HSE for inspecting damaged gaskets. Tray gaskets need to be closely inspected to ensure they are in good condition. Torn or damaged gaskets (e.g., Figure 10.12) should be replaced, and the pieces removed from the column.

When shear clips are used, they should not be welded to the tray support ring (288).

Figure 10.13 shows a sample of miscellaneous issues with nuts, bolts, and clamps that were observed during inspections and that inspectors should be on the look for. The diagram is self-explanatory, and there are more examples in Ref. 42.

10.3.7   Tray Assembly

Tray Panels   Leakage from the tray active area is less critical than leakage from panels under the downcomers, but still should be minimized, especially where high efficiency at turndown is important. Bolles (in Ref. 192) recommended that cracks and crevices should not take up more than 2% of the tray area. If they do, excessive weeping and/or channeling may result. Figure 10.14

(a)   (b)

Figure 10.12 Missing and damaged gaskets (a) on a tray support ring (b) in a packed tower downpipe. (Part (b): From Krishnamoorthy, S. and L. M. Yang, "Troubleshooting EB/SM Splitters: How Can a Maldistribution Analysis Help?," in Kister Distillation Symposium, Topical Conference Proceedings, p. 43, AIChE Spring Meeting, San Antonio, Texas, March 26–30, 2017. Reprinted with permission.)

shows a sample of cracks in the tray floor observed during inspections that the inspector should be on the look for.

Time pressure during turnarounds often breeds the question of whether to fix the cracks or not. As a general rule, if it is easy to seal the cracks, then it is best to seal them. If sealing the crack is costly and/or time-consuming, the magnitude of the crack, the cost and schedule impact of sealing it, and the consequences of not fixing it should be evaluated. As stated in Section 10.3.6, cracks in the panels under the downcomers (e.g., Figure 10.10), as well as cracks at the tray inlet, from which the weep will bypass two trays, need to be sealed. Large cracks (e.g., Figure 10.14a,d) also need to be sealed. In contrast, one can get away with leaving smaller cracks (e.g., Figure 10.14b,c) unsealed, assuming their area is small compared with the tray hole area and the trays operate near maximum rates most of the time.

With new trays, and also in existing trays where this has not been done before, the hole area or number of valves should be checked, together with the diameters of the sieve holes or the dimensions of the valves (particularly the valve slot height or open float lift above the deck). This open area should be within 3–5% of the design (192). Care should also be taken to ensure that the holes or valves are of consistent dimensions and that any blanking strips are correctly positioned. This is especially important when the hole area or dimensions change from section to section, as panels are often interchanged. Premature flood due to panel interchange is common (e.g., 112). The author is familiar with one case of a tower performing below the design expectation, but the panel interchange was so extensive that the plant accepted the substandard performance rather than attempting to correct the problem in the short turnaround. The author is also familiar with major capacity losses incurred by grossly diminished hole areas, in one case on only one tray and in others in a section of the tower. Similar issues were reported by others (41). Finally, the author is familiar with many cases in which the sieve tray hole area increased due to corrosion and fewer cases in which it decreased due to scaling.

Counting holes or valves is best performed outside of the tower. In new installations or revamps, it is best to install one or more trays on the ground, which makes valve or hole counting and checking easy. If practical, this should be done before the turnaround so that any corrective action can be taken in time.

Most fixed valves, as well as some moving valves, are directional (e.g., Figure 10.15a,b). Suppliers' drawings show the required direction, and the direction is often also etched on the valve cap. The inspectors should ensure that the panel installation adheres to these requirements. As a general rule, unless the drawings show otherwise, the wide legs of the fixed valves should face the liquid flow (as in Figure 10.15b) to minimize weep. Nonetheless, misdirection of panels

Figure 10.13   Miscellaneous fastening issues. (a) Nuts and bolts incorrectly installed. (b) Nuts and bolts incorrectly fabricated. (c) Unfastened nuts and bolts. (d) Loose clamps. (e) Light duty bolting provides poor mechanical strength, left panel uplifted during service. (f) manway clip with a permanent bend. (Parts (a–c): Courtesy of Karl Kolmetz, Consultant. Part (f): From Bouck, D., Chemical Engineering Progress, p. 26, September 2018. Reprinted Courtesy of Chemical Engineering Progress (CEP). Copyright © American Institute of Chemical Engineers (AIChE).)

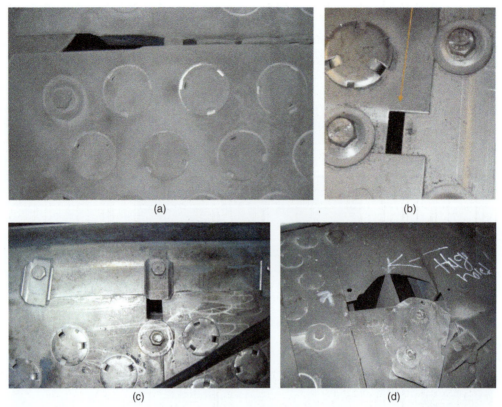

Figure 10.14   Cracks in tray active areas. (a) Due to misalignment of tray panels. (b) Due to incorrect fabrication. (c) Due to poor assembly. (d) Due to damage.

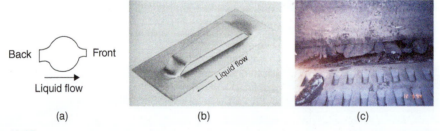

Figure 10.15   Checks of layout of fixed valves. (a) Directional round fixed valve. (b) Directional trapezoidal fixed valve. (c) Fixed valves installed upside down. (Part (b): Courtesy of Sulzer Chemtech.)

has been common, especially on manways. Misdirecting may lead to an increased weeping and a slight loss in capacity and efficiency. In most cases, misdirecting only leads to a relatively small loss of performance. Should redirecting a large number of panels threaten to prolong the turnaround, or be costly, redirection may not be justified.

The opposite applies to trays installed upside down (e.g., Figure 10.15c). Such installation leads to a major loss of performance and must be corrected. In one case (254), bubble caps were installed under the tray panels; this column flooded at 30–40% of the design.

Another situation where misdirection cannot be tolerated is when it affects the functionality of devices like weirs or downcomers. A case history by Golden cited in Ref. 192 is of tray panels in a low-liquid-rate amine contactor that were rotated 180° to the desired orientation. The inlet weirs were therefore installed near the tray outlet and were ineffective for sealing the downcomers. Vapor entered the downcomers and interfered with liquid descent, which caused excessive liquid carryover from the tower. To minimize the carryover, the liquid rate was lowered to 20% of the design rate, which provided poor tray washing. The tray valves plugged, and frequent cleaning became necessary. After the fault was corrected, normal liquid rates were reinstated and the plugging problem disappeared.

It is important to closely audit any plugging, corrosion, or damage (Figure 10.16). Deposit samples are invaluable in shedding light on the foulants, even if these are known. There have been many instances in which deposits of a new foulant looked the same as those of a past foulant, but were entirely different and required a different mitigation strategy. Explore for bent, warped, or bowing tray parts. Good photographs should be taken of corroded regions, damaged regions, warping, and cracks. These should be forwarded to personnel who have expertise in the relevant discipline to see whether any action is needed to prevent recurrence. For instance, a crack shown in Figure 10.16d indicates possible fatigue or vibrations that need to be further investigated.

Tray levelness is normally not critical in conventional one- and two-pass trays (192, 303). Since tray-by-tray levelness checks are tedious, it is often satisfactory only to spot-check their levelness. Checking tray levelness becomes important in multipass trays, where it may affect the liquid distribution to the passes, in dual-flow trays where vapor can preferentially channel through shallower parts, and in specialty trays.

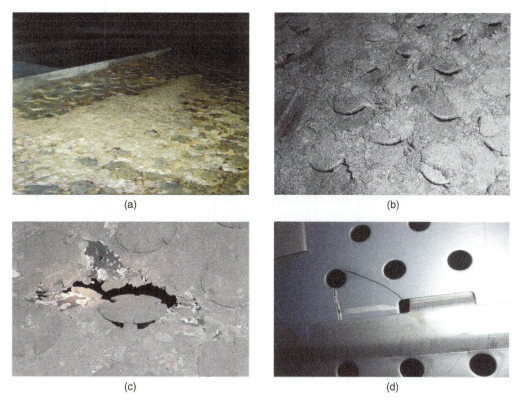

(a)

(b)

(c)

(d)

Figure 10.16 Plugging (a–c), under-deposit corrosion (c) and cracks (d) on trays. (Part (c), courtesy of Karl Kolmetz, Consultant. Part (d), courtesy of Sulzer Chemtech.)

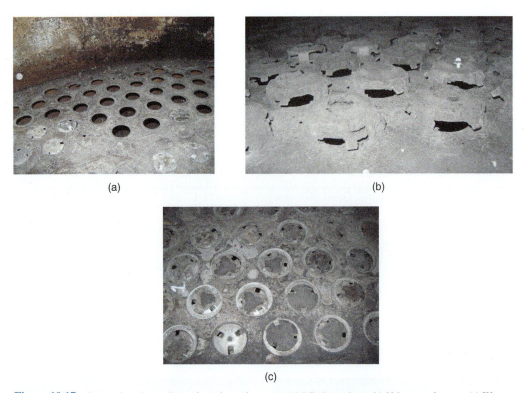

(a)

(b)

(c)

Figure 10.17 Scenes from inspections of moving valve trays. (a) Missing valves. (b) Valves stuck open. (c) Watermarks showing valve spins.

With moving valve trays, inspectors should also look for popped-out valve floats (Figure 10.17a). This is very common. The popped-out floats should be properly reinstalled (if the floats are found) or replaced. If only a few floats are randomly missing (<3%) and timely replacement is impractical, then "do nothing" is generally acceptable. Some experts accept even a higher percentage of missing floats, up to 5–10% (293). While replacing valve tray floats, attention should be paid to replace by the same type whenever possible. Many valve trays contain alternate rows of light-gauge and heavy-gauge floats. The light floats should be replaced by light floats and the heavy with heavy floats. Failing to do so (very common) may detract from the optimum performance, but seldom causes a major operating problem. What is critical is that the installers correctly replace the floats. There have been incidents where the new floats were a loose fit and all popped out again in the next run (42); the author had similar experiences. In another case, the installers bent back the legs of the floats so the floats could not rise above the tray floor, causing premature flooding and forcing an unplanned outage to fix the problem. The symptoms were like severe plugging of the bottom trays.

Popped-out valve caps are also experienced in large-opening fixed valves where the valve caps are fastened to the tray floor by tabs (rather than an integral part of the tray floor). This problem is far less severe and less frequent than in moving valves and is experienced when the tabs are not properly bent (a manufacturing issue) or corrode.

The inspection should determine what caused the floats to pop. Are the legs worn or corroded? Are the holes expanded due to wear or corrosion (if so, adding light-gauge retainer rings is needed before replacing the floats)? Were the floats blown out by a process upset?

By impingement by high-velocity flashing feeds (in this case, the feed distributor should be modified, like in the case study in Refs. 60, 61)?

Valve floats may stick to deposits on the tray floor, and their opening may be restricted. Check that the floats can move freely. Float movement may also be restricted by deposits at the valve legs; check for these too. Floats in moving valves also stick open (Figure 10.17b) or spin (watermarks in Figure 10.17c). Sticking open reduces the turndown. Spinning may damage the tray floor and is usually not a big issue, but may justify using valves with stops (a small tongue projecting into the hole to stop the spin) during the next turnaround. It is important also to ensure that moving valves do not easily come out of their orifices. If they do during inspection, they are likely to do the same in service.

With bubble caps, at least a sample of the caps should be removed, and the clearance between the riser and cap measured and inspected for cleanliness. Dirt accumulating in this area can cause premature flooding (293 describes one case). When replacing the caps, check that the nuts that secure the caps in place on the threaded nut protruding from the bubble cap are not overtightened (288). One case was reported (144) in which overtightening these nuts reduced the clearance between the caps and the risers, inducing premature flood. In another case, not knowing the clearance between the caps and the risers, which was not on the drawings and not in decades of inspection reports, led to an expensive replacement of welded-in bubble-cap trays that was completely unnecessary. Also check that the outlet weirs are shorter than the risers but are taller than the bottom edge of the caps.

Downcomer Clearances and Weir Heights   The only tray internal issue that specifically made it to the top 10 assembly mishaps (Table 10.2) is the downcomer clearance. This is both because it is common to find incorrectly installed clearances and because incorrectly set clearances generally have a greater adverse impact on tower performance than most other tray internals installation errors. Clearances that are too small can cause premature downcomer backup flooding or can easily plug and cause the same. Clearances that are too large can allow vapor entry into the downcomer with possible downcomer unsealing flood (Section 2.11) or just a loss of tray efficiency. Clearances that vary along the downcomer length maldistribute liquid to the tray. With multipass and specialty trays, incorrectly set clearances can lead to liquid and vapor maldistribution to the passes with efficiency and capacity penalties.

In one case (161), a column flooded prematurely after replacing valve trays by sieve trays. The cause was scale left on the support rings raised many panels beneath the downcomers, reducing the clearances from $1''$ to $5/8''$–$3/4''$. In another case (292), reducing clearances from $2''$ to $1''$ in a stripping section caused premature flooding. In another case (286), a stripping tray was installed with zero clearance, flooding the entire stripping section and propagating to the rectifying section above, causing off-spec product. In another case (93), premature flooding started three trays from the bottom due to vapor entering the downcomer. The latter two cases teach that it is sufficient to have one incorrectly set clearance to bottleneck an entire tower. In one more case (254), the downcomer clearance was about 7 to 8 in. at the feed tray due to miscommunication, and premature flooding (due to lack of downcomer seal) resulted. In another case (9), excessive downcomer clearances caused downcomer seal loss accompanied by the loss of tower capacity and product purity.

Figure 10.18a shows an invaluable tool for inspecting downcomer clearances. This template should be inserted under the downcomer clearance in a number of locations along the downcomer length. The template is marked at quarter-inch intervals, which permits clearance measurement within $\pm 1/8''$, which is almost always satisfactory. Most major tray suppliers are delighted to provide templates like this (upon request) as they want their trays to work. Of course, short rulers or tape measures (Figures 10.18b,c) can also be used to improve accuracy, but are much slower.

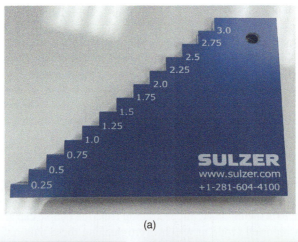

(a)

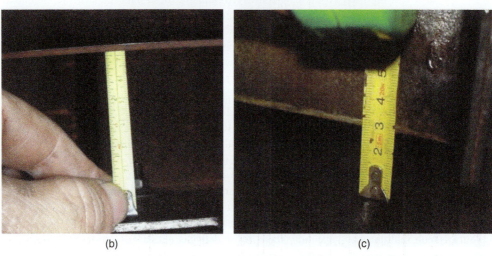

(b)   (c)

Figure 10.18 Inspections of downcomer clearances. (a) Template for inserting under the downcomer plate to determine the downcomer clearance. (b) Clearance too large, 72 mm where it should have been 40 mm. (c) Clearance too small, 29 mm where it should have been 44 mm. (Part (a): Courtesy of Sulzer Chemtech. Part (b): Courtesy of Karl Kolmetz, Consultant. Part (c): From Sanchez, J. M., A. Valverde, C. Di Marco, and E. Carosio, Chemical Engineering, p. 44, July 2011. Reprinted with permission.)

Some inspectors use wooden blocks (of different sizes) that are inserted under downcomers. They work, but the blocks often slip out of the inspector's pocket, leaving the inspector without a tool until the slipped block is found or replaced.

In large-diameter towers (>10 ft), it was recommended (140) to take at least three measurement points along the clearance and allow variations of no more than 1/8″ along the length.

Special caution is required when the downcomer clearances change from one tower section to another. A common error is to interchange tray parts, and this may cause incorrect clearances to be set. One incorrectly set clearance is sufficient to bottleneck an entire tower.

If the column was previously in service and no changes were made, clearance measurements need only be spot-checked to ensure the absence of corrosion, scaling, or fouling effects.

The guidelines above extend to outlet weir heights. Fortunately, in conventional one- or two-pass trays, incorrectly set outlet weir heights are less likely to cause a spectacular tower

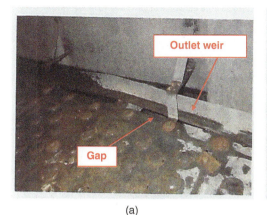

(a)   (b)

Figure 10.19   Outlet weir installation issues. (a) Weir displaced and pushed away from under its retaining washers; some tray liquid can flow through the gap under the weir. (b) A missing piece in a swept-back weir, causing liquid maldistribution on the tray. (From Bouck, D., Chemical Engineering Progress, p. 26, September 2018. Reprinted Courtesy of Chemical Engineering Progress (CEP). Copyright © American Institute of Chemical Engineers (AIChE).)

failure like those described above. However, outlet weir heights are frequently incorrectly set, which may reduce tray efficiency and should be avoided. Correct setting of outlet weir heights is critical for multipass and specialty trays, as variations in weir heights will maldistribute the liquid to the passes, accompanied by efficiency and capacity penalties.

Other issues to look for include excessive gaps near the walls, weir pieces displaced and pushed away from their retaining bolting (Figure 10.19a), or missing weir pieces (Figure 10.19b). As above, this is critical for multipass and specialty trays.

Keep an eye open for trusses and draw sumps mounted over outlet weirs that the restrict downcomer entrance area, especially in foaming or high-pressure systems. These have led to premature flooding (Section 8.5).

Picket-fence weirs (192, 245) contain pickets or weir blocks to reduce the effective weir length. In the spray regime, this design is used to keep the liquid on the trays and, in multipass trays, to balance the liquid flow between passes. The dimensions of the pickets should be per the drawings, but reasonable deviations can usually be tolerated when approved by the supplier. The pickets should be held together by stiffening brackets and not "flap in the wind."

Incorrect installation of inlet weirs can be a lot more troublesome. In services where they are installed on every tray, usually low-liquid load services, their function is to give adequate seal to the downcomer and good liquid distribution to the tray. Incorrect setting may allow vapor into the downcomer, causing a premature downcomer unsealing flood, and/or maldistribute liquid to the tray, which at low liquid loads may lead to drying of tray sections, liquid entrainment, and poor efficiency. If installed too close to the downcomers or if the downcomer plate bows onto them, they may cause restriction and premature flood. In one case (486), bowing of several downcomers over the inlet weirs led to a premature flood and capacity restriction.

In most services, inlet weirs are only installed on selected trays, typically the top tray, to distribute liquid to the tray. A common error is to have the inlet weir installed in some other location in the tower. In one case (254), an inlet weir intended for the top tray was located five trays below, effectively closing off the downcomer. In one more case (112), an inlet weir on a tray blocked the flow from the tray above, restricting capacity. In another case (61), during the replacement of a corroded feed tray by an in-kind, an unintended inlet weir was installed on the tray, resulting from the supplier's misinterpretation of the tray drawing that called for

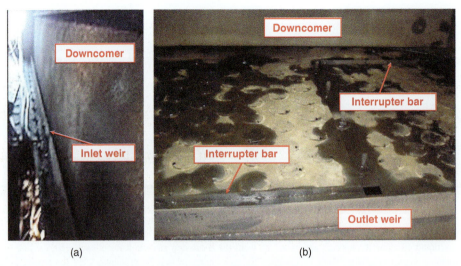

(a)   (b)

Figure 10.20 Misplaced inlet weirs. (a) Incorrectly installed on feed tray, causing narrow gap for liquid exiting the downcomer. (b) Due to panel reversal, interrupter bar becoming ineffective. (Part (b): From Sanchez, J. M., A. Valverde, C. Di Marco, and E. Carosio, Chemical Engineering, p. 44, July 2011. Reprinted with permission.)

an inlet weir on the top tray only. The narrow $<0.5''$ gap between the downcomer and inlet weir (Figure 10.20a) precluded the installation of a downcomer brace, leaving the downcomer dangling against the weir. Fortunately, due to the low liquid load and the extended tray spacing at the feed, no operational issue resulted.

With circular moving valve trays, a short inlet weir (about $½''–¾''$), referred to as "interrupter bar" or "breaker bar," is often installed to prevent liquid from opening the front row of valves and weeping through them. In one case (409, Figure 10.19b), a panel was rotated, rendering the interrupter bar ineffective.

The inspection tools for weir heights are the same as those described above for downcomer clearances, except that the template in Figure 10.18a is turned upside down to fit over the weirs and marked accordingly.

Seal Pans A seal pan is provided below the bottom tray, often also at the tray just above the feed, to stop vapor ascending the downcomer. In some chemical towers, especially with smaller tray spacings $(<20'')$, a seal pan is provided at the bottom of each downcomer to permit a larger clearance to be used without excessive downcomer backup.

Seal pans are often awkward to access and, as a shortcut, left uninspected or only partially inspected. The author learned in the school of hard knocks that this can be fatal to the tower performance (216). Incorrect seal pan installation is one of the most common installation errors in tray towers and often leads to very poor performance. Troubleshooting maestro Lieberman (288) concurs, stating "If you have concluded that I've had lots of bad experiences with flooding due to seal pan malfunctions, you have drawn the correct conclusion." The author recommends expanding every possible effort to ensure proper inspection of the seal pans, both during initial installation and again in every turnaround.

It is amazing how often seal pans are not installed at all or installed at an incorrect location. The diagram on the left in Figure 10.21 shows a seal pan that needs to be installed above the feed tray; the photo on the right shows what the photographer caught following the installation. The seal pan simply was not there. In another case (69), the bottom seal pan was inadvertently

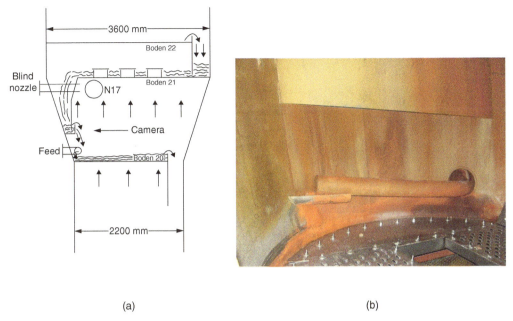

(a)   (b)

Figure 10.21   Anything missing? (Courtesy of Mr. Andre Ohligschläger, Nobian, Frankfurt, Germany.)

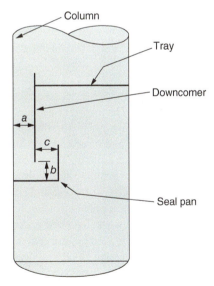

Figure 10.22   Clearances to be checked for a typical bottom seal
pan arrangement. (From Kister, H. Z., Chemical Engineering, p. 107,
February 9, 1981. Reprinted with permission.)

blocked off during a revamp that replaced a reboiler by steam injection. In one more case (489),
an unsealed overflow pipe from a chimney tray prevented liquid descent, causing liquid buildup
above the chimneys and entrainment (Section 8.6 and Figure 8.6b).

Figure 10.22 shows the clearances that need to be checked (a, b, and c). Clearance b should
be less than c and less than the overflow weir height. It is also essential to check that the stiffener
lip at the bottom of the downcomer plate immersed in the seal pan liquid does not significantly

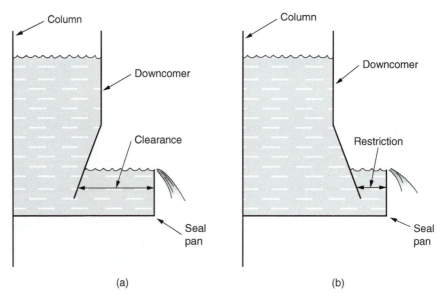

Figure 10.23 Incorrectly installed downcomer causes premature flooding. (a) Correct. (b) Incorrect. (From Kister, H. Z., Chemical Engineering, p. 107, February 9, 1981. Reprinted with permission.)

reduce dimension c to interfere with the flow (140) and that the downcomer plate is solidly supported. If this plate bends under liquid hydrostatic force, liquid downflow will be impeded and flooding of the bottom tray may initiate. Lieberman et al. (144, 288, 292) describe experiences with restricted "*c*" clearance. In one case (288), a discrepancy between the tower and tray fabricators caused dimension *c* to be 1″ when it should have been 3″, leading to restriction and premature flood.

Figure 10.23 shows a sloped downcomer from the bottom tray to the seal pan installed back to front (191, 216). The incorrect arrangement (Figure 10.23b) restricted the liquid flow at the bottom seal pan, producing cyclic flooding in service. This tower was the first that the author took from design to startup and turned success into failure until the flaw was diagnosed and repaired. This hard-earned lesson from the school of hard knocks taught the author, and hopefully the industry, the importance of avoiding shortcuts in seal pan inspections.

Corrosion, water trapping, and collection of debris or deposits are common issues affecting both new and operational units (Figure 10.24). The photographs are self-explanatory. Ensure that the weep hole(s) work. Trapped water from commissioning can lead to operating problems. In one case (342), plugged bottom seal pan weep holes trapped water, which later caused a pressure surge and tray damage. Good practices for weep hole sizes are in Ref. 192.

When the tower base contains a preferential baffle dividing it into a reboiler draw and a bottoms product draw compartments, it is imperative that all the seal pan liquid overflows into the reboiler draw compartment. Failure is likely to lead to starving the reboiler of liquid and to lights in the tower bottoms.

Odd Features   It is important to look for and question any odd features observed during inspections. Some odd features can bottleneck or adversely affect performance; others are quite harmless. In all cases, they should be noted and evaluated. Figure 10.25a shows an obstruction at the downcomer exit. No one knew the reason, and the tower operation was fine. Figure 10.25b shows

Figure 10.24   Checking seal pan condition and integrity. (a) Fouled. (b) Corroded. (c) Trapping water.

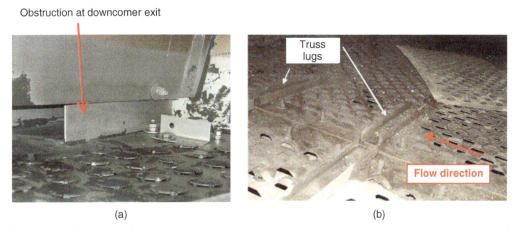

Figure 10.25   Some odd features. (a) Obstruction at the downcomer exit. (b) Truss lugs perpendicular to tray liquid flow.

truss lugs. These are used for tray support in large towers but stick out above the tray floor, impeding liquid flow and in fouling services potentially collecting solids. Changing the truss lugs can be costly and, in the absence of operating issues, also unnecessary.

Specialty Trays   These contain intricate features, many of which are omitted from the tray drawings, often due to confidentiality concerns by the supplier. When shown, they are often in insufficient detail. It is best to install a couple of trays from each section on the ground well before the turnaround, ideally by the crew that will be installing them in the tower. Often, tray parts can be installed in different ways, with the drawings providing insufficient guidance to the correct way. The supplier should then be consulted and provide guidance. Photographs and videos, emphasizing the issues in question, should be taken for future reference.

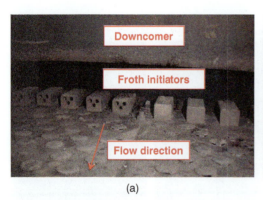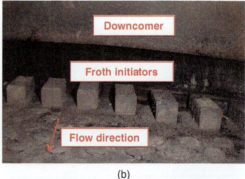

(a)   (b)

Figure 10.26 Incorrectly installed froth initiators on specialty trays. (a) The five froth initiators on the left correctly installed, the three on the right installed backward. (b) All froth initiators installed backward.

Many specialty trays contain "push valves." These are fixed or moving valves with a long front leg and a short back leg (or closed at the back), inducing horizontal vapor flow. These push valves should be installed per the supplier drawings. If the drawings are not clear, the supplier should be questioned. As a general rule, the openings should direct the vapor issuing from them either in the direction of flow or toward the stagnant regions on the sides of the main flow path (or both).

Figure 10.26 shows incorrectly installed froth initiators (or bubbling promoters) on specialty trays. Froth initiators inject some of the rising vapor into the liquid issuing from the downcomers in order to aerate the liquid upon tray entry and also to impart this liquid a push in the direction of flow. Three of the froth initiators in Figure 10.26a, and all the froth initiators in Figure 10.26b, were installed backward, sending this vapor into the downcomer. The author has seen cases in which vapor entering downcomers by a similar mechanism choked the downcomers and incurred a tower capacity limitation. It is also important to check that froth initiators are firmly fixed to the tray and do not come loose.

Some specialty trays feature a multitude of downcomers and passes. Good liquid split to each of these passes is essential for achieving the design tray efficiency. It is therefore critical to positively ensure that tray levelness, weir heights, and downcomer clearances are within the (usually very tight) tolerances set by the proprietor. Tower sway may render these inspections difficult on windy days. The author is familiar with cases of poor separation in tall towers containing specialty trays with multiple downcomers resulting from a relatively small degree of out-of-levelness (which significantly exceeded the manufacturer's specifications).

Materials of Construction   It is essential to ensure that all tray parts, including clamps, bolts, nuts, and washers, are installed according to the materials of construction specifications. If lower-grade materials are arbitrarily substituted (even in only one tray or one section of the column), severe corrosion may occur. It is not uncommon to find that all tray parts, except for a very few, are fabricated from the correct materials, but these few are usually sufficient to cause problems.

The materials of construction are often marked on hardware and tray parts. A magnetic rod can be used to distinguish austenitic stainless steel (300 series) from carbon and martensitic stainless steel (400 series).

Particular attention must be given to nuts, washers, and bolts. Often, those used are of an inferior grade. This could be caused by inadvertent mixing of nuts and bolts or installers running

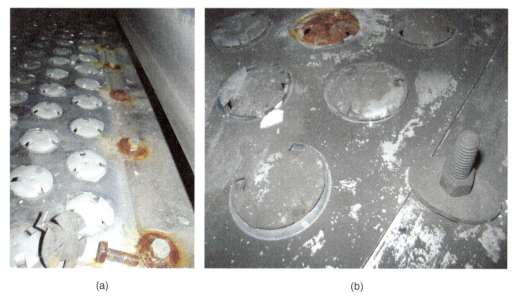

(a)   (b)

Figure 10.27   Inspection pinpoints incorrect materials of construction: (a) washers; (b) a valve float.

out of an item and using an inferior substitute they find in their toolbox. The substitute corrodes in service (e.g., Figure 10.27), causing a variety of problems such as mechanical failure, leakage, rusting, and inability to undo when required. One inspection (329) identified four installed trays as well as many nuts and bolts fabricated from 304 SS where 316 SS was specified; these would have failed in service. Another inspection (291) revealed stripping trays and bolts in a refinery vacuum tower fabricated from carbon steel when 410 SS was specified.

If different materials are used in different parts of a tower, care must be taken to ensure that no interchanging of materials occurs between sections. Internals of existing towers should be inspected for corrosion and operation damage. The materials of any changed parts should also be checked.

Damage   If damage occurred, the inspection should explore and identify the mode. By "mode" we mean how it happened, not why. The reason is sometimes well known and at other times obscure. Refs. 192, 201 discuss many possible mechanisms of damage, which can be identified or at least narrowed in on once the mode of damage is determined.

Upward push on the trays is consistent with a pressure surge below the trays, high liquid level at the tower base, fast depressuring from the tower top or from a location above the damage, or a sudden step-up in condensation. Downward push is consistent with a vapor gap problem, subcooled liquid entering into the bottom sump, liquid slugging from above, or rapid depressuring from the tower bottom. Trays above the feed bent up while those below the feed bent down may suggest rapid flashing at the feed. Trays above the feed bent down while those below the feed bent up may suggest rapid quenching at the feed due to cold liquid entry into a hot column. Bending or cracking can suggest heavy stepping, overtorquing, or vibration. Loose hardware and cracking can also indicate vibrations.

Of course, there can be many other explanations for various modes of tray damage. Identifying the mode of damage can guide the user to the most likely mechanism.

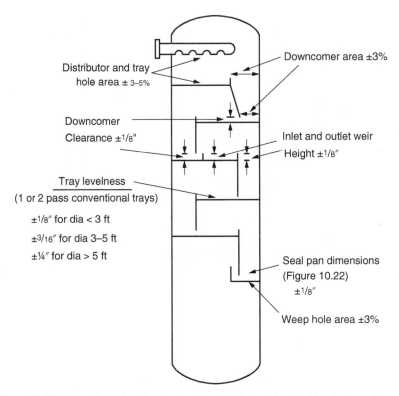

Figure 10.28 Some dimensions that should be inspected and their typical installation tolerances.

Tolerances Figure 10.28 depicts typical tolerances recommended in many literature sources surveyed in *Distillation Operation* (192). These are shown only as a general guide. When company or supplier specifications differ, they should prevail over those shown in Figure 10.28. The inspector should be aware of the possible adverse effects of departures from the specified or recommended values, as described in sections discussing the relevant parts.

10.3.8 Feeds/Draws Obstruction, Misorientation, and Poor Assembly

What to Look for Incorrect orientation of internal pipes, baffles, and other removables is one of the most common installation errors. These are often installed in situ by people who have little understanding of column operation.

Figure 10.29 is an excellent illustration. It shows correct and incorrect ways to assemble the inlet bend of a liquid drawoff line. The correct arrangement (Figure 10.29a) was intended to prevent drawing solids into the liquid outlet line. The incorrect arrangement installed (Figure 10.29b) possibly made more sense to the installer, thinking that the pipe was to overflow liquid or to catch the falling liquid. A good, but uneducated logic. This arrangement caused 160 psig vapor rather than liquid to escape through the outlet pipe (191) into an atmospheric storage tank, lifting its roof; fortunately, no one was hurt. Hence, it is essential to ensure that such pipes have not been installed upside down or sideways in the column.

In another excellent illustration (164), a revamp added product draw sumps beneath the downcomers a few trays below the top of a tower. The product was drawn from the sumps, with tower reflux descending onto the tray below via the clearances between the downcomers and the sumps.

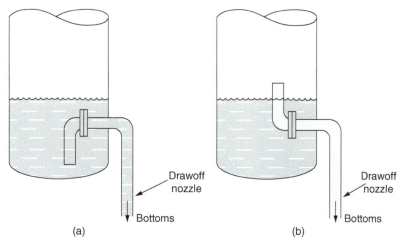

Figure 10.29   Orientation of a bottoms drawoff. (a) Correctly installed (liquid in drawoff). (b) Incorrectly installed (vapor in drawoff). (From Kister, H. Z., Chemical Engineering, p. 107, February 9, 1981. Reprinted with permission.)

The installers presumably did not understand the need for the clearances and welded them shut. The reflux was unable to descend, causing flooding above.

For liquid feed, critically review the possibility of vapor presence or vapor flash upon entry, especially with packed towers. Sections 4.9 and 4.13 provide detailed discussion with case studies.

When checking the location and orientation of internals during inspection, note should be taken of the possible interference of internals with instruments or other parts. It is common to have different internals shown on different drawings, and interferences between these remain unappreciated until the parts are assembled.

Explore the possibility of plugging of any internal piping and vapor channels, especially those that are difficult to check. In one case (281), plugging of internal pipes bottlenecked a tower for 12 years and evaded several inspections.

Incorrect location or orientation of instruments is a common flaw. Thermocouples may be installed in a stagnant region or in a region where the feed or weep from above hits the thermocouple, therefore giving erroneous readings. Pressure taps are sometimes installed in turbulent regions, or at the wrong nozzles, and consequently give incorrect indications.

Strategy   A very useful technique, recommended by the author and others (409), for critically inspecting any point of transition (feed, draw, chimney tray, change in diameter or number of passes), is to imagine yourself as a pocket of liquid or vapor. This pocket enters at a given velocity (which you should calculate) and travels in an initial direction dictated by the entrance geometry. Once inside the tower, the pocket will seek the easiest path, be deflected by walls or baffles in its path, and change direction as it hits them. Baffles and walls also break and/or redirect its momentum. Your challenge: "Would it work? Would you (the pocket of liquid or vapor) get to your intended destination without obstruction and without upsetting or adversely interfering with the tray action?"

Closely look for watermarks. They are invaluable in identifying mysteries, often unexpected ones. The watermarks in Figure 4.26 identify a horizontal direction of liquid coming out of distributor openings. The watermarks in Figure 10.17c show that the valve floats have been spinning. The watermarks in Figures 10.30a and b show a 7-ft vortex in a tower base and leaks from a reboiler draw pan. None of these issues was expected and could have remained undetected were it not for the watermarks.

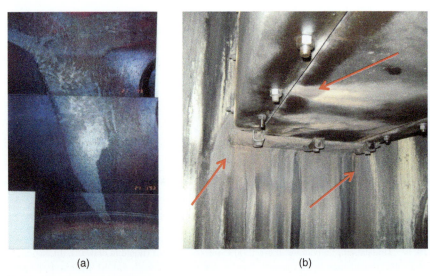

(a)   (b)

Figure 10.30 Watermarks reveal unexpected phenomena. (a) A 7-ft vortex in the bottom of a tower. (b) A leaking reboiler collection pan, with arrows pointing to leak locations. (Part (a): Contributed by R. F. Olsson. Part (b): From Cardoso, R., and H. Z. Kister, Hydrocarbon Processing, p. 27, May 2020. Reprinted with permission.)

Items that Frequently Go Wrong Below is a list of items that commonly go wrong and need to be closely watched by a process/operation inspector:

- Misoriented feed pipes (e.g., Figure 10.31a) and misoriented pipe distributor holes. Also check that the number and size of the holes conform to the drawings. In one case (41, 43), a feed pipe was welded in place upside down, which could have been circumvented by a timely inspection. In another case, the feed pipe holes pointed at 45° toward the outlet weir when they should have pointed at 45° toward the inlet downcomer. Luckily, this did not cause a bottleneck.

- Plugged holes or slots (Figure 10.31b), corroded holes or slots, unintended holes (Figure 10.31c,d), loose nuts and bolts, and damage. Corrosion tends to target welds and areas near welds (43); inspect the bottom corners and all welded seams.

- Flashing feeds finding a path into the downcomers. The flash vapor is likely to choke the downcomer. Section 8.2 and Figure 8.1b show a feed pipe bringing a flashing feed into a downcomer, leading to a capacity bottleneck. Section 5.5.4 and Figure 5.15b show a high-velocity flashing feed directed at a tray floor, causing a tower capacity bottleneck.

- Flashing feeds causing damage upon entry. Baffles, false downcomers, and other internals in contact with a flashing feed may be subjected to severe hydraulic pounding and vibrations. They need to be secured to their supports with brackets, bracing, through bolts, and double nuts. The supports should allow pipe movement due to thermal expansion. Check for deflection, wear, broken brackets, or other damage. Audit the possibility of velocity components impinging on the tray above or below, the tower wall, and the seal pan above. These should generally be avoided.

- Feeds significantly hotter than the tray liquid getting into the downcomers and causing vaporization there. In one revamp (167), a feed 80°F hotter than the tray liquid was introduced a short distance upstream of a downcomer, causing erratic operation and poor separation in a column.

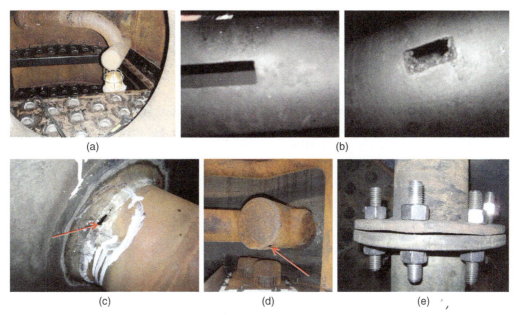

Figure 10.31 Some pipe distributor issues identified by inspection. (a) Feed pipe incorrectly installed; (b) clear slot (left) versus fouled slot (right); (c, d) arrows point at unintended holes in pipe; (e) warped and leaking plate flanges in pipe. (Part (a): Courtesy of Karl Kolmetz, Consultant. Part (e): From Cardoso, R., and H. Z. Kister, "Refinery Tower Inspections: Discovering Problems and Preventing Malfunctions," Distillation 2013, Kister Distillation Symposium, p. 91, AIChE Spring Meeting, San Antonio, Texas, April 28–May 2, 2013. Reprinted with permission.)

- Feed pipes obstructing liquid entry into the feed tray downcomer. Section 8.4 and Figures 8.3 and 8.4 are illustrations.

- Center downcomers, supports, or baffles obstructing the movement of vapor or flashing feed or reboiler return and channeling the vapor into one section of the tray. Some case studies are in Section 8.12 and Figure 8.13.

- Plugged, blocked, or missing drain or vent holes on internal pipes. Inability to drain or vent internal pipes can trap toxic or pyrophoric materials and be hazardous to personnel entering the tower in the next turnaround.

- Draw boxes or draw sumps choking downcomer entrance, as illustrated in Section 8.5 and Figure 8.5 as well as in Case 9.9 in *Distillation Troubleshooting* (201).

- Substandard flanges/gaskets in internal pipes, missing gaskets, incorrect gasket sizes, and materials. Using incorrect gaskets in internal flanges has been a common problem. The temperature rating of the gaskets should conform to both normal and abnormal operation (e.g., steamout). In many cases, incorrect gasket materials were attacked by the chemicals in the tower, leading to severe leakage. In one case (61), metallic spacers were found in lieu of gaskets in the flanges of the reflux distributor, causing a large leak, liquid maldistribution to the three-pass trays, and poor tray efficiency. This was missed in past inspections. In another case (60), many shop-fabricated plate flanges were found leaking and needed replacing (Figure 10.31e). Other cases of leakage from such flanges were described (198, 285).

- Impingement of high-velocity feeds (especially those containing vapor) on the tray liquid, tray floor, distributor liquid, chimney tray liquid (Section 8.13, Figure 8.14), downcomer entry, tower walls (causing corrosion or erosion), or instrument taps in the feed region. In one case (60), high discharge velocity of flashing feed from downward-pointing pipe distributor perforations damaged and destroyed valve floats located right beneath the distributor near its closed end, with weeping and penetration through the holes with the missing floats. In another case (144), a high-velocity hot flashing feed to a refinery atmospheric crude tower was preferentially deflected by a center downcomer into one quadrant of the tray above where it caused severe fouling.

• Impingement of high-velocity reboiler return feeds on the base liquid level, seal pan overflow, tower walls (causing corrosion or erosion), or instrument taps at the tower base region. Figure 10.32 shows reboiler return entries that impinged on the sump liquid level and caused severe bottlenecks in towers. Sections 8.14–8.16, and Figures 8.15–8.17 have additional good examples.

• Maldistribution of feed liquid or vapor in multipass trays, favoring some passes to others. Section 8.3 and Figure 8.2 are good examples.

• Poor mixing between reflux from the tray above and the feed. This poor mixing can lead to poor separation in the section below and is fatal for extractive distillation. Sections 8.18 and 8.19 and Figures 8.24 through 8.27 have discussion and illustrations.

• Misoriented or mislocated draws, seal pans, and inlet weirs (at feed or reflux entry).

• Leaking draws and chimney trays. These are discussed in this Section under Water Testing.

• On chimney trays and draws, features that obstruct liquid flow or cause excessive hydraulic gradients that lead to overflows. Figure 10.33a,b shows supports installed during a revamp inside the downcomer from which a pumparound was drawn. The supports impeded liquid flow toward the draw nozzle, causing cavitation of the draw pump. The problem was solved

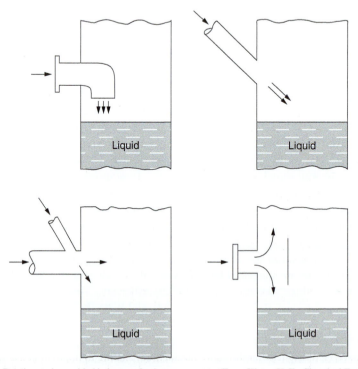

Figure 10.32   Practices to be avoided in bottom feed arrangements. (From Kister, H. Z., Chemical Engineering, p. 138, May 19, 1980. Reprinted with permission.)

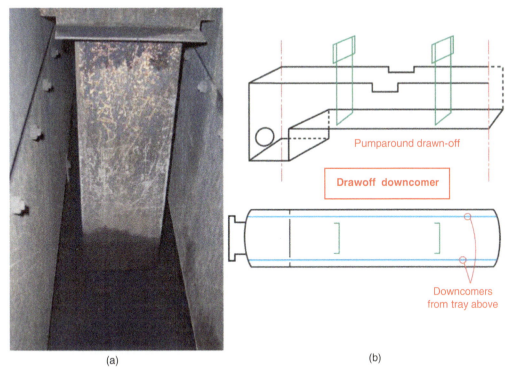

(a)   (b)

Figure 10.33 Looking for obstruction to liquid flow toward the draw nozzle. (a, b) Supports perpendicular to liquid flow inside a draw downcomer impede liquid flow toward the draw nozzle.

by cutting holes in the supports to allow the liquid to drain (in this case, this could be done without compromising the mechanical strength). In Section 8.10 Case 1 and Figure 8.11a (236), rectangular chimneys perpendicular to the liquid flow restricted the liquid flow area toward the outlet nozzle, causing a steep hydraulic gradient, an overflow, and excessive lights in the bottom product.

- Chimney trays and draws are often intended to be total draws or to permit only liquid overflow to descend. During inspection, attempt to visualize the flow patterns on the tray, as well as the vapor–liquid interaction, and look for hidden paths for the liquid to bypass the tray. Section 8.8 and Figure 8.8b describe vapor issuing from chimneys, horizontally deflected by chimney hats, blowing some of the liquid descending from the seal pans above into overflow downcomers that were meant to have no flow. Case 1 and Figure 8.9b in Section 8.9 describe a situation experienced by the author several times. In all these, only liquid overflowing into the downcomer was meant to descend. However, while the chimneys had hats, the overflow downcomer did not, so liquid from the packed bed (or tray weep) rained into it. Case 2 in Section 8.9 describes two other mechanisms for liquid bypassing a chimney tray and raining into the risers: by pouring into an open chimney and by wrapping itself around hats that had no weirs. In another case (56), a support for the seal pan of the tray above did not permit installing one of the chimney hats; seal pan liquid poured into the chimney. Finally, a case was described (144) in which adding manways to a chimney tray led to leakage as the manways could not be seal-welded.

- Oversized hats may restrict the vapor ascent area, leading to flood, or cause vapor maldistribution to the bed or trays above. During inspection, visualize the vapor rise path and evaluate whether there is an oversized hat issue. Review Case 3 in Section 8.11 and Figure 8.12b where oversized hats led to premature flood. The author has seen other similar cases. Review the case studies in Section 8.12 and Figure 8.13, where oversized hats and their interaction with support beams led to vapor maldistribution. In another case (140), a very small vapor space between chimney hats was identified by inspection and corrected prior to startup.

- Errors in the dimensions on the chimney tray. The chimney area and the area between the chimneys and the hats should be within 3–5% of the design. Look for any signs of bowing in long chimneys. They may need straightening and bracing to prevent recurrence. Look for signs of buckling at the tray floor, especially in large-diameter hot services. Buckling is a sign of expansion (usually thermal) between rigid supports and is a common cause of leakage.

- No path for liquid to go from the chimney tray to the tower section below. This path can be external to the tower or internal via a properly designed overflow pipe. In one case (56), the intended overflow was not installed; in another (41, 43), redesign of a draw tray in a revamp overlooked replacing the original downcomer to the tray below, causing flooding above the chimney tray. In one more case (41, 42) a block valve added in the draw to isolate a side reboiler for maintenance was unusable because the chimney tray had no overflow. The extra pressure drop of the block valves caused excessive backup in the tower. In another case (292), the chimney tray feeding the tower reboiler had an overflow, but the top of the overflow was 18 in. above the bottom seal pan, causing tower flooding when the reboiler fouled up.

- Chimney hats incorrectly installed. Upside-down installation of chimney hats is a common error that can bottleneck the tower, similar to Case 11.12 in *Distillation Troubleshooting* (201). The author has had a similar experience with a chimney tray. For hats equipped with weirs, the liquid collected between the weirs should pour out over a closed portion of the chimney and not descend over an open chimney window. Figure 10.34 shows a good arrangement.

- Chimney hats, especially long hats, need to be adequately supported; if not, additional support brackets may need to be added. The author is familiar with many incidents in which a hat breaking loose blocked or partially blocked the liquid outlet. In one case (256), a broken hat suddenly restricted the circulation of olefin plant quench water, forcing a shutdown.

- With very tall chimneys (>36″ height), holes found at the base of the chimneys indicate chimney sway as described in one case study (56). Stabilizing struts to connect the risers together and prevent the sway solved the problem.

- Vapor from chimneys impinging on the chimney tray liquid level or instrument taps.

- Unsealed chimney tray overflow pipes or downcomers (e.g., Section 8.6, Case 2).

- When the chimney tray includes water decanting and removal features, check that there is residence time for water–organic separation, that the water path to the sump is unobstructed, and that the sump and water collection regions do not leak. One case of an obstructed path to the water draws was reported (41).

- Tower bottom sumps are often split into a reboiler draw compartment and a bottoms product draw compartment as described in Section 8.17. The intention is to divert the cold liquid from the tray above to the reboiler. The hot liquid also enters the reboiler draw compartment and overflows into the product draw compartment. Critically look at the arrangement in the tower and convince yourself that it will work as intended. The case studies in Section 8.17 show several arrangements that worked poorly. Lieberman and Lieberman

Figure 10.34   Chimney tray with well-mounted hats. Liquid collected between the weirs on the hats is directed toward the middle of the tray over closed lips that prevent interference with the rising vapor. The middle of the tray provides a path for the liquid toward a sump (not shown).

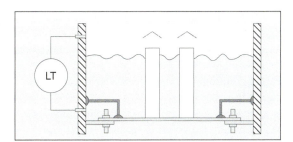

Figure 10.35   Lower level tap installed behind an angle iron intended to keep the tray support area on a chimney tray dry to prevent leakage from the tray. This disabled the level transmitter. (From Kister, H. Z., B. Blum, and T. Rosenzweig, Hydrocarbon Processing, p. 101, April 2001. Reprinted with permission.)

(292) list errors in baffle arrangement among the most common refinery distillation design errors. A similar check is needed on a chimney tray that has hot and cold compartments (Section 7.3.4).

- Mislocated level taps on draw trays and tower base. Figure 10.35 shows a level tap on a chimney tray installed in a dry spot that does not see the liquid level. This is a common error and, if not picked by inspection, will leave the chimney tray with no level measurement and be very embarrassing, as it was one time to the author. A similar incident was reported by Sloley (440). Verify that each tower tap is connected to the correct instrument.

- Other instrument taps in regions where modifications took place. In one case (440) where single-pass trays were replaced by two-pass trays, a pressure tap that used to be in the vapor space found itself in the middle of a downcomer. In Case 25.3 in Ref. 201, pressure transmitters were installed below their taps during a revamp, giving false high readings due to condensate buildup above the taps.

- The upper level tap of the tower level transmitter should be below the bottom of the reboiler return nozzle.

- Vortex breakers are installed, firm, and clear of trash and plugging materials. When gratings are used, each grating layer is rotated 90°. The gratings support should allow free flow to the liquid outlet.

Water Testing Visual inspection for leakage can be misleading. The author has seen many situations where a pan "did not look like it would leak," but water testing revealed a significant leak. For chimney trays, total drawoff pans, liquid outlet pans, bubble-cap trays, and seal areas behind inlet weirs, it is imperative to perform a leakage test. A leakage test is conducted by plugging weep holes and then water-filling the chimney tray or draw or seal area up to near the top of the weir, marking this level, and then monitoring how fast the water leaks down. A leakage rate of 1″ per 20 minutes for normal services (192), or as little as 1″ per hour (even less) for services where leakage is to be positively avoided, are common criteria used. Any leaking joints need to be tightened or repaired, and the test then repeated. Watch the repair work. There have been cases where the leaking joints were repaired by silica cement instead of seal welding. At the end of the test, all temporary plugs must be removed.

At the conclusion of the water test, once the temporary plugs are removed, all the water should drain out. Remaining puddles indicate low points trapping pockets of water, which may cause operating problems and, in hot towers, also pressure surges.

It has been argued that a tray that does not leak under test conditions may leak under operating conditions, and vice versa, but experience has taught that the water test generally provides a reliable indication of leakage under operating conditions.

Water testing should be conducted with solid-free, noncorrosive water and, in the case of stainless steel, also water low in chlorides. The quantity of water needed for the test should be estimated ahead of time, and additional pumps ordered as needed. Work underneath the tested pan should be stopped during the tests.

Figure 10.36 shows a massive leak from a once-through thermosiphon reboiler draw pan (60, 61). The leak induced lights into the bottoms product. Initial visual inspection did not reveal any integrity problems, but identified the watermarks in Figure 10.30b. Upon water introduction using hoses, the leak was so massive that it was impossible to build a liquid level in the draw pan. Gasketing gave some improvement, but the pan still failed the leakage test. Seal-welding the entire draw pan eliminated the leakage and the lights losses in the bottoms and reduced steam consumption, saving the refinery $1.5 million per year. In another case (168), "no amount of care in installation of the gaskets (of a chimney tray collector) prevented distortion of the gaskets in service, and liquid bypassing."

References (60, 61) describe a second case, where the draw pans had big gaps and no gaskets. A water test proved that the draw pans were incapable of holding liquid level. Seal welding eliminated the leakage and improved product recovery and purity.

Case study 10.4 in Ref. 201 describes another case in which leakage from a diesel draw pan in an atmospheric crude tower caused a large loss of diesel product to the much lower value residue. That leakage lasted for at least 11 years, with the hidden flaw becoming the norm. When a leakage test was conducted, the water could not even fill the draw pan. Seal welding eliminated most of the leakage, giving a major improvement in diesel yield. An atmospheric crude tower in another refinery experienced a similar loss in diesel yield due to incorrect seal pan installation, which would have been detected by a water test. To improve the diesel yield, high-capacity trays

Figure 10.36   Water testing a reboiler draw pan shows massive leaks from the pan. (From Cardoso, R., and H. Z. Kister, "Refinery Tower Inspections: Discovering Problems and Preventing Malfunctions," Distillation 2013, Kister Distillation Symposium, p. 91, AIChE Spring Meeting, San Antonio, Texas, April 28–May 2, 2013. Reprinted with permission.)

were installed. During their installation, the seal pan problem was detected and corrected, which by itself would have improved the yield.

Draw sumps should drain into a drawoff nozzle flush with the sump floor, usually centrally located in the sump. With this arrangement, weep holes are not needed and should be sealed off if found.

When a preferential baffle divides the tower bottom base into a reboiler draw and a bottoms product draw compartments, there is usually no need to water test. An exception is when the bottoms product stream is small compared with the reboiler draw stream. In this case, leakage across the baffle can starve the reboiler of liquid, leading to lights in the tower bottoms. In this case, water testing of the reboiler draw compartment is a good idea.

In situ water testing of packings distributors is invaluable, which is further discussed in Sections 4.15 for spray distributors and in Sections 4.16.2 and 9.2.8 for gravity distributors.

10.3.9   Cleanliness of Internals

The column should be inspected for proper cleanliness and absence of debris. While cleaning, steps 1, 4, 5, 9, 13 in Section 10.2.3 and step 10 in Section 10.2.1 are important. Chemical cleaning is sometimes practiced. Washing and chemical cleaning are discussed elsewhere (192).

The inspection should pay special attention to the cleanliness of distributors, downcomers, seal pans, and nozzles. Debris left in the column is a very common experience, as listed in Table 10.2. Debris commonly found in columns can consist of working tools, clamps, nuts, bolts, gaskets, shipping covers, gloves, rags, ear plugs, paper, boards, masking tape, cups, and beverage containers. Each can cause bottlenecks or severe operating problems if not removed. Some troublesome experiences have been reported (88, 222, 288, 319).

In packed towers, dirt and sand may be lodged in narrow distributor or parting box troughs. These can often be sucked out by handheld vacuum cleaners. Plugged small holes can be reamed out with thin rods or metal wire.

Temporary plugs are commonly used during construction and cleaning at tower outlets to prevent small parts such as clamps, nuts, and bolts from entering outlet lines and in weep holes (for leak testing). These plugs must be removed prior to startup. All weep holes must be cleared of possible plugging; in one case (342), weep hole plugging resulted in tray damage.

Trash often travels past the bottom nozzle and finds its way to the bottom line. This line needs to be adequately flushed or blown during commissioning. Also, closely inspect draw nozzles and vortex breakers for trapped gloves and rags (288).

When the tower was previously in service, inspection by backlighting may reveal internal blockages in pipes, tubes, mist eliminators, and other internals, even when the device appears visually clean (392).

Lieberman and Lieberman (293) describe experiences where poor column performance resulted from a variety of debris left inside, including a rag caught in a vortex breaker in a jet fuel draw causing the shutdown of an entire refinery, and a case in which a complete scaffold including boards and piles was left inside a tower.

10.3.10   Final Inspection

Safety considerations permitting, immediately upon completion of column inspection, manhole doors should be shut and from then on should only be kept open while work (e.g., tray reassembly) is being performed inside. This would keep rain, sand, dust, and animals from entering the column. There have been cases (e.g., 284) where animal carcasses lodged in tower internals and caused premature flooding.

It is essential to maintain a continuous close watch of activities around the column during the period preceding the final bolting up of the column. Common errors during this period are (i) manways left loosely placed and unbolted or loosely bolted to the trays, (ii) debris reintroduced, (iii) pipe scaffolds left in the tower, and (iv) bottom baffle hatchways put into place but left unbolted.

Leaving manways unbolted or loosely bolted is a very common experience. Lieberman and Lieberman (293) state that the problem is not just common, it is universal. The author has experienced numerous cases of missing or unbolted manways, and many more are described in the literature. In the author's experience, missing manways deserves a much higher spot in Table 10.2 than the ninth spot that it received. A few of the many cases are described below.

One 10-ft ID chemical tower (226) was returned to service after a routine turnaround in which no modifications to the trays were performed. The tray manways were dismantled for inspection but were not reinstalled following the inspection. The tower flooded at 70% of the rated jet flood, which did not happen previously.

In the most common cases (123, as well as some that the author is familiar with), poor separation and excessive entrainment were experienced due to missing or unbolted manways. In one case, separation was poor, but everything looked good on the gamma scans; in fact, the trays with properly installed manways were thought to be entraining (112).

One classic case (287, 294) showed how leaving manways uninstalled on only four trays in a 30-tray atmospheric crude tower was enough to bottleneck production and lead to building a new vacuum feed heater (with adverse carbon footprints) that would have been unnecessary had the manways been installed. The four uninstalled manways were in the section separating diesel from residue. Their absence led to more diesel in the residue and reduced bottoms temperature, both of which increased the heat duty on the downstream vacuum tower feed heater, which in turn led to building a new vacuum heater to overcome the bottleneck, which could have been circumvented by bolting the four manways.

Sometimes good luck is on your side; one of the author's experiences had been finding a few manways left sitting on nearby tray decks from the last turnaround, leaving a large gap in the tray floor; surprisingly, the column still functioned.

There are different supervision procedures to circumvent such incidents. The level of effort expended varies with the procedure. Procedure selection should be based on the safety, environmental, and economic consequences of having to shut down the unit because of the unbolted manways, the criticality of the service, and the degree of confidence in circumventing such malfunction.

One low-effort component that is invaluable in minimizing, even avoiding, such malfunctions is to clearly explain to the supervisors and work crews the importance of properly bolting up the manways and the consequences of unbolted manways. In the author's experience, as well as a few other experts he talked to, workers want to do a good job and will pay much closer attention when they understand why they are doing it and what is at stake. At the same time, inform the work crews that plant personnel will perform spot checks.

It is important to alert the work crews to potential miscommunication upon shift changes. In one case, the leaving shift bolted the top manway, leaving several unbolted manways below in an endeavor to prevent unauthorized entry. The next shift thought the work was complete and wanted to leave, but fortunately were stopped by the process inspector. One lesson is that it is important to make sure that all manways are accounted for. There have been other cases (293) where only the manways near the tower external manholes were bolted.

Some of the common techniques are listed below:

- As each manway is being installed, having trained responsible persons directly reporting to plant (or project) management inspect it and sign off on it. This is also recommended by Lieberman (288). Very often, process engineers are utilized for such tasks. This is probably the most reliable method, but also the most laborious. It is justified when the stakes are high. For instance, in one giant tower with 4-pass trays (242), there were more than 1000 manways, and even a few unbolted ones could have led to maldistribution and costly operating problems. The installation was very closely inspected, and the next time when the tower was entered, not one manway was found unbolted.

  One benefit of this procedure is affording one final check of fastening all the nuts and bolts, being able to fix any issues on the spot, and positively preventing the reintroduction of debris.

- Request the construction supervisor to number all the manways and take a photo as well as sign off as each manway is bolted in. This is reliable most of the time, but there is always a possibility that some manways inadvertently remain unphotographed and not signed off, and by the time it is realized, it may be too late to check.

- Rely on the work crew, with spot checks by plant personnel. This can be satisfactory when contracting a well-trained work crew (e.g., from a tray or packing supplier). With this technique, it is imperative to make sure that the supervisors as well as the work crew fully understand the importance of properly bolting up the manways and the consequences of unbolted manways – even when the crews are from a tray or packing supplier, as some of their workers may not be well trained.

- "Do nothing." This is the one technique that the author does not recommend, as it has been the root cause of many malfunctions.

While the main issue is manways not being installed or adequately bolted, there are smaller issues that also need to be watched. When the tray contains directional valves, manways should be installed at the desired direction. Manway locking clamps need to be oriented in the correct direction (25). Many older installations of Nutter trays have unique manway hardware that requires correct orientation and installation, as discussed in detail elsewhere (42). When tray design varies from section to section, interchanging needs to be avoided. A good practice is to check this ahead of time and to mark the manways and the desired direction before their installation commences.

In towers that have a preferential baffle that divides the tower base into separate reboiler draw and bottoms product draw compartments, a check should be made that the baffle hatchway has been bolted. In one case (304), poor column performance resulted from a failure to detect (during inspection) that a bottoms hatchway had not been bolted.

10.3.11   Externals Inspection

The previous sections emphasized tower internals inspection, but the inspectors must not lose sight of some external units that need to be inspected while the tower is down for maintenance. These include:

- Heat exchanges: reboilers, condensers, and preheaters. These should be inspected for fouling and corrosion and leak tested.

- Filters, strainers, and coalescers. It is not uncommon to find filter baskets damaged or missing altogether. Experiences with missing and deformed meshes (Figure 10.37) were reported (409). Also look for watermarks. Poor mounting of filter baskets can give rise to liquid bypassing the baskets and deposits at the bottom of the filter casing (outside the baskets). The baskets should be installed securely and inspected by mechanical personnel to prevent bypassing. Filter elements are only as good as the sealing surface between the elements and the vessel (273, 274), so very close seals should be ensured. A check should

Figure 10.37   Deformed mesh taken from a Y-strainer which would allow solids to flow to spray nozzles and potentially cause plugging. (From Sanchez, J. M., A. Valverde, C. Di Marco, and E. Carosio, Chemical Engineering, p. 44, July 2011. Reprinted with permission.)

also be made that all components of the filter are fabricated from materials compatible with the service.

- Solids in the lines leading to the tower, especially those leading to packing distributors, are likely to lead to plugging. They should be adequately flushed, blown, or otherwise cleaned before startup.

- Any maintenance items in lines connecting to the tower. In one case (216), a new under-sized replacement orifice plate installed in the tower reflux line caused an off-spec product. In another case (Section 5.5.2) another improperly-installed new orifice plate led to pre-mature flooding. In one more case, adding block valves in a side reboiler circuit to permit isolation introduced excessive pressure drop and premature flooding (42).

References

1. Achoundong, C., C. Ribeiro, and L. Liu, "Troubleshooting and Redesign of a Solvent Stripper in a Chemical Plant," in *Kister Distillation Symposium, Topical Conference Proceedings*, p. 605, AIChE Spring Meeting, New Orleans, Louisiana, March 31 – April 4, 2019.

2. Achoundong, C., C. Brasington, M. Sukopp, and C. Ribeiro, "Troubleshooting of a Caustic Scrubber in a Chemical Plant," in *Kister Distillation Symposium, Topical Conference Proceedings*, p. 517, AIChE Spring Meeting, Virtual, April 18–22, 2021.

3. Aggrawal, R., and M. Shah, "Conventional Ammonia Plant Revamp with Purifier™," *Ammonia Plant Saf. Relat. Facil.*, 54, p. 31, 2013.

4. AIChE Equipment Testing Procedure, *"Trayed and Packed Columns, A Guide to Performance Evaluation,"* 3rd ed., American Institute of Chemical Engineers, September 2014.

5. Ali, A. M., P. J. Jansens, and Z. Olujic, "Experimental Characterization and Computational Fluid Dynamics Simulation of Gas Distribution Performance of Liquid (Re)Distributors and Collectors in Packed Columns," *Trans. IChemE*, 81, Part A, p. 108, January 2003.

6. Al-Mudaibegh, S. H., M. A. Al-Harbi, and M. R. Tariq, "Feed Gas Quality Limits Acid Gas Removal Unit Capacity," *Gas Processing and LNG*, p. 15, January-February 2022.

7. American National Standards International, "Standard Test Methods for Measuring and Compensating for Reflected Temperature Using Infrared Imaging Radiometers," *ASTM E 1862-97*, American Society for Testing and Materials, West Conshohocken, Pennsylvania, April 1997.

8. Andersen, A. E., and J. C. Jubin, "Case Histories of the Distillation Practitioner," *Chem. Eng. Prog.*, 60(10), p. 60, 1964.

9. Anderson, C. F., "Simultaneous Scanning Techniques for More Accurate Analysis," Paper presented at the *AIChE Spring National Meeting*, New Orleans, Louisiana, March 29-April 2, 1992.

10. Anon, "Miscellaneous Case Histories," in C. H. Vervalin (ed.), *Fire Protection Manual*, vol. 2, Gulf Publishing, Houston, Texas, 1981, p. 29.

11. Anon, "Absorber Changes Solve Offshore High -H_2S Problems," *Oil Gas J.*, 86, May 23, p. 40, 1988.

12. Anon, "The New Schoepentoeter Plus," *Sulzer Tech. Rev.*, p. 6, 3/2010.

13. Arora, V., "Cost-Effective Revamp of CO_2 Removal Systems," *Hydrocarbon Process.*, 97, p. 65, March 2018.

14. Averill, W. P., "Conceptual Troubleshooting Training for Refinery Operators," *Hydrocarbon Process.*, 96, p. 43, October 2017.

15. Ballard, D., "Cut Energy, Chemical, and Corrosion Costs in Amine Units," *Energy Prog.*, 6(2), p. 112, 1986

16. Balmert, P., and D. Glaser, "Learning how to Troubleshoot: the Overlooked Operating Competency," *Hydrocarbon Process.*, 82, p. 89, May 2003.

17. Banik, S., "Non-Uniform Weeping and Its Effect on Tray Efficiency," Paper presented at the *AIChE Spring National Meeting*, Houston, Texas, April 1989.

18. Barletta, T., "Pump Cavitation Caused by Entrained Gas," *Hydrocarbon Process.*, 82, p. 69, November 2003.

19. Barletta, T., and S. Golden, "Deep-Cut Vacuum Unit Design," *PTQ*, Q4, p. 91, 2005.

20. Barletta, T., and K. J. Kurzym, "Consider Retrofits to Handle High-Viscosity Crudes," *Hydrocarbon Process.*, 83, p. 49, September 2004.

21. Barletta, T., J. Nigg, S. Ruoss, J. Mayfield, and W. Landry, "Diagnose Column Flooding Efficiently," *Hydrocarbon Process.*, 80, p. 71, July 2001.

22. Barnard, H., "Use Thermography to Expose What's Hidden," *Chem. Eng. Prog.*, 113(1), p. 21, January 2017.

23. Barnard, H., and E. P. du Toit, "Quick and Easy Troubleshooting of a Packed Column: Thermal Imaging as a Novel Method," in *Kister Distillation Symposium, Topical Conference Proceedings*, AIChE Spring Meeting, Austin, Texas, April 27 – 30, 2015.

24. Bellner, S. P.,W. Ege, and H. Z. Kister, "Hydraulic Analysis is Key to Effective, Low-Cost Demethanizer Debottleneck," *Oil Gas J.*, 102(44), p. 56, November 22, 2004.

25. Bennet, T., "Avoiding Operating Problems in the Distillation Column," *Hydrocarbon Technology International '89/'90*, p. 83, Sterling Publishing International, London, UK, 1990.

26. Bernard, A., "Loss of Downcomer Seal on a Depropanizer During Plant Startup," in *Distillation 2011: The Dr. Jim Fair Heritage Distillation Symposium, Topical Conference Proceedings*, p. 283, AIChE Spring National meeting, Chicago, Illinois, March 13–17, 2011.

27. Bernard, A., "Evaluation of Fouling Precursors Residence Time in Distillation Tower," in *Distillation 2013: The Kister Distillation Symposium, Topical Conference Proceedings*, p. 577, the AIChE National Spring Meeting, San Antonio, Texas, April 29–May 2, 2013.

28. Bernard, A., "The Operating Window – Description and Application," in *Kister Distillation Symposium, Topical Conference Proceedings*, p. 264, AIChE Spring Meeting, San Antonio, Texas, March 26–30, 2017.

29. Bernard, A., "Practical Approach to Distillation Tower Revamp," in *Kister Distillation Symposium, Topical Conference Proceedings*, p. 652, AIChE Spring Meeting, New Orleans, Louisiana, March 31 – April 4, 2019.

30. Bernard, A., "Distillation Tower Startup Following a Plant Revamp with Focus on Reboiler Operation," *Paper 62e, Topical Conference Proceedings, Kister Distillation Symposium*, AIChE Spring Meeting, Houston, Texas, March 12–16, 2023.

31. Biales, G. A., "How Not to Pack a Packed Column," *Chem. Eng. Prog.*, 60(10), p. 71, 1964.

32. Bickerman, J. J., "*Foams*," Springer-Verlag, New York, 1973.

33. Biddulph, M. W., "Tray Efficiency is Not Constant," *Hydrocarbon Proc.*, 56(10), p. 145, October, 1977.

34. Billet, R., "*Distillation Engineering*," Chemical Publishing Company, New York, 1979.

35. Billet, R., "*Packed Column Analysis and Design*," Ruhr University, Bochum, 1989.

36. Billingham, J. F., and M. J. Lockett, "A Simple Method to Assess the Sensitivity of Packed Distillation Columns to Maldistribution," *Trans. IChemE*, 80, Part A, p. 373, May 2002.

37. Blum, B., H. Kister, and R. Tsang, "Good Distributor Design for High-Velocity Feed Debottlenecks a Crude Preflash Tower," *Hydrocarbon Process.*, 100, p. 29, June 2021.

38. Bolles, W. L., "Optimum Bubble-Cap Tray Design," *Pet. Proc.*, p. 65, February 1956; p. 82, March 1956; p. 72, April 1956; p. 109, May 1956.

39. Bolles, W. L., "The Solution of a Foam Problem," *Chem. Eng. Prog.*, 63(9), p. 48, 1967.

40. Bonilla, J., "Don't Neglect Liquid Distributors," *Chem. Eng. Prog.*, 89, p. 47, March 1993.

41. Bouck, D., "10 Distillation Revamp Pitfalls to Avoid," *Chem. Eng. Prog.*, 110, p. 31, February 2014.

42. Bouck, D., "Inspecting Distillation Towers Part 1: Turnarounds," *Chem. Eng. Prog.*, 114, p. 26, September 2018.

43. Bouck, D., "Inspecting Distillation Towers Part 2: Revamps and Other Inspections," *Chem. Eng. Prog.*, 114, p. 42, December 2018.

44. Bouck, D., and C. J. Erickson, "Gamma Scans: A Look Into Troubled Towers," Paper presented at the *AIChE National Meeting*, Miami Beach, Florida, November 2–7, 1986.

45. Bowman, J. D., "Use Column Scanning for Predictive Maintenance," *Chem. Eng. Prog.*, 87, p. 25, February 1991.

46. Bowman, J., "Troubleshoot Packed Towers with Radioisotopes," *Chem. Eng. Prog.*, 89, p. 34, September 1993.

47. Branan, C., "*The Fractionator Analysis Pocket Handbook*," Gulf Publishing, Houston, Texas, 1978.

48. Braswell, Z., B. F. Hanley, and H. Ragsdale, "Reconsidering the Reliability of Chemical Process Simulators I: Implementation of Murphree and Vaporization Efficiencies," *Ind. Eng. Chem. Res.*, 63, 4519, 2024.

49. Brown Burns, J., D. Hanson, C. Riley, and C. Wicklow, "Targeted Injection Improves Shed Deck Performance," in *The Kister Distillation Symposium, Topical Conference Proceedings*," p. 119, the AIChE National Spring Meeting, San Antonio, Texas, April 29–May 2, 2013.

50. Buckley, P. S., W. L. Luyben, and J. P. Shunta, "*Design of Distillation Column Control Systems*," Instrument Society of America, Research Triangle Park, North Carolina, 1985.

51. Bu-Naiyan, A., and K. Kamali, "Troubleshooting Debutanizer at Ras Tanura Refinery's Hydrocracking Unit," in *Asian Refining Technology Conference (ARTC) 6th Annual Meeting*, Singapore, 2003.

52. Brown Burns, J., K. Becht, and B. Mueller, "Troubleshooting Crude Tower Constraints," *PTQ*, Q3, p. 15, 2017.

53. Cai, T. J., "Column Performance Testing Procedures," Chapter 3 in A. Gorak and H. Schoenmakers (ed.), *Distillation Operation and Applications*, p. 103, Elsevier, 2014.

54. Cai, T. J., and M. R. Resetarits, "Pressure Drop Measurements on Distillation Columns (in English)," *Chin. J. Chem. Eng.*, 19(5), p. 1, 2011.

55. Cai, T. J., G. X. Chen, C. W. Fitz, and J. G. Kunesh, "Effect of Bed Length and Vapor Maldistribution on Structured Packing Performance," *Chem. Eng. Res. Des.*, 81, Part A, p. 85, January 2003

56. Cantley, G. A., "Inspection War Stories – Part 1," Presented at the *Distillation Topical Conference*, AIChE Spring National Meeting, Houston, Texas, April 1–5, 2012.

57. Cantley, G. A., "Coking Mechanism in Refinery Fractionators Wash Beds," in *Kister Distillation Symposium, Topical Conference Proceedings*, p. 96, AIChE Spring Meeting, Austin, Texas, 26–30 April, 2015.

58. Cantley, G. A., and L. Pless, "Managing Wash Beds," *Hydrocarbon Eng.*, April 2017.

59. Cardoso, R., "Aqueous Chemistry Principles Applied to Refinery Processes," *Hydrocarbon Proc.*, 97, p. 55, April 2018.

60. Cardoso, R., and H. Z. Kister, "Refinery Tower Inspections: Discovering Problems and Preventing Malfunctions," in *Distillation 2013, Kister Distillation Symposium*, p. 91, AIChE Spring Meeting, San Antonio, Texas, April 28–May 2, 2013.

61. Cardoso, R., and H. Z. Kister, "Refinery Tower Inspections: Discovering Problems and Preventing Malfunctions," *Hydrocarbon Proc.*, 99, p. 27, May 2020.

62. Chabris, C., and D. Simons, *"The Invisible Gorilla and Other Ways Our Intuitions Deceive Us,"* Random House, New York, 2010.

63. Chakraborty, A., S. Mokhatab, and W. A. Poe, "Solving Gas Plant Operational Problems," *PTQ*, Q1, p. 111, 2007.

64. Chambers, S., L. Pless, R. Carlson, and M. Schultes, "Gamma Scan and CAT-Scan Data from a Distillation Column at Various Loadings," in *The Kister Distillation Symposium, Topical Conference Proceedings*, p. 27, the AIChE National Spring Meeting, San Antonio, Texas, April 29–May 2, 2013.

65. Chambers, S., L. Pless, and R. Carlson, "TRU-SCAN^R and TRU-CAT^{TM} Scanning of Structured Packing Operating in Deep Vacuum," in "Distillation & Absorption 2014," *Proceedings of the 10th International Conference on Distillation and Absorption*, p. 995, EFCE Event No. 705, Friedrichshafen, Germany, September 14–17, 2014.

66. Charlton, J. S. (ed.), *"Radioisotope Techniques for Problem Solving in Industrial Process Plants,"* Gulf Publishing, Houston, Texas, 1986.

67. Charlton, J. S., and M. Polarski, "Radioisotope Techniques Solve CPI Problems," *Chem. Eng.*, p. 125, January 24, 1983, and p. 93, February 21, 1983.

68. Chen, G. K., "Packed Column Internals," *Chem. Eng.*, 91, p. 40, March 5, 1984.

69. Chen, G. K., "Troubleshooting Distribution Problems in Packed Columns," *The Chem. Engineer (Suppl.) (London)*, p. 10, September 1987.

70. Chen, G. K., and K. T. Chuang, "Recent Developments in Distillation," *Hydrocarbon Proc.*, 68(2), p. 37, 1989.

71. Chen, G. K., T. L. Holmes, and J. H. Shieh, "Effects of Subcooled or Flashing Feed on Packed Column Performance," *IChemE Symp. Ser.*, 94, p. 185, 1985.

72. Clark, D., and S. Golden, "Improving FCCU's: Bottoms System Upgrades," *PTQ*, p. 65, Summer 2003.

73. Colwell, C. J., "Clear Liquid Height and Froth Density on Sieve Trays," *Ind. Eng. Chem. Process Des. Dev.*, 20(2), p. 298, 1981.

74. Colwell, C. J., and J. T. O'Bara, "Estimate Sieve Tray efficiency in the Weeping Region," Paper presented at the *AIChE Spring National Meeting*, Houston, Texas, April 1989.

75. Costanzo, S., S. M. Wong, and M. Pilling, "Improve Vacuum Tower Revamp Projects," *Hydrocarbon Proc.*, 89, p. 81, September 2010.

76. Cox, K. R., R. N. French, and G. J. Koplos, "Limiting Lemmas and Lemons: Some Common Pitfalls of Modeling Gibbs Excess Energy Data," in *Distillation Tools for the Practicing Engineer, Topical Conference Proceedings*, p. 212, AIChE Spring National Meeting, New Orleans, Louisiana, March 10–14, 2002.

77. Cunha, J. A. de C., and P. R. de M. Freitas, "Acid Gases Absorber Improvements in Copene's Ethylene Plant," Paper presented at the *AIChE Spring National Meeting*, Houston, Texas, March 9–13, 1997.

78. Curry, R. N., *"Fundamentals of Natural Gas Conditioning,"* PennWell, Tulsa, Oklahoma, 1981.

79. Custer, R. S., "Case Histories of Distillation Columns," *Chem. Eng. Prog.*, 61(9), 1965, p. 89.

80. Davies, J. A., "Bubble Trays – Design and Layout," *Pet. Refin.*, 29(8), p. 93, 1950, and 29(9), p. 121, 1950.

81. De Souza, L. L. G., A. L. R. S. M. Smiderle, I. S. Militão, and M. de Alcântara, "Modeling Ethane Absorption in MEA Solution," *Gas*, p. 29, 2018.

82. Deibele, L., and H. W. Brandt, "Fehlerbetrachtung bei der Messung der Theoretischen Bodenzahl von Destillations-kolonnen," *Chem. Ing. Tech.*, 57(5), p. 439, 1985.

83. Dhabalia, D., and M. Pilling, "Distributor Design and Testing," *Revamps*, p. 9, 2006.

84. Diaz, A., R. Husfeld, and C. Vaughan, "Case Study: Utilizing Process Modeling to Optimize a Batch Column Retrofit," *Paper 124c, Topical Conference Proceedings, Kister Distillation Symposium*, AIChE Spring Meeting, Houston, Texas, March 12–16, 2023.

85. Diehl, J. E., and C. R. Koppany, "Flooding Velocity Correlation for Gas-Liquid Counterflow in Vertical Tubes," *Chem. Eng. Prog. Symp. Ser.*, 65(92), p. 77, 1969.

86. Doyle, W. H., "Industrial Explosions and Insurance," *Loss Prevent.*, 3, p. 11, 1969.

87. Doyle, W. H., "Instrument-Connected Losses in the CPI," *Instrum. Technol.*, 19(10), p. 38, 1972.

88. Drew, J. W., "Distillation Column Startup," *Chem. Eng.*, 90, p. 221, November 14, 1983.

89. Duarte Pinto, R., M. Perez, and H. Z. Kister, "Combine Temperature Surveys, Field Tests and Gamma Scans for Effective Troubleshooting," *Hydrocarbon Process.*, 82, p. 69, April 2003.

90. Dumas, B., "*Monitoring of Atmospheric Crude Column Detects Fouling Build-up*," Tracerco Insight, Volume 6, 3rd ed., p. 2, 2016.

91. Dzyacky, G., and S. Carlson, "Improve Column Performance: Operate Closer to the Hydraulic Limit without Flooding," in *Distillation 2009: Proceedings of Topical Conference*, p. 369, AIChE National Spring Meeting, Tampa, Florida, April 26–30, 2009.

92. Dzyacky, G. E., A. F. Siebert, and J. C. Lewis, "Increasing Distillation Column Throughput," *PTQ*, Q4, p. 99, 2005.

93. Eagle, R. S., "Trouble-shooting with Gamma Radiation," *Chem. Eng. Prog.*, 60(10), p. 69, 1964.

94. Eckert, J. S., "Selecting the Proper Distillation Column Packing," *Chem. Eng. Prog.*, 66 (3), p. 39, 1970.

95. Ellingsen, W. R., "Diagnosing and Preventing Tray Damage in Distillation Columns," *DYCORD 86, IFAC Proceedings of International Symposium on Dynamics and Control of Chemical Reactors and Distillation Columns*, Bournemouth, U.K., December 8–10, 1986.

96. Emiliano, J. B. F., Quadro, E., H. Z. Kister, and D. R. Summers, "War Strategies to Achieve Peace in a Primary Fractionator," *Proceedings of the 22nd Annual Ethylene Producers Conference*, San Antonio, Texas, March 2010.

97. Ender, C., and D. Laird, "Minimize the Risk of Fire During Distillation Column Maintenance," *Chem. Eng. Prog.*, 99, p. 54, September 2003.

98. Engel, D., "Manage Contaminants in Amine Treating Units – Part 2: Rich Amine Filtration and Foaming," *Hydrocarbon Process.*, 97, p. 41, July 2018.

99. Engel, V., "How to Design and Optimize (1) Sieve Trays, (3) Float Valve Trays (4) Fixed Valve Trays" *Eng. Pract.*, (1), p. 4, January 2020; (3), p. 7, July 2020; (4), p. 27, October 2020.

100. Engel, D., and M. Sheilan, "The 'Seven Deadly Sins' of Filtration and Separation Systems," *Gas Process.*, p. 35, March/April 2016.

101. Ferguson, D., "Radioisotope Techniques for Troubleshooting Olefins Plants," in *7th Annual Ethylene Producers Conference Proceedings*, AIChE, 1995.

102. Ferguson, D., and S. Xu, "Improving Gas Flow Measurement," *PTQ*, Q1, p. 47, 2006.

103. Fitz, C. W., D. W. King, and J. G. Kunesh, "Controlled Liquid Maldistribution Studies on Structured Packings," *Trans. IChemE*, 77, Part A, p. 482, 1999

104. Fleming, B., G. R. Martin, and E. L. Hartman, "Pay Attention to Reflux/Feed Entry Design," *Chem. Eng. Prog.*, 92, p. 56, January 1996.

105. FLIR Systems, "FLIR 600 Series User Manual," FLIR, Wilsonville, Oregon.

106. Fractionation Research Inc., "The Performance of Trays with Downcomers," Movie A, revised Edition, 2009, available to the public from FRI, Stillwater, Oklahoma, www.fri.org.

107. Fractionation Research Inc., "Spray Collapse Study," Motion Picture 919, Stillwater, Oklahoma (available to the public www.fri.org.).

108. Fractionation Research Inc., Design Practices Committee, "Causes and Prevention of Packing Fires," *Chem. Eng.*, 114, p. 34, July 2007.

109. Fractionation Research Inc., Design Practices Committee, "Reboiler Circuits for Trayed Columns," *Chem. Eng.*, 118, p. 26, January 2011.

110. Fradgley, D., (Tracerco, UK), Private Communication, November 1994.

111. France, J. J., "Avoiding Pitfalls in Distillation System Revamps," Paper presented at the *AIChE Spring Meeting*, Houston, Texas, March 1993.

112. France, J. J., "Troubleshooting Distillation Columns," Paper presented at the *AIChE Spring Meeting*, Houston, Texas, March 1993.

113. Freeman, L., and J. D. Bowman, "Use of Column Scanning to Troubleshoot Demethanizer Operation," Paper presented at the *AIChE Spring National Meeting*, Houston, Texas, March 31, 1993.

114. Fulham, M. J., and V. G. Hulbert, "Gamma Scanning of Large Towers," *Chem. Eng. Prog.*, 71 (6), p. 73, June 1975.

115. Gamble, C., "Maintaining Heat Transfer Fluid Quality," *Chem. Eng.*, 125, p. 32, July 2018.

116. Gans, M., S. A. Kiorpes, and F. A. Fitzgerald, "Plant Startup – Step by Step," *Chem. Eng.*, 90, p. 74, October 3, 1983.

117. Gans, M., D. Kohan, and B. Palmer, "Systematize Troubleshooting Techniques," *Chem. Eng. Prog.*, 87, p. 25, April 1991.

118. Garcia, J. A., B. Ballance, and J. Hatfield, "Debottlenecking of a Large Industrial Deethanizer by Gravity Driven Two-Phase Flow Analysis," Paper presented at the *AIChE Spring National Meeting*, Orlando, Florida, April 23rd, 2018.

119. Garvin, R. G., and E. R. Norton, "Sieve Tray Performance under GS Process Conditions," *Chem. Eng. Prog.*, 64(3), 1968, p. 99.

120. Geipel, C., "Troubleshooting a Methanol Recovery Unit," *Paper 156a, Topical Conference Proceedings, Kister Distillation Symposium*, AIChE Spring Meeting, Houston, Texas, March 12–16, 2023.

121. Gibson G. J., "Efficient Test Runs," *Chem. Eng.*, 94, p. 75, May 11, 1987.

122. Glausser, W. E., "Foaming in a Natural Gasoline Absorber," *Chem. Eng. Prog.*, 60(10), p. 67, 1964.

123. Golden, S. W., Glitsch at the time, now PCS, Private communication, August 1987.

124. Golden, S. W., "Revamping FCC's – Process and Reliability," *Pet. Tech Quarterly*, p. 85, Summer 1996.

125. Golden, S. W., "Case Studies Reveal Common Design, Equipment Errors in Revamps," *Oil Gas J.*, 95, p. 62, April 7, 1997 and April 14, 1997.

126. Golden, S. W., "Pushing Plant Limits, Test Runs, Plant Expectations, and Performance Confidence," *World Refin.*, p. 1, March/April 1999.

127. Golden, S. W., and A. W. Sloley, "Simple Methods Solve Vacuum Column Problems Using Plant Data," *Oil Gas J.*, 90, p. 74, September 14, 1992.

128. Golden, S. W., A. W. Sloley, and B. Fleming, "Refinery Vacuum Column Troubleshooting," Paper presented at the *AIChE Spring National Meeting*, Houston, TX, March 31, 1993.

129. Golden, S. W., G. R. Martin, and K. D. Schmidt, "Field Data, New Design Correct Faulty FCC Tower Revamp," *Oil Gas J.*, 91, p. 54, May 31, 1993.

130. Golden, S. W., N. P. Lieberman, and E. T. Lieberman, "Troubleshooting Vacuum Columns with Low-Capital Methods," *Hydrocarbon Process.*, 72, p. 81, July 1993.

131. Golden, S. W., D. C. Villalanti, and G. R. Martin, "Feed Characterization and Deepcut Vacuum Columns: Simulation and Design," Paper presented at the *AIChE Spring National Meeting*, Atlanta, Georgia, April 18–20, 1994.

132. Golden, S. W., N. P. Lieberman, and G. R. Martin, "Correcting Design Errors Can Prevent Coking in Main Fractionators," *Oil Gas J.*, 92, p. 72, November 21, 1994.

133. Golden, S. W., J. Moore, and J. Nigg, "Optimize Revamp Projects with a Logic-Based Approach," *Hydrocarbon Process.*, 82, p. 75, September 2003.

134. Golden, S. W., D. W. Hanson, J. Hansen, and M. Brown, "Revamp Improved FCC Performance at BP's Texas City Refinery," *Oil Gas J.*, 102, March 15, 2004.

135. Golden, S., T. Barletta, and S. White, "Vacuum Unit Performance," *Sour Heavy*, p. 11, 2012.

136. González, R., and J. M. Ferrer, "Analyzing the Value of First-Principles Dynamic Simulation," *Hydrocarbon Process.*, 85, September, p. 69, 2006.

137. Gorrochotegui, E., and J. Portillo, "Natural Gas Cryogenic Demethanizer Troubleshooting: An Internal Look at the Column Operation," Paper presented at the *Distillation Symposium*, AIChE Spring Meeting, San Antonio, Texas, April 2022.

138. Goyal, O. P., "Guidelines Aid Troubleshooting," *Hydrocarbon Process.*, 79, p. 69, January 2000.

139. Graham, G., P. Pednbinkekar, and D. Bunning, "Distillation Part 1: Experimental Validation of Column Simulations," *Chem. Eng.*, 125, p. 30, February 2018.

140. Grave, E., and P. Tanaka, "The Final Step to Success – Tower Internals Inspection," *Part 1 and Part 2*, in *Proceedings of at the Topical Conference on Distillation*, p. 533 and 547, the AIChE National Spring Meeting, Houston, Texas, April 22–26, 2007.

141. Grave, E., N. Yeh, and J. C. Juarez, "Troubleshooting Tower Carryover," in *Kister Distillation Symposium, Topical Conference Proceedings*, p. 155, AIChE Spring Meeting, San Antonio, Texas, March 26–30, 2017.

142. Groberichter, D., and J. Stichlmair, "Crystallization Fouling in Packed Columns," *Chem. Eng. Res. Des.*, 81, Part A, p. 68, January 2003.

143. Grover, B. S., and E. S. Holmes, "The Benfield Process for High Efficiency and Reliability in Ammonia Plant Acid Gas Removal – Four Case Studies," in *Nitrogen 1986*, The British Sulphur Corp. Ltd., Amsterdam, p. 101, April 20–23, 1986.

144. Guarda, C. F., E. Lieberman, and N. Lieberman, "The Lost Art of Tower Internal Inspection: A Practical Guide for the New Generation," in *Kister Distillation Symposium, Topical Conference Proceedings*, p. 64, AIChE Spring Meeting, Virtual, April 18–22, 2021.

145. Hanson, D., "Details Matter When Diagnosing Hot Vapor Bypasses," in *Kister Distillation Symposium, Topical Conference Proceedings*, p. 402, AIChE Spring Meeting, San Antonio, Texas, March 26–30, 2017.

146. Hanson, D., and I. Buttridge, "Revamp, Troubleshooting Optimize NGL Depropanizer Operations," *Oil Gas J.*, 101, p. 88, August 25, 2003.

147. Hanson, D. W., and E. L. Hartman, "Process Design Pitfalls Hinder Successful Atmospheric Tower Revamp," Paper presented at the *AIChE Spring National Meeting*, Houston, Texas, March 9–12, 1997.

148. Hanson, D. W., and S. H. Lee, "Reducing FCC Main Fractionator Operating Risks," *PTQ*, Q1, p. 19, 2021.

149. Hanson, D. W., and D. Lin, "Crude Unit Optimization," *Hydrocarbon Eng.*, p. 2, July/August 1998.

150. Hanson, D., and K. Subramanian, "Side-Reboiler Redesign Improves Alkylation Isostripper Operation," Paper presented at the *AIChE Spring Distillation Conference*, San Antonio, Texas, April 29-May 1, 2013.

151. Hanson, D., and P. Williams, "Innovative Stripper Revamp Improved Hydrocracker Operation," in *Distillation 2011: The Dr. James Fair Heritage Distillation Symposium, Topical Conference Proceedings*, p. 87, AIChE Spring Meeting, Chicago, Illinois, March 13–17, 2011.

152. Hanson, D. W., and C. Winfield, "Advanced Scan Techniques Aide in Column Diagnostics," in *Kister Distillation Symposium, Topical Conference Proceedings*, p. 27, AIChE Spring Meeting, Virtual, April 18–22, 2021.

153. Hanson, D. W., S. W. Golden, and G. R. Martin, "Make the Most of Distillation Test Runs," *Chem. Eng. Prog.*, 92, p. 72, February 1996.

154. Hanson, D. W., E. L. Hartman, and S. Costanzo, "Modifying Crude Units to Deepcut Operation," *Hydrocarbon Eng.*, p. 2, January-February 1997.

155. Hanson, D. W., N. P. Lieberman, and E. T. Lieberman, "De-entrainment and Washing of Flash Zone Vapors in Heavy Oil Fractionators," *Hydrocarbon Process.*, 78, p. 55, July 1999.

156. Hanson, D. W., T. Barletta, and J. Bernickas, "An Atmospheric Crude Tower Revamp," *PTQ*, Q3, p. 61, 2005.

157. Hanson, D. W., J. Brown Burns, and M. Teders, "How Does a Collector Tray Leak Impact Column Operation?" in *Kister Distillation Symposium, Topical Conference Proceedings*, p. 105, AIChE Spring Meeting, Austin, Texas, April 27–30, 2015.

158. Hanson, D. W., J. Brown Burns, and M. Teders, "FCC Collector Tray Leaks Impact Column Operation," *Hydrocarbon Process.*, 96, p. 37, September 2017.

159. Harper, G., "Clamp-on Flow Meter Gains Firm Hold," *Chem. Proc.*, 73, p. 42, March 2011.

160. Harrison, M. E., "Gamma Scan Evaluation for Distillation Column Debottlenecking," *Chem. Eng. Prog.*, 86, p. 37, March 1990.

161. Harrison, M. E., and J. J. France, "Distillation Column Troubleshooting," *Chem. Eng.*, 96, p. 116, March 1989; p. 121, April 1989; p. 126, May 1989; and p. 139, June 1989.

162. Hartman, E. L., "New Millennium, Old Problems: Vapor Cross Flow Channeling on Valve Trays," in *Distillation 2001: Frontiers in a New Millennium, Proceedings of Topical Conference*, p. 108, AIChE Spring National Meeting, Houston, Texas, April 22–26, 2001.

163. Hartman, E. L., and T. Barletta, "Reboiler and Condenser Operating Problems," *PTQ*, p. 47, Summer 2003.

164. Hartman, E. L., and E. Menzes, "Successful Column Revamps Require Careful Design," *Oil Gas J.*, 98, p. 42, January 24, 2000.

165. Hartman, E., and B. White, "Maximizing Stripping Section Performance," *PTQ*, Q3, p. 55, 2021.

166. Hasbrouck, J. F., J. G. Kunesh, and V. C. Smith, "Successfully Troubleshoot Distillation Columns," *Chem. Eng. Prog.*, 89, p. 63, March 1993.

167. Hausch, D. C., "How Flooding Can Affect Tower Operation," *Chem. Eng. Prog.*, 60(10), 1964, p. 55.

168. Helling, R. K., and M. A. Des-Jardin, "Get the Best Performance from Structured Packing," *Chem. Eng. Prog.*, 90, p. 62, October 1994.

169. Hennigan, S., "Full Scale Plant Column Efficiency – Measurement and Management," in *Distillation 2011: The Dr. Jim Fair Heritage Distillation Symposium, Topical Conference Proceedings*, p. 151, AIChE Spring National meeting, Chicago, Illinois, March 13–17, 2011.

170. Hesselink, W. H., and A. Van Huuksloot, "Foaming of Amine Solutions," *IChemE Symp. Ser.*, 94, p. 193, 1985.

171. Hiller, C., R. Meier, G. Nigglemann, and R. Rix, "Distillation in Special Chemistry," *Chem. Eng. Res. Des.*, 99, p. 220, 2015.

172. Holden, B. S., P. H. Au-Yeung, and T. W. Kajdan, "Watch Out for Trapped Components in Towers," *Chem. Proc.*, 74, p. 38, May 2012.

173. Horner, G., "How to Select Internals for Packed Columns," *Proc. Eng.*, p. 79, May 1985.

174. Horner, G. V., "Tips for Installing Structured Packings and Internals," *The Chem. Engineer (Suppl.) London*, p. 8, September 1987.

175. Horwitz, B. A., "Don't Let Startup or Debugging Problems Bug You," *Chem. Eng. Prog.*, 90, p. 62, November 1994.

176. Horwitz, B. A., "Hardware, Software, Nowhere," *Chem. Eng. Prog.*, 94, p. 69, September 1998.

177. Hower, T. C. Jr., and H. Z. Kister, "Solve Column Process Problems, Part 1," *Hydrocarbon Process.*, 70, p. 89, May 1991.

178. Hsiao, H. H., C. T. Hong, and H. C. Shen, "Troubleshoot of Column Containing Structured Packing," Paper presented at the *AIChE Spring National Meeting*, San Antonio, Texas, March 24, 2010.

179. Hsieh, C. L., and K. J. McNulty, "Predict Weeping of Sieve and Valve Trays," *Chem. Eng. Prog.*, 89, p. 71, July 1993.

180. Hussman, D., G. Bruce, A. Abufara, K. Chandi, and H. Z. Kister, "Diagnosing a Premature Flood Near the Diesel Draw of an Atmospheric Crude Tower," *Paper 25e, Topical Conference Proceedings, Kister Distillation Symposium*, AIChE Spring Meeting, Houston, Texas, March 12–16, 2023.

181. Iyengar, J. N., P. W. Sibal, and D. S. Clarke, "Operations and Recovery Improvement Via Heavy Hydrocarbon Extraction," in *48th Lawrence Reid Gas Conditioning Conference*, p. 161, Norman, Oklahoma, March 1–4, 1998.

182. Jacobs, J. K., "Reboiler Selection Simplified," *Hydrocarbon Process. Pet. Refin.*, 40(7), p. 190, 1961.

183. Jain, A. D., "Avoid Stress Corrosion Cracking of Stainless Steel," *Hydrocarbon Process.*, 91, p. 39, March 2012.

184. Jarvis, H. C., "Butadiene Explosion at Texas City," *Loss Prevent.*, 5, p. 57, 1971; R. H. Freeman and M. P. McCready, *ibid.*, p. 61; R. G. Keister, B. I. Pesetsky, and S. W. Clark, *ibid.*, p. 67.

185. Jones, D. W., and J. B. Jones, "Tray Performance Evaluation," *Chem. Eng. Prog.*, 71(6), p. 65, 1975.

186. Kabakov, M. I., and A. M. Rozen, "Hydrodynamic Inhomogeneities in Large-Diameter Packed Columns and Ways to Eliminate Them," *Khim. Prom.*, (8), p. 496, 1984; *Sov. Chem. Ind.*, 16(8), p. 1059, 1984.

187. Kelley, R. E., T. W. Pickel, and G. W. Wilson, "How to Test Fractionators," *Pet. Refin.*, 34(1), p. 110, 1955, and 34(2), p. 159, 1955.

188. Killat, G. R., and D. Perry, "High Performance Distributor for Low Liquid Rates," Paper presented at the *AIChE Annual Meeting*, Los Angeles, California, November 17–22, 1991.

189. Kirmse, B., K. Krase, and D. Ferguson, "Gamma Scanning Fractionators," *PTQ*, Q1, p. 69, 2007.

190. Kister, H. Z., "When Tower Startup Has Problems," *Hydrocarbon Process.*, 58(2), p. 89, 1979.

191. Kister, H. Z., "Inspection Assures Trouble-Free Operation," *Chem. Eng.*, p. 107, February 9, 1981.

192. Kister, H. Z., "*Distillation Operation*," McGraw-Hill, New York, 1990.

193. Kister, H. Z., "*Distillation Design*," McGraw-Hill, New York, 1992.

194. Kister, H. Z., "Can Vapor Cross Flow Channeling Occur on Valve Trays?," *The Chem. Engineer (London)*, 544, p. 18, June 24, 1993.

195. Kister, H. Z., "Troubleshooting Distillation Simulations," *Chem. Eng. Prog.*, 91, p. 63, June 1995.

196. Kister, H. Z., "Are Column Malfunctions Becoming Extinct—or Will They Persist in the 21st Century," *Trans. IChemE*, 75, Part A, p. 563, September 1997.

197. Kister, H. Z., "Can We Believe the Simulation Results," *Chem. Eng. Prog.*, 98, p. 52, October 2002.

198. Kister, H. Z., "What Caused Tower Malfunctions in the Last 50 Years?," *Trans. IChemE*, 81, Part A, p. 5, January 2003.

199. Kister, H. Z., "Component Trapping in Distillation Towers: Causes, Symptoms, and Cures," *Chem. Eng. Prog.*, 100, p. 22, August 2004.

200. Kister, H. Z., "Distillation: Introducing Reboiler Return," *Chem. Eng. Prog.*, 102, p. 16, March 2006.

201. Kister, H. Z., "*Distillation Troubleshooting*," John Wiley and Sons, New Jersey, 2006.

202. Kister, H. Z., "Effect of Design on Tray Efficiency in Commercial Columns," *Chem. Eng. Prog.*, 104, p. 39, June 2008.

203. Kister, H. Z., "Is the Hydraulic Gradient on Sieve and Valve Trays Negligible?" Paper presented at the *Topical Conference on Distillation*, the AIChE Meeting, Houston, Texas, April 2012.

204. Kister, H. Z., "Use Quantitative Gamma Scans to Troubleshoot Maldistribution on Trays," *Chem. Eng. Prog.*, 109, p. 33, February 2013.

205. Kister, H. Z., "Apply Quantitative Gamma Scanning to High-Capacity Trays," *Chem. Eng. Prog.*, 109, p. 45, April 2013.

206. Kister, H. Z., "Flooded Condenser Controls: Principles and Troubleshooting," *Chem. Eng.*, 123, p. 37, January 2016.

207. Kister, H. Z., "Reboiler Circuit Debottleneck with No Hardware Changes," *Chem. Eng.*, 124, p. 30, August 2017.

208. Kister, H. Z., "Controlling Reboilers Heated by Condensing Steam or Vapor," *Chem. Eng.*, 127, p. 22, July 2020.

209. Kister, H. Z., "Light Tales from the Tower Doctor's Casebook," in *Kister Distillation Symposium, Topical Conference Proceedings*, p. 683, AIChE Spring Meeting, Virtual, April 18–22, 2021.

210. Kister, H. Z., "Gas Trapping Can Unsettle Towers," *Chem. Eng.*, 129, April, p. 47, 2022.

211. Kister, H. Z., "Tower Doctor: A Quick Diagnosis Makes a Short Assignment," *Chem. Eng.*, 129, p. 37, July 2022.

212. Kister, H. Z., "Tower Doctor: Downcomer Unsealing: A Current Diagnosis Brings a Correct Cure," *Chem. Eng.*, 129, p. 32, September 2022.

213. Kister, H. Z., "Tower Doctor: Should the Doctor believe a Flowmeter?," *Chem. Eng.*, 129, p. 29, May 2022.

214. Kister, H. Z., "Tower Doctor: X-Raying The Patient: Gamma Scanning Vapor Cross-Flow Channeling," *Chem. Eng.*, 129, p. 24, August 2022.

215. Kister, H. Z., "Practical Distillation Technology," continuing education course, sponsored by *IChemE*, 2023.

216. Kister, H. Z., "Light Tales from the Tower Doctor's Casebook II," *Paper 157f, Topical Conference Proceedings, Kister Distillation Symposium*, AIChE Spring Meeting, Houston, Texas, March 12–16, 2023.

217. Kister, H. Z., and M. A. Chaves, "Kettle Troubleshooting," *Chem. Eng.*, 117, p. 26, February 2010.

218. Kister, H. Z., and S. Chen, "Solving a Tower's Salt Plugging Problem," *Chem. Eng.*, 107, p. 129, August 2000.

219. Kister, H. Z., and I. D. Doig, "Distillation Pressure Ups Throughput," *Hydrocarbon Process.*, 56, p. 132, July 1977.

220. Kister, H. Z., and D. R. Gill, "Flooding and Pressure Drop in Structured Packings," *Distillation and Absorption 1992, IChemE Symposium Series 128*, p. A109, IChemE/EFCE, 1992.

221. Kister, H. Z., and D. W. Hanson, "Control Column Pressure via Hot Vapor Bypass," *Chem. Eng. Prog.*, 111, p. 35, February 2015.

222. Kister, H. Z., and T. C. Hower, Jr., "Unusual Case Histories of Gas Processing and Olefins Plant Columns," *Plant/Oper. Prog.*, 6(3), p. 151, 1987.

223. Kister, H. Z., and R. Kumar, "Premature Foam Flood in an Amine Absorber Parts 1 and 2," *PTQ*, Q3, p. 13, and Q4, p. 16, 2021.

224. Kister, H. Z., and N. M. Mohamed, "Troubleshoot and Solve a Gas Treater Downcomer Unsealing Problem: Part 2," *Gas Process.*, p. 49, March/April 2015.

225. Kister, H. Z., and N. M. Mohamed, "Troubleshoot and Solve a Gas Treater Downcomer Unsealing Problem: Part 1," *Gas Process.*, p. 39, January/February 2015.

226. Kister, H. Z., and M. Olsson, "An Investigation of Premature Flooding in a Distillation Column," *Chem. Eng.*, 126, p. 29, January 2019.

227. Kister, H. Z., and C. Winfield, "Use Downcomer Gamma Scans to Troubleshoot Your Process," *Chem. Eng. Prog.*, 113, p. 28, January 2017.

228. Kister, H. Z., K. F. Larson, and P. E. Madsen, "Vapor Cross Flow Channeling on Sieve Trays: Fact or Myth?," *Chem. Eng. Prog.*, 88(11), p. 86, November 1992.

229. Kister, H. Z., S. G. Chellappan, and C. E. Spivey, "Debottleneck and Performance of a Packed Demethanizer," *Proceedings of the 4th Ethylene Producers Conference*, New Orleans, Louisiana, p. 283, 1992.

230. Kister, H. Z., H. Pathak, M. Korst, D. Strangmeier, and R. Carlson, "Troubleshoot Reboilers by Neutron Backscatter," *Chem. Eng.*, 102, p. 145, September 1995.

231. Kister, H. Z., K. F. Larson, J. M. Burke, R. J. Callejas, and F. Dunbar, "Troubleshooting a Water Quench Tower," *7th Annual Ethylene Producers Conference*, AIChE, Houston, Texas, March 20, 1995.

232. Kister, H. Z., R. Rhoad, and K. A. Hoyt, "Improve Vacuum-Tower Performance," *Chem. Eng. Prog.*, 92, p. 36, September 1996.

233. Kister, H. Z., T. C. Hower Jr., P. R. de M. Freitas, and J. Nery, "Problems and Solutions in Demethanizers with Interreboilers," *Proceedings of the 8th Ethylene Producers Conference*, New Orleans, Louisiana, 1996.

234. Kister, H. Z., S. Bello Neves, R. C. Siles, and R. da Costa Lima, "Does Your Distillation Simulation Reflect the Real World?," *Hydrocarbon Process.*, 76, p. 103, August 1997.

235. Kister, H. Z., E. Brown, and K. Sorensen, "Sensitivity Analysis Key to Successful DC_5 Simulation," *Hydrocarbon Process.*, 77, p. 124, October 1998.

236. Kister, H. Z., B. Blum, and T. Rosenzweig, "Troubleshoot Chimney Trays Effectively," *Hydrocarbon Process.*, 80, p. 101, April 2001.

237. Kister, H. Z., D. W. Hanson, and T. Morrison, "California Refiner Identifies Crude Tower Instability Using Root Cause Analysis," *Oil Gas J.*, 100, p. 42, February 18, 2002.

238. Kister, H. Z., D. E. Grich, and R. Yeley, "Better Feed Entry Ups Debutanizer Capacity," *PTQ Revamps and Operation*, p. 31, 2003.

239. Kister, H. Z., W. J. Stupin, and S. Stupin, "Pack up Your Troubles," *TCE*, p. 44, December 2006/January 2007.

240. Kister, H. Z., W. J. Stupin, J. E. O. Lenferink, and S. W. Stupin, "Troubleshooting a Packing Maldistribution Upset," *Trans. IChemE*, 85, Part A, p. 136, January 2007.

241. Kister, H. Z., R. W. Dionne, W. J. Stupin, and M. Olsson, "Preventing Maldistribution in Four-Pass Trays," *Chem. Eng. Prog.*, 106, p. 32, April 2010

242. Kister, H. Z., B. Clancy-Jundt, and R. Miller, "Troubleshooting a C3 Splitter Tower Part 1: Evaluation," *PTQ*, Q4, p. 97, 2014.
243. Kister, H. Z., N. O'Shea, and D. Cronin, "Loss into Gain in High Capacity Trays, Part 1: Excursion," *PTQ*, Q2, p. 43, 2016.
244. Kister, H. Z., N. O'Shea, and D. Cronin, "Loss into Gain in High Capacity Trays, Part 2: Reverse Vapour Cross Flow Channeling," *PTQ*, Q3, p. 27, 2016.
245. Kister, H. Z., P. M. Mathias, D. E. Steinmeyer, W. R. Penney, V. S. Monical, and J. R. Fair, "Equipment for Distillation, Gas Absorption, Phase Dispersion, and Phase Separation," Section 14, in R. H. Perry and D. Green (ed.), *Chemical Engineers' Handbook*," 9th ed., McGraw-Hill, New York, 2018.
246. Kister, H. Z., C. Trompiz, and B. Clancy-Jundt, "Troubleshooting Instability in a Debutanizer Tower," *Revamps*, p. 11, 2019.
247. Kister, H. Z., G. Jacobs, and A. A. Kister, "Gaining Insight into Parting Box Hydraulics," in *Proceedings of the Kister Distillation Symposium, Topical Conference Proceedings*, p. 463, AIChE Spring Meeting, New Orleans, Louisiana, March 31–April 4, 2019.
248. Kister, H. Z., G. Jacobs, and A. A. Kister, "Parting Box Can Make or Break Packed Tower Performance Part1: Perforated Sparger Feeds," *Chem. Eng. Prog.*, 117, p. 22, May 2021.
249. Kister, H. Z., G. Jacobs, and A. A. Kister, "Parting Box Can Make or Break Packed Tower Performance Part 2: Spargers with Dip Tubes," *Chem. Eng. Prog.*, 117, p. 53, July 2021.
250. Kister, H. Z., G. Jacobs, and A. A. Kister, "Gas Entrainment by Liquid Jets and Cascades Can Unsettle Towers," *Chem. Eng. Res. Des.*, 191, p. 313, 2023.
251. Kister, H. Z., J. Li, S. Radovcich, and O. Wilding, "Troubleshooting and Evaluation of Air Condenser Limitation," *Paper 147f, Topical Conference Proceedings, Kister Distillation Symposium*, AIChE Spring Meeting, Houston, Texas, March 12–16, 2023.
252. Kister, H. Z., B. Clancy-Jundt, and R. Miller, "Troubleshooting a C3 Splitter Tower Part 2: Root Cause and Solution," *PTQ*, Q1, p. 39, 2015.
253. Kitterman, L., "Tower internals and accessories," *Congresso Brasileiro de Petroquimica*, Rio de Janeiro, November 8–12, 1976.
254. Kitterman, L., "Things I Have Seen," unpublished paper, 1988.
255. Kletz, T. A., "Fires and Explosions of Hydrocarbon Oxidation Plants," *Plant/Oper. Prog.*, 7(4), p. 226, 1988.
256. Kolmetz, K., and T. M. Zygula, "Guidelines for Ethylene Quench Towers," *Eng. Pract.*, p. 5, April 2018.
257. Kolmetz, K., M. Chua, R. Desai, J. Gray, and A. W. Sloley, "Staged Modifications Improve BTX Extractive Distillation Unit Capacity," *Oil Gas J.*, 101, p. 60, October 13, 2003.
258. Kolmetz, K., A. W. Sloley, T. M. Zygula, W. K. Ng, and P. W. Faesseler, "Design Guidelines for Distillation Columns in Fouling Service," *Proceedings of the 16th Ethylene Producers Conference*, New Orleans, Louisiana, 2004.
259. Kolmetz, K., W. K. Ng, P. W. Faessler, A. W. Sloley, and T. M. Zygula, "Case Studies Demonstrate Guidelines for Reducing Fouling in Distillation Columns," *Oil Gas J.*, 102, p. 46, August 23, 2004.
260. Kolmetz, K., W. K. Ng, P. W. Faessler, K. Senthil, Y. T. Lim, A. W. Sloley, and T. M. Zygula, "Optimize Process Operation with New Vacuum Distillation Methods," *Hydrocarbon Process.*, 84, p. 77, May 2005.
261. Kolmetz, K., J. C. Gentry, and J. N. Gray, "Guidelines for BTX Distillation Revamps," in *Distillation 2007: Proceedings of Topical Conference*, p. 59, AIChE National Spring Meeting, Houston, Texas, April 22–26, 2007.

262. Kooijman, H. A., and R. Taylor, "Distillation of Bulk Chemicals," Chapter 5, in A. Gorak and H. Schoenmakers (ed.), *Distillation Operation and Application*, p. 191, Academic Press, 2014.

263. Kouloheris, A. P., "Foam: Friend or Foe?," *Chem. Eng.*, 94, p. 88, October 26, 1987.

264. Krishnamoorthy, S., and L. M. Yang, "Troubleshooting EB/SM Splitters: How Can a Maldistribution Analysis Help?," in *Kister Distillation Symposium, Topical Conference Proceedings*, p. 43, AIChE Spring Meeting, San Antonio, Texas, March 26–30, 2017.

265. Krishnamoorthy, S., and T. M. Zygula, "Condensate Stripper Revamp to Mitigate Flooding," *Paper 62d, Topical Conference Proceedings, Kister Distillation Symposium*, AIChE Spring Meeting, Houston, Texas, March 12–16, 2023.

266. Kumar, S. M., A. Al-qahtani, I. Al-juhani, T. A. Al-seraihi, and H. Z. Kister, "Overcoming a Challenging C_2 Splitter Instability Problem," *Paper 62g, Topical Conference Proceedings, Kister Distillation Symposium*, AIChE Spring Meeting, Houston, Texas, March 12–16, 2023.

267. Kunesh, J. G., "Practical Tips on Tower Packing," *Chem. Eng.*, 94, p. 101, December 7, 1987.

268. Kunesh, J. G., L. Lahm, and T. Yanagi, "Commercial Scale Experiments that Provide Insight on Packed Tower Distributors," *Ind. Eng. Chem. Res.*, 26(9), 1845, 1987.

269. Kunesh, J. G., L. Lahm, and T. Yanagi, "Controlled Maldistribution Studies on Random Packing at a Commercial Scale," *IChemE Symp. Ser.*, 104, p. A233, 1987.

270. Kunz, R., "Simulation of Batch Distillation with Varying Alpha (α): the Ethanol Water and Methanol Water Systems at 1 ATM," in *Kister Distillation Symposium*, p. 96, AIChE Spring Meeting, Austin, Texas, April 26–30, 2015.

271. Laird, D., "Packed FCCU Main Fractionator Upgrades for Performance and Reliability," Paper presented at the *NPRA Annual Meeting*, San Antonio, Texas, March 21–23, 2004.

272. Le Grange, P., M. Sheilan, and B. Spooner, "How to limit Amine System Failures," *The Chem. Engineer*, p. 50, October 2017.

273. Le Grange, P., M. Sheilan, and B. Spooner, "Why Amine Systems Fail in Sour Service," *Sulphur*, 368, January-February, p. 42, 2017.

274. Le Grange, P., M. Sheilan, and B. Spooner, "Trends in Tragedy – An In-Depth Study of Amine System Failures," Paper presented in *Nitrogen + Syngas Conference*, The Hague, February 17–19, 2020.

275. Lee, S. H., "Enhancing Fractionator Efficiency," *PTQ*, Q2, p. 59, 2016,

276. Lee, S. H., "Optimizing Fouled Distillation Unit," *Paper 25f, Topical Conference Proceedings, Kister Distillation Symposium*, AIChE Spring Meeting, Houston, Texas, March 12–16, 2023.

277. Lee, S. H., and M. J. Binkley, "Optimize Design for Distillation Feed," *Hydrocarbon Process.*, 90, p. 101, June 2011.

278. Lee, S. H., and K. G. Min, "Improving Distillation Energy Network Using High Performance Distillation Equipment,"in *The Kister Distillation Symposium, Topical Conference Proceedings*," p. 637, the AIChE National Spring Meeting, San Antonio, Texas, April 29-May 2, 2013.

279. Lenfeld, P., and I. Buttridge, "High Costs of Quick Turnarounds and Erroneous Procedures," in *Distillation 2013, Kister Distillation Symposium*, p. 455, AIChE Spring Meeting, San Antonio, Texas, April 28-May 2, 2013.

280. Lenfeld, P., and I. Buttridge, "Plant Revamps and Turnarounds: Some Lessons Learned," *Chem. Eng.*, 120, p. 45, December 2013.

281. Lenfeld, P., and I. Buttridge, "Troubleshooting Tower Internals Foulings," Paper presented at the *Distillation Symposium*, AIChE Spring Meeting, San Antonio, Texas, April 10–14, 2022.

282. Lenfeld, P., and D. Lin, "Tricky Liquid Distributors," in *Kister Distillation Symposium, Topical Conference Proceedings*, p. 428, AIChE Spring Meeting, San Antonio, Texas, March 26–30, 2017.

283. Leslie, V. J., and D. Ferguson, "Radioactive Techniques for Solving Ammonia Plant Problems," *Plant/Oper. Prog.*, 4(3), p. 144, 1985.

284. Lieberman, N. P., *Troubleshooting Natural Gas Processing*, PennWell Publishing, Tulsa, Oklahoma, 1987.

285. Lieberman, N. P., *Process Design for Reliable Operation*, 2nd ed., Gulf Publishing, Houston, Texas, 1988.

286. Lieberman, N. P., *Troubleshooting Process Operations*, 4th ed., PennWell Books, Tulsa, Oklahoma, 2009.

287. Lieberman, N. P., *"Process Engineering for a Small Planet,"* Wiley, New Jersey, 2010.

288. Lieberman, N. P., *Process Equipment Malfunctions*, McGraw-Hill, New York, 2011.

289. Lieberman, N. P., "Performance Testing: the Key to Successful Revamps," *Hydrocarbon Process.*, 96, p. 57, February 2017.

290. Lieberman, N. P., *Understanding Process Equipment for Operators and Engineers*, Elsevier, Amsterdam, 2019.

291. Lieberman, N. P., and E. T. Lieberman, "Design, Installation Pitfalls Appear in Vac Tower Retrofit," *Oil Gas J.*, 89, p. 57, August 26, 1991.

292. Lieberman, E., and N. Lieberman, "Common Refinery Distillation Design Errors," in *Kister Distillation Symposium, Topical Conference Proceedings*, p. 562, AIChE Spring Meeting, New Orleans, Louisiana, March 31–April 4, 2019.

293. Lieberman, N. P., and E. T. Lieberman, *A Working Guide to Process Equipment*, 5th ed., McGraw-Hill, New York, 2022.

294. Lieberman, N. P., and E. T. Lieberman, "Reduction of Process Plant CO_2 Emissions – Short Term Steps," *Paper 156h, Topical Conference Proceedings, Kister Distillation Symposium*, AIChE Spring Meeting, Houston, Texas, March 12–16, 2023.

295. Lieberman, N. P., and E. T. Lieberman, "Troubleshooting Distillation Problems," *Paper 25b, Topical Conference Proceedings, Kister Distillation Symposium*, AIChE Spring Meeting, Houston, Texas, March 12–16, 2023.

296. Lin, D., A. T. Lee, S. D. Williams, J. N. Lee, and J. Poon, "A Case Study of Crude/Vacuum Unit Revamp," Paper presented at the *AIChE Annual Meeting*, New Orleans, Louisiana, February 25–29, 1996.

297. Lin, D., K. Wu, A. Yanoma, and S. Costanzo, "Entrainment Limits and Operating Capacity of Large-Size Structured Packing with Sprayed-type Distributor," Paper presented at the *AIChE Spring National Meeting*, Houston, Texas, March 9–13, 1997.

298. Litzen, D. B., and J. L. Bravo, "Uncover Low-Cost Debottlenecking Opportunities," *Chem. Eng. Prog.*, 95, p. 25, March 1999.

299. Lockett, M. J., *Distillation Tray Fundamentals*, Cambridge University Press, Cambridge, UK, 1986.

300. Lockett, M. J., and S. Banik, "Weeping from Sieve Trays," *Ind. Eng. Chem. Process Des. Dev.*, 25, p. 561, 1986.

301. Lockett, M. J., and J. F. Billingham, "The Effect of Maldistribution on Separation in Packed Distillation Columns," *Trans. IChemE*, 81, Part A, p. 131, January 2003.

302. Lockett, M. J., and A. A. W. Gharani, "Downcomer Hydraulics at High Liquid Flowrates," *IChemE Symp. Ser.*, 56, p. 2.3/43, London, 1979.

303. Lockwood, D. C., and W. E. Glausser, "Are Level Trays Worth their Cost?," *Pet. Refin.*, 38(9), p. 281, 1959.

304. Love, F. S., "Troubleshooting Distillation Problems," *Chem. Eng. Prog.*, 71(6), 1975, p. 61.

305. Ludwig, E. E., *Applied Process Design for Chemical and Petrochemical Plants*, vol. 2, 2nd ed., Gulf Publishing, 1979.

306. LumaSense Technologies Inc., "Mikron Infrared MikroScan 7600PRO Users' Manual," LumaSense, Santa Clara, California.

307. Manifould, D., "Distillation Tray Maintenance," in *1977 NPRA Refinery and Petrochemical Plant Maintenance Conference*, Petroleum Publishing Co., Tulsa, Oklahoma, p. 114, 1977.

308. Mannan, S., *Loss Prevention in the Process Industries*, 4th ed., vols. 1–3, Elsevier, Amsterdam, 2012.

309. Manufacturing Chemists' Association Inc., *Case Histories of Accidents in the Chemical Industry*, vol. 1, Washington, District of Columbia, 1962; vol. 2, 1966; vol. 3, 1970.

310. Marchiori, A., A. L. Wild, A. Y. Saito, A. L. de Souza, C. Mittmann, C. C. Anton, F. S. Duarte, S. L. A. Pereira, and S. Waintraub, "More Tower Damages Caused by Water-Induced Pressure Surge: Unprecedented Sequence of Events," in *Distillation 2013: The Kister Distillation Symposium, Topical Conference Proceedings*, p. 421, the AIChE National Spring Meeting, San Antonio, Texas, April 29–May 2, 2013.

311. Martin, H. W., "Scale-up Problems in a Solvent-Water Fractionator," *Chem. Eng. Prog.*, 60(10), p. 50, 1964.

312. Martin, G. R., E. Luque, and R. Rodriguez, "Revamping Crude Unit Increases Reliability and Operability," *Hydrocarbon Process.*, 79, p. 45, June 2000.

313. Mathias, P. M., "110th Anniversary: A Case Study on developing Accurate and Reliable Excess Gibbs Energy Correlations for Industrial Application," *Ind. Eng. Chem. Res.*, p. 12465, June 24, 2019.

314. Mathias, P., "Effective Use of Physical Properties in Distillation Simulations," *Paper 124b, Topical Conference Proceedings, Kister Distillation Symposium*, AIChE Spring Meeting, Houston, Texas, March 12–16, 2023.

315. Mathur, U., "Experience in Using Group Contribution Methods for Phase Equilibria in Industrial Distillation Column Design," in *Paper 124e, Topical Conference Proceedings, Kister Distillation Symposium*, AIChE Spring Meeting, Houston, Texas, March 12–16, 2023.

316. McCallum, V., "New Mexico Gas Processing Plants Add Filtration to Improve Amine Operations," *Oil Gas J.*, 103, p. 60, April 25, 2005.

317. McCormick, D., "Seeing Mechanical," *Mech. Eng.*, p. 35, September 2007.

318. McGowan, T., and D. J. Coughlin, "Field Troubleshooting 101 and How to Get the Job Done," *Chem. Eng.*, 123, p. 62, June 2016.

319. McLaren, D. B., and J. C. Upchurch, "Guide to Trouble-Free Distillation," *Chem. Eng.*, 77, p. 139, June 1, 1970.

320. McMullan, B. D., A. E. Ravicz, and S. Y. J. Wei, "Troubleshooting a Packed Vacuum Column - A Success Story," *Chem. Eng. Prog.*, 87, p. 69, July 1991.

321. Meier, W., R. Hunkeler, and D. Stöcker, "Performance of New Regular Packing Mellapak," *IChemE Symp. Ser.*, 56, p. 3.3/1, 1979.

322. Miller, J. E., "Include Tech Service Engineers in Turnaround Inspections," *Hydrocarbon Process.*, 66(5), p. 53, 1987.

323. Mixon, W., and L. Pless, "Utilizing Gamma Scans to Monitor Fouling Accumulation in Packed Towers," in *Kister Distillation Symposium, Topical Conference Proceedings*, p. 550, AIChE Spring Meeting, New Orleans, Louisiana, March 31 – April 4, 2019.

324. Mixon, W., and S. X. Xu, "Identify Liquid Maldistribution in Packed Distillation Towers by CAT-Scan Technology," in *Distillation 2005: Learning from the Past and Advancing the Future, Topical Conference Proceedings*, AIChE Spring National Meeting, Atlanta, Georgia, April 10–13, p. 275, 2005.

325. Mohan, R., R. Kumar, H. K. Gupta, and B. Zohail N., "Amine Anomaly in a Mild Hydrocracker," *PTQ*, Q2, p. 39, 2021.

326. Moore, F., and F. Rukovena, "Liquid and Gas Distribution in Commercial Packed Towers," Paper presented at the *36th Canadian Chemical Engineering Conference*, October 5–8, 1986.

327. Mosca, G., E. Tacchini, S. Sander, D. Röttger, A. Wolff, A. Rix, and G. Schilling, "Upgrading a Super-Fractionator C4s Splitter with 4-Pass High Performance Trays at Evonik Degussa, Marl, Germany," in *Distillation 2009: Topical Conference Proceedings*, p. 229, AIChE Spring National Meeting, Tampa, Florida, April 26–30, 2009.

328. Mostia, W. Jr., "Don't Let Cognitive Bias Steer You Wrong," *Control*, 34(2), p. 32, February 2021.

329. Murray, R. M., and J. E. Wright, "Trouble-Free Startup of Distillation Columns," *Chem. Eng. Prog.*, 63(12), p. 40, 1967.

330. Musumeci, J., J. Estill, and G. Mitchell, "Taking a Holistic Approach to a Revamp," *PTQ*, Q2, p. 55, 2021.

331. Mykitta, R. S., "Operating Experiences of Shell's Yellowhammer Gas Plant," *49th Laurence Reid Gas Conditioning Conference*, Norman, Oklahoma, February 1999.

332. Naklie, M. M., L. Pless, T. P. Gurning, and M. Ilyasak, "Radiation Scanning Aids Tower Diagnosis at Arun LNG Plant," *Oil Gas J.*, 88, March 26, 1990.

333. Nath, B. K., "Methanol Plant Applies SPC: A Case Study," *Hydrocarbon Process.*, 75, p. 116-B, January 1996.

334. Niccum, G., and S. White, "Troubleshooting Refinery Equipment with Multiphase CFD Modeling," *PTQ*, Q1, p. 133, 2014.

335. Niccum, G., and S. White, "How Dynamic Pressure Can Affect Vacuum Column Readings," *PTQ*, Q2, p. 75, 2016.

336. Niggeman, G., A. Rix, and R. Meier, "Distillation of Specialty Chemicals," Chapter 7, in A. Gorak and H. Schoenmakers (ed.), *"Distillation Operation and Application,"* p. 297, Academic Press, 2014.

337. Norton Company, "Norton Ceramic Intalox Saddles," Bulletin CI-78, Akron, Ohio, 1973.

338. Norton Company (now Koch-Glitsch), "Packed Tower Internals", Bulletin TA-80R, 1974.

339. Norton Company (now Koch-Glitsch), "Design Information for Packed Towers," Bulletin DC-11, Akron, Ohio, 1977.

340. NPRA, Q & A Session on Refining and Petrochemical Technology, p. 17, 59, 60, 1988.

341. NPRA, Q & A Session on Refining and Petrochemical Technology, p. 25, 26, 29, 177, 178, 1994.

342. NPRA, Panel Discussion, "Refiners Respond to Distillation Queries," *Oil Gas J.*, 78, p. 189, July 28, 1980.

343. O'Brien, N. G., and H. F. Porter, "Design of Gas-Liquid Contactors," *Chem. Eng. Prog.*, 80, p. 44, May 1984.

344. O'Connell, H. E., "Plate Efficiency of Fractionating Columns and Absorbers," *Trans. AIChE*, 42, 741, 1946.

345. Ognisty, T. P., and M. Sakata, "Multicomponent Diffusion: Theory vs. Industrial Data," *Chem. Eng. Prog.*, 83, p. 60, March 1987.

346. Olsson, F. R., "Detect Distributor Defects Before They Cripple Columns," *Chem. Eng. Prog.*, 95, p. 57, October 1999.

347. Olsson, F. R., "Distributors that Worked and those that Did Not," Paper presented at the *AIChE Spring National Meeting*, Houston, Texas, March 14–18, 1999.

348. Olsson, M. R., and H. Z. Kister, "Can We Count on Good Turndown in Two-Pass Moving Valve Trays?," *Chem. Eng. Prog.*, 114 p. 43, November 2018.

349. Olujic, Z., and H. Jansen, "Large Diameter Experimental Evidence on Liquid (Mal)distribution Properties of Structured Packings," *Chem. Eng. Res. Des.*, 99, 2, 2015.

350. Olujic Z., A. M. Ali, and P. J. Jansens, "Effect of Initial Gas Maldistribution on the Pressure Drop of Structured Packings," *Chem. Eng. Process.*, 43, p. 465, 2004.

351. Opong, S., and D. R. Short, "Troubleshooting Columns Using Steady State Models," in *Distillation: Horizons for the New Millennium, Topical Conference Preprints*, p. 129, AIChE Spring National Meeting, Houston, Texas, March 14–18, 1999.

352. Ouni, T., K-M Hansson, S. Welin, C. Hedvall, and H. Z. Kister, "Troubleshooting a Thermosiphon Reboiler – Why New is Not Always Better than Old," in *Kister Distillation Symposium, Topical Conference Proceedings*, p. 747, AIChE Spring Meeting, New Orleans, Louisiana, March 31–April 4, 2019.

353. Özkan, L., I. Aydin, and G. Öztürk, "Cleaning Distillation Column and Heat Exchangers through Cutter Stock Circulation," *Hydrocarbon Process.*, 100, p. 19, May 2021.

354. Öztürk, G., S. Kibar, and G. Akyildiz, "Troubleshoot Flooding Problems in a Crude Distillation Tower," *Hydrocarbon Process.*, 100, p. 57, February 2021.

355. Paladino, E. E., D. Ribeiro, M. V. Reis, W. Geraldelli, G. Torres, A. Saito, and C. R. Maliska, "CFD Simulation of Vacuum CFD Device Using the Two-Fluid Model," in *Distillation 2003: Topical Conference Proceedings*, AIChE Spring National Meeting, New Orleans, Louisiana, p. 241, 2003.

356. Palluzi, R., "Effectively Troubleshoot Pilot Plant Equipment," *Chem. Process.*, 84, p. 17, May 2022.

357. Parthasarthy, M., and T. Scanlan, "Effects of Cellulose Filter Media on Alkanolamines," *PTQ*, Q4, p. 5, 2009.

358. Patel, J. P., V. Desai, D. Mehta, and S. Patil, "Absorber Optimization: Employing Process Simulation Software," *Chem. Eng.*, 120, p. 46, August 2013.

359. Payne, D., "Caustic Tower Operation at Westlake Chemical," *21st Annual Ethylene Producers Conference*, AIChE, Tampa, Florida, April 26–29, 2009.

360. Peng, A. C., T. H. Min, M. Chalakova, L. Bauer, and C. Arguelles, "Revamp of a Methanol Wash Column," *PTQ*, Q2, p. 33, 2022.

361. Peramanu, S., and J. C. Wah, "Improve Material Balance by Using Proper Flowmeter Corrections," *Hydrocarbon Process.*, 90, p. 77, October 2011.

362. Perry, D., D. E. Nutter, and A. Hale, "Liquid Distribution for Optimum Packing Performance," *Chem. Eng. Prog.*, 86, p. 30, January 1990.

363. Perschmann, A., D. Bruder, T. Walter, and F. Schroeder, "Root Cause Analysis of a Premature Flooded Debutanizer Using Gamma Scanning, Engineering and Installation of a Bypass Line by Hot Tapping," in *Kister Distillation Symposium, Topical Conference Proceedings*, p. 275, AIChE Spring Meeting, New Orleans, Louisiana, March 31–April 4, 2019.

364. Pigford, R. L., and C. Pyle, "Performance Characteristics of Spray-Type Absorption Equipment," *Ind. Eng. Chem.*, 43, p. 1649, 1951.

365. Pihlaja, R. K., "Is Your Process Whispering to You?," *Chem. Process.*, 73, p. 30, September 2011.

366. Pilling, M., "Foaming in Fractionation Columns," *PTQ*, Q4, p. 105, 2015.

367. Pilling, M., and Mannion, P., "Process Simulation Aids Expansion of Alkylation Unit Main Fractionator," Paper presented at the *AIChE Spring National Meeting*, Houston, Texas, March 9–13, 1997.

368. Pinczewski, W. V., and C. J. D. Fell, "Nature of the Two-Phase Dispersion on Sieve Plates Operating in the Spray Regime," *Trans. Inst. Chem. Eng. (London)*, 52, p. 294, 1974.

369. Pitzer, A., "Foam Decay with Knitted Wire Mesh," in *The Kister Distillation Symposium, Topical Conference Proceedings*, p. 295, the AIChE National Spring Meeting, San Antonio, Texas, April 29–May 2, 2013.

370. Pless, L., "Tower Scanning as a Quantitative Tool – PackView," in *The Kister Distillation Symposium, Topical Conference Proceedings*," p. 471, the AIChE National Spring Meeting, San Antonio, Texas, April 29–May 2, 2013.

371. Pless, L., "Distillation Tower Gamma Scanning – Now a Quantitative Tool for Measuring Useful Capacity," in *Distillation & Absorption 2014, Proceedings of the 10th International Conference on Distillation and Absorption*, p. 988, EFCE Event No. 705, Friedrichshafen, Germany, September 14–17, 2014.

372. Pless, L., "Monitor Fouling in Distillation Towers with Gamma Scans," in *Distillation 2015: Kister Distillation Symposium, Topical Conference Proceedings*, p. 385, AIChE Spring Meeting, Austin, Texas, April 26–30, 2015.

373. Pless, L., Tru-Grid™ Scans Confirms Need to Repair Damaged Packing," *Tracerco Insight*, Volume 5, Edition 2, p. 3, 2015

374. Pless, L., "Identify the Cause of Liquid Maldistribution Inside Packed Columns," *Hydrocarbon Process.*, 96, p. 65, August 2017.

375. Pless, L., "Crude Vacuum Tower Wash Bed Optimization," *Tracerco Insight*, Volume 8, Edition 2, p. 3, 2018.

376. Pless, L., "Packed Bed Performance Analytics," *PTQ*, Q2, 2018.

377. Pless, L., and B. Asseln, "Using Gamma Scans to Plan Maintenance of Columns," *PTQ*, p. 115, Spring 2002.

378. Pless, L., and D. Ferguson, "Tracer Technology Provides Insight into Leaking Trays and Entrainment," *Tracerco Insight*, Volume 5, Edition 1, p. 1, 2016.

379. Pless, L., and S. Lacasse, "Troubleshooting and Improving the Operation of a Main Fractionating Tower," Presented at the *AIChE Distillation Topical Conference*, Houston, Texas, April 1–5, 2012.

380. Pless, L., and I. Nieuwoudt, "Tower Scanning as a Quantitative Tool," in *Distillation 2013: The Kister Distillation Symposium, Topical Conference Proceedings*, p. 323, the AIChE National Spring Meeting, San Antonio, Texas, April 29–May 2, 2013.

381. Pless, L., C. McKeown, and S. Xu, "Diagnostics of High-Pressure Depropanizer," *Distillation and Absorption 2010*, Eindhoven, The Netherlands, p. 687, 2010.

382. Pless, L., R. Carlson, M. Schultes, and S. Chambers, "Gamma Scan and Tomography Scan Data from a Distillation Column at Various Loadings," in *Distillation & Absorption 2014, Proceedings of the 10th International Conference on Distillation and Absorption*, p. 339, EFCE Event No. 705, Friedrichshafen, Germany, September 14–17, 2014.

383. Pless, L., A. Perschmann, D. Bruder, and T. Walter, "Troubleshooting Premature Tower Flooding," *PTQ*, Q3, p. 99, 2020.

384. Polverini, D., C. Cucinelli, and M. Ferrara, "Online Cleaning and Decontamination of a Butadiene Unit," *PTQ*, Q2, p. 35, 2010.

385. Ponting, J., Kister, H. Z., and R. B. Nielsen, "Troubleshooting and Solving a Sour Water Stripper Problem," *Chem. Eng.*, 120, p. 28, November 2013.

386. Pugh, T., B. York, G. Laxton, and G. D. Simpson, "Improve Heat Exchanger Leak Detection," *Hydrocarbon Process.*, 72, p. 85, November 1993.

387. Quotson, P., H. Peña, and C. Tippman, "Tower Internals Modifications Improve Performance of Atmospheric Crude Tower," in *Distillation 2007: Proceedings of Topical Conference*, p. 59, AIChE National Spring Meeting, Houston, Texas, April 22–26, 2007.

388. Rahman, A. A., A. A. Yusof, J. D. Wilkinson, and L. D. Tyler, "Improving Ethane Extraction at the Petronas Gas GPP-A Facilities in Malaysia," Paper presented at the *83rd Annual Convention of the Gas Processors Association*, March 15, 2004.

389. Ramchandran, S., "Minimize Trapped Components in Distillation Towers," *Chem. Eng.*, 113, p. 65, March 2006.

390. Rao, M. J., K. S. Ganesh, R. Madhavan, and S. Yadav, "Tray Revamp Restores Crude Tower Performance," *Revamps*, p. 15, 2018.

391. Ray, D., A. Arora, and A. Phanikumar, "CFD Study of a VDU Feed Inlet Device and Wash Bed," *Revamps*, p. 40, 2009.

392. Reay, D. W., "Vacuum Distillation of Heavy Residues – Meeting Changing Refinery Requirements," *IChemE Symp. Ser.*, 73, p. D51, 1982.

393. Reay, D. W. (ret. BP Oil), private communication, 1988.

394. Remesat, D., and Z. Riha, "Consider CFD Analysis to Support Critical Separation Operations," *Hydrocarbon Process.*, 89, p. 67, May 2010.

395. Rendón, G. T., L. A. Quintero Rivera, J. A. Recio Espinosa, and R. A. Loubet Guzmán, "Troubleshooting Jet-Fuel Production on Crude Distillation Unit in Pemex Cadereyta Refinery," in *The Kister Distillation Symposium, Topical Conference Proceedings*, p. 79, the AIChE National Spring Meeting, San Antonio, Texas, April 29–May 2, 2013.

396. Resetarits, M., "Believe Your Instruments Unless . . . ," *Chem. Eng.*, 118, p. 28, September 2011.

397. Resetarits, M. R., and N. Pappademos, "Factors Influencing Vapor Crossflow Channeling," Paper presented at the *AIChE Fall Annual Meeting*, Reno, Nevada, November 2001.

398. Resetarits, M. R., J. Agnello, M. J. Lockett, and H. L. Kirkpatrick, "Retraying Increases C3 Splitter Column Capacity," *Oil Gas J.*, 86, p. 54, June 6, 1988.

399. Risko, J. R., "Steam Quality Considerations," *Chem. Eng.*, 127(5), p. 42, May 2020.

400. Risko, J. R., "Optimize Reboiler Performance via Effective Condensate Drainage," *Chem. Eng. Prog.*, 117, p. 43, July 2021.

401. Rose, L. M., *Distillation Design in Practice*, Elsevier, Amsterdam, 1985.

402. Ross, S., "Mechanisms of Foam Stabilization and Antifoaming Action," *Chem. Eng. Prog.*, 63(9), p. 41, 1967.

403. Ross, S., and G. Nishioka, "Foaminess of Binary and Ternary Solutions," *J. Phys. Chem.*, 79(15), p. 1761, 1975.

404. Roy, P., and G. K. Hobson, "The Selection and Use of VLE Methods and Data," *IChemE Symp. Ser.*, 104, p. A273, 1987.

405. Roy, P., and A. C. Mercer, "The Use of Structured Packings in Crude Oil Atmospheric Distillation Column," *IChemE Symp. Ser.*, 104, p. A103, 1987.

406. Ruffert, G., K. Hallenberger, and J. Bausa, "Debottlenecking of a Hexane-Distillation," in *Distillation 2007: Proceedings of Topical Conference*, p. 565, AIChE National Spring Meeting, Houston, Texas, April 22–26, 2007.

407. Rukovena, F., "Properly Install Column Internals," *Chem. Proc.*, 69, p. 19, February 2007.

408. Saletan, D., "Avoid Trapping Minor Components in Distillation," *Chem. Eng. Prog.*, 91, p. 76, September 1995.

409. Sanchez, J. M., Valverde, A., Di Marco, C., and E. Carosio, "Inspecting Fractionation Towers," *Chem. Eng.*, 118, p. 44, July 2011.

410. Sands, R., "Column Instrumentation Basics," *Chem. Eng.*, 115, p. 48, March 2008.

411. Sands, R., "Thoughts on Troubleshooting: Tips from Those Who Do It Best," in *Distillation 2011: The Dr. James Fair Heritage Distillation Symposium, Topical Conference Proceedings*, p. 315, AIChE Spring National Meeting, Chicago, Illinois, March 13–17, 2011.

412. Sattler, F. J., "Nondestructive Testing Methods Can Aid Plant Operation," *Chem. Eng.*, 97, p. 177, October 1990.

413. Sauter, J. R., and W. E. Younts III, "Tower Packings Cut Olefin-Plant Energy Needs," *Oil Gas J.*, 84, p. 45, September 1, 1986.

414. Schad, R. C., "Make the Most of Process Simulation," *Chem. Eng. Prog.*, 94, p. 21, January 1998.

415. Schafer, T. A., Y. S. Lam, and S. Vragolic, "Organic Strippers: Vapor Maldistribution Problems and Solutions," Paper presented at the *AIChE Spring National Meeting*, New Orleans, Louisiana, March 29–April 2, 1992.

416. Schnaibel, G., and M. E. P. L. Marsiglia, "The Recurring Jet Fuel Problem (and what it took to solve it)," Paper presented at the *AIChE Spring National Meeting*, New Orleans, Louisiana, April 25–29, 2004.

417. Schnaibel, G., A. C. Bellote, A. V. S. Castro, C. N. Fonseca, M. E. P. L. Marsiglia, G. A. S. Torres, K. Ropelato, A. S. Moraes, and F. Feitosa, "The Jet Fuel Problem Strikes Again," in *Distillation 2009: Topical Conference Proceedings*, p. 89, AIChE Spring National Meeting, Tampa, Florida, April 26–30, 2009.

418. Schneider, D., "Build a Better Process Model," *Chem. Eng. Prog.*, 94, p. 75, April 1998.

419. Schon, S. G., "A New technique to Prevent Minor Component Trapping in Distillation," Paper presented at the *Distillation Symposium*, AIChE Spring Meeting, San Antonio, Texas, April 2022.

420. Schultes, M., "The Raschig Super-Ring is Pushing the FRI Test Column to New Limits," in *The Kister Distillation Symposium, Topical Conference Proceedings*, p. 381, the AIChE National Spring Meeting, San Antonio, Texas, April 29–May 2, 2013.

421. Seidel, R. O., "Experience in the Operation of Activated Hot Potassium Carbonate Acid Gas Removal Plants (U.S.)," in *Seminar on Raising Productivity in Fertilizer Plants*, Baghdad, Iraq, March 23–25, 1978.

422. Senger, G., and G. Wozney, "Investigation of Foam in Distillation and Absorption Columns," in *Distillation 2011: The Dr. Jim Fair Heritage Distillation Symposium, Topical Conference Proceedings*, p. 337, AIChE Spring National meeting, Chicago, Illinois, March 13–17, 2011.

423. Severance, W. A. N., "Advances in Radiation Scanning of Distillation Columns," *Chem. Eng. Prog.*, 77(9), p. 38, 1981.

424. Severance, W. A. N., "Differential Radiation Scanning Improves the Visibility of Liquid Distribution," *Chem. Eng. Prog.*, 81(4), p. 48, 1985.

425. Sewell, A., "Practical Aspects of Distillation Column Design," *The Chem. Engineer*, 299/300, p. 442, 1975.

426. Shah, K., and B. Stucky, "Operating Eunice, La., Demethanizer at High Pressures Yields Unique Data," *Oil Gas J.*, 102(43), p. 54, November 15, 2004.

427. Shakur, S. M., N. F. Urbanski, M. R. Resitarits, D. R. Monkelbaan, and D. A. Bucior, "Gamma Scanning a Column Containing Closely Spaced Trays," Paper presented at the *AIChE Annual Meeting*, Dallas, Texas, November 4, 1999.

428. Shariat, A., and J. G. Kunesh, "Packing Efficiency Testing on Commercial Scale with Good (and Not So Good) Reflux Distribution," *Ind. Eng. Chem. Res.*, 34(4), 1273, 1995.

429. Sharma, A., "Caustic Tower Design Guidelines and Recommendations," *Hydrocarbon Process.*, 101, p. 63, July 2022.

430. Sheilan, M. H., and R. H. Weiland, "Troubleshooting AGRUs in an LNG Train," *LNG Industry*, August 2015.

431. Shinskey, F. G., *Distillation Control for Productivity and Energy Conservation*, 2nd ed., McGraw-Hill, New York, 1984.

432. Shiveler, G., and H. Wandke, "Steps for Troubleshooting Amine Sweetening Plants," in *Kister Distillation Symposium*, p. 96, AIChE Spring Meeting, Austin, Texas, April 26–30, 2015.

433. Shiveler, G., D. Love, and D. Pierce, "Retrofit of Depropanizer and Debutanizer Packed Column," Paper presented at the *AIChE Spring Meeting*, New Orleans, Louisiana, April 25–29, 2004.

434. Shtayieh, S., C. A. Durr, J. C. McMillan, and C. Collins, "Successful Operation of a Large LPG Plant," *Oil Gas J.*, 80, p. 79, March 1, 1982.

435. Siddiqui, D., and N. Hedrick, "Measuring and Controlling Flow of Dangerous Materials," *Chem. Eng.*, 125, p. 44, July 2018.

436. Silverstein, J. L., "Use of Distillation and Process Analysis Techniques in Toll Manufacturing: A Case Study," in *The Kister Distillation Symposium, Topical Conference Proceedings*, p. 249, the AIChE National Spring Meeting, San Antonio, Texas, April 29–May 2, 2013.

437. Silvey, F. C., and G. J. Keller, "Testing on a Commercial Scale," *Chem. Eng. Prog.*, 62(1), p. 68, 1966.

438. Simmons, C. V., Jr., "Avoiding Excessive Glycol Costs in Operation of Gas Dehydrators," *Oil Gas J.*, 79, p. 121, September 21, 1981.

439. Simpson, L. L., "Sizing Piping for Process Plants," *Chem. Eng.*, 75, p. 192, June 17, 1968.

440. Sloley, A. W., "Avoid Problems During Distillation Column Startups," *Chem. Eng. Prog.*, 92, p. 30, July 1996.

441. Sloley, A. W., "Don't Get Drawn into Distillation Difficulties," *Chem. Eng. Prog.*, 94, p. 63, June 1998.

442. Sloley, A. W., "Instrumentation: the Key to Revamp Benefits," *PTQ*, p. 127, Spring 2001.

443. Sloley, A. W., "Troubleshooting Refinery Vacuum Towers," in *Distillation 2001: Frontiers in a New Millennium, Proceedings of Topical Conference*, p. 251, AIChE Spring National Meeting, Houston, Texas, April 22–26, 2001.

444. Sloley, A. W., "Reuse and Relocation: A Case history Part III: Troubleshooting Approach and Technique," in *Kister Distillation Symposium, Topical Conference Proceedings*, p. 46, AIChE Spring Meeting, Virtual, April 18–22, 2021.

445. Sloley, A. W., and B. Fleming, "Successfully Downsize Trayed Columns," *Chem. Eng. Prog.*, 90, p. 39, March 1994.

446. Sloley, A. W., and A. C. S. Fraser, "Consider Modeling Tools to Revamp Existing Process Units," *Hydrocarbon Process.*, 79, p. 57, June 2000.

447. Sloley, A. W., and C. K. Hoepner, "Setting the Revamp Basis: What is the Current Operation?," Paper presented at the *AIChE Spring National Meeting*, San Antonio, Texas, April 28–May 2, 2013.

448. Sloley, A. W., and G. R. Martin, "Subdue Solids in Towers," *Chem. Eng. Prog.*, 91, p. 64, January 1995.

449. Sloley, A. W., S. W. Golden, and E. L. Hartman, "Why Towers Do Not Work," *Nat. Eng.*, Part 1, p. 19, August 1995, and Part 2, p. 16, September, 1995.

450. Sloley, A. W., T. M. Zygula, and K. Kolmetz, "Troubleshooting Practice in the Refinery," Paper presented at the *AIChE Spring National Meeting*, Houston, Texas, April 2001.

451. Smith, B. D., *Design of Equilibrium Stage Processes*, McGraw-Hill, 1963.

452. Smith, R. F., "Curing Foam Problems in Gas Processing," *Oil Gas J.*, 77, p. 186, July 30, 1979.

453. Smith, C. L., *Distillation Control*, Wiley, New Jersey, 2012.

454. Sofronas, A., "Case 70: Twenty Rules for Troubleshooting," *Hydrocarbon Process.*, 91, p. 35, September 2012.

455. Sofronas, A., "Case 72: Interaction between Disciplines when Troubleshooting," *Hydrocarbon Process.*, 92, p. 114, May 2013.

456. Sofronas, A., "Case 97: Beware of Confirmation Bias in Industry," *Hydrocarbon Process.*, 96, p. 25, July 2017.

457. Spiegel, L., "New Method to Assess Liquid Distributor Quality," *Chem. Eng. Process.*, p. 1011, 2006.

458. Spiegel, L., and M. Duss, "Structured Packings," Chapter 4, in A. Gorak and Z. Olujic (ed.), *Distillation Equipment and Processes*, p. 145, Academic Press, 2014.

459. Stichlmair, J. G., and J. R. Fair, *Distillation Principles and Practice*, Wiley, 1998.

460. Stoter, F., Z. Olujic, and J. de Graauw, "Modelling of Hydraulic and Separation Performance of Large Diameter Columns Containing Structured Packings," *IChemE Symp. Ser.*, 128, A201, 1992.

461. Strigle, R. F. Jr., *Random Packings and Packed Towers*, Gulf Publishing, Houston, Texas, 1987.

462. Strigle, R. F. Jr., *Packed Tower Design and Applications*, 2nd ed., Gulf Publishing, Houston, Texas, 1994.

463. Stupin, W. J., and H. Z. Kister, "System Limit – The Ultimate Capacity of Fractionators," *Trans. IChemE*, 81, Part A, p. 136, January 2003.

464. Summers, D. R., *Performance Diagrams - All Your Tray Hydraulics in One Place*, Paper presented at the AIChE Annual Meeting, Austin, Texas, 2004.

465. Summers, D. R., "Evaluating and Documenting Tower Performance," *Chem. Eng. Prog.*, 106, p. 38, February 2010.

466. Summers, D. R., "Distillation Tray Design Through a Better Understanding of Operating Windows," Paper presented at the *AIChE Annual Meeting*, San Francisco, California, November 14, 2016.

467. Summers, D. R., "Sieve Tray Stability Factor Based on Efficiency Data," *Chem. Eng.*, 128, p. 26, July 2021.

468. Summers, D. R., L. Spiegel, and E. Kolesnikov, "Tray Stability at Low Vapour Load," in *Distillation and Absorption 2010*, p. 611, Einhoven University of Technology, The Netherlands, 2010.

469. Summers, D. R., W. Mônaco, and M. A. Britto, "Butyraldehyde Tower Capacity Limitations," in *Distillation 2013, Kister Distillation Symposium, Topical Conference Proceedings*, p. 441, AIChE Spring Meeting, San Antonio, Texas, April 28–May 2, 2013.

470. Swain, A. D., *Accident Sequence Evaluation Procedure/Human Reliability Analysis Procedure*, Sandia National Laboratories, US Nuclear Regulatory Commission, Washington, District of Columbia, 1987.

471. Swisher, C., "On the Trail of Hidden Design Flaws," *Chem. Eng.*, 102, p. 133, February 1995.

472. Talley, D. L., "Startup of a Sour Gas Plant," *Hydrocarbon Process.*, 55, p. 90, April 1976.

473. Tariq, M. R., and M. Al Harbi, "Importance of Lean Amine Quality to Avoid H2S Breakthrough from Gas Treat Unit," *Presentation at SulGas 2022 (Virtual)*, India, February 3, 2022.

474. Taylor, R., R. Krishna, and H. Kooijman, "Real-World Modeling of Distillation," *Chem. Eng. Prog.*, 99, p. 28, July 2003.

475. Teletzke, E, and C. Bickham, "Troubleshoot Acid-Gas Removal Systems," *Chem. Eng. Prog.*, 110, p. 47, July 2014.

476. Teletzke, E, and B. Madhyani, "Minimize Amine Losses in Gas and Liquid Treating," *Gas*, p. 35, 2018.

477. Teletzke, E., and B. Roberts, "Infrared Scans Improve Permian Acid-Gas Removal Operations," *Oil & Gas J.*, 219, p. 53, 2021.

478. The Pfaudler Company, "Column Packings and Design Data," Bulletin SB-14-600-2, 1983.

479. Theunick, G., "Composition Change Impacts Pressure Control," in *Kister Distillation Symposium, Topical Conference Proceedings*, p. 597, AIChE Spring Meeting, New Orleans, Louisiana, March 31–April 4, 2019.

480. Theunick, G., "Solutions for Fouled Towers that Made Things Worse," in *Kister Distillation Symposium, Topical Conference Proceedings, Paper 156d*, AIChE Spring Meeting, Houston, Texas, March 12–16, 2023.

481. Theunick, G., J. Molloy, M. Ludwick, D. Theimer, and L. Jones, "Troubleshooting Tunnel Vision: Packing Change Blamed for Premature Flooding," in *Kister Distillation Symposium, Topical Conference Proceedings*, p. 563, AIChE Spring Meeting, Virtual, April 18–22, 2021.

482. Torres, G., A. da Silva, C. N. da Fonesca, N. M. Pinto, and S. L. de Araujo, "How a Simple Modification Resulted in a Great and Unexpected Change in the Operation of a Vacuum Tower Wash Section," in *Distillation 2003: On the Path to High Capacity, Efficient Splits, Topical Conference Proceedings*, p. 284, AIChE Spring National Meeting, New Orleans, Louisiana, March 31–April 3, 2003.

483. Tracerco, "Process Diagnostics - TRU-SCAN[R] Case Studies," 2010. Available from Tracerco, Pasadena, Texas.

484. Trompiz, C. J., and J. R. Fair, "Entrainment from Spray Distributors for Packed Columns," *Ind. Eng. Chem. Res.*, 39, 1797, 2000.

485. Tru-Tec Division, "Sample Scan Brochure," 1998, Available from Tracerco, Pasadena, Texas.

486. Tru-Tec Division of Koch Engineering Company Inc., "Scans Identify Vapor Bypassing in a Debutanizer," Tru-News, Vol. 4, Ed. 3, p. 1, La Porte, Texas, 1996, Available from Tracerco, Pasadena, Texas.

487. Vaidyanathan, S., M. Wehrli, H. Ulrich, and R. Hunkeler, "New Design Techniques for Vacuum Column Flash Zone Design," in *Distillation 2001, Topical Conference Proceedings*, AIChE Spring National Meeting, Houston, Texas, p. 287, April 22–26, 2001

488. Vail, R., "Gas Con Recovery and Stripping Issues at Turndown Conditions," in *Kister Distillation Symposium, Topical Conference Proceedings*, p. 661, AIChE Spring Meeting, Virtual, April 18–22, 2021.

489. Vail, R., and G. Cantley, "Pinch Points: Commonly Overlooked Pieces of Equipment which Result in Flooding and Unit Constraints," *Paper 25c, Topical Conference Proceedings, Kister Distillation Symposium*, AIChE Spring Meeting, Houston, Texas, March 12–16, 2023.

490. Van der Meer, D., F. J. Zuiderweg, and H. J. Scheffer, "Foam Suppression in Extract Purification and Recovery Trains," *Proceedings of the International Solvent Extraction Conference (ISEC 71)*, Society of Chemical Industry, London, p. 350, 1971.

491. Van der Merwe, J., "Distillation Pressure Control Troubleshooting – The Hidden Pitfalls of Overdesign," *Chem. Eng. Res. Des.*, 89, p. 1377, 2011.

492. Van Winkle, M., *Distillation*, McGraw-Hill, 1967.

493. Vidrine, S., and P. Hewitt, "Radioisotope Technology - Benefits and Limitations in Packed Bed Tower Diagnostics," Paper presented at the *AIChE Spring Meeting*, New Orleans, Louisiana, April 25–29, 2004.

494. de Villiers, W. E., R. N. French, and G. J. Koplos, "Navigate Phase Equilibria via Residue Curve Maps," *Chem. Eng. Prog.*, 98, p. 66, November 2002.

495. Vollmer, M., and K. P. Mollmann, *Infrared Thermal Imaging*, 2nd ed., Wiley-VCH, Weinheim, Germany, 2018.

496. Von Phul, S. A., "Sweetening Process Foaming and Abatement," *Laurence Reid Gas Conditioning Conference*, p. 251, Norman, Oklahoma, February 25–28, 2001.

497. Waintraub, S, G. A. da S. Torres, A. L. W. Serpa, R. L. Waschburger, and A. G. Oliveira, "Removing Packings from Heat Transfer Sections of Vacuum Towers," in *Distillation 2005: Topical Conference Proceedings*, AIChE Spring National Meeting, Atlanta, Georgia, p. 79, 2005.

498. Wallsgrove, C. S., and J. C. Butler, "Process Plant Startup," *Continuing Education Seminar*, The Center for Professional Advancement, East Brunswick, New Jersey.

499. Wehrli, M., S. Hischberg, and R. Scweitzer, "Influence of Vapour Feed Design on the Flow Distribution below Packings," *Chem. Eng. Res. Des.*, 81, Part A, p. 116, January 2003.

500. Wehrli, M., P. Schaeffer, U. Marti, F. Muggli, and H. Kooijman, "Mixed Phase Feeds in Mass Transfer Columns and Liquid Separation," *IChemE Symp. Ser.*, 152, p. 230, London, 2006.

501. Weiland, R. H., and J. C. Dingman, "Column Design Using Mass Transfer Rate Simulation," *PTQ*, p. 139, Spring 2002.

502. Weiland, R. H., and A. Hatcher, "What are the Benefits from Mass Transfer Rate-Based Simulation?," *Hydrocarbon Process.*, 90, p. 43, July 2011.

503. White, R. L., "On-Line Troubleshooting of Chemical Plants," *Chem. Eng. Prog.*, 83(5), p. 33, 1987.

504. Wieslaw, T., and N. P. Lieberman, "Amine Unit Startup Illustrates Commissioning Considerations," *Oil Gas J.*, 115, p. 62, August 7, 2017.

505. Willard, D., "Ease Packed-Column Commissioning," *Chem. Proc.*, 75, p. 10, January 2013.

506. Williamson, J. C., "Improve Vapor and Mixed-Phase Feed Distribution," *World Refin.*, p. 22, May 2000.

507. Xu, S. X., and W. Mixon, "Diagnosing Maldistribution in Towers," *Chem. Eng. Prog.*, 103, p. 28, May 2007.

508. Xu, S. X., and L. Pless, "Understand More Fundamentals of Distillation Column Operation from Gamma Scans," in *Distillation 2001: Frontiers in a New Millennium, Proceedings of at the Topical Conference on Distillation*, the AIChE National Spring Meeting, Houston, Texas, April 22–26, 2001.

509. Xu, S. X., and L. Pless, "Distillation Tower Flooding – More Complex Than You Think," *Chem. Eng.*, 109, p. 60, June 2002.

510. Xu, S. X., C. Winfield, and J. D. Bowman, "Distillation – How to Push a Tower to Its Maximum Capacity," *Chem. Eng.*, 105, p. 100, August 1998.

511. Xu, S. X., G. Kennedy, C. Conforti, T. Marut, and J. Dusseault, "Troubleshooting Industrial Packed Columns by Gamma-Ray Tomography," Paper presented at the *Chemical Engineering (Magazine) Expo*, Houston, Texas, June 9–10, 1999.

512. Xu, F. J., F. H. Zhou, Z. M. Cao, Q. Yang, L. Zuber, and J. J. Dong, "Investigations of High Pressure Drop Observed in ASU Columns," *Gas Process. LNG*, p. 19, March-April 2022; *Hydrocarbon Process.*, 101, p. 27, June 2022.

513. Yanagi, T., and M. Sakata, "Performance of a Commercial Scale 14% Hole Area Sieve Tray," *Ind. Eng. Chem. Process Des. Dev.*, 21, p. 712, 1982.

514. Yang, Q., J. J. Dong, F. J. Xu, and X. D. Zhang, "Revamping an Air Separation Unit," *Revamps*, p. 19, 2019.

515. Yarborough, L., L. E. Petty, and R. H. Wilson, "Using Performance Data to Improve Plant Operations," in *Proceedings of the 59th Annual Convention of the Gas Processors Associations*, p. 86, Houston, Texas, March 17–19, 1980.

516. Yarbrough, B., T. Cooper, and C. Davidson, "Extractive Column Troubleshooting after Internals Revamp," Paper presented in the *Distillation Symposium, AIChE Spring Meeting*, Virtual, August 18, 2020.

517. Zanetti, R., "Boosting Packed Tower Performance by More than a Trickle," *Chem. Eng.*, 92, p. 22, May 27, 1985.

518. Zenz, F. A., "Designing Gas Absorption Columns," *Chem. Eng.*, 79, p. 120, November 13, 1972.

519. ZipLevel Corp., https://ziplevel.com/pro-2000.

520. Zuiderweg, F. J., P. A. M. Hofhuis, and J. Kuzniar, "Flow Regimes on Sieve Trays: The Significance of the Emulsion Flow regime," *Chem. Eng. Res. Des.*, 62, p. 39, 1984.

521. Zuiderweg, F. J., P. J. Hoek, and L. Lahm, Jr., "The Effect of Liquid Distribution and Redistribution on the Separating Efficiency of Packed Columns," *IChemE Symp. Ser.*, 104, p. A217, 1987.

522. Zygula, T., E. Roy, and P. C. Dautenhahn, "The Importance of Thermodynamics on Process Simulation Modeling," in *Distillation 2001: Frontiers in a New Millennium, Proceedings of at the Topical Conference on Distillation*, the AIChE National Spring Meeting, Houston, Texas, April 22–26, 2001.

523. Zygula, T. M., K. Kolmetz, and R. Sommerfeldt, "Troubleshooting an Ethylene Feed Saturator Column," *Oil Gas J.*, 101, p. 88, August 25, 2003.

Index

Bold page numbers signify major references on indexed item.

Distillation Diagnostics: An Engineer's Guidebook, First Edition. Henry Z. Kister.
© 2025 American Institute of Chemical Engineers, Inc. Published 2025 by John Wiley & Sons, Inc.